# BMW 3-Series Automotive Repair Manual

by Robert Rooney,
Mike Stubblefield
and John H Haynes

Member of the Guild of Motoring Writers

**Models covered:**

BMW 3-Series models (1992 through 1998)
and Z3 models (1996 through 1998)
*Does not include information specific to M3 models*

(18021-3K2)

Haynes Group Limited
Haynes North America, Inc.
www.haynes.com

## Disclaimer

There are risks associated with automotive repairs. The ability to make repairs depends on the individual's skill, experience and proper tools. Individuals should act with due care and acknowledge and assume the risk of performing automotive repairs.

The purpose of this manual is to provide comprehensive, useful and accessible automotive repair information, to help you get the best value from your vehicle. However, this manual is not a substitute for a professional certified technician or mechanic.

This repair manual is produced by a third party and is not associated with an individual vehicle manufacturer. If there is any doubt or discrepancy between this manual and the owner's manual or the factory service manual, please refer to the factory service manual or seek assistance from a professional certified technician or mechanic.

Even though we have prepared this manual with extreme care and every attempt is made to ensure that the information in this manual is correct, neither the publisher nor the author can accept responsibility for loss, damage or injury caused by any errors in, or omissions from, the information given.

The manufacturer's authorised
representative in the EU for product safety is:

HaynesPro BV
Stationsstraat 79 F, 3811MH Amersfoort
The Netherlands
gpsr@haynes.co.uk

© **Haynes North America, Inc. 2000**
With permission from Haynes Group Limited

**A book in the Haynes Automotive Repair Manual Series**

All rights reserved. No part of this book may be reproduced or transmitted in any form or by any means, electronic or mechanical, including photocopying, recording or by any information storage or retrieval system, without permission in writing from the copyright holder.

**ISBN-10: 1 56392 376 9**
**ISBN-13: 978 1 56392 376 0**

**Library of Congress Control Number 00-102053**

While every attempt is made to ensure that the information in this manual is correct, no liability can be accepted by the authors or publishers for loss, damage or injury caused by any errors in, or omissions from, the information given.

# Contents

**Introductory pages**

| | |
|---|---|
| About this manual | 0-5 |
| Introduction to the BMW 3-Series | 0-5 |
| Vehicle identification numbers | 0-6 |
| Booster battery (jump) starting | 0-6 |
| Buying parts | 0-7 |
| Maintenance techniques, tools and working facilities | 0-7 |
| Jacking and towing | 0-13 |
| Anti-theft audio system | 0-13 |
| Automotive chemicals and lubricants | 0-14 |
| Conversion factors | 0-15 |
| Fraction/decimal/millimeter equivalents | 0-16 |
| Safety first! | 0-17 |
| Troubleshooting | 0-18 |

**Chapter 1**
Tune-up and routine maintenance — 1-1

**Chapter 2    Part A**
Four-cylinder engines — 2A-1

**Chapter 2    Part B**
Six-cylinder engines — 2B-1

**Chapter 2    Part C**
General engine overhaul procedures — 2C-1

**Chapter 3**
Cooling, heating and air conditioning systems — 3-1

**Chapter 4    Part A**
Fuel and exhaust systems - four-cylinder engines — 4A-1

**Chapter 4    Part B**
Fuel and exhaust systems - six-cylinder engines — 4B-1

**Chapter 5    Part A**
Starting and charging systems — 5A-1

**Chapter 5    Part B**
Ignition system — 5B-1

**Chapter 6**
Emissions control systems — 6-1

**Chapter 7    Part A**
Manual transmission — 7A-1

**Chapter 7    Part B**
Automatic transmission — 7B-1

**Chapter 8**
Clutch and driveline — 8-1

**Chapter 9**
Brakes — 9-1

**Chapter 10**
Suspension and steering systems — 10-1

**Chapter 11**
Body — 11-1

**Chapter 12**
Chassis electrical system — 12-1

**Wiring diagrams** — 12-32

**Index** — IND-1

BMW 325is Coupe

# About this manual

## *Its purpose*

The purpose of this manual is to help you get the best value from your vehicle. It can do so in several ways. It can help you decide what work must be done, even if you choose to have it done by a dealer service department or a repair shop; it provides information and procedures for routine maintenance and servicing; and it offers diagnostic and repair procedures to follow when trouble occurs.

We hope you use the manual to tackle the work yourself. For many simpler jobs, doing it yourself may be quicker than arranging an appointment to get the vehicle into a shop and making the trips to leave it and pick it up. More importantly, a lot of money can be saved by avoiding the expense the shop must pass on to you to cover its labor and overhead costs. An added benefit is the sense of satisfaction and accomplishment that you feel after doing the job yourself.

## *Using the manual*

The manual is divided into Chapters. Each Chapter is divided into numbered Sections, which are headed in bold type between horizontal lines. Each Section consists of consecutively numbered paragraphs.

At the beginning of each numbered Section you will be referred to any illustrations which apply to the procedures in that Section. The reference numbers used in illustration captions pinpoint the pertinent Section and the Step within that Section. That is, illustration 3.2 means the illustration refers to Section 3 and Step (or paragraph) 2 within that Section.

Procedures, once described in the text, are not normally repeated. When it's necessary to refer to another Chapter, the reference will be given as Chapter and Section number. Cross references given without use of the word "Chapter" apply to Sections and/or paragraphs in the same Chapter. For example, "see Section 8" means in the same Chapter.

References to the left or right side of the vehicle assume you are sitting in the driver's seat, facing forward.

Even though we have prepared this manual with extreme care, neither the publisher nor the author can accept responsibility for any errors in, or omissions from, the information given.

**NOTE**

A **Note** provides information necessary to properly complete a procedure or information which will make the procedure easier to understand.

**CAUTION**

A **Caution** provides a special procedure or special steps which must be taken while completing the procedure where the Caution is found. Not heeding a Caution can result in damage to the assembly being worked on.

**WARNING**

A **Warning** provides a special procedure or special steps which must be taken while completing the procedure where the Warning is found. Not heeding a Warning can result in personal injury.

# Introduction to the BMW 3-Series

The models covered by this manual are equipped with either a 1.8L or 1.9L four-cylinder engine or a 2.5L or 2.8L six-cylinder engine. All engines are of the Dual Overhead Camshaft (DOHC) design. These engines feature a computer-controlled ignition system and electronic fuel injection.

Transmissions are either a five-speed manual or a four-speed automatic. Power is transmitted from the transmission to the fully independent rear axle through a two-piece driveshaft. The differential is bolted to a crossmember and drives the wheels through driveaxles equipped with inner and outer constant velocity joints.

The front suspension is of the MacPherson strut design, with the coil spring/shock absorber unit making up the upper suspension link. The rear suspension on all models except the 318ti and the Z3 is of the multi-link design, while the 318ti and the Z3 use a semi-trailing arm setup. Power-assisted rack-and-pinion steering is standard.

Power-assisted front and rear disc brakes are standard. An Anti-lock Braking System (ABS) is also standard equipment.

# Vehicle identification numbers

### Vehicle Identification Number

The Vehicle Identification Number (VIN) is located on the on the firewall, just below the windshield, on the driver's side door and on the Manufacturer's Plate **(see illustrations)**. It contains valuable information such as where and when the vehicle was manufactured, the model year and the body style. This number can be used to cross-check the registration and license.

### Engine serial number

The engine serial number is stamped on a machined pad, above the starter on the left rear side of the engine.

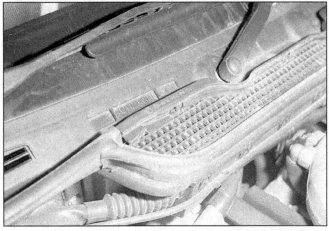

The VIN is stamped into the firewall, just below the windshield

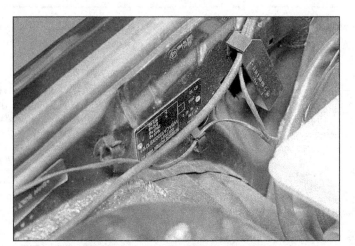

The VIN is also present on the Manufacturer's Plate, located under the hood on the right inner fender panel

# Booster battery (jump) starting

**Note:** *On all models except four-cylinder 3-series models and the 1996 Z3, the battery is located in the trunk. A remote positive terminal is provided in the engine compartment on these models.*

Observe the following precautions when using a booster battery to start a vehicle:

a) Before connecting the booster battery, make sure the ignition switch is in the Off position.
b) Turn off the lights, heater and other electrical loads.
c) Your eyes should be shielded. Safety goggles are a good idea.
d) Make sure the booster battery is the same voltage as the dead one in the vehicle.
e) The two vehicles MUST NOT TOUCH each other.
f) Make sure the transmission is in Neutral (manual transaxle) or Park (automatic transaxle).

Connect the red jumper cable to the positive (+) terminals of each battery (or to the remote positive terminal in the engine compartment). Connect one end of the black cable to the negative (-) terminal of the booster battery. The other end of this cable should be connected to a good ground on the engine block **(see illustration)**. Make sure the cable will not come into contact with the fan, drivebelts or other moving parts of the engine.

Start the engine using the booster battery, then, with the engine running at idle speed, disconnect the jumper cables in the reverse order of connection.

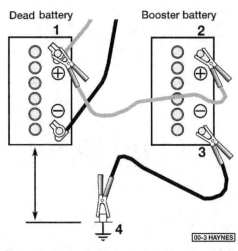

Make the booster battery cable connections in the numerical order shown (note that the negative cable of the booster battery is NOT attached to the negative terminal of the dead battery)

# Buying parts

Replacement parts are available from many sources, which generally fall into one of two categories - authorized dealer parts departments and independent retail auto parts stores. Our advice concerning these parts is as follows:

***Retail auto parts stores:*** Good auto parts stores will stock frequently needed components which wear out relatively fast, such as clutch components, exhaust systems, brake parts, tune-up parts, etc. These stores often supply new or reconditioned parts on an exchange basis, which can save a considerable amount of money. Discount auto parts stores are often very good places to buy materials and parts needed for general vehicle maintenance such as oil, grease, filters, spark plugs, belts, touch-up paint, bulbs, etc. They also usually sell tools and general accessories, have convenient hours, charge lower prices and can often be found not far from home.

***Authorized dealer parts department:*** This is the best source for parts which are unique to the vehicle and not generally available elsewhere (such as major engine parts, transmission parts, trim pieces, etc.).

***Warranty information:*** If the vehicle is still covered under warranty, be sure that any replacement parts purchased - regardless of the source - do not invalidate the warranty!

To be sure of obtaining the correct parts, have engine and chassis numbers available and, if possible, take the old parts along for positive identification.

# Maintenance techniques, tools and working facilities

## *Maintenance techniques*

There are a number of techniques involved in maintenance and repair that will be referred to throughout this manual. Application of these techniques will enable the home mechanic to be more efficient, better organized and capable of performing the various tasks properly, which will ensure that the repair job is thorough and complete.

### Fasteners

Fasteners are nuts, bolts, studs and screws used to hold two or more parts together. There are a few things to keep in mind when working with fasteners. Almost all of them use a locking device of some type, either a lockwasher, locknut, locking tab or thread adhesive. All threaded fasteners should be clean and straight, with undamaged threads and undamaged corners on the hex head where the wrench fits. Develop the habit of replacing all damaged nuts and bolts with new ones. Special locknuts with nylon or fiber inserts can only be used once. If they are removed, they lose their locking ability and must be replaced with new ones.

Rusted nuts and bolts should be treated with a penetrating fluid to ease removal and prevent breakage. Some mechanics use turpentine in a spout-type oil can, which works quite well. After applying the rust penetrant, let it work for a few minutes before trying to loosen the nut or bolt. Badly rusted fasteners may have to be chiseled or sawed off or removed with a special nut breaker, available at tool stores.

If a bolt or stud breaks off in an assembly, it can be drilled and removed with a special tool commonly available for this purpose. Most automotive machine shops can perform this task, as well as other repair procedures, such as the repair of threaded holes that have been stripped out.

Flat washers and lockwashers, when removed from an assembly, should always be replaced exactly as removed. Replace any damaged washers with new ones. Never use a lockwasher on any soft metal surface (such as aluminum), thin sheet metal or plastic.

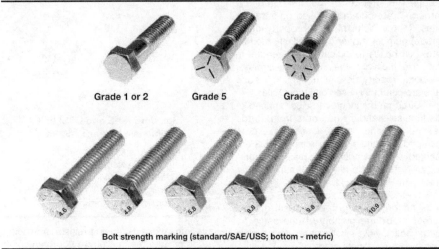

## Fastener sizes

For a number of reasons, automobile manufacturers are making wider and wider use of metric fasteners. Therefore, it is important to be able to tell the difference between standard (sometimes called U.S. or SAE) and metric hardware, since they cannot be interchanged.

All bolts, whether standard or metric, are sized according to diameter, thread pitch and length. For example, a standard 1/2 - 13 x 1 bolt is 1/2 inch in diameter, has 13 threads per inch and is 1 inch long. An M12 - 1.75 x 25 metric bolt is 12 mm in diameter, has a thread pitch of 1.75 mm (the distance between threads) and is 25 mm long. The two bolts are nearly identical, and easily confused, but they are not interchangeable.

In addition to the differences in diameter, thread pitch and length, metric and standard bolts can also be distinguished by examining the bolt heads. To begin with, the distance across the flats on a standard bolt head is measured in inches, while the same dimension on a metric bolt is sized in millimeters (the same is true for nuts). As a result, a standard wrench should not be used on a metric bolt and a metric wrench should not be used on a standard bolt. Also, most standard bolts have slashes radiating out from the center of the head to denote the grade or strength of the bolt, which is an indication of the amount of torque that can be applied to it. The greater the number of slashes, the greater the strength of the bolt. Grades 0 through 5 are commonly used on automobiles. Metric bolts have a property class (grade) number, rather than a slash, molded into their heads to indicate bolt strength. In this case, the higher the number, the stronger the bolt. Property class numbers 8.8, 9.8 and 10.9 are commonly used on automobiles.

Strength markings can also be used to distinguish standard hex nuts from metric hex nuts. Many standard nuts have dots stamped into one side, while metric nuts are marked with a number. The greater the number of dots, or the higher the number, the greater the strength of the nut.

Metric studs are also marked on their ends according to property class (grade). Larger studs are numbered (the same as metric bolts), while smaller studs carry a geometric code to denote grade.

It should be noted that many fasteners, especially Grades 0 through 2, have no distinguishing marks on them. When such is the case, the only way to determine whether it is standard or metric is to measure the thread pitch or compare it to a known fastener of the same size.

Standard fasteners are often referred to as SAE, as opposed to metric. However, it should be noted that SAE technically refers to a non-metric fine thread fastener only. Coarse thread non-metric fasteners are referred to as USS sizes.

Since fasteners of the same size (both standard and metric) may have different

| | Ft-lbs | Nm |
|---|---|---|
| **Metric thread sizes** | | |
| M-6 | 6 to 9 | 9 to 12 |
| M-8 | 14 to 21 | 19 to 28 |
| M-10 | 28 to 40 | 38 to 54 |
| M-12 | 50 to 71 | 68 to 96 |
| M-14 | 80 to 140 | 109 to 154 |
| **Pipe thread sizes** | | |
| 1/8 | 5 to 8 | 7 to 10 |
| 1/4 | 12 to 18 | 17 to 24 |
| 3/8 | 22 to 33 | 30 to 44 |
| 1/2 | 25 to 35 | 34 to 47 |
| **U.S. thread sizes** | | |
| 1/4 - 20 | 6 to 9 | 9 to 12 |
| 5/16 - 18 | 12 to 18 | 17 to 24 |
| 5/16 - 24 | 14 to 20 | 19 to 27 |
| 3/8 - 16 | 22 to 32 | 30 to 43 |
| 3/8 - 24 | 27 to 38 | 37 to 51 |
| 7/16 - 14 | 40 to 55 | 55 to 74 |
| 7/16 - 20 | 40 to 60 | 55 to 81 |
| 1/2 - 13 | 55 to 80 | 75 to 108 |

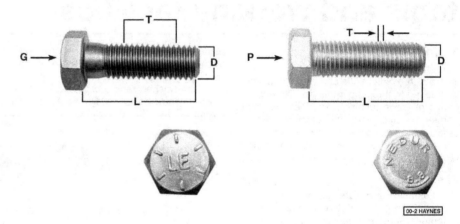

**Standard (SAE and USS) bolt dimensions/grade marks**

- G  Grade marks (bolt strength)
- L  Length (in inches)
- T  Thread pitch (number of threads per inch)
- D  Nominal diameter (in inches)

**Metric bolt dimensions/grade marks**

- P  Property class (bolt strength)
- L  Length (in millimeters)
- T  Thread pitch (distance between threads in millimeters)
- D  Diameter

strength ratings, be sure to reinstall any bolts, studs or nuts removed from your vehicle in their original locations. Also, when replacing a fastener with a new one, make sure that the new one has a strength rating equal to or greater than the original.

## Tightening sequences and procedures

Most threaded fasteners should be tightened to a specific torque value (torque is the twisting force applied to a threaded component such as a nut or bolt). Overtightening the fastener can weaken it and cause it to break, while undertightening can cause it to eventually come loose. Bolts, screws and studs, depending on the material they are made of and their thread diameters, have specific torque values, many of which are noted in the Specifications at the beginning of each Chapter. Be sure to follow the torque recommendations closely. For fasteners not assigned a specific torque, a general torque value chart is presented here as a guide. These torque values are for dry (unlubricated) fasteners threaded into steel or cast iron (not aluminum). As was previously mentioned, the size and grade of a fastener determine the amount of torque that can safely be applied to it. The figures listed here are approximate for Grade 2 and Grade 3 fasteners. Higher grades can tolerate higher torque values.

Fasteners laid out in a pattern, such as cylinder head bolts, oil pan bolts, differential cover bolts, etc., must be loosened or tightened in sequence to avoid warping the com-

# Maintenance techniques, tools and working facilities

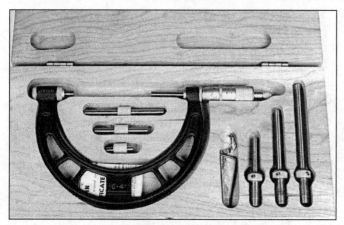

**Micrometer set**

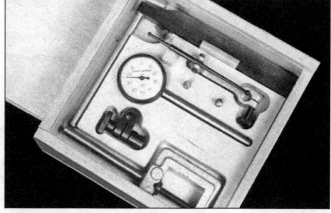

**Dial indicator set**

ponent. This sequence will normally be shown in the appropriate Chapter. If a specific pattern is not given, the following procedures can be used to prevent warping.

Initially, the bolts or nuts should be assembled finger-tight only. Next, they should be tightened one full turn each, in a criss-cross or diagonal pattern. After each one has been tightened one full turn, return to the first one and tighten them all one-half turn, following the same pattern. Finally, tighten each of them one-quarter turn at a time until each fastener has been tightened to the proper torque. To loosen and remove the fasteners, the procedure would be reversed.

## Component disassembly

Component disassembly should be done with care and purpose to help ensure that the parts go back together properly. Always keep track of the sequence in which parts are removed. Make note of special characteristics or marks on parts that can be installed more than one way, such as a grooved thrust washer on a shaft. It is a good idea to lay the disassembled parts out on a clean surface in the order that they were removed. It may also be helpful to make sketches or take instant photos of components before removal.

When removing fasteners from a component, keep track of their locations. Sometimes threading a bolt back in a part, or putting the washers and nut back on a stud, can prevent mix-ups later. If nuts and bolts cannot be returned to their original locations, they should be kept in a compartmented box or a series of small boxes. A cupcake or muffin tin is ideal for this purpose, since each cavity can hold the bolts and nuts from a particular area (i.e. oil pan bolts, valve cover bolts, engine mount bolts, etc.). A pan of this type is especially helpful when working on assemblies with very small parts, such as the carburetor, alternator, valve train or interior dash and trim pieces. The cavities can be marked with paint or tape to identify the contents.

Whenever wiring looms, harnesses or connectors are separated, it is a good idea to identify the two halves with numbered pieces of masking tape so they can be easily reconnected.

## Gasket sealing surfaces

Throughout any vehicle, gaskets are used to seal the mating surfaces between two parts and keep lubricants, fluids, vacuum or pressure contained in an assembly.

Many times these gaskets are coated with a liquid or paste-type gasket sealing compound before assembly. Age, heat and pressure can sometimes cause the two parts to stick together so tightly that they are very difficult to separate. Often, the assembly can be loosened by striking it with a soft-face hammer near the mating surfaces. A regular hammer can be used if a block of wood is placed between the hammer and the part. Do not hammer on cast parts or parts that could be easily damaged. With any particularly stubborn part, always recheck to make sure that every fastener has been removed.

Avoid using a screwdriver or bar to pry apart an assembly, as they can easily mar the gasket sealing surfaces of the parts, which must remain smooth. If prying is absolutely necessary, use an old broom handle, but keep in mind that extra clean up will be necessary if the wood splinters.

After the parts are separated, the old gasket must be carefully scraped off and the gasket surfaces cleaned. Stubborn gasket material can be soaked with rust penetrant or treated with a special chemical to soften it so it can be easily scraped off. A scraper can be fashioned from a piece of copper tubing by flattening and sharpening one end. Copper is recommended because it is usually softer than the surfaces to be scraped, which reduces the chance of gouging the part. Some gaskets can be removed with a wire brush, but regardless of the method used, the mating surfaces must be left clean and smooth. If for some reason the gasket surface is gouged, then a gasket sealer thick enough to fill scratches will have to be used during reassembly of the components. For most applications, a non-drying (or semi-drying) gasket sealer should be used.

## Hose removal tips

**Warning:** *If the vehicle is equipped with air conditioning, do not disconnect any of the A/C hoses without first having the system depressurized by a dealer service department or a service station.*

Hose removal precautions closely parallel gasket removal precautions. Avoid scratching or gouging the surface that the hose mates against or the connection may leak. This is especially true for radiator hoses. Because of various chemical reactions, the rubber in hoses can bond itself to the metal spigot that the hose fits over. To remove a hose, first loosen the hose clamps that secure it to the spigot. Then, with slip-joint pliers, grab the hose at the clamp and rotate it around the spigot. Work it back and forth until it is completely free, then pull it off. Silicone or other lubricants will ease removal if they can be applied between the hose and the outside of the spigot. Apply the same lubricant to the inside of the hose and the outside of the spigot to simplify installation.

As a last resort (and if the hose is to be replaced with a new one anyway), the rubber can be slit with a knife and the hose peeled from the spigot. If this must be done, be careful that the metal connection is not damaged.

If a hose clamp is broken or damaged, do not reuse it. Wire-type clamps usually weaken with age, so it is a good idea to replace them with screw-type clamps whenever a hose is removed.

## Tools

A selection of good tools is a basic requirement for anyone who plans to maintain and repair his or her own vehicle. For the owner who has few tools, the initial investment might seem high, but when compared to the spiraling costs of professional auto maintenance and repair, it is a wise one.

To help the owner decide which tools are needed to perform the tasks detailed in this manual, the following tool lists are offered: *Maintenance and minor repair, Repair/overhaul* and *Special.*

The newcomer to practical mechanics

## Maintenance techniques, tools and working facilities

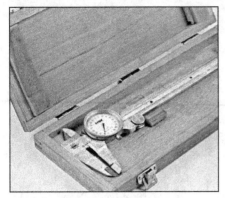
Dial caliper

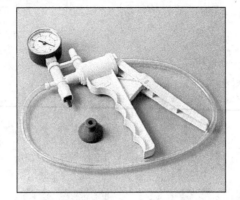
Hand-operated vacuum pump

Timing light

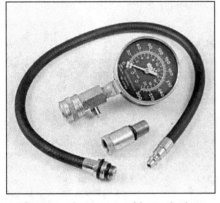
Compression gauge with spark plug hole adapter

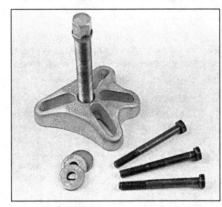
Damper/steering wheel puller

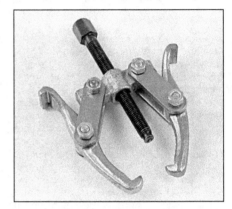
General purpose puller

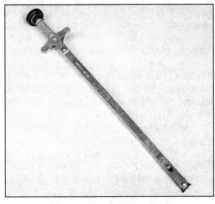
Hydraulic lifter removal tool

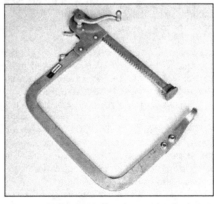
Valve spring compressor

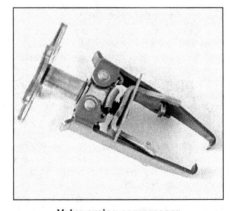
Valve spring compressor

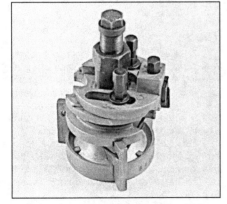
Ridge reamer

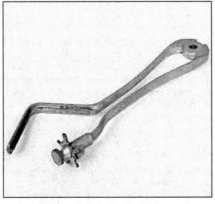
Piston ring groove cleaning tool

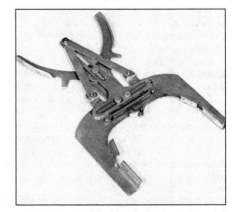
Ring removal/installation tool

## Maintenance techniques, tools and working facilities 0-11

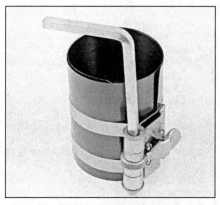
Ring compressor

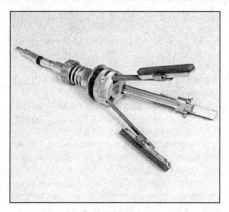
Cylinder hone

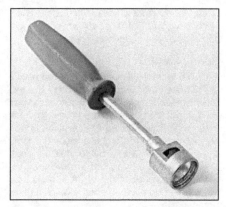
Brake hold-down spring tool

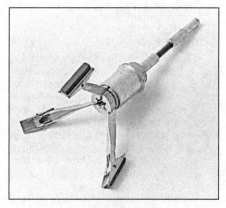
Brake cylinder hone

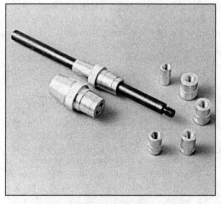
Clutch plate alignment tool

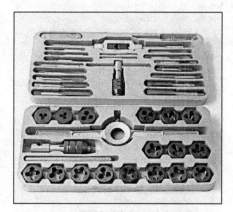
Tap and die set

should start off with the *maintenance and minor repair* tool kit, which is adequate for the simpler jobs performed on a vehicle. Then, as confidence and experience grow, the owner can tackle more difficult tasks, buying additional tools as they are needed. Eventually the basic kit will be expanded into the *repair and overhaul* tool set. Over a period of time, the experienced do-it-yourselfer will assemble a tool set complete enough for most repair and overhaul procedures and will add tools from the special category when it is felt that the expense is justified by the frequency of use.

### Maintenance and minor repair tool kit

The tools in this list should be considered the minimum required for performance of routine maintenance, servicing and minor repair work. We recommend the purchase of combination wrenches (box-end and open-end combined in one wrench). While more expensive than open end wrenches, they offer the advantages of both types of wrench.

*Combination wrench set (1/4-inch to 1 inch or 6 mm to 19 mm)*
*Adjustable wrench, 8 inch*
*Spark plug wrench with rubber insert*
*Spark plug gap adjusting tool*
*Feeler gauge set*
*Brake bleeder wrench*
*Standard screwdriver (5/16-inch x 6 inch)*
*Phillips screwdriver (No. 2 x 6 inch)*
*Combination pliers - 6 inch*
*Hacksaw and assortment of blades*
*Tire pressure gauge*
*Grease gun*
*Oil can*
*Fine emery cloth*
*Wire brush*
*Battery post and cable cleaning tool*
*Oil filter wrench*
*Funnel (medium size)*
*Safety goggles*
*Jackstands (2)*
*Drain pan*

**Note:** *If basic tune-ups are going to be part of routine maintenance, it will be necessary to purchase a good quality stroboscopic timing light and combination tachometer/dwell meter. Although they are included in the list of special tools, it is mentioned here because they are absolutely necessary for tuning most vehicles properly.*

### Repair and overhaul tool set

These tools are essential for anyone who plans to perform major repairs and are in addition to those in the maintenance and minor repair tool kit. Included is a comprehensive set of sockets which, though expensive, are invaluable because of their versatility, especially when various extensions and drives are available. We recommend the 1/2-inch drive over the 3/8-inch drive. Although the larger drive is bulky and more expensive, it has the capacity of accepting a very wide range of large sockets. Ideally, however, the mechanic should have a 3/8-inch drive set and a 1/2-inch drive set.

*Socket set(s)*
*Reversible ratchet*
*Extension - 10 inch*
*Universal joint*
*Torque wrench (same size drive as sockets)*
*Ball peen hammer - 8 ounce*
*Soft-face hammer (plastic/rubber)*
*Standard screwdriver (1/4-inch x 6 inch)*
*Standard screwdriver (stubby - 5/16-inch)*
*Phillips screwdriver (No. 3 x 8 inch)*
*Phillips screwdriver (stubby - No. 2)*
*Pliers - vise grip*
*Pliers - lineman's*
*Pliers - needle nose*
*Pliers - snap-ring (internal and external)*
*Cold chisel - 1/2-inch*
*Scribe*
*Scraper (made from flattened copper tubing)*
*Centerpunch*
*Pin punches (1/16, 1/8, 3/16-inch)*
*Steel rule/straightedge - 12 inch*
*Allen wrench set (1/8 to 3/8-inch or 4 mm to 10 mm)*
*A selection of files*
*Wire brush (large)*
*Jackstands (second set)*
*Jack (scissor or hydraulic type)*

**Note:** *Another tool which is often useful is an electric drill with a chuck capacity of 3/8-inch and a set of good quality drill bits.*

### Special tools

The tools in this list include those which are not used regularly, are expensive to buy, or which need to be used in accordance with their manufacturer's instructions. Unless these tools will be used frequently, it is not very economical to purchase many of them. A consideration would be to split the cost and use between yourself and a friend or friends. In addition, most of these tools can be obtained from a tool rental shop on a temporary basis.

This list primarily contains only those tools and instruments widely available to the public, and not those special tools produced by the vehicle manufacturer for distribution to dealer service departments. Occasionally, references to the manufacturer's special tools are included in the text of this manual. Generally, an alternative method of doing the job without the special tool is offered. However, sometimes there is no alternative to their use. Where this is the case, and the tool cannot be purchased or borrowed, the work should be turned over to the dealer service department or an automotive repair shop.

*Valve spring compressor*
*Piston ring groove cleaning tool*
*Piston ring compressor*
*Piston ring installation tool*
*Cylinder compression gauge*
*Cylinder ridge reamer*
*Cylinder surfacing hone*
*Cylinder bore gauge*
*Micrometers and/or dial calipers*
*Hydraulic lifter removal tool*
*Balljoint separator*
*Universal-type puller*
*Impact screwdriver*
*Dial indicator set*
*Stroboscopic timing light (inductive pick-up)*
*Hand operated vacuum/pressure pump*
*Tachometer/dwell meter*
*Universal electrical multimeter*
*Cable hoist*
*Brake spring removal and installation tools*
*Floor jack*

### Buying tools

For the do-it-yourselfer who is just starting to get involved in vehicle maintenance and repair, there are a number of options available when purchasing tools. If maintenance and minor repair is the extent of the work to be done, the purchase of individual tools is satisfactory. If, on the other hand, extensive work is planned, it would be a good idea to purchase a modest tool set from one of the large retail chain stores. A set can usually be bought at a substantial savings over the individual tool prices, and they often come with a tool box. As additional tools are needed, add-on sets, individual tools and a larger tool box can be purchased to expand the tool selection. Building a tool set gradually allows the cost of the tools to be spread over a longer period of time and gives the mechanic the freedom to choose only those tools that will actually be used.

Tool stores will often be the only source of some of the special tools that are needed, but regardless of where tools are bought, try to avoid cheap ones, especially when buying screwdrivers and sockets, because they won't last very long. The expense involved in replacing cheap tools will eventually be greater than the initial cost of quality tools.

### Care and maintenance of tools

Good tools are expensive, so it makes sense to treat them with respect. Keep them clean and in usable condition and store them properly when not in use. Always wipe off any dirt, grease or metal chips before putting them away. Never leave tools lying around in the work area. Upon completion of a job, always check closely under the hood for tools that may have been left there so they won't get lost during a test drive.

Some tools, such as screwdrivers, pliers, wrenches and sockets, can be hung on a panel mounted on the garage or workshop wall, while others should be kept in a tool box or tray. Measuring instruments, gauges, meters, etc. must be carefully stored where they cannot be damaged by weather or impact from other tools.

When tools are used with care and stored properly, they will last a very long time. Even with the best of care, though, tools will wear out if used frequently. When a tool is damaged or worn out, replace it. Subsequent jobs will be safer and more enjoyable if you do.

### How to repair damaged threads

Sometimes, the internal threads of a nut or bolt hole can become stripped, usually from overtightening. Stripping threads is an all-too-common occurrence, especially when working with aluminum parts, because aluminum is so soft that it easily strips out.

Usually, external or internal threads are only partially stripped. After they've been cleaned up with a tap or die, they'll still work. Sometimes, however, threads are badly damaged. When this happens, you've got three choices:

1) *Drill and tap the hole to the next suitable oversize and install a larger diameter bolt, screw or stud.*
2) *Drill and tap the hole to accept a threaded plug, then drill and tap the plug to the original screw size. You can also buy a plug already threaded to the original size. Then you simply drill a hole to the specified size, then run the threaded plug into the hole with a bolt and jam nut. Once the plug is fully seated, remove the jam nut and bolt.*
3) *The third method uses a patented thread repair kit like Heli-Coil or Slimsert. These easy-to-use kits are designed to repair damaged threads in straight-through holes and blind holes. Both are available as kits which can handle a variety of sizes and thread patterns. Drill the hole, then tap it with the special included tap. Install the Heli-Coil and the hole is back to its original diameter and thread pitch.*

Regardless of which method you use, be sure to proceed calmly and carefully. A little impatience or carelessness during one of these relatively simple procedures can ruin your whole day's work and cost you a bundle if you wreck an expensive part.

### Working facilities

Not to be overlooked when discussing tools is the workshop. If anything more than routine maintenance is to be carried out, some sort of suitable work area is essential.

It is understood, and appreciated, that many home mechanics do not have a good workshop or garage available, and end up removing an engine or doing major repairs outside. It is recommended, however, that the overhaul or repair be completed under the cover of a roof.

A clean, flat workbench or table of comfortable working height is an absolute necessity. The workbench should be equipped with a vise that has a jaw opening of at least four inches.

As mentioned previously, some clean, dry storage space is also required for tools, as well as the lubricants, fluids, cleaning solvents, etc. which soon become necessary.

Sometimes waste oil and fluids, drained from the engine or cooling system during normal maintenance or repairs, present a disposal problem. To avoid pouring them on the ground or into a sewage system, pour the used fluids into large containers, seal them with caps and take them to an authorized disposal site or recycling center. Plastic jugs, such as old antifreeze containers, are ideal for this purpose.

Always keep a supply of old newspapers and clean rags available. Old towels are excellent for mopping up spills. Many mechanics use rolls of paper towels for most work because they are readily available and disposable. To help keep the area under the vehicle clean, a large cardboard box can be cut open and flattened to protect the garage or shop floor.

Whenever working over a painted surface, such as when leaning over a fender to service something under the hood, always cover it with an old blanket or bedspread to protect the finish. Vinyl covered pads, made especially for this purpose, are available at auto parts stores.

# Jacking and towing

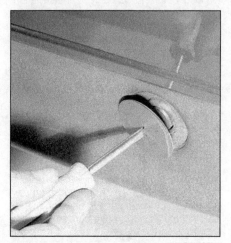

Unscrew the access cover from the rocker panel . . .

. . . and insert the jack into the hole (all except Z3 models)

## Jacking

The jack supplied with the vehicle should be used only for raising the vehicle when changing a tire or placing jackstands under the frame. **Warning:** *Never crawl under the vehicle or start the engine when this jack is being used as the only means of support.*

The vehicle should be on level ground with the wheels blocked and the transmission in Park (automatic) or Reverse (manual). Pry off the hub cap (if equipped) using the tapered end of the lug wrench. Loosen the lug bolts one-half turn and leave them in place until the wheel is raised off the ground.

On all except Z3 models, unscrew the access cover in the rocker panel nearest the wheel to be changed **(see illustration)**. Engage the head of the jack with the hole in the rocker panel **(see illustration)**. If you're working on a Z3 model, engage the head of the jack with the jacking point on the underside of the vehicle, directly behind the front wheel of right in front of the rear wheel. Block the wheel opposite the one being changed. Make sure the jack is located on firm ground; if not, place a block of wood under it. Turn the jack handle clockwise until the wheel is raised off the ground, then unscrew the lug bolts and remove the wheel.

Install the spare wheel and tighten the lug bolts until they are snug. Lower the vehicle by turning the jack handle counterclockwise, then remove the jack and install the access cover (3-series models). Tighten the bolts in a diagonal pattern to the torque listed in the Chapter 1 Specifications. If a torque wrench is not available, have the torque checked by a service station as soon as possible. Reinstall the hub cap, if equipped.

## Towing

Manual transmission-equipped vehicles can be towed with all four wheels on the ground. Automatic transmission-equipped vehicles can be towed with all four wheels on the ground if speeds do not exceed 35 mph and the distance is not over 50 miles, otherwise transmission damage can result.

Towing equipment specifically designed for this purpose should be used and should be attached to the main structural members of the vehicle, not the bumper or brackets. For towing with a strap or cable (and for pulling the vehicle onto a trailer with a winch), a towing eye is provided in the tool kit (located in the luggage compartment lid) which can be screwed into the front or rear of the vehicle, as necessary, after prying open a cover on the bumper **(see illustration)**.

The towing eye is supplied as part of the vehicle's tool kit. To install the eye, pry out the access cover from the front or rear bumper, then screw the eye into position and tighten it securely

# Anti-theft audio system

The stereo unit installed as standard equipment by BMW is equipped with a built-in security code to deter thieves. If the power source to the unit is cut, the anti-theft system will activate. Even if the power source is immediately reconnected, the stereo unit will not function until the correct security code has been entered. Therefore, if you do not know the correct security code for the unit, **do not** disconnect the battery negative cable or remove the radio/cassette unit from the vehicle.

The procedure for reprogramming a unit that has been disconnected from its power supply varies from model to model - consult the owners manual supplied with the unit for specific details or refer to your BMW dealer.

# Automotive chemicals and lubricants

A number of automotive chemicals and lubricants are available for use during vehicle maintenance and repair. They include a wide variety of products ranging from cleaning solvents and degreasers to lubricants and protective sprays for rubber, plastic and vinyl.

## Cleaners

**Carburetor cleaner and choke cleaner** is a strong solvent for gum, varnish and carbon. Most carburetor cleaners leave a dry-type lubricant film which will not harden or gum up. Because of this film it is not recommended for use on electrical components.

**Brake system cleaner** is used to remove grease and brake fluid from the brake system, where clean surfaces are absolutely necessary. It leaves no residue and often eliminates brake squeal caused by contaminants.

**Electrical cleaner** removes oxidation, corrosion and carbon deposits from electrical contacts, restoring full current flow. It can also be used to clean spark plugs, carburetor jets, voltage regulators and other parts where an oil-free surface is desired.

**Demoisturants** remove water and moisture from electrical components such as alternators, voltage regulators, electrical connectors and fuse blocks. They are non-conductive, non-corrosive and non-flammable.

**Degreasers** are heavy-duty solvents used to remove grease from the outside of the engine and from chassis components. They can be sprayed or brushed on and, depending on the type, are rinsed off either with water or solvent.

## Lubricants

**Motor oil** is the lubricant formulated for use in engines. It normally contains a wide variety of additives to prevent corrosion and reduce foaming and wear. Motor oil comes in various weights (viscosity ratings) from 0 to 50. The recommended weight of the oil depends on the season, temperature and the demands on the engine. Light oil is used in cold climates and under light load conditions. Heavy oil is used in hot climates and where high loads are encountered. Multi-viscosity oils are designed to have characteristics of both light and heavy oils and are available in a number of weights from 5W-20 to 20W-50.

**Gear oil** is designed to be used in differentials, manual transmissions and other areas where high-temperature lubrication is required.

**Chassis and wheel bearing grease** is a heavy grease used where increased loads and friction are encountered, such as for wheel bearings, balljoints, tie-rod ends and universal joints.

**High-temperature wheel bearing grease** is designed to withstand the extreme temperatures encountered by wheel bearings in disc brake equipped vehicles. It usually contains molybdenum disulfide (moly), which is a dry-type lubricant.

**White grease** is a heavy grease for metal-to-metal applications where water is a problem. White grease stays soft under both low and high temperatures (usually from -100 to +190-degrees F), and will not wash off or dilute in the presence of water.

**Assembly lube** is a special extreme pressure lubricant, usually containing moly, used to lubricate high-load parts (such as main and rod bearings and cam lobes) for initial start-up of a new engine. The assembly lube lubricates the parts without being squeezed out or washed away until the engine oiling system begins to function.

**Silicone lubricants** are used to protect rubber, plastic, vinyl and nylon parts.

**Graphite lubricants** are used where oils cannot be used due to contamination problems, such as in locks. The dry graphite will lubricate metal parts while remaining uncontaminated by dirt, water, oil or acids. It is electrically conductive and will not foul electrical contacts in locks such as the ignition switch.

**Moly penetrants** loosen and lubricate frozen, rusted and corroded fasteners and prevent future rusting or freezing.

**Heat-sink grease** is a special electrically non-conductive grease that is used for mounting electronic ignition modules where it is essential that heat is transferred away from the module.

## Sealants

**RTV sealant** is one of the most widely used gasket compounds. Made from silicone, RTV is air curing, it seals, bonds, waterproofs, fills surface irregularities, remains flexible, doesn't shrink, is relatively easy to remove, and is used as a supplementary sealer with almost all low and medium temperature gaskets.

**Anaerobic sealant** is much like RTV in that it can be used either to seal gaskets or to form gaskets by itself. It remains flexible, is solvent resistant and fills surface imperfections. The difference between an anaerobic sealant and an RTV-type sealant is in the curing. RTV cures when exposed to air, while an anaerobic sealant cures only in the absence of air. This means that an anaerobic sealant cures only after the assembly of parts, sealing them together.

**Thread and pipe sealant** is used for sealing hydraulic and pneumatic fittings and vacuum lines. It is usually made from a Teflon compound, and comes in a spray, a paint-on liquid and as a wrap-around tape.

## Chemicals

**Anti-seize compound** prevents seizing, galling, cold welding, rust and corrosion in fasteners. High-temperature anti-seize, usually made with copper and graphite lubricants, is used for exhaust system and exhaust manifold bolts.

**Anaerobic locking compounds** are used to keep fasteners from vibrating or working loose and cure only after installation, in the absence of air. Medium strength locking compound is used for small nuts, bolts and screws that may be removed later. High-strength locking compound is for large nuts, bolts and studs which aren't removed on a regular basis.

**Oil additives** range from viscosity index improvers to chemical treatments that claim to reduce internal engine friction. It should be noted that most oil manufacturers caution against using additives with their oils.

**Gas additives** perform several functions, depending on their chemical makeup. They usually contain solvents that help dissolve gum and varnish that build up on carburetor, fuel injection and intake parts. They also serve to break down carbon deposits that form on the inside surfaces of the combustion chambers. Some additives contain upper cylinder lubricants for valves and piston rings, and others contain chemicals to remove condensation from the gas tank.

## Miscellaneous

**Brake fluid** is specially formulated hydraulic fluid that can withstand the heat and pressure encountered in brake systems. Care must be taken so this fluid does not come in contact with painted surfaces or plastics. An opened container should always be resealed to prevent contamination by water or dirt.

**Weatherstrip adhesive** is used to bond weatherstripping around doors, windows and trunk lids. It is sometimes used to attach trim pieces.

**Undercoating** is a petroleum-based, tar-like substance that is designed to protect metal surfaces on the underside of the vehicle from corrosion. It also acts as a sound-deadening agent by insulating the bottom of the vehicle.

**Waxes and polishes** are used to help protect painted and plated surfaces from the weather. Different types of paint may require the use of different types of wax and polish. Some polishes utilize a chemical or abrasive cleaner to help remove the top layer of oxidized (dull) paint on older vehicles. In recent years many non-wax polishes that contain a wide variety of chemicals such as polymers and silicones have been introduced. These non-wax polishes are usually easier to apply and last longer than conventional waxes and polishes.

# Conversion factors

### Length (distance)
| | | | | |
|---|---|---|---|---|
| Inches (in) | X 25.4 | = Millimetres (mm) | X 0.0394 | = Inches (in) |
| Feet (ft) | X 0.305 | = Metres (m) | X 3.281 | = Feet (ft) |
| Miles | X 1.609 | = Kilometres (km) | X 0.621 | = Miles |

### Volume (capacity)
| | | | | |
|---|---|---|---|---|
| Cubic inches (cu in; in$^3$) | X 16.387 | = Cubic centimetres (cc; cm$^3$) | X 0.061 | = Cubic inches (cu in; in$^3$) |
| Imperial pints (Imp pt) | X 0.568 | = Litres (l) | X 1.76 | = Imperial pints (Imp pt) |
| Imperial quarts (Imp qt) | X 1.137 | = Litres (l) | X 0.88 | = Imperial quarts (Imp qt) |
| Imperial quarts (Imp qt) | X 1.201 | = US quarts (US qt) | X 0.833 | = Imperial quarts (Imp qt) |
| US quarts (US qt) | X 0.946 | = Litres (l) | X 1.057 | = US quarts (US qt) |
| Imperial gallons (Imp gal) | X 4.546 | = Litres (l) | X 0.22 | = Imperial gallons (Imp gal) |
| Imperial gallons (Imp gal) | X 1.201 | = US gallons (US gal) | X 0.833 | = Imperial gallons (Imp gal) |
| US gallons (US gal) | X 3.785 | = Litres (l) | X 0.264 | = US gallons (US gal) |

### Mass (weight)
| | | | | |
|---|---|---|---|---|
| Ounces (oz) | X 28.35 | = Grams (g) | X 0.035 | = Ounces (oz) |
| Pounds (lb) | X 0.454 | = Kilograms (kg) | X 2.205 | = Pounds (lb) |

### Force
| | | | | |
|---|---|---|---|---|
| Ounces-force (ozf; oz) | X 0.278 | = Newtons (N) | X 3.6 | = Ounces-force (ozf; oz) |
| Pounds-force (lbf; lb) | X 4.448 | = Newtons (N) | X 0.225 | = Pounds-force (lbf; lb) |
| Newtons (N) | X 0.1 | = Kilograms-force (kgf; kg) | X 9.81 | = Newtons (N) |

### Pressure
| | | | | |
|---|---|---|---|---|
| Pounds-force per square inch (psi; lbf/in$^2$; lb/in$^2$) | X 0.070 | = Kilograms-force per square centimetre (kgf/cm$^2$; kg/cm$^2$) | X 14.223 | = Pounds-force per square inch (psi; lbf/in$^2$; lb/in$^2$) |
| Pounds-force per square inch (psi; lbf/in$^2$; lb/in$^2$) | X 0.068 | = Atmospheres (atm) | X 14.696 | = Pounds-force per square inch (psi; lbf/in$^2$; lb/in$^2$) |
| Pounds-force per square inch (psi; lbf/in$^2$; lb/in$^2$) | X 0.069 | = Bars | X 14.5 | = Pounds-force per square inch (psi; lbf/in$^2$; lb/in$^2$) |
| Pounds-force per square inch (psi; lbf/in$^2$; lb/in$^2$) | X 6.895 | = Kilopascals (kPa) | X 0.145 | = Pounds-force per square inch (psi; lbf/in$^2$; lb/in$^2$) |
| Kilopascals (kPa) | X 0.01 | = Kilograms-force per square centimetre (kgf/cm$^2$; kg/cm$^2$) | X 98.1 | = Kilopascals (kPa) |

### Torque (moment of force)
| | | | | |
|---|---|---|---|---|
| Pounds-force inches (lbf in; lb in) | X 1.152 | = Kilograms-force centimetre (kgf cm; kg cm) | X 0.868 | = Pounds-force inches (lbf in; lb in) |
| Pounds-force inches (lbf in; lb in) | X 0.113 | = Newton metres (Nm) | X 8.85 | = Pounds-force inches (lbf in; lb in) |
| Pounds-force inches (lbf in; lb in) | X 0.083 | = Pounds-force feet (lbf ft; lb ft) | X 12 | = Pounds-force inches (lbf in; lb in) |
| Pounds-force feet (lbf ft; lb ft) | X 0.138 | = Kilograms-force metres (kgf m; kg m) | X 7.233 | = Pounds-force feet (lbf ft; lb ft) |
| Pounds-force feet (lbf ft; lb ft) | X 1.356 | = Newton metres (Nm) | X 0.738 | = Pounds-force feet (lbf ft; lb ft) |
| Newton metres (Nm) | X 0.102 | = Kilograms-force metres (kgf m; kg m) | X 9.804 | = Newton metres (Nm) |

### Vacuum
| | | | | |
|---|---|---|---|---|
| Inches mercury (in. Hg) | X 3.377 | = Kilopascals (kPa) | X 0.2961 | = Inches mercury |
| Inches mercury (in. Hg) | X 25.4 | = Millimeters mercury (mm Hg) | X 0.0394 | = Inches mercury |

### Power
| | | | | |
|---|---|---|---|---|
| Horsepower (hp) | X 745.7 | = Watts (W) | X 0.0013 | = Horsepower (hp) |

### Velocity (speed)
| | | | | |
|---|---|---|---|---|
| Miles per hour (miles/hr; mph) | X 1.609 | = Kilometres per hour (km/hr; kph) | X 0.621 | = Miles per hour (miles/hr; mph) |

### Fuel consumption*
| | | | | |
|---|---|---|---|---|
| Miles per gallon, Imperial (mpg) | X 0.354 | = Kilometres per litre (km/l) | X 2.825 | = Miles per gallon, Imperial (mpg) |
| Miles per gallon, US (mpg) | X 0.425 | = Kilometres per litre (km/l) | X 2.352 | = Miles per gallon, US (mpg) |

### Temperature
Degrees Fahrenheit = (°C x 1.8) + 32         Degrees Celsius (Degrees Centigrade; °C) = (°F - 32) x 0.56

*It is common practice to convert from miles per gallon (mpg) to litres/100 kilometres (l/100km), where mpg (Imperial) x l/100 km = 282 and mpg (US) x l/100 km = 235

## Fraction/Decimal/Millimeter Equivalents

**DECIMALS TO MILLIMETERS**  **FRACTIONS TO DECIMALS TO MILLIMETERS**

| Decimal | mm | Decimal | mm | Fraction | Decimal | mm | Fraction | Decimal | mm |
|---|---|---|---|---|---|---|---|---|---|
| 0.001 | 0.0254 | 0.500 | 12.7000 | 1/64 | 0.0156 | 0.3969 | 33/64 | 0.5156 | 13.0969 |
| 0.002 | 0.0508 | 0.510 | 12.9540 | 1/32 | 0.0312 | 0.7938 | 17/32 | 0.5312 | 13.4938 |
| 0.003 | 0.0762 | 0.520 | 13.2080 | 3/64 | 0.0469 | 1.1906 | 35/64 | 0.5469 | 13.8906 |
| 0.004 | 0.1016 | 0.530 | 13.4620 | | | | | | |
| 0.005 | 0.1270 | 0.540 | 13.7160 | | | | | | |
| 0.006 | 0.1524 | 0.550 | 13.9700 | 1/16 | 0.0625 | 1.5875 | 9/16 | 0.5625 | 14.2875 |
| 0.007 | 0.1778 | 0.560 | 14.2240 | | | | | | |
| 0.008 | 0.2032 | 0.570 | 14.4780 | | | | | | |
| 0.009 | 0.2286 | 0.580 | 14.7320 | 5/64 | 0.0781 | 1.9844 | 37/64 | 0.5781 | 14.6844 |
| | | 0.590 | 14.9860 | 3/32 | 0.0938 | 2.3812 | 19/32 | 0.5938 | 15.0812 |
| 0.010 | 0.2540 | | | 7/64 | 0.1094 | 2.7781 | 39/64 | 0.6094 | 15.4781 |
| 0.020 | 0.5080 | | | | | | | | |
| 0.030 | 0.7620 | | | | | | | | |
| 0.040 | 1.0160 | 0.600 | 15.2400 | 1/8 | 0.1250 | 3.1750 | 5/8 | 0.6250 | 15.8750 |
| 0.050 | 1.2700 | 0.610 | 15.4940 | | | | | | |
| 0.060 | 1.5240 | 0.620 | 15.7480 | | | | | | |
| 0.070 | 1.7780 | 0.630 | 16.0020 | 9/64 | 0.1406 | 3.5719 | 41/64 | 0.6406 | 16.2719 |
| 0.080 | 2.0320 | 0.640 | 16.2560 | 5/32 | 0.1562 | 3.9688 | 21/32 | 0.6562 | 16.6688 |
| 0.090 | 2.2860 | 0.650 | 16.5100 | 11/64 | 0.1719 | 4.3656 | 43/64 | 0.6719 | 17.0656 |
| | | 0.660 | 16.7640 | | | | | | |
| 0.100 | 2.5400 | 0.670 | 17.0180 | | | | | | |
| 0.110 | 2.7940 | 0.680 | 17.2720 | | | | | | |
| 0.120 | 3.0480 | 0.690 | 17.5260 | 3/16 | 0.1875 | 4.7625 | 11/16 | 0.6875 | 17.4625 |
| 0.130 | 3.3020 | | | | | | | | |
| 0.140 | 3.5560 | | | | | | | | |
| 0.150 | 3.8100 | | | 13/64 | 0.2031 | 5.1594 | 45/64 | 0.7031 | 17.8594 |
| 0.160 | 4.0640 | 0.700 | 17.7800 | 7/32 | 0.2188 | 5.5562 | 23/32 | 0.7188 | 18.2562 |
| 0.170 | 4.3180 | 0.710 | 18.0340 | 15/64 | 0.2344 | 5.9531 | 47/64 | 0.7344 | 18.6531 |
| 0.180 | 4.5720 | 0.720 | 18.2880 | | | | | | |
| 0.190 | 4.8260 | 0.730 | 18.5420 | | | | | | |
| | | 0.740 | 18.7960 | 1/4 | 0.2500 | 6.3500 | 3/4 | 0.7500 | 19.0500 |
| 0.200 | 5.0800 | 0.750 | 19.0500 | | | | | | |
| 0.210 | 5.3340 | 0.760 | 19.3040 | | | | | | |
| 0.220 | 5.5880 | 0.770 | 19.5580 | 17/64 | 0.2656 | 6.7469 | 49/64 | 0.7656 | 19.4469 |
| 0.230 | 5.8420 | 0.780 | 19.8120 | 9/32 | 0.2812 | 7.1438 | 25/32 | 0.7812 | 19.8438 |
| 0.240 | 6.0960 | 0.790 | 20.0660 | 19/64 | 0.2969 | 7.5406 | 51/64 | 0.7969 | 20.2406 |
| 0.250 | 6.3500 | | | | | | | | |
| 0.260 | 6.6040 | 0.800 | 20.3200 | | | | | | |
| 0.270 | 6.8580 | 0.810 | 20.5740 | 5/16 | 0.3125 | 7.9375 | 13/16 | 0.8125 | 20.6375 |
| 0.280 | 7.1120 | 0.820 | 21.8280 | | | | | | |
| 0.290 | 7.3660 | 0.830 | 21.0820 | | | | | | |
| | | 0.840 | 21.3360 | 21/64 | 0.3281 | 8.3344 | 53/64 | 0.8281 | 21.0344 |
| 0.300 | 7.6200 | 0.850 | 21.5900 | 11/32 | 0.3438 | 8.7312 | 27/32 | 0.8438 | 21.4312 |
| 0.310 | 7.8740 | 0.860 | 21.8440 | 23/64 | 0.3594 | 9.1281 | 55/64 | 0.8594 | 21.8281 |
| 0.320 | 8.1280 | 0.870 | 22.0980 | | | | | | |
| 0.330 | 8.3820 | 0.880 | 22.3520 | | | | | | |
| 0.340 | 8.6360 | 0.890 | 22.6060 | 3/8 | 0.3750 | 9.5250 | 7/8 | 0.8750 | 22.2250 |
| 0.350 | 8.8900 | | | | | | | | |
| 0.360 | 9.1440 | | | | | | | | |
| 0.370 | 9.3980 | | | 25/64 | 0.3906 | 9.9219 | 57/64 | 0.8906 | 22.6219 |
| 0.380 | 9.6520 | | | 13/32 | 0.4062 | 10.3188 | 29/32 | 0.9062 | 23.0188 |
| 0.390 | 9.9060 | | | 27/64 | 0.4219 | 10.7156 | 59/64 | 0.9219 | 23.4156 |
| | | 0.900 | 22.8600 | | | | | | |
| 0.400 | 10.1600 | 0.910 | 23.1140 | | | | | | |
| 0.410 | 10.4140 | 0.920 | 23.3680 | | | | | | |
| 0.420 | 10.6680 | 0.930 | 23.6220 | 7/16 | 0.4375 | 11.1125 | 15/16 | 0.9375 | 23.8125 |
| 0.430 | 10.9220 | 0.940 | 23.8760 | | | | | | |
| 0.440 | 11.1760 | 0.950 | 24.1300 | | | | | | |
| 0.450 | 11.4300 | 0.960 | 24.3840 | 29/64 | 0.4531 | 11.5094 | 61/64 | 0.9531 | 24.2094 |
| 0.460 | 11.6840 | 0.970 | 24.6380 | 15/32 | 0.4688 | 11.9062 | 31/32 | 0.9688 | 24.6062 |
| 0.470 | 11.9380 | 0.980 | 24.8920 | 31/64 | 0.4844 | 12.3031 | 63/64 | 0.9844 | 25.0031 |
| 0.480 | 12.1920 | 0.990 | 25.1460 | | | | | | |
| 0.490 | 12.4460 | 1.000 | 25.4000 | 1/2 | 0.5000 | 12.7000 | 1 | 1.0000 | 25.4000 |

# Safety first!

Regardless of how enthusiastic you may be about getting on with the job at hand, take the time to ensure that your safety is not jeopardized. A moment's lack of attention can result in an accident, as can failure to observe certain simple safety precautions. The possibility of an accident will always exist, and the following points should not be considered a comprehensive list of all dangers. Rather, they are intended to make you aware of the risks and to encourage a safety conscious approach to all work you carry out on your vehicle.

## Essential DOs and DON'Ts

**DON'T** rely on a jack when working under the vehicle. Always use approved jackstands to support the weight of the vehicle and place them under the recommended lift or support points.

**DON'T** attempt to loosen extremely tight fasteners (i.e. wheel lug nuts) while the vehicle is on a jack - it may fall.

**DON'T** start the engine without first making sure that the transmission is in Neutral (or Park where applicable) and the parking brake is set.

**DON'T** remove the radiator cap from a hot cooling system - let it cool or cover it with a cloth and release the pressure gradually.

**DON'T** attempt to drain the engine oil until you are sure it has cooled to the point that it will not burn you.

**DON'T** touch any part of the engine or exhaust system until it has cooled sufficiently to avoid burns.

**DON'T** siphon toxic liquids such as gasoline, antifreeze and brake fluid by mouth, or allow them to remain on your skin.

**DON'T** inhale brake lining dust - it is potentially hazardous (see *Asbestos* below).

**DON'T** allow spilled oil or grease to remain on the floor - wipe it up before someone slips on it.

**DON'T** use loose fitting wrenches or other tools which may slip and cause injury.

**DON'T** push on wrenches when loosening or tightening nuts or bolts. Always try to pull the wrench toward you. If the situation calls for pushing the wrench away, push with an open hand to avoid scraped knuckles if the wrench should slip.

**DON'T** attempt to lift a heavy component alone - get someone to help you.

**DON'T** rush or take unsafe shortcuts to finish a job.

**DON'T** allow children or animals in or around the vehicle while you are working on it.

**DO** wear eye protection when using power tools such as a drill, sander, bench grinder, etc. and when working under a vehicle.

**DO** keep loose clothing and long hair well out of the way of moving parts.

**DO** make sure that any hoist used has a safe working load rating adequate for the job.

**DO** get someone to check on you periodically when working alone on a vehicle.

**DO** carry out work in a logical sequence and make sure that everything is correctly assembled and tightened.

**DO** keep chemicals and fluids tightly capped and out of the reach of children and pets.

**DO** remember that your vehicle's safety affects that of yourself and others. If in doubt on any point, get professional advice.

## Asbestos

Certain friction, insulating, sealing, and other products - such as brake linings, brake bands, clutch linings, torque converters, gaskets, etc. - may contain asbestos. Extreme care must be taken to avoid inhalation of dust from such products, since it is hazardous to health. If in doubt, assume that they do contain asbestos.

## Fire

Remember at all times that gasoline is highly flammable. Never smoke or have any kind of open flame around when working on a vehicle. But the risk does not end there. A spark caused by an electrical short circuit, by two metal surfaces contacting each other, or even by static electricity built up in your body under certain conditions, can ignite gasoline vapors, which in a confined space are highly explosive. Do not, under any circumstances, use gasoline for cleaning parts. Use an approved safety solvent.

Always disconnect the battery ground (-) cable at the battery before working on any part of the fuel system or electrical system. Never risk spilling fuel on a hot engine or exhaust component. It is strongly recommended that a fire extinguisher suitable for use on fuel and electrical fires be kept handy in the garage or workshop at all times. Never try to extinguish a fuel or electrical fire with water.

## Fumes

Certain fumes are highly toxic and can quickly cause unconsciousness and even death if inhaled to any extent. Gasoline vapor falls into this category, as do the vapors from some cleaning solvents. Any draining or pouring of such volatile fluids should be done in a well ventilated area.

When using cleaning fluids and solvents, read the instructions on the container carefully. Never use materials from unmarked containers.

Never run the engine in an enclosed space, such as a garage. Exhaust fumes contain carbon monoxide, which is extremely poisonous. If you need to run the engine, always do so in the open air, or at least have the rear of the vehicle outside the work area.

If you are fortunate enough to have the use of an inspection pit, never drain or pour gasoline and never run the engine while the vehicle is over the pit. The fumes, being heavier than air, will concentrate in the pit with possibly lethal results.

## The battery

Never create a spark or allow a bare light bulb near a battery. They normally give off a certain amount of hydrogen gas, which is highly explosive.

Always disconnect the battery ground (-) cable at the battery before working on the fuel or electrical systems.

If possible, loosen the filler caps or cover when charging the battery from an external source (this does not apply to sealed or maintenance-free batteries). Do not charge at an excessive rate or the battery may burst.

Take care when adding water to a non maintenance-free battery and when carrying a battery. The electrolyte, even when diluted, is very corrosive and should not be allowed to contact clothing or skin.

Always wear eye protection when cleaning the battery to prevent the caustic deposits from entering your eyes.

## Household current

When using an electric power tool, inspection light, etc., which operates on household current, always make sure that the tool is correctly connected to its plug and that, where necessary, it is properly grounded. Do not use such items in damp conditions and, again, do not create a spark or apply excessive heat in the vicinity of fuel or fuel vapor.

## Secondary ignition system voltage

A severe electric shock can result from touching certain parts of the ignition system (such as the spark plug wires) when the engine is running or being cranked, particularly if components are damp or the insulation is defective. In the case of an electronic ignition system, the secondary system voltage is much higher and could prove fatal.

# Troubleshooting

## Contents

### Engine

| Symptom | Section |
|---|---|
| Engine fails to rotate when attempting to start | 1 |
| Engine rotates, but will not start | 2 |
| Engine difficult to start when cold | 3 |
| Engine difficult to start when hot | 4 |
| Starter motor noisy or excessively rough in engagement | 5 |
| Starter motor turns engine slowly | 6 |
| Engine starts, but stops immediately | 7 |
| Engine idles erratically | 8 |
| Engine misfires at idle speed | 9 |
| Engine misfires throughout the driving speed range | 10 |
| Engine stalls | 11 |
| Engine lacks power | 12 |
| Engine backfires | 13 |
| Oil pressure warning light illuminated with engine running | 14 |
| Engine runs-on after switching off | 15 |
| Engine noises | 16 |

### Cooling system

| Symptom | Section |
|---|---|
| Overheating | 17 |
| Overcooling | 18 |
| External coolant leakage | 19 |
| Internal coolant leakage | 20 |
| Corrosion | 21 |

### Fuel and exhaust systems

| Symptom | Section |
|---|---|
| Excessive fuel consumption | 22 |
| Fuel leakage and/or fuel odor | 23 |
| Excessive noise or fumes from exhaust system | 24 |

### Clutch

| Symptom | Section |
|---|---|
| Pedal travels to floor - no pressure or very little resistance | 25 |
| Clutch fails to disengage (unable to select gears) | 26 |
| Clutch slips (engine speed increases, with no increase in vehicle speed) | 27 |
| Shudder as clutch is engaged | 28 |
| Noise when depressing or releasing clutch pedal | 29 |

### Manual transmission

| Symptom | Section |
|---|---|
| Noisy in neutral with engine running | 30 |
| Noisy in one particular gear | 31 |
| Difficulty engaging gears | 32 |
| Jumps out of gear | 33 |
| Vibration | 34 |
| Lubricant leaks | 35 |

### Automatic transmission

| Symptom | Section |
|---|---|
| Fluid leakage | 36 |
| Transmission fluid brown, or has burned smell | 37 |
| General gear selection problems | 38 |
| Transmission will not downshift (kickdown) with accelerator pedal fully depressed | 39 |
| Engine will not start in any gear, or starts in gears other than Park or Neutral | 40 |
| Transmission slips, shifts roughly, is noisy, or has no drive in forward or reverse gears | 41 |

### Differential and driveshaft

| Symptom | Section |
|---|---|
| Vibration when accelerating or decelerating | 42 |
| Low pitched whining; increasing with road speed | 43 |

### Braking system

| Symptom | Section |
|---|---|
| Vehicle pulls to one side under braking | 44 |
| Noise (grinding or high-pitched squeal) when brakes applied | 45 |
| Excessive brake pedal travel | 46 |
| Brake pedal feels spongy when depressed | 47 |
| Excessive brake pedal effort required to stop vehicle | 48 |
| Shudder felt through brake pedal or steering wheel when braking | 49 |
| Brakes binding | 50 |
| Rear wheels locking under normal braking | 51 |

### Suspension and steering

| Symptom | Section |
|---|---|
| Vehicle pulls to one side | 52 |
| Wheel wobble and vibration | 53 |
| Excessive pitching and/or rolling around corners, or during braking | 54 |
| Wandering or general instability | 55 |
| Excessively stiff steering | 56 |
| Excessive play in steering | 57 |
| Lack of power assistance | 58 |
| Tire wear excessive | 59 |

### Electrical system

| Symptom | Section |
|---|---|
| Battery will not hold a charge for more than a few days | 60 |
| Ignition/no-charge warning light remains illuminated with engine running | 61 |
| Ignition/no-charge warning light fails to come on | 62 |
| Lights inoperative | 63 |
| Instrument readings inaccurate or erratic | 64 |
| Horn inoperative, or unsatisfactory in operation | 65 |
| Windshield wipers inoperative, or unsatisfactory in operation | 66 |
| Windshield washers inoperative, or unsatisfactory in operation | 67 |
| Electric windows inoperative, or unsatisfactory in operation | 68 |
| Window glass fails to move | 69 |
| Central locking system inoperative, or unsatisfactory in operation | 70 |

# Troubleshooting 0-19

The vehicle owner who does his or her own maintenance according to the recommended service schedules should not have to use this section of the manual very often. Modern component reliability is such that, provided those items subject to wear or deterioration are inspected or replaced at the specified intervals, sudden failure is comparatively rare. Problems do not usually just happen as a result of sudden failure, but develop over a period of time. Major mechanical failures in particular are usually preceded by characteristic symptoms over hundreds or even thousands of miles. Those components which do occasionally fail without warning are often small and easily carried in the vehicle.

With any troubleshooting, the first step is to decide where to begin investigations. Sometimes this is obvious, but on other occasions, a little detective work will be necessary. The owner who makes half a dozen haphazard adjustments or replacements may be successful in curing a problem (or its symptoms), but will be none the wiser if it recurs, and ultimately may have spent more time and money than was necessary. A calm and logical approach will be found to be more satisfactory in the long run. Always take into account any warning signs or abnormalities that may have been noticed in the period preceding the trouble - power loss, high or low gauge readings, unusual smells, etc. - and remember that failure of components such as fuses or spark plugs may only be pointers to some underlying fault.

The pages which follow provide an easy-reference guide to the more common problems which may occur during the operation of the vehicle. These problems and their possible causes are grouped under headings denoting various components or systems, such as Engine, Cooling system, etc. The Chapter and/or Section which deals with the problem is also shown in brackets. Whatever the trouble, certain basic principles apply. These are as follows:

*Verify the problem.* This is simply a matter of being sure that you know what the symptoms are before starting work. This is particularly important if you are investigating a fault for someone else, who may not have described it very accurately.

*Don't overlook the obvious.* For example, if the vehicle won't start, is there fuel in the tank? (Don't take anyone else's word on this particular point, and don't trust the fuel gauge either!) If an electrical problem is indicated, look for loose or broken wires before digging out the test gear.

*Cure the disease, not the symptom.* Substituting a flat battery with a fully charged one will get you off the hard shoulder, but if the underlying cause is not attended to, the new battery will go the same way. Similarly, changing oil-fouled spark plugs for a new set will get you moving again, but remember that the reason for the fouling (if it wasn't simply an incorrect grade of plug) will have to be established and corrected.

*Don't take anything for granted.* Particularly, don't forget that a "new" component may itself be defective (especially if it's been rattling around in the luggage compartment for months), and don't leave components out of a troubleshooting sequence just because they are new or recently installed. When you do finally diagnose a difficult problem, you'll probably realize that all the evidence was there from the start.

## Engine

### 1 Engine fails to rotate when attempting to start

1 Battery terminal connections loose or corroded (Chapter 1)
2 Battery discharged or faulty (Chapter 5)
3 Broken, loose or disconnected wiring in the starting circuit (Chapter 5)
4 Defective starter solenoid or switch (Chapter 5)
5 Defective starter motor (Chapter 5)
6 Starter pinion or flywheel ring gear teeth loose or broken (Chapters 2 or 5)
7 Engine ground strap broken or disconnected.

### 2 Engine rotates, but will not start

1 Fuel tank empty
2 Battery discharged (engine rotates slowly) (Chapter 5)
3 Battery terminal connections loose or corroded (Chapter 1)
4 Air filter element dirty or clogged (Chapter 1)
5 Low cylinder compressions (Chapter 2)
6 Major mechanical failure (broken timing chain, for example) (Chapter 2)
7 Ignition components damp or damaged (Chapter 5)
8 Fuel injection system fault (Chapter 4)
9 Worn, faulty or incorrectly gapped spark plugs (Chapter 1)
10 Broken, loose or disconnected wiring in ignition circuit (Chapter 5)

### 3 Engine difficult to start when cold

1 Battery discharged (Chapter 5)
2 Battery terminal connections loose or corroded (Chapter 1)
3 Air filter element dirty or clogged (Chapter 1)
4 Worn, faulty or incorrectly gapped spark plugs (Chapter 1)
5 Low cylinder compressions (Chapter 2)
6 Fuel injection system fault (Chapter 4)
7 Ignition system fault (Chapter 5)

### 4 Engine difficult to start when hot

1 Battery discharged (Chapter 5)
2 Battery terminal connections loose or corroded (Chapter 1)
3 Air filter element dirty or clogged (Chapter 1)
4 Fuel injection system fault (Chapter 4)

### 5 Starter motor noisy or excessively rough in engagement

1 Starter pinion or flywheel ring gear teeth loose or broken (Chapter 2 or 5)
2 Starter motor mounting bolts loose or missing (Chapter 5)
3 Starter motor internal components worn or damaged (Chapter 5)

### 6 Starter motor turns engine slowly

1 Battery discharged (Chapter 5)
2 Battery terminal connections loose or corroded (Chapter 1)
3 Ground strap broken or disconnected (Chapter 5)
4 Starter motor wiring loose (Chapter 5)
5 Starter motor internal fault (Chapter 5)

### 7 Engine starts, but stops immediately

1 Loose ignition system wiring (Chapter 5)
2 Dirt in fuel system (Chapter 4)
3 Fuel injector fault (Chapter 4)
4 Fuel pump or pressure regulator fault (Chapter 4)
5 Vacuum leak at throttle body, intake manifold or hoses (Chapters 2 and 4)

### 8 Engine idles erratically

1 Air filter element clogged (Chapter 1)
2 Air in fuel system (Chapter 4)
3 Worn, faulty or incorrectly gapped spark plugs (Chapter 1)
4 Vacuum leak at throttle body, intake manifold or hoses (Chapters 2 and 4)
5 Uneven or low cylinder compressions (Chapter 2)
6 Timing chain incorrectly installed or tensioned (Chapter 2)
7 Camshaft lobes worn (Chapter 2)
8 Faulty fuel injector(s) (Chapter 4)

### 9 Engine misfires at idle speed

1 Distributor cap cracked or tracking internally (Chapter 5)
2 Faulty fuel injector(s) (Chapter 4)
3 Uneven or low cylinder compressions (Chapter 2)

## Troubleshooting

4    Disconnected, leaking, or deteriorated crankcase ventilation hoses (Chapter 4)
5    Vacuum leak at the throttle body, intake manifold or associated hoses (Chapter 4)

### 10  Engine misfires throughout the driving speed range

1    Fuel filter plugged (Chapter 1)
2    Fuel pump faulty, or delivery pressure low (Chapter 4)
3    Fuel tank vent blocked, or fuel pipes restricted (Chapter 4)
4    Uneven or low cylinder compressions (Chapter 2)
5    Worn, faulty or incorrectly gapped spark plugs (Chapter 1)
6    Faulty spark plug or spark plug wires (Chapter 1)

### 11  Engine stalls

1    Fuel filter plugged (Chapter 1)
2    Blocked injector/fuel injection system fault (Chapter 4)
3    Fuel pump faulty, or delivery pressure low (Chapter 4)
4    Vacuum leak at the throttle body, intake manifold or associated hoses (Chapter 4)
5    Fuel tank vent blocked, or fuel pipes restricted (Chapter 4)

### 12  Engine lacks power

1    Fuel filter plugged (Chapter 1)
2    Timing chain incorrectly installed or tensioned (Chapter 2)
3    Fuel pump faulty, or delivery pressure low (Chapter 4)
4    Worn, faulty or incorrectly gapped spark plugs (Chapter 1)
5    Vacuum leak at the throttle body, intake manifold or associated hoses (Chapter 4)
6    Uneven or low cylinder compressions (Chapter 2)
7    Brakes binding (Chapters 1 and 9)
8    Clutch slipping (Chapter 8)
9    Blocked injector/fuel injection system fault (Chapter 4)

### 13  Engine backfires

1    Timing chain incorrectly installed (Chapter 2)
2    Faulty injector/fuel injection system fault (Chapter 4)

### 14  Oil pressure warning light illuminated with engine running

1    Low oil level, or incorrect oil grade (Chapter 1)
2    Faulty oil pressure sensor (Chapter 5)
3    Worn engine bearings and/or oil pump (Chapter 2)
4    Excessively high engine operating temperature (Chapter 3)
5    Oil pressure relief valve defective (Chapter 2)
6    Oil pick-up strainer clogged (Chapter 2)

**Note:** *Low oil pressure in a high-mileage engine at idle is not necessarily a cause for concern. Sudden pressure loss at speed is far more significant. In any event, check the gauge or warning light sender before condemning the engine.*

### 15  Engine runs-on after switching off

1    Excessive carbon build-up in engine (Chapter 2)
2    Excessively high engine operating temperature (Chapter 3)

### 16  Engine noises

1    Pre-ignition (pinging) or knocking during acceleration or under load
  a) *Excessive carbon build-up in engine (Chapter 2)*
  b) *Faulty fuel injector(s) (Chapter 4)*
  c) *Ignition system fault (Chapter 5)*
2    Whistling or wheezing noises
  a) *Leaking exhaust manifold gasket (Chapter 4)*
  b) *Leaking vacuum hose (Chapter 4 or 9)*
  c) *Blowing cylinder head gasket (Chapter 2)*
3    Tapping or rattling noises
  a) *Worn valve gear or camshaft (Chapter 2)*
  b) *Ancillary component fault (water pump, alternator, etc.) (Chapters 3, 5, etc.)*
4    Knocking or thumping noises
  a) *Worn connecting rod bearings (regular heavy knocking, perhaps less under load) (Chapter 2)*
  b) *Worn main bearings (rumbling and knocking, perhaps worsening under load) (Chapter 2)*
  c) *Piston slap (most noticeable when cold) (Chapter 2)*
  d) *Ancillary component fault (water pump, alternator, etc.) (Chapters 3, 5, etc.)*

## Cooling system

### 17  Overheating

1    Insufficient coolant in system (Chapter 1)
2    Thermostat faulty (Chapter 3)
3    Radiator core blocked, or grille restricted (Chapter 3)
4    Cooling fan or viscous coupling faulty (Chapter 3)
5    Inaccurate temperature gauge sender unit (Chapter 3)
6    Airlock in cooling system (Chapter 3)
7    Expansion tank pressure cap faulty (Chapter 3)

### 18  Overcooling

1    Thermostat faulty (Chapter 3)
2    Inaccurate temperature gauge sender unit (Chapter 3)
3    Viscous coupling faulty (Chapter 3)

### 19  External coolant leakage

1    Deteriorated or damaged hoses or hose clips (Chapter 1)
2    Radiator core or heater matrix leaking (Chapter 3)
3    Pressure cap faulty (Chapter 3)
4    Water pump internal seal leaking (Chapter 3)
5    Water pump-to-block seal leaking (Chapter 3)
6    Boiling due to overheating (Chapter 3)
7    Core plug leaking (Chapter 2)

### 20  Internal coolant leakage

1    Leaking cylinder head gasket (Chapter 2)
2    Cracked cylinder head or cylinder block (Chapter 2)

### 21  Corrosion

1    Infrequent draining and flushing (Chapter 1)
2    Incorrect coolant mixture or inappropriate coolant type (Chapter 1)

## Fuel and exhaust systems

### 22  Excessive fuel consumption

1    Air filter element dirty or clogged (Chapter 1)
2    Fuel injection system fault (Chapter 4)
3    Ignition timing incorrect/ignition system fault (Chapters 1 and 5)
4    Tires under-inflated (Chapter 1)

### 23  Fuel leakage and/or fuel odor

    Damaged or corroded fuel tank, pipes or connections (Chapter 4)

# Troubleshooting 0-21

## 24 Excessive noise or fumes from exhaust system

1 Leaking exhaust system or manifold joints (Chapters 1 and 4)
2 Leaking, corroded or damaged mufflers or pipe (Chapters 1 and 4)
3 Broken mountings causing body or suspension contact (Chapter 1)

## Clutch

## 25 Pedal travels to floor - no pressure or very little resistance

1 Brake fluid level low/air in the hydraulic system (Chapters 1 and 8)
2 Broken clutch release bearing or fork (Chapter 8)
3 Broken diaphragm spring in clutch pressure plate (Chapter 8)

## 26 Clutch fails to disengage (unable to select gears)

1 Brake fluid level low/air in the hydraulic system (Chapters 1 and 8)
2 Clutch disc sticking on transmission input shaft splines (Chapter 8)
3 Clutch disc sticking to flywheel or pressure plate (Chapter 8)
4 Faulty pressure plate assembly (Chapter 8)
5 Clutch release mechanism worn or incorrectly assembled (Chapter 8)

## 27 Clutch slips (engine speed increases, with no increase in vehicle speed)

1 Clutch disc linings excessively worn (Chapter 8)
2 Clutch disc linings contaminated with oil or grease (Chapter 8)
3 Faulty pressure plate or weak diaphragm spring (Chapter 8)

## 28 Shudder as clutch is engaged

1 Clutch disc linings contaminated with oil or grease (Chapter 8)
2 Clutch disc linings excessively worn (Chapter 8)
3 Faulty or distorted pressure plate or diaphragm spring (Chapter 8)
4 Worn or loose engine or transmission mountings (Chapter 2A or 2B)
5 Clutch disc hub or transmission input shaft splines worn (Chapter 8)

## 29 Noise when depressing or releasing clutch pedal

1 Worn clutch release bearing (Chapter 8)
2 Worn or dry clutch pedal bushings (Chapter 8)
3 Faulty pressure plate assembly (Chapter 8)
4 Pressure plate diaphragm spring broken (Chapter 8)
5 Broken clutch disc cushioning springs (Chapter 8)

## Manual transmission

## 30 Noisy in neutral with engine running

1 Input shaft bearings worn (noise apparent with clutch pedal released, but not when depressed) (Chapter 7)*
2 Clutch release bearing worn (noise apparent with clutch pedal depressed, possibly less when released) (Chapter 8)

## 31 Noisy in one particular gear

Worn, damaged or chipped gear teeth (Chapter 7)*

## 32 Difficulty engaging gears

1 Clutch fault (Chapter 8)
2 Worn or damaged shift linkage/cable (Chapter 7)
3 Incorrectly adjusted shift linkage/cable (Chapter 7)
4 Worn synchronizer units (Chapter 7)*

## 33 Jumps out of gear

1 Worn or damaged shift linkage/cable (Chapter 7)
2 Incorrectly adjusted shift linkage/cable (Chapter 7)
3 Worn synchronizer units (Chapter 7)*
4 Worn selector forks (Chapter 7)*

## 34 Vibration

1 Lack of oil (Chapter 1)
2 Worn bearings (Chapter 7)*

## 35 Lubricant leaks

1 Leaking differential output oil seal (Chapter 7)
2 Leaking housing joint (Chapter 7)*
3 Leaking input shaft oil seal (Chapter 7)*

*Although the corrective action necessary to remedy the symptoms described is beyond the scope of the home mechanic, the above information should be helpful in isolating the cause of the condition, so that the owner can communicate clearly with a professional mechanic.*

## Automatic transmission

**Note:** *Due to the complexity of the automatic transmission, it is difficult for the home mechanic to properly diagnose and service this unit. For problems other than the following, the vehicle should be taken to a dealer service department or automatic transmission specialist. Do not be too hasty in removing the transmission if a fault is suspected, as most of the testing is carried out with the unit installed.*

## 36 Fluid leakage

Automatic transmission fluid is usually dark in color. Fluid leaks should not be confused with engine oil, which can easily be blown onto the transmission by airflow.
To determine the source of a leak, first remove all built-up dirt and grime from the transmission housing and surrounding areas using a degreasing agent, or by steam-cleaning. Drive the vehicle at low speed, so airflow will not blow the leak far from its source. Raise and support the vehicle, and determine where the leak is coming from. The following are common areas of leakage:

a) *Oil pan (Chapters 1 and 7)*
b) *Dipstick tube (Chapter 1 and 7)*
c) *Transmission-to-fluid cooler pipes/unions (Chapter 7)*

## 37 Transmission fluid brown, or has burned smell

Transmission fluid level low, or fluid in need of replacement (Chapter 1)

## 38 General gear selection problems

Chapter 7B deals with checking and adjusting the selector cable on automatic transmissions. The following are common problems which may be caused by a poorly adjusted cable:

a) *Engine starting in gears other than Park or Neutral*
b) *Indicator panel indicating a gear other than the one actually being used*
c) *Vehicle moves when in Park or Neutral*
d) *Poor gear shift quality or erratic gear changes*

Refer to Chapter 7B for the selector cable adjustment procedure.

## 39 Transmission will not downshift (kickdown) with accelerator pedal fully depressed

1 Low transmission fluid level (Chapter 1)
2 Incorrect selector cable adjustment (Chapter 7)

## 40 Engine will not start in any gear, or starts in gears other than Park or Neutral

1 Incorrect starter/inhibitor switch adjustment (Chapter 7)
2 Incorrect selector cable adjustment (Chapter 7)

## 41 Transmission slips, shifts roughly, is noisy, or has no drive in forward or reverse gears

There are many probable causes for the above problems, but the home mechanic should be concerned with only one possibility - fluid level. Before taking the vehicle to a dealer or transmission specialist, check the fluid level and condition of the fluid as described in Chapter 1. Correct the fluid level as necessary, or change the fluid and filter if needed. If the problem persists, professional help will be necessary.

## Differential and driveshaft

## 42 Vibration when accelerating or decelerating

1 Worn universal joint (Chapter 8)
2 Bent or distorted driveshaft (Chapter 8)

## 43 Low pitched whining; increasing with road speed

Worn differential (Chapter 8)

## Brakes

**Note:** *Before assuming that a brake problem exists, make sure that the tires are in good condition and correctly inflated, that the front wheel alignment is correct, and that the vehicle is not loaded with weight in an unequal manner. Apart from checking the condition of all pipe and hose connections, any faults occurring on the anti-lock braking system should be referred to a BMW dealer or other qualified repair shop for diagnosis.*

## 44 Vehicle pulls to one side under braking

1 Worn, defective, damaged or contaminated brake pads/shoes on one side (Chapters 1 and 9)
2 Seized or partially seized front brake caliper/wheel cylinder piston (Chapters 1 and 9)
3 A mixture of brake pad/shoe lining materials (Chapters 1 and 9)
4 Brake caliper or backing plate mounting bolts loose (Chapter 9)
5 Worn or damaged steering or suspension components (Chapters 1 and 10)

## 45 Noise (grinding or high-pitched squeal) when brakes applied

1 Brake pad or shoe friction lining material worn down to metal backing (Chapters 1 and 9)
2 Excessive corrosion of brake disc or drum. (May be apparent after the vehicle has been standing for some time (Chapters 1 and 9)
3 Foreign object (stone chipping, etc.) trapped between brake disc and shield (Chapters 1 and 9)

## 46 Excessive brake pedal travel

1 Inoperative rear brake self-adjust mechanism - drum brakes (Chapters 1 and 9)
2 Faulty master cylinder (Chapter 9)
3 Air in hydraulic system (Chapters 1 and 9)

## 47 Brake pedal feels spongy when depressed

1 Air in hydraulic system (Chapters 1 and 9)
2 Deteriorated flexible rubber brake hoses (Chapters 1 and 9)
3 Master cylinder mounting nuts loose (Chapter 9)
4 Faulty master cylinder (Chapter 9)

## 48 Excessive brake pedal effort required to stop vehicle

1 Faulty brake booster (Chapter 9)
2 Disconnected, damaged or insecure brake booster vacuum hose (Chapter 9)
3 Primary or secondary hydraulic circuit failure (Chapter 9)
4 Seized brake caliper or wheel cylinder piston(s) (Chapter 9)
5 Brake pads or brake shoes incorrectly installed (Chapters 1 and 9)
6 Incorrect grade of brake pads or brake shoes installed (Chapters 1 and 9)
7 Brake pads or brake shoe linings contaminated (Chapters 1 and 9)

## 49 Shudder felt through brake pedal or steering wheel when braking

1 Excessive run-out or distortion of discs/drums (Chapters 1 and 9)
2 Brake pad or brake shoe linings worn (Chapters 1 and 9)
3 Brake caliper or brake backing plate mounting bolts loose (Chapter 9)
4 Wear in suspension or steering components or mountings (Chapters 1 and 10)

## 50 Brakes binding

1 Seized brake caliper or wheel cylinder piston(s) (Chapter 9)
2 Incorrectly adjusted parking brake mechanism (Chapter 9)
3 Faulty master cylinder (Chapter 9)

## 51 Rear wheels locking under normal braking

1 Rear brake shoe linings contaminated (Chapters 1 and 9)
2 Faulty brake pressure regulator (Chapter 9)

## Suspension and steering

**Note:** *Before diagnosing suspension or steering faults, be sure that the trouble is not due to incorrect tire pressures, mixtures of tire types, or binding brakes.*

## 52 Vehicle pulls to one side

1 Defective tire (Chapter 1)
2 Excessive wear in suspension or steering components (Chapters 1 and 10)
3 Incorrect front wheel alignment (Chapter 10)
4 Accident damage to steering or suspension components (Chapter 1)

## 53 Wheel wobble and vibration

1 Front wheels out of balance (vibration felt mainly through the steering wheel) (Chapters 1 and 10)
2 Rear wheels out of balance (vibration felt throughout the vehicle) (Chapters 1 and 10)
3 Wheels damaged or distorted (Chapters 1 and 10)
4 Faulty or damaged tire (Chapter 1)
5 Worn steering or suspension joints, bushings or components (Chapters 1 and 10)
6 Wheel bolts loose (Chapters 1 and 10)

# Troubleshooting 0-23

## 54 Excessive pitching and/or rolling around corners, or during braking

1. Defective shock absorbers (Chapters 1 and 10)
2. Broken or weak spring and/or suspension component (Chapters 1 and 10)
3. Worn or damaged anti-roll bar or mountings (Chapter 10)

## 55 Wandering or general instability

1. Incorrect front wheel alignment (Chapter 10)
2. Worn steering or suspension joints, bushings or components (Chapters 1 and 10)
3. Wheels out of balance (Chapters 1 and 10)
4. Faulty or damaged tire (Chapter 1)
5. Wheel bolts loose (Chapters 1 and 10)
6. Defective shock absorbers (Chapters 1 and 10)

## 56 Excessively stiff steering

1. Lack of power steering fluid (Chapter 10)
2. Seized tie-rod end balljoint or suspension balljoint (Chapters 1 and 10)
3. Broken or incorrectly adjusted drivebelt - power steering (Chapter 1)
4. Incorrect front wheel alignment (Chapter 10)
5. Steering rack or column bent or damaged (Chapter 10)

## 57 Excessive play in steering

1. Worn steering column intermediate shaft universal joint (Chapter 10)
2. Worn steering track rod end balljoints (Chapters 1 and 10)
3. Worn rack-and-pinion steering gear (Chapter 10)
4. Worn steering or suspension joints, bushings or components (Chapters 1 and 10)

## 58 Lack of power assistance

1. Broken or incorrectly adjusted drivebelt (Chapter 1)
2. Incorrect power steering fluid level (Chapter 1)
3. Restriction in power steering fluid hoses (Chapter 1)
4. Faulty power steering pump (Chapter 10)
5. Faulty rack-and-pinion steering gear (Chapter 10)

## 59 Tire wear excessive

1. Tires worn on inside or outside edges
   a) *Tires under-inflated (wear on both edges) (Chapter 1)*
   b) *Incorrect camber or caster angles (wear on one edge only) (Chapter 10)*
   c) *Worn steering or suspension joints, bushings or components (Chapters 1 and 10)*
   d) *Excessively hard cornering.*
   e) *Accident damage.*
2. Tire treads exhibit feathered edges
   *Incorrect toe setting (Chapter 10)*
3. Tires worn in center of tread
   *Tires over-inflated (Chapter 1)*
4. Tires worn on inside and outside edges
   *Tires under-inflated (Chapter 1)*
5. Tires worn unevenly
   a) *Tires/wheels out of balance (Chapter 1)*
   b) *Excessive wheel or tire run-out (Chapter 1)*
   c) *Worn shock absorbers (Chapters 1 and 10)*
   d) *Faulty tire (Chapter 1)*

## Electrical system

**Note:** *For problems associated with the starting system, refer to the faults listed under "Engine" earlier in this Section.*

## 60 Battery will not hold a charge for more than a few days

1. Battery defective internally (Chapter 5)
2. Battery terminal connections loose or corroded (Chapter 1)
3. Drivebelt worn or incorrectly adjusted (Chapter 1)
4. Alternator not charging at correct output (Chapter 5)
5. Alternator or voltage regulator faulty (Chapter 5)
6. Short-circuit causing continual battery drain (Chapters 5 and 12)

## 61 Ignition/no-charge warning light remains illuminated with engine running

1. Drivebelt broken, worn, or incorrectly adjusted (Chapter 1)
2. Alternator brushes worn, sticking, or dirty (Chapter 5)
3. Alternator brush springs weak or broken (Chapter 5)
4. Internal fault in alternator or voltage regulator (Chapter 5)
5. Broken, disconnected, or loose wiring in charging circuit (Chapter 5)

## 62 Ignition/no-charge warning light fails to come on

1. Warning light bulb blown (Chapter 12)
2. Broken, disconnected, or loose wiring in warning light circuit (Chapter 12)
3. Alternator faulty (Chapter 5)

## 63 Lights inoperative

1. Bulb blown (Chapter 12)
2. Corrosion of bulb or bulb holder contacts (Chapter 12)
3. Blown fuse (Chapter 12)
4. Faulty relay (Chapter 12)
5. Broken, loose, or disconnected wiring (Chapter 12)
6. Faulty switch (Chapter 12)

## 64 Instrument readings inaccurate or erratic

1. Instrument readings increase with engine speed
   *Faulty voltage regulator (Chapter 12)*
2. Fuel or temperature gauges give no reading
   a) *Faulty gauge sender unit (Chapters 3 and 4)*
   b) *Wiring open-circuit (Chapter 12)*
   c) *Faulty gauge (Chapter 12)*
3. Fuel or temperature gauges give continuous maximum reading
   a) *Faulty gauge sender unit (Chapters 3 and 4)*
   b) *Wiring short-circuit (Chapter 12)*
   c) *Faulty gauge (Chapter 12)*

## 65 Horn inoperative, or unsatisfactory in operation

1. Horn operates all the time
   a) *Horn push either grounded or stuck down (Chapter 12)*
   b) *Horn cable-to-horn switch grounded (Chapter 12)*
2. Horn fails to operate
   a) *Blown fuse (Chapter 12)*
   b) *Cable or cable connections loose, broken or disconnected (Chapter 12)*
   c) *Faulty horn (Chapter 12)*
3. Horn emits intermittent or unsatisfactory sound
   a) *Cable connections loose (Chapter 12)*
   b) *Horn mountings loose (Chapter 12)*
   c) *Faulty horn (Chapter 12)*

## 66 Windshield wipers inoperative, or unsatisfactory in operation

1. Wipers fail to operate, or operate very slowly
   a) *Wiper blades stuck to screen, or linkage seized or binding (Chapters 1 and 12)*
   b) *Blown fuse (Chapter 12)*

c) Cable or cable connections loose, broken or disconnected (Chapter 12)
d) Faulty relay (Chapter 12)
e) Faulty wiper motor (Chapter 12)

2 Wiper blades sweep over too large or too small an area of the glass
a) Wiper arms incorrectly positioned on spindles (Chapter 1)
b) Excessive wear of wiper linkage (Chapter 12)
c) Wiper motor or linkage mountings loose or insecure (Chapter 12)

3 Wiper blades fail to clean the glass effectively
a) Wiper blade inserts worn or deteriorated (Chapter 1)
b) Wiper arm tension springs broken, or arm pivots seized (Chapter 12)
c) Insufficient windshield washer additive to adequately remove road film (Chapter 1)

## 67 Windshield washers inoperative, or unsatisfactory in operation

1 One or more washer jets inoperative
a) Blocked washer jet (Chapter 1)
b) Disconnected, kinked or restricted fluid hose (Chapter 12)
c) Insufficient fluid in washer reservoir (Chapter 1)

2 Washer pump fails to operate
a) Broken or disconnected wiring or connections (Chapter 12)
b) Blown fuse (Chapter 12)
c) Faulty washer switch (Chapter 12)
d) Faulty washer pump (Chapter 12)

3 Washer pump runs for some time before fluid is emitted from jets
Faulty one-way valve in fluid supply hose (Chapter 12)

## 68 Electric windows inoperative, or unsatisfactory in operation

1 Window glass will only move in one direction
Faulty switch (Chapter 12)

2 Window glass slow to move
a) Regulator seized or damaged, or in need of lubrication (Chapter 11)
b) Door internal components or trim fouling regulator (Chapter 11)
c) Faulty motor (Chapter 11)

## 69 Window glass fails to move

1 Blown fuse (Chapter 12)
2 Faulty relay (Chapter 12)
3 Broken or disconnected wiring or connections (Chapter 12)
4 Faulty motor (Chapter 11)

## 70 Central locking system inoperative, or unsatisfactory in operation

1 Complete system failure
a) Blown fuse (Chapter 12)
b) Faulty relay (Chapter 12)
c) Broken or disconnected wiring or connections (Chapter 12)
d) Faulty motor (Chapter 11)

2 Latch locks but will not unlock, or unlocks but will not lock
a) Faulty master switch (Chapter 12)
b) Broken or disconnected latch operating rods or levers (Chapter 11)
c) Faulty relay (Chapter 12)
d) Faulty motor (Chapter 11)

3 One solenoid/motor fails to operate
a) Broken or disconnected wiring or connections (Chapter 12)
b) Faulty operating assembly (Chapter 11)
c) Broken, binding or disconnected latch operating rods or levers (Chapter 11)
d) Fault in door latch (Chapter 11)

# Chapter 1
# Tune-up and routine maintenance

## Contents

| | Section | | Section |
|---|---|---|---|
| Airbag system inspection | 35 | Hinge and lock lubrication | 15 |
| Air filter element replacement | 24 | Hose and fluid leak check | 6 |
| Automatic transmission fluid level check | 5 | Introduction | 1 |
| Automatic transmission fluid replacement | 25 | Maintenance schedule | Page 1-6 |
| Battery check, maintenance and charging | 21 | Manual transmission oil replacement | 29 |
| Brake fluid replacement | 32 | Oxygen sensor replacement | 34 |
| Brake hoses and lines - inspection | 10 | Parking brake check | 12 |
| Brake pad check | 9 | Parking brake shoe check | 28 |
| Clutch check | 31 | Pollen filter replacement | 16 |
| Coolant replacement | 33 | Resetting the service interval display | 4 |
| Differential oil replacement | 26 | Road test | 20 |
| Driveaxle boot check | 27 | Seat belt check | 14 |
| Drivebelts - check and replacement | 7 | Spark plug replacement | 23 |
| Engine management system check | 19 | Steering and suspension check | 8 |
| Engine oil and filter replacement | 3 | Throttle linkage - lubrication | 11 |
| Exhaust system check | 13 | Weekly checks | 2 |
| Fuel filter replacement | 30 | Windshield/headlight washer system(s) check | 18 |
| Headlight beam alignment check | 17 | Wiper blade check and replacement | 22 |

## Specifications

### Recommended lubricants and fluids

**Note:** *Listed here are manufacturer recommendations at the time this manual was written. Manufacturers occasionally upgrade their fluid and lubricant specifications, so check with your local auto parts store for current recommendations.*

Engine oil
    Type .................................................................................. API "certified for gasoline engines"
    Viscosity ............................................................................ SAE 10W/40 to 20W/50
Cooling system ....................................................................... 50/50 mixture of ethylene glycol based antifreeze
Manual transmission
    1992
        Green label .................................................................. Gear Lubricant Special (GLS)*
        Red label ...................................................................... DEXRON III type ATF*
        No label ........................................................................ 80W GL-4*
    1993 and later .................................................................. DEXRON III type ATF*
Automatic transmission
    1992 through 1996 ........................................................... DEXRON III type ATF*
    1997 and later .................................................................. Special Lubricant Fluid (SLF)*
Final drive unit ......................................................................... Hypoid gear oil*
Braking system ........................................................................ DOT 4 brake fluid
Power steering ......................................................................... DEXRON II type ATF*

*Refer to your BMW dealer for brand name and type recommendations.

## Capacities*

### Engine oil (including filter)
| | |
|---|---|
| Four-cylinder engines | 4.7 quarts |
| Six-cylinder engines | 6.8 quarts |

### Cooling system
| | |
|---|---|
| Four-cylinder engines | 6.8 quarts |
| Six-cylinder engines | 11 quarts |

### Transmission
| | |
|---|---|
| Manual transmission | |
| 1992 | 1.2 quarts |
| 1993 and later | |
| 318, 320 and 325 | 1.2 quarts |
| 328 and Z3 | 1.3 quarts |
| Automatic transmission | 3.2 quarts |

### Differential
| | |
|---|---|
| Four-cylinder models | 1.2 quarts |
| Six-cylinder models | 1.8 quarts |

### Power-assisted steering
| | |
|---|---|
| All models (approximate) | 1.5 quarts |

*All capacities approximate. Add as necessary to bring to the appropriate level.

## Ignition system

| | |
|---|---|
| Spark plug type | |
| Four-cylinder engines | Champion RC9YC |
| Six-cylinder engines | |
| 2.5L | Champion RC9YC |
| 2.8L | Champion R12YC |
| Spark plug electrode gap (all) | 0.035 inch |

*The spark plug gap quoted is that recommended by Champion for their specified plugs listed above. If spark plugs of any other type are to be installed, refer to their manufacturer's recommendations.

## Drivebelts

| | |
|---|---|
| Air conditioning compressor drivebelt tensioning torque (see Section 7): | |
| Used drivebelt | 5.0 to 6.0 ft-lbs |
| New drivebelt | 4.0 to 5.0 ft-lbs |

## Brakes

| | |
|---|---|
| Brake pad friction material minimum thickness | 5/64 inch |
| Parking brake shoe friction material minimum thickness | 1/16 inch |

## Torque specifications

**Ft-lbs** (unless otherwise indicated)

| | |
|---|---|
| Engine oil pan oil drain plug: | |
| M12 plug | 26 |
| M22 plug | 44 |
| Cylinder block coolant drain plug | 21 |
| Automatic transmission oil drain plug | 133 to 150 in-lbs |
| Automatic transmission oil filler/level plug | |
| 1994 and earlier | Not available |
| 1995 on | 74 |
| Automatic transmission fluid pan bolts | 62 to 71 in-lbs |
| Automatic transmission oil screen screws | 44 in-lbs |
| Manual transmission oil drain plug | 37 |
| Manual transmission oil filler/level plug | 37 |
| Differential filler/level and drain plugs | 52 |
| Spark plugs | 22 |
| Wheel bolts | 74 |

# Chapter 1  Tune-up and routine maintenance      1-3

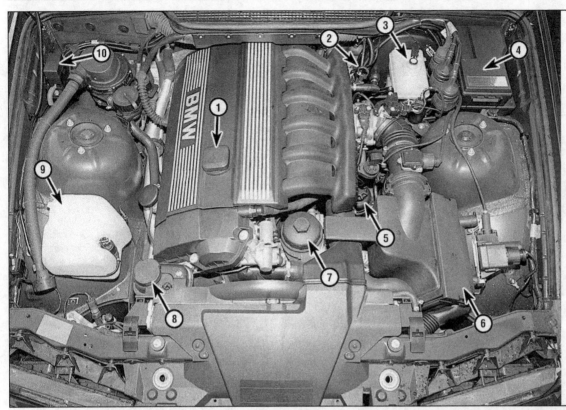

**Typical 3-Series engine compartment components (2.8L six-cylinder shown, others similar)**

1. Engine oil filler cap
2. Engine oil dipstick
3. Brake/clutch fluid reservoir
4. Fuse box
5. Power steering fluid reservoir
6. Air cleaner housing
7. Engine oil filter
8. Coolant expansion tank
9. Windshield washer fluid reservoir
10. Battery jump starting terminal

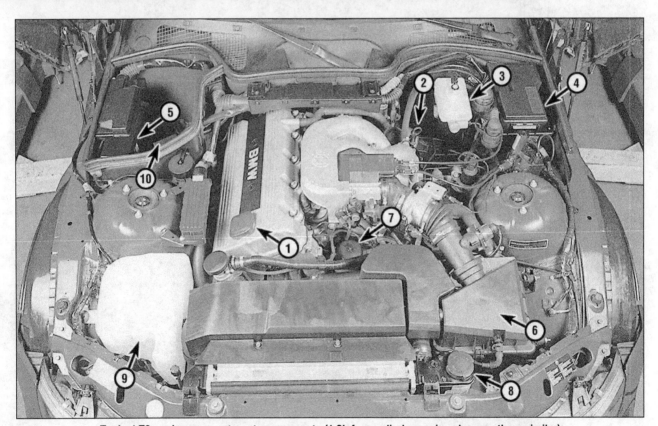

**Typical Z3 engine compartment components (1.9L four-cylinder engine shown, others similar)**

1. Engine oil filler cap
2. Engine oil dipstick
3. Brake/clutch fluid reservoir
4. Fuse box
5. Fuse box
6. Air cleaner housing
7. Engine oil filter
8. Coolant expansion tank
9. Windshield washer fluid reservoir
10. Battery jump starting terminal (1997 and later models)

# Chapter 1 Tune-up and routine maintenance

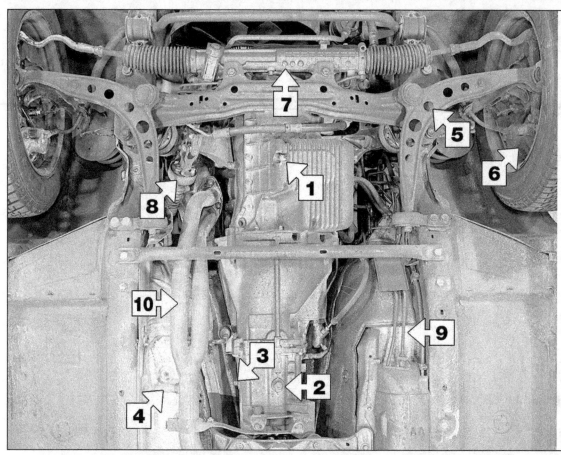

**Typical 3-Series front underside components**

1. Engine oil drain plug
2. Manual transmission drain plug
3. Manual transmission check/fill plug
4. Oxygen sensor
5. Lower arm
6. Brake caliper
7. Steering gear
8. Steering column intermediate shaft
9. Fuel lines
10. Exhaust pipe

**Typical Z3 front underside components**

1. Engine oil drain plug
2. Manual transmission drain plug
3. Manual transmission check/fill plug
4. Oxygen sensor
5. Lower arm
6. Brake caliper
7. Steering gear
8. Steering column intermediate shaft
9. Fuel lines
10. Exhaust pipe

# Chapter 1  Tune-up and routine maintenance

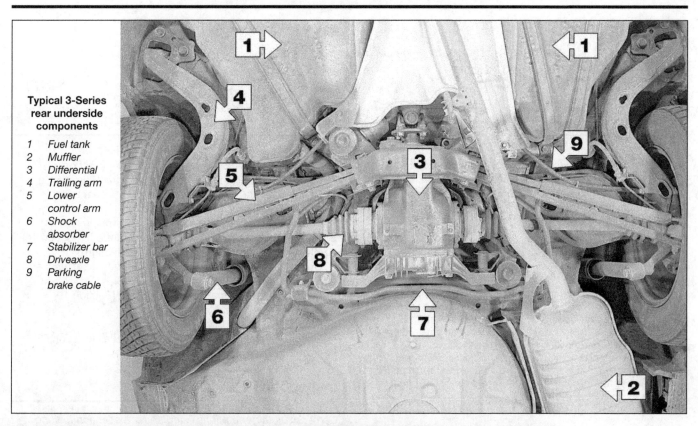

**Typical 3-Series rear underside components**

1. Fuel tank
2. Muffler
3. Differential
4. Trailing arm
5. Lower control arm
6. Shock absorber
7. Stabilizer bar
8. Driveaxle
9. Parking brake cable

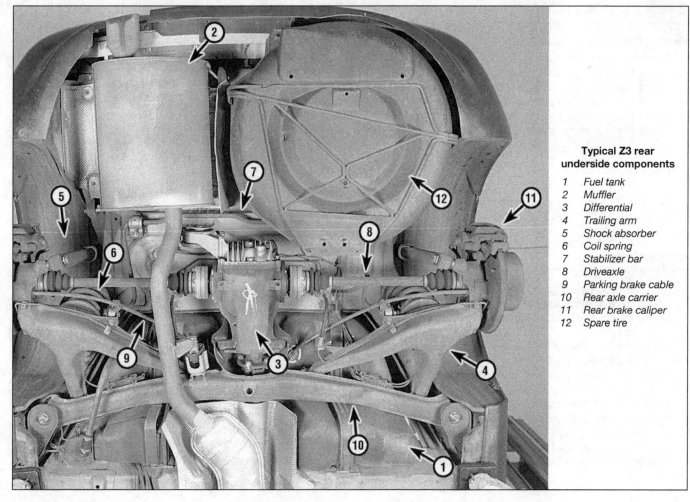

**Typical Z3 rear underside components**

1. Fuel tank
2. Muffler
3. Differential
4. Trailing arm
5. Shock absorber
6. Coil spring
7. Stabilizer bar
8. Driveaxle
9. Parking brake cable
10. Rear axle carrier
11. Rear brake caliper
12. Spare tire

# BMW 3-Series Maintenance schedule

All BMW 3-series models are equipped with a service interval display indicator and a row of LEDs in the instrument panel.

When the ignition is switched on, the panel and LEDs will illuminate for a few seconds and then go out. The green LEDs indicate the amount of time until the next service, the fewer LEDs lit then the nearer the next service interval is. When the yellow LED is lit, a service is due and the display will show whether an oil service or inspection is needed; the clock symbol will be illuminated if the additional annual inspection operations are required. If the red LED is illuminated then the service is overdue.

There are two different inspection services, Inspection I and Inspection II, these should be carried out alternately with some additional items to be included every second Inspection II. If you are unclear as to which inspection schedule was carried out last time start with Inspection II (including the additional items).

To reset the service interval display indicator a BMW service tool is required. Aftermarket alternatives to the BMW tool are produced by several leading tool manufacturers and should be available from larger car accessory shops.

## Every 250 miles (400 km) or weekly

Refer to *Weekly checks* (Section 2)
    Check the engine oil level
    Check the engine coolant level
    Check the brake and clutch fluid levels
    Check the power steering fluid level
    Check the windshield washer fluid level
    Check the tires and tire pressures

## Oil service

Replace the engine oil and filter (Section 3)
Reset the service interval display (Section 4)

## Inspection I

Replace the engine oil and filter (Section 3)
Check the automatic transmission fluid level (Section 5)
Check all underhood components and hoses for fluid leaks (Section 6)
Check the condition of the drivebelt(s), and adjust/replace if necessary (Section 7)
Check the steering and suspension components for condition and security (Section 8)
Check the front brake pad thickness (Section 9)
Check the rear brake pad thickness - rear disc brake models (Section 9)
Check the brake hoses and lines (Section 10)
Check and lubricate the throttle linkage
Check the operation of the parking brake (Section 12)
Check the exhaust system and mountings (Section 13)
Check the condition and operation of the seat belts (Section 14)
Lubricate all hinges and locks (Section 15)
Replace the pollen filter element (if equipped) (Section 16)
Check the headlight beam alignment (Section 17)
Check the operation of the windshield/headlight washer system(s) (as applicable) (Section 18)
Check the engine management system (Section 19)
Carry out a road test (Section 20)
Check/service the battery (Section 21)
Check/replace the windshield wiper blades (Section 22)
Reset service interval display (Section 4)

## Inspection II

*Carry out all the operations listed under Inspection I, along with the following:*

    Replace the spark plugs (Section 23)
    Replace the air filter element (Section 24)
    Replace the automatic transmission fluid (Section 25)
    Replace the differential oil (Section 26)
    Check the condition of the driveaxle boots (Section 27)
    Check the condition of the parking brake shoe linings (Section 28)
    Reset the service interval display (Section 4)

Additional work to be carried out every second inspection II:
    Replace the manual transmission oil (Section 29)
    Replace the fuel filter (Section 30)
    Check the clutch (Section 31)

## Annual service

**Note:** *BMW specify that the following should be carried out whenever the service display clock illuminates or at least every 2 years*

    Replace the brake fluid (Section 32)
    Replace the coolant (Section 33)
    Reset the service interval display (Section 4)

## Every 50,000 miles

Replace the oxygen sensor(s) (Section 34)

## Every three years

Inspect the airbag system (Section 35)

# Chapter 1 Tune-up and routine maintenance

1-7

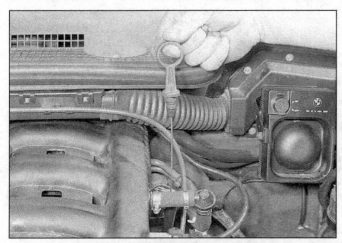

2.1 The dipstick top is often brightly colored for easy identification

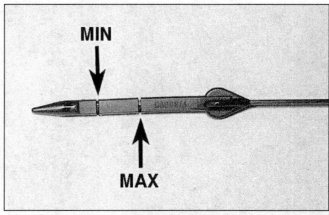

2.3 Note the oil level on the end of the dipstick, which should be between the upper (MAX) and lower (MIN) marks (approximately one quart of oil will raise the oil level from the lower mark to the upper mark)

## 1 Introduction

### General information

1  This Chapter is designed to help the home mechanic maintain his/her vehicle for safety, economy, long life and peak performance.

2  The Chapter contains a master maintenance schedule (page 1-6), followed by Sections dealing specifically with each task in the schedule. Visual checks, adjustments, component replacement and other helpful items are included. Refer to the illustrations of the engine compartment and the underside of the vehicle at the beginning of this Chapter for the locations of the various components.

3  Servicing your vehicle in accordance with the service indicator display and the following Sections will provide a planned maintenance program, which should result in a long and reliable service life. This is a comprehensive plan, so maintaining some items but not others at the specified service intervals, will not produce the same results.

4  As you service your vehicle, you will discover that many of the procedures can - and should - be grouped together, because of the particular procedure being performed, or because of the proximity of two otherwise-unrelated components to one another. For example, if the vehicle is raised for any reason, the exhaust can be inspected at the same time as the suspension and steering components.

5  The first step in this maintenance program is to prepare yourself before the actual work begins. Read through all the Sections relevant to the work to be carried out, then make a list and gather all the parts and tools required. If a problem is encountered, seek advice from a parts specialist, or a dealer service department.

### Intensive maintenance

6  If, from the time the vehicle is new, the routine maintenance schedule is followed closely, and frequent checks are made of fluid levels and high-wear items, as suggested throughout this manual, the engine will be kept in relatively good running condition, and the need for additional work will be minimized.

7  It is possible that there will be times when the engine is running poorly due to the lack of regular maintenance. This is even more likely if a used vehicle, which has not received regular and frequent maintenance checks, is purchased. In such cases, additional work may need to be carried out, outside of the regular maintenance intervals.

8  If engine wear is suspected, a compression test (refer to the relevant Part of Chapter 2) will provide valuable information regarding the overall performance of the main internal components. Such a test can be used as a basis to decide on the extent of the work to be carried out. If, for example, a compression test indicates serious internal engine wear, conventional maintenance as described in this Chapter will not greatly improve the performance of the engine, and may prove a waste of time and money, unless extensive overhaul work is carried out first.

9  The following series of operations are those most often required to improve the performance of a generally poor-running engine:

### Primary operations

a) Clean, inspect and test the battery (See Section 2).
b) Check all the engine-related fluids (See Section 2).
c) Check the condition and tension of the drivebelt (Section 7).
d) Replace the spark plugs (Section 23).
e) Check the condition of the air filter, and replace if necessary (Section 24).
f) Check the fuel filter (Section 30).
g) Check the condition of all hoses, and check for fluid leaks (Section 6).

5  If the above operations do not prove fully effective, carry out the following secondary operations:

### Secondary operations

All items listed under "Primary operations", plus the following:

a) Check the charging system (see Chapter 5A).
b) Check the ignition system (see Chapter 5B).
c) Check the fuel system (see Chapter 4).

## 2 Weekly checks

**Note:** *See Recommended lubricants and fluids at the beginning of this Chapter before adding fluid to any of the following components. The vehicle must be on level ground when fluid levels are checked.*

### Engine oil level check

*Refer to illustrations 2.1 and 2.3*

1  Engine oil is checked with a dipstick, which is located on the side of the engine **(see illustration)**. The dipstick extends through a metal tube down into the oil pan.

2  The engine oil should be checked before the vehicle has been driven, or about 5 minutes after the engine has been shut off. If the oil is checked immediately after driving the vehicle, some of the oil will remain in the upper part of the engine, resulting in an inaccurate reading on the dipstick.

3  Pull the dipstick out of the tube and wipe all of the oil away from the end with a clean rag or paper towel. Insert the clean dipstick all the way back into the tube and pull it out again. Note the oil at the end of the dipstick. At its highest point, the oil should be between the two **(see illustration)**.

4  It takes approximately one quart of oil to raise the level from the lower mark to the upper mark on the dipstick. Do not allow the level to drop below the lower mark or oil starvation may cause engine damage. Conversely, overfilling the engine (adding oil above the upper mark) may cause oil fouled spark plugs, oil leaks or oil seal failures.

# 1-8  Chapter 1  Tune-up and routine maintenance

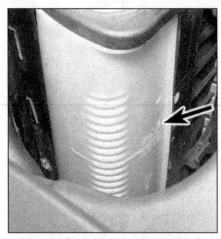

2.8  The coolant level varies with the temperature of the engine. When the engine is cold, the coolant level should be between the "KALT/COLD" mark on the expansion tank. When the engine hot, the level will rise above the "KALT/COLD" mark

2.11  If it is necessary to add coolant, wait until the engine is cold, then *slowly* unscrew the expansion tank cap to release any pressure present in the cooling system

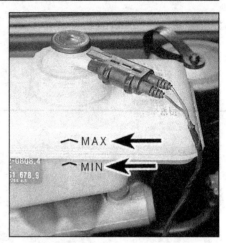

2.14  The MAX and MIN marks are indicated on the side of the brake fluid reservoir (the fluid level must be maintained between these marks

5  To add oil, remove the filler cap located on the valve cover. After adding oil, wait a few minutes to allow the level to stabilize, then pull the dipstick out and check the level again. Add more oil if required. Install the filler cap and tighten it by hand only.

6  Checking the oil level is an important preventive maintenance step. A consistently low oil level indicates oil leakage through damaged seals, defective gaskets or past worn rings or valve guides. The condition of the oil should also be noted. If the oil looks milky in color or has water droplets in it, the cylinder head gasket may be blown or the head or block may be cracked. The engine should be repaired immediately. Whenever you check the oil level, slide your thumb and index finger up the dipstick before wiping off the oil. If you see small dirt or metal particles clinging to the dipstick, the oil should be changed (see Section 3).

## Engine coolant level check

*Refer to illustrations 2.8 and 2.11*
**Warning:** *Do not allow antifreeze to come in contact with your skin or painted surfaces of the vehicle. Rinse off spills immediately with plenty of water. Antifreeze is highly toxic if ingested. Never leave antifreeze lying around in an open container or in puddles on the floor; children and pets are attracted by it's sweet smell and may drink it. Check with local authorities about disposing of used antifreeze. Many communities have collection centers which will see that antifreeze is disposed of safely.*

7  All vehicles covered by this manual are equipped with a pressurized coolant recovery system. A plastic expansion tank (or coolant reservoir) is mounted alongside the radiator. As the engine heats up during operation, the expanding coolant fills the tank. As the engine cools, the coolant is automatically drawn back into the cooling system to maintain the correct level.

8  The coolant level in the reservoir **(see illustration)** should be checked regularly. **Warning:** *Do not remove the expansion tank cap to check the coolant level when the engine is warm!* The level in the reservoir varies with the temperature of the engine. When the engine is cold, the coolant level should be above the LOW mark on the reservoir. Once the engine has warmed up, the level should be at or near the FULL mark. If it isn't, allow the engine to cool, then remove the cap from the reservoir and add a 50/50 mixture of ethylene glycol based antifreeze and water. Don't use rust inhibitors or additives.

9  Drive the vehicle and recheck the coolant level. If only a small amount of coolant is required to bring the system up to the proper level, water can be used. However, repeated additions of water will dilute the antifreeze and water solution. In order to maintain the proper ratio of antifreeze and water, always top up the coolant level with the correct mixture. An empty plastic milk jug or bleach bottle makes an excellent container for mixing coolant.

10  If the coolant level drops consistently, there may be a leak in the system. Inspect the radiator, hoses, filler cap, drain plugs and water pump. If no leaks are noted, have the expansion tank cap pressure tested by a service station.

11  If you have to remove the cap, wait until the engine has cooled completely, then slowly unscrew it **(see illustration)**. If coolant or steam escapes, let the engine cool down longer, then remove the cap.

12  Check the condition of the coolant as well. It should be relatively clear. If it's brown or rust colored, the system should be drained, flushed and refilled. Even if the coolant appears to be normal, the corrosion inhibitors wear out, so it must be replaced at the specified intervals.

## Brake and clutch fluid level check

*Refer to illustration 2.14*
**Warning:** *Brake fluid can harm your eyes and damage painted surfaces, so use extreme caution when handling or pouring it. Do not use brake fluid that has been standing open or fluid that is more than one year old. Brake fluid absorbs moisture from the air, which can cause a dangerous loss of brake effectiveness. Use only the specified type of brake fluid. Mixing different types (such as DOT 3 or 4 and DOT 5) can cause brake failure.*

13  The brake master cylinder is mounted at the left (driver's side) rear corner of the engine compartment. The clutch hydraulic system is also served by this reservoir.

14  The fluid level is checked by looking through the plastic reservoir mounted on the brake master cylinder **(see illustration)**. The fluid level should be between the MAX and MIN lines on the reservoir. If the fluid level is low, wipe the top of the reservoir and the cap with a clean rag to prevent contamination of the system as the cap is unscrewed. Top up with the recommended brake fluid, but do not overfill.

15  While the reservoir cap is off, check the master cylinder reservoir for contamination. If rust deposits, dirt particles or water droplets are present, the system should be drained, flushed and refilled by a dealer service department or other repair shop.

16  After filling the reservoir to the proper level, make sure the cap is seated to prevent fluid leakage and/or contamination.

17  The fluid level in the master cylinder will drop slightly as the disc brake pads wear. A very low level may indicate worn brake pads. Check for wear (see Section 9).

18  If the brake fluid level drops consistently, check the entire system for leaks immediately. Examine all brake lines, hoses and connections, along with the calipers, wheel cylinders and master cylinder.

19  When checking the fluid level, if you dis-

# Chapter 1 Tune-up and routine maintenance

2.21 The power steering fluid reservoir is located near the front of the engine compartment. Wipe off the area around the cap, then unscrew the cap/dipstick from the reservoir

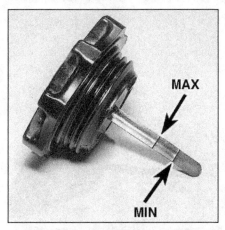

2.24 Check the fluid level with the dipstick (rest the cap on the reservoir's filler neck - don't screw it into place). The fluid level should be between the MIN and MAX marks

2.28 The windshield washer fluid reservoir is located in the front right corner of the engine compartment

cover one or both reservoirs empty or nearly empty, the brake or clutch hydraulic system should be checked for leaks and bled (see Chapters 8 and 9).

## Power steering fluid level check

*Refer to illustrations 2.21 and 2.24*

20  Check the power steering fluid level periodically to avoid steering system problems, such as damage to the pump. **Caution:** *DO NOT hold the steering wheel against either stop (extreme left or right turn) for more than five seconds. If you do, the power steering pump could be damaged.*
21  The power steering fluid reservoir is located on the left side of the engine compartment, and is equipped with a twist-off cap with an integral fluid level dipstick **(see illustration)**.
22  Park the vehicle on level ground and apply the parking brake.
23  Run the engine until it has reached normal operating temperature. With the engine at idle, turn the steering wheel back-and-forth several times to get any air out of the steering system. Shut the engine off, remove the cap by turning it counterclockwise, wipe the dipstick clean and reinstall the cap without screwing it down.
24  Remove the cap again and note the fluid level on the dipstick. It must be between the two lines **(see illustration)**.
25  Add small amounts of fluid until the level is correct. **Caution:** *Do not overfill the reservoir. If too much fluid is added, remove the excess with a clean syringe or suction pump.* Reinstall the cap.
26  Check the power steering hoses and connections for leaks and wear.
27  Check the condition and tension of the drivebelt.

## Windshield washer fluid level check

*Refer to illustrations 2.28 and 2.30*

28  Fluid for the windshield washer system is stored in a plastic reservoir in the engine compartment **(see illustration)**.
29  In milder climates, plain water can be used in the reservoir, but it should be kept no more than two-thirds full to allow for expansion if the water freezes. In colder climates, use windshield washer system antifreeze, available at any auto parts store, to lower the freezing point of the fluid. This comes in concentrated or pre-mixed form. If you purchase concentrated antifreeze, mix the antifreeze with water in accordance with the manufacturer's directions on the container. **Caution:** *Do not use cooling system antifreeze - it will damage the vehicle's paint.*
30  Before installing the cap, make sure the filter screen is clean **(see illustration)**.

## Tire and tire pressure checks

*Refer to illustrations 2.32, 2.33, 2.34a, 2.34b and 2.38*

31  Periodic inspection of the tires may save you the inconvenience of being stranded with a flat tire. It can also provide you with vital information regarding possible problems in the steering and suspension systems before major damage occurs.
32  Tires are equipped with 1/2-inch wide bands that will appear when tread depth

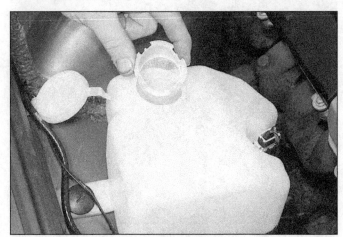

2.30 Before adding fluid to the washer reservoir, make sure the filter screen is clean

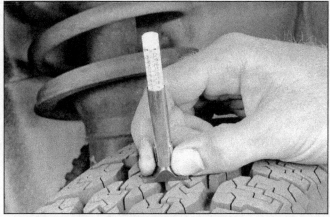

2.32 Use a tire tread depth indicator to monitor tire wear - they are available at auto parts stores and service stations and cost very little

**UNDERINFLATION**

**CUPPING**

Cupping may be caused by:
- Underinflation and/or mechanical irregularities such as out-of-balance condition of wheel and/or tire, and bent or damaged wheel.
- Loose or worn steering tie-rod or steering idler arm.
- Loose, damaged or worn front suspension parts.

**OVERINFLATION**

**INCORRECT TOE-IN OR EXTREME CAMBER**

**FEATHERING DUE TO MISALIGNMENT**

2.33 This chart will help you determine the condition of the tires, the probable cause(s) of abnormal wear and the corrective action necessary

reaches 1/16-inch, at which time the tires can be considered worn out. Tread wear can be monitored with a simple, inexpensive device known as a tread depth indicator **(see illustration)**.

33 Note any abnormal tire wear **(see illustration)**. Tread pattern irregularities such as cupping, flat spots and more wear on one side than the other are indications of front end alignment and/or balance problems. If any of these conditions are noted, take the vehicle to a tire shop or service station to correct the problem.

34 Look closely for cuts, punctures and embedded nails or tacks. Sometimes a tire will hold air pressure for a short time or leak down very slowly after a nail has embedded itself in the tread. If a slow leak persists, check the valve stem core to make sure it is tight **(see illustration)**. Examine the tread for an object that may have embedded itself in the tire or for a "plug" that may have begun to leak (radial tire punctures are repaired with a plug that is installed in the puncture). If a puncture is suspected, it can be easily verified by spraying a solution of soapy water onto the puncture **(see illustration)**. The

2.34a If a tire loses air on a steady basis, check the valve core first to make sure it's snug (special inexpensive wrenches are commonly available at auto parts stores)

2.34b If the valve core is tight, raise the corner of the vehicle with the low tire and spray a soapy water solution onto the tread as the tire is turned slowly - leaks will cause small bubbles to appear

# Chapter 1 Tune-up and routine maintenance

soapy solution will bubble if there is a leak. Unless the puncture is unusually large, a tire shop or service station can usually repair the tire.

35 Carefully inspect the inner sidewall of each tire for evidence of brake fluid leakage. If you see any, inspect the brakes immediately.

36 Correct air pressure adds miles to the lifespan of the tires, improves mileage and enhances overall ride quality. Tire pressure cannot be accurately estimated by looking at a tire, especially if it's a radial. A tire pressure gauge is essential. Keep an accurate gauge in the glove compartment. The pressure gauges attached to the nozzles of air hoses at gas stations are often inaccurate.

37 Always check tire pressure when the tires are cold. Cold, in this case, means the vehicle has not been driven over a mile in the three hours preceding a tire pressure check. A pressure rise of four to eight pounds is not uncommon once the tires are warm.

38 Unscrew the valve cap protruding from the wheel or hubcap and push the gauge firmly onto the valve stem **(see illustration)**. Note the reading on the gauge and compare the figure to the recommended tire pressure shown in your owner's manual or on the tire placard on the passenger side door or door pillar. Be sure to reinstall the valve cap to keep dirt and moisture out of the valve stem mechanism. Check all four tires and, if necessary, add enough air to bring them to the recommended pressure.

39 Don't forget to keep the spare tire inflated to the specified pressure (refer to your owner's manual).

2.38 To extend the life of the tires, check the air pressure at least once a week with an accurate gauge (don't forget the spare!)

# Oil service

### 3  Engine oil and filter replacement

*Refer to illustrations 3.5a, 3.5b, 3.8a, 3.8b and 3.11*

1 Frequent oil and filter changes are the most important preventative maintenance procedures which can be undertaken by the home mechanic. As engine oil ages, it becomes diluted and contaminated, which leads to premature engine wear.

2 Before starting this procedure, gather together all the necessary tools and materials. Also make sure that you have plenty of clean rags and newspapers handy, to mop up any spills. Ideally, the engine oil should be warm, as it will drain better, and more built-up sludge will be removed with it. Take care, however, not to touch the exhaust or any other hot parts of the engine when working under the vehicle. To avoid any possibility of scalding, and to protect yourself from possible skin irritants and other harmful contaminants in used engine oils, it is advisable to wear gloves when carrying out this work. Access to the underside of the vehicle will be greatly improved if it can be raised on a lift, driven onto ramps, or jacked up and supported with jackstands. Whichever method is chosen, make sure that the vehicle remains level, or if it is at an angle, so that the drain plug is at the lowest point. Where necessary remove the splash guard from under the engine.

3 Working in the engine compartment, locate the oil filter housing on the left-hand side of the engine, in front of the intake manifold.

4 Place a bundle of rags around the bottom of the housing to absorb any spilled oil.

5 Unscrew the through-bolt, then slowly remove the cover, and lift the filter cartridge out **(see illustrations)**. The oil will drain from the housing back into the oil pan as the cover is removed.

6 Remove the O-rings from the cover and from the bottom of the through-bolt.

7 Using a clean rag, wipe the mating faces

3.5a Unscrew the through-bolt . . .

of the housing and cover.

8 Install new O-rings to the cover and the through-bolt **(see illustrations)**.

9 Lower the new filter cartridge into the housing.

3.5b . . . and lift out the filter cartridge

3.8a Install a new O-ring on the cover . . .

3.8b . . . and on the through-bolt

## Chapter 1 Tune-up and routine maintenance

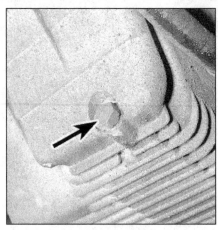

3.11 Oil pan drain plug (arrow) - four-cylinder engine shown

10  Smear a little clean engine oil on the O-rings, then install the cover. Install and tighten the through-bolt, ensuring that the washer is in place under the bolt head.
11  Working under the vehicle, loosen the oil pan drain plug about half a turn (see illustration). Position the draining container under the drain plug, then remove the plug completely. If possible, try to keep the plug pressed into the oil pan while unscrewing it by hand the last couple of turns.
12  Recover the sealing ring from the drain plug.
13  Allow some time for the old oil to drain, noting that it may be necessary to reposition the container as the oil flow slows to a trickle.
14  After all the oil has drained, wipe off the drain plug with a clean rag. Check the sealing washer for condition, and replace it if necessary. Clean the area around the drain plug opening, then install and tighten the plug.
15  Remove the old oil and all tools from under the car, then lower the car to the ground (if applicable).
16  Remove the dipstick then unscrew the oil filler cap from the valve cover. Fill the engine, using the correct grade and type of oil (see Section 2). A funnel may help to reduce spillage. Pour in half the specified quantity of oil first, then wait a few minutes for the oil to fall to the oil pan. Continue adding oil a little at a time until the level is up to the lower mark on the dipstick. Finally, bring the level up to the upper mark on the dipstick. Insert the dipstick, and install the filler cap.
17  Start the engine and run it for a few minutes; check for leaks around the oil filter seal and the oil pan drain plug. Note that there may be a delay of a few seconds before the oil pressure warning light goes out when the engine is first started, as the oil circulates through the engine oil galleries and the new oil filter, before the pressure builds up.
18  Switch off the engine, and wait a few minutes for the oil to settle in the oil pan once more. With the new oil circulated and the filter completely full, recheck the level on the dipstick, and add more oil as necessary.
19  The old oil drained from the engine cannot be reused in its present state and should be disposed of. Check with your local refuse disposal company, disposal facility or environmental agency to see if they will accept the oil for recycling. Don't pour used oil into drains or onto the ground. After the oil has cooled, it can be drained into a suitable container (capped plastic jugs, topped bottles, milk cartons, etc.) for transport to one of these disposal sites.

## 4  Resetting the service interval display

Note: *The following is for use with the special BMW service tool and adapter. If an aftermarket tool is being used, refer to the instructions supplied by the manufacturer.*
1  Turn the ignition off and plug the BMW service interval resetting tool (No. 62 1 110) and adapter (No. 62 1 140) into the diagnostic socket.
2  Ensure that all electrical items are switched off then turn on the ignition switch.
Note: *Do not start the engine.*
3  Press and hold the red Inspection button; the green (function check) light will illuminate. After approximately 3 seconds the red lamp should also light, remain on for approximately 12 seconds, and then go out. Release the Inspection button and the green (function check) light will go out.
4  If the clock (annual service) symbol was illuminated at the same time as the oil service or inspection indicator, wait 10 seconds then repeat the operation in Step 3.
5  Turn off the ignition switch and disconnect the resetting tool and adapter from the diagnostic connector.
6  Turn the ignition switch on and check that the all service interval display green LEDs (the yellow and red LEDs may also light) and indicator illuminate and then go out.

# Inspection I

## 5  Automatic transmission fluid level check

*Refer to illustration 5.3*
Note: *A new filler/level plug sealing ring will be required.*
1  The fluid level is checked by removing the filler/level plug from the transmission fluid pan. If desired, jack up the vehicle and support on jack stands to improve access, but make sure that the vehicle is level (all four wheels should be off the ground).
2  The fluid level should only be checked when the transmission is neither cold nor at normal operating temperature. Fluid temperature should be between 86 and 130-degrees F (the automatic transmission fluid pan will be warm to the touch, but not hot). Starting the vehicle when cold, allowing it to run for a few minutes and shifting the transmission into all of its gear ranges should bring the fluid up to the proper temperature for checking.
3  Turn off the engine. Working under the vehicle, place a container under the transmission fluid pan, then unscrew the filler/level plug (see illustration). Recover the sealing ring.
4  Restart the engine and shift the transmission into Neutral; the fluid level should be up to the lower edge of the filler/level plug hole.

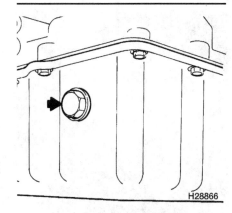

5.3 Automatic transmission filler/level plug (arrow)

5  If necessary, top-up the fluid until it overflows from the plug hole.
6  The condition of the fluid should also be checked along with the level. If the fluid is black or a dark reddish-brown color, or if it smells burned, it should also be replaced (see Section 25).
7  Install the filler/level plug, using a new sealing ring, and tighten it to the specified torque.
8  Stop the engine and, if raised, lower the vehicle to the ground.

## 6  Hose and fluid leak check

*Refer to illustration 6.1*
1  Visually inspect the engine joint faces, gaskets and seals for any signs of water or oil leaks. Pay particular attention to the areas around the camshaft cover, cylinder head, oil filter and oil pan joint faces. Bear in mind that, over a period of time, some very slight seepage from these areas is to be expected - what you are really looking for is any indica-

# Chapter 1 Tune-up and routine maintenance

1-13

6.1 A leak in the cooling system will usually show up as white - or rust - colored deposits on the area adjoining the leak

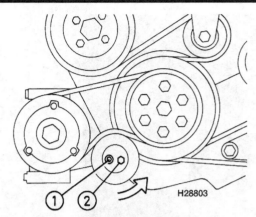

7.8 Air conditioning compressor drivebelt tensioner - four-cylinder engines

1  Pulley bolt  2  Hexagon cut-out

tion of a serious leak **(see illustration)**. Should a leak be found, replace the offending gasket or oil seal by referring to the appropriate Chapters in this manual.

2   Also check the security and condition of all the engine-related pipes and hoses. Ensure that all cable-ties or securing clips are in place and in good condition. Clips which are broken or missing can lead to chafing of the hoses, pipes or wiring, which could cause more serious problems in the future.

3   Carefully check the radiator hoses and heater hoses along their entire length. Replace any hose which is cracked, swollen or deteriorated. Cracks will show up better if the hose is squeezed. Pay close attention to the hose clamps that secure the hoses to the cooling system components. Hose clamps can pinch and puncture hoses, resulting in cooling system leaks.

4   Inspect all the cooling system components (hoses, joint faces etc.) for leaks. A leak in the cooling system will usually show up as white- or rust-colored deposits on the area adjoining the leak. Where any problems of this nature are found on system components, replace the component or gasket with reference to Chapter 3.

5   Where applicable, inspect the automatic transmission fluid cooler hoses for leaks or deterioration.

6   With the vehicle raised, inspect the gasoline tank and filler neck for punctures, cracks and other damage. The connection between the filler neck and tank is especially critical. Sometimes a rubber filler neck or connecting hose will leak due to loose retaining clamps or deteriorated rubber.

7   Carefully check all rubber hoses and metal fuel lines leading away from the gasoline tank. Check for loose connections, deteriorated hoses, crimped lines, and other damage. Pay particular attention to the vent pipes and hoses, which often loop up around the filler neck and can become blocked or crimped. Follow the lines to the front of the vehicle, carefully inspecting them all the way. Replace damaged sections as necessary.

8   Closely inspect the metal brake pipes which run along the vehicle underbody. If they show signs of excessive corrosion or damage they must be replaced.

9   From within the engine compartment, check the security of all fuel hose attachments and pipe unions, and inspect the fuel hoses and vacuum hoses for kinks, chafing and deterioration.

10  Where applicable, check the condition of the power steering fluid hoses and pipes.

## 7  Drivebelts - check and replacement

### Check

1   Due to their function and construction, the belts are prone to failure after a period of time, and should be inspected periodically to prevent problems.

2   The number of belts used on a particular vehicle depends on the accessories installed. Drivebelts are used to drive the water pump, alternator, power steering pump and air conditioning compressor.

3   To improve access for belt inspection, if desired, remove the viscous cooling fan and cowl (see Chapter 3).

4   With the engine stopped, using your fingers (and a flashlight if necessary), move along the belts, checking for cracks and separation of the belt plies. Also check for fraying and glazing, which gives the belt a shiny appearance. Both sides of the belts should be inspected, which means the belt will have to be twisted to check the underside. If necessary turn the engine using a wrench or socket on the crankshaft pulley bolt to that the whole of the belt can be inspected.

### Replacement

#### Four-cylinder engines (1992 and 1993 models)

**Air conditioning compressor drivebelt**

*Refer to illustration 7.8*

5   Access is most easily obtained from under the vehicle. If desired, jack up the front of the vehicle and support securely on axle stands.

6   Loosen the tensioner pulley bolt, and slide the drivebelt from the pulleys.

7   Loosen the tensioner pulley bolt until the tensioner roller rotates smoothly with no friction.

8   Install the drivebelt around the pulleys, then engage a hexagon bit and torque wrench with the hexagon cut-out in the tensioner pulley and apply the specified torque (see Specifications) to the pulley. Tighten the tensioner pulley bolt **(see illustration)**.

9   Where applicable, lower the vehicle to the ground.

**Power steering pump drivebelt**

*Refer to illustration 7.12*

10  Access is most easily obtained from under the vehicle. If desired, jack up the front of the vehicle and support securely on axle stands.

11  Where applicable, remove the air conditioning compressor drivebelt as described previously in this Section.

12  Loosen the power steering pump tensioner locknut, then turn the tensioner nut to loosen the drivebelt until it can be slid from

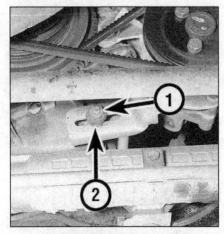

7.12 Power steering pump drivebelt tensioner locknut (1) and tensioner nut (2) - four-cylinder engine

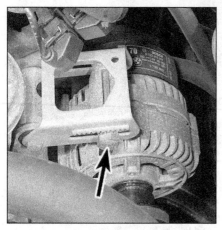

7.20 Alternator/water pump drivebelt tensioner bolt (arrow) - four-cylinder engine

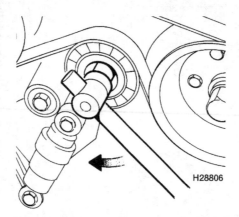

7.30 Lever the air conditioning drivebelt tensioner clockwise - six-cylinder engine

7.36 Removing the water pump/alternator/power steering pump drivebelt - six-cylinder engine

the pulleys **(see illustration)**. If necessary, loosen the power steering pump mounting bolts to allow the pump to pivot.
13  Engage the belt with the pulleys, then turn the tensioner nut to tension the drivebelt. As a guide, it should be possible to push the belt at the center of the belt run between the power steering pump and crankshaft pulleys to give a deflection of approximately 3/32-inch under moderate finger pressure.
14  When the drivebelt is correctly tensioned, tighten the tensioner locknut, and where applicable tighten the power steering pump mounting bolts.
15  Where applicable, lower the vehicle to the ground.

### Alternator/water pump drivebelt

*Refer to illustration 7.20*
16  Where applicable, remove the air conditioning compressor and/or power steering pump drivebelt(s), as described previously in this Section.
17  If the drivebelt is to be re-used, mark the running direction of the belt before removal.
18  To improve access, remove the viscous cooling fan and shroud (see Chapter 3).
19  Make a careful note of the routing of the drivebelt before removal.
20  Loosen the nut at the rear of the alternator tensioner bolt, then turn the bolt as necessary to loosen the belt **(see illustration)**. If necessary loosen the lower alternator mounting bolt to allow the alternator to pivot.
21  Slide the drivebelt from the pulleys.
22  If the original belt is being installed, observe the running direction mark made before removal.
23  Engage the belt with the pulleys, ensuring that it is routed as noted before removal.
24  Lever the tensioner bolt to tension the drivebelt. As a guide, it should be possible to push the belt at the center of the belt run between the alternator and water pump pulleys to give a deflection of approximately 13/64-inch under moderate finger pressure.
25  When the drivebelt is correctly tensioned, tighten the tensioner nut, and where applicable tighten the lower alternator mounting bolt.
26  Install the viscous cooling fan and shroud (see Chapter 3).
27  Where applicable, install the air conditioning compressor and/or power steering pump drivebelt(s), as described previously in this Section.

### Four-cylinder engines (1994 and later models) and all six-cylinder engines

#### Air conditioning compressor drivebelt

*Refer to illustration 7.30*
28  Access is most easily obtained from under the vehicle. If desired, jack up the front of the vehicle and support it securely on jackstands.
29  Where applicable, pry the cover from the center of the tensioner pulley.
30  Engage a hexagon bit and extension bar with the tensioner bolt, and lever the tensioner clockwise **(see illustration)**. Slide the belt from the pulleys.
31  Lever the tensioner until the drivebelt can be routed around the pulleys, then release the tensioner. Ensure that the belt is engaged with the grooves in the pulleys.
32  Where applicable, install the pulley cover and lower the vehicle to the ground.

#### Water pump/alternator/power steering pump drivebelt

*Refer to illustration 7.36*
33  Where applicable, remove the air conditioning compressor drivebelt as described previously in this Section.
34  Proceed as described in Steps 17 to 19.
35  Pry the cover from the center of the tensioner pulley.
36  Engage a hexagon key and extension bar with the pulley bolt, then lever the pulley (counterclockwise) to compress the tensioner, and slide the drivebelt from the pulleys **(see illustration)**.
37  If the original belt is being installed, observe the running direction mark made before removal.
38  Compress the tensioner, and engage the belt with the pulleys, ensuring that it is routed as noted before removal. Make sure that the belt engages correctly with the grooves in the pulleys.
39  Install the viscous cooling fan and shroud with reference to Chapter 3.
40  Where applicable, install the air conditioning compressor drivebelt as described previously in this Section.

## 8  Steering and suspension check

### Front suspension and steering check

*Refer to illustration 8.4*
1  Raise the front of the vehicle and securely support it on jackstands.
2  Visually inspect the balljoint dust covers and the steering rack-and-pinion boots for splits, chafing or deterioration. Any wear of these components will cause loss of lubricant, together with dirt and water entry, resulting in rapid deterioration of the balljoints or steering gear.
3  On vehicles with power steering, check the fluid hoses for chafing or deterioration, and the pipe and hose connections for fluid leaks. Also check for signs of fluid leakage under pressure from the steering gear rubber boots, which would indicate failed fluid seals within the steering gear.
4  Grasp the wheel at the 12 o'clock and 6 o'clock positions, and try to rock it **(see illustration)**. Very slight free play may be felt, but if the movement is appreciable, further investigation is necessary to determine the source. Continue rocking the wheel while an assistant depresses the brake pedal. If the movement is now eliminated or significantly reduced, it is likely that the hub bearings are at fault. If the free play is still evident with the brake pedal depressed, then there is wear in the suspension joints or mountings.
5  Now grasp the wheel at the 9 o'clock

# Chapter 1  Tune-up and routine maintenance

8.4 Check for wear in the hub bearings by grasping the wheel and trying to rock it

9.3 For a quick check, the thickness of the friction material of the brake pad can be measured through the aperture in the caliper body

13.2 Check the condition of the exhaust system mounts (arrows)

and 3 o'clock positions, and try to rock it as before. Any movement felt now may again be caused by wear in the hub bearings or the steering tie-rod balljoints. If the inner or outer balljoint is worn, the visual movement will be obvious.

6  Using a large screwdriver or prybar, check for wear in the suspension mounting bushings by levering between the relevant suspension component and its attachment point. Some movement is to be expected as the mountings are made of rubber, but excessive wear should be obvious. Also check the condition of any visible rubber bushings, looking for splits, cracks or contamination of the rubber.

7  With the car standing on its wheels, have an assistant turn the steering wheel back and forth about an eighth of a turn each way. There should be very little, if any, lost movement between the steering wheel and wheels. If this is not the case, closely observe the joints and mountings previously described, but in addition, check the steering column universal joints for wear, and the rack-and-pinion steering gear itself.

## Suspension strut/shock absorber check

8  Check for any signs of fluid leakage around the suspension strut/shock absorber body, or from the rubber boot around the piston rod. Should any fluid be noticed, the suspension strut/shock absorber is defective internally, and should be replaced. **Note:** *Suspension struts/shock absorbers should always be replaced in pairs on the same axle.*

9  The efficiency of the suspension strut/shock absorber may be checked by bouncing the vehicle at each corner. Generally speaking, the body will return to its normal position and stop after being depressed. If it rises and returns on a rebound, the suspension strut/shock absorber is probably suspect. Examine also the suspension strut/shock absorber upper and lower mountings for any signs of wear.

## 9  Brake pad check

*Refer to illustration 9.3*

**Warning:** *The dust created by the brake system is harmful to your health. Never blow it out with compressed air and don't inhale any of it. An approved filtering mask should be worn when working on the brakes. Do not, under any circumstances, use petroleum-based solvents to clean brake parts. Use brake system cleaner only!*

1  Loosen the wheel lug bolts. Jack up the car and support it securely on jackstands. Remove the wheels.

2  For a comprehensive check, the brake pads should be removed and cleaned. The condition of the calipers can then also be checked, and the condition of the brake discs can be fully examined on both sides. Refer to Chapter 9 for further information.

3  Look through the cutout in each caliper and measure the amount of friction material remaining on the brake pads **(see illustration)**.

4  If any front or rear pad's friction material is worn to the specified thickness or less, *all four front or rear pads must be replaced as a set.*

## 10  Brake hoses and lines - inspection

1  Raise the vehicle and place it securely on jackstands.

2  Inspect all flexible brake hoses. Verify that they're in good condition and correctly routed. If a hose is cracked or otherwise damaged, replace it (see Chapter 9). And make sure that the brake hose-to-brake line fittings are tight and free of corrosion.

3  Inspect all rigid metal brake lines. Verify that they're not corroded, dented or otherwise damaged. Replace worn or damaged lines (see Chapter 9).

## 11  Throttle linkage - lubrication

1  Periodically, lubricate the accelerator and throttle linkage with a general purpose oil.

2  Lubricate all joints and bearings in the linkage.

3  Lubricate the throttle plate shaft with multipurpose grease.

## 12  Parking brake check

Check and, if necessary, adjust the parking brake (see Chapter 9). Check that the parking brake cables are free to move easily and lubricate all exposed linkages/cable pivots.

## 13  Exhaust system check

*Refer to illustration 13.2*

1  With the engine cold (at least an hour after the vehicle has been driven), check the complete exhaust system from the engine to the end of the tailpipe. The exhaust system is most easily checked with the vehicle raised on a hoist, or suitably supported on axle stands, so that the exhaust components are readily visible and accessible.

2  Check the exhaust pipes and connections for evidence of leaks, severe corrosion and damage. Make sure that all brackets and mountings are in good condition, and that all relevant nuts and bolts are tight **(see illustration)**. Leakage at any of the joints or other parts of the system will usually show up as a black sooty stain in the vicinity of the leak.

3  Rattles and other noises can often be traced to the exhaust system, especially the brackets and mountings. Try to move the pipes and silencers. If the components are able to come into contact with the body or suspension parts, secure the system with new mountings. Otherwise separate the joints (if possible) and twist the pipes as necessary to provide additional clearance.

## 14 Seat belt check

1 Carefully examine the seat belt webbing for cuts or any signs of serious fraying or deterioration. If the seat belt is of the retractable type, pull the belt all the way out, and examine the full extent of the webbing.
2 Fasten and unfasten the belt, ensuring that the locking mechanism holds securely and releases properly when intended. If the belt is of the retractable type, check also that the retracting mechanism operates correctly when the belt is released.
3 Check the security of all seat belt mountings and attachments which are accessible, without removing any trim or other components, from inside the vehicle.

## 15 Hinge and lock lubrication

Lubricate the hinges of the hood, doors and tailgate with a light general-purpose oil. Similarly, lubricate all latches, locks and lock strikers. At the same time, check the security and operation of all the locks, adjusting them if necessary (see Chapter 11).
Lightly lubricate the hood release mechanism and cable with a suitable grease.

## 16 Pollen filter replacement

### Models without air conditioning

1 Working in the engine compartment, remove the rubber seal from the top of the heating/ventilation system inlet and release the grille from the inlet. On models where the grille is an integral part of the windshield wiper motor cover panel, to improve access remove the wiper arms and remove the one-piece cover panel (see Chapter 12).
2 Remove the retaining screws and free the wiring harness duct from the inlet duct.
3 Loosen and remove the retaining screws and retaining plate and remove the inlet from the firewall. Note: *On six-cylinder engines it may be necessary to remove the injector and spark plug covers from the engine to enable the inlet to be removed.*
4 Depress the retaining clips and remove the pollen filter(s) from the side of the blower motor housing.
5 Clip the new filter(s) onto the housing and install all disturbed components by reversing the removal procedure.

### Models with air conditioning

6 Remove the heating/ventilation system control unit (see Chapter 3).
7 Remove the retaining screws then slide the driver's side lower dash panel to the side, to release its retaining clips, and remove it from the vehicle.

8 Unclip the air duct from the side of the air distribution housing.
9 Remove the retaining screws and pivot the control unit mounting bracket forwards to gain access to the pollen filter housing.
10 Release the retaining clip by rotating it counterclockwise then remove the cover from the side of the air distribution housing and slide out the pollen filter.
11 Install the new pollen filter then install the cover, securing it in position with the retaining clip.
12 Install the control unit mounting bracket screws and tighten them securely.
13 Install the air duct and lower dash panel.
14 Install the control unit (see Chapter 3).

## 17 Headlight beam alignment check

Accurate adjustment of the headlight beam is only possible using optical beam-setting equipment, and this work should therefore be carried out by a BMW dealer or service station with the necessary facilities.
Basic adjustments can be carried out in an emergency, and further details are given in Chapter 12.

## 18 Windshield/headlight washer system(s) check

Check that each of the washer jet nozzles are clear and that each nozzle provides a strong jet of washer fluid. The headlight jets should be aimed to spray at a point slightly above the center of the headlight. On the windshield washer nozzles where there are two jets, aim one of the jets slightly above then center of the windshield and aim the other just below to ensure complete coverage of the windshield. If necessary, adjust the jets using a pin.

## 19 Engine management system check

1 This check is part of the manufacturer's maintenance schedule, and involves testing the engine management system using special dedicated test equipment. Such testing will allow the test equipment to read any fault codes stored in the electronic control unit memory.
2 Unless a fault is suspected, this test is not essential, although it should be noted that it is recommended by the manufacturers.
3 If access to suitable test equipment is not possible, make a thorough check of all ignition, fuel and emission control system components, hoses, and wiring, for security and obvious signs of damage. Further details of the fuel system, emission control system and ignition system can be found in Chapters 4, 5 and 6.

## 20 Road test

### Instruments and electrical equipment

1 Check the operation of all instruments and electrical equipment.
2 Make sure that all instruments read correctly, and switch on all electrical equipment in turn, to check that it functions properly.

### Steering and suspension

3 Check for any abnormalities in the steering, suspension, handling or road "feel".
4 Drive the vehicle, and check that there are no unusual vibrations or noises.
5 Check that the steering feels positive, with no excessive "sloppiness", or roughness, and check for any suspension noises when cornering and driving over bumps.

### Drivetrain

6 Check the performance of the engine, clutch (where applicable), transmission and driveaxles.
7 Listen for any unusual noises from the engine, clutch and transmission.
8 Make sure that the engine runs smoothly when idling, and that there is no hesitation when accelerating.
9 Check that, where applicable, the clutch action is smooth and progressive, that the drive is taken up smoothly, and that the pedal travel is not excessive. Also listen for any noises when the clutch pedal is depressed.
10 On manual transmission models, check that all gears can be engaged smoothly without noise, and that the gear lever action is not abnormally vague or "notchy".
11 On automatic transmission models, make sure that all gearchanges occur smoothly, without harshness, and without an increase in engine speed between changes. Check that all the gear positions can be selected with the vehicle at rest. If any problems are found, they should be referred to a BMW dealer.

### Check the operation and performance of the braking system

12 Make sure that the vehicle does not pull to one side when braking, and that the wheels do not lock prematurely when braking hard.
13 Check that there is no vibration through the steering when braking.
14 Check that the parking brake operates correctly without excessive movement of the lever, and that it holds the vehicle stationary on a slope.
15 Test the operation of the brake booster unit as follows. With the engine off, depress the brake pedal four or five times to exhaust the vacuum. Hold the brake pedal depressed, then start the engine. As the engine starts, there should be a noticeable "give" in the brake pedal as vacuum builds up. Allow the engine to run for at least two minutes, and then switch it off. If the brake pedal is

# Chapter 1  Tune-up and routine maintenance

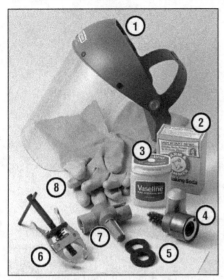

**21.1  Tools and materials required for battery maintenance**

1. **Face shield/safety goggles** - When removing corrosion with a brush, the acidic particles can easily fly up into your eyes
2. **Baking soda** - A solution of baking soda and water can be used to neutralize corrosion
3. **Petroleum jelly** - A layer of this on the battery posts will help prevent corrosion
4. **Battery post/cable cleaner** - This wire brush cleaning tool will remove all traces of corrosion from the battery posts and cable clamps
5. **Treated felt washers** - Placing one of these on each post, directly under the cable clamps, will help prevent corrosion
6. **Puller** - Sometimes the cable clamps are very difficult to pull off the posts, even after the nut/bolt has been completely loosened. This tool pulls the clamp straight up and off the post without damage
7. **Battery post/cable cleaner** - Here is another cleaning tool which is a slightly different version of Number 4 above, but it does the same thing
8. **Rubber gloves** - Another safety item to consider when servicing the battery; remember that's acid inside the battery!

depressed now, it should be possible to detect a hiss from the booster as the pedal is depressed. After about four or five applications, no further hissing should be heard, and the pedal should feel considerably harder.

## 21  Battery check, maintenance and charging

### Check and maintenance

Refer to illustrations 21.1, 21.2, 21.4, 21.7, 21.8a, 21.8b and 12.8c
**Warning:** Certain precautions must be followed when checking and servicing the bat-

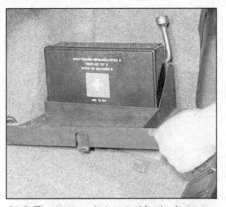

**21.2  The battery is located in the luggage compartment on six-cylinder 3-Series models and on 1997 and later Z3 models; it's under the first aid kit (shown) on 3-Series models, under a battery cover on Z3 models**

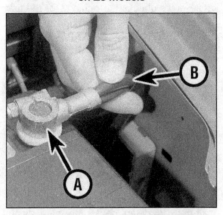

**21.7  Check the tightness of the battery cable clamps (A) to ensure good electrical connections. You should not be able to move them. Also check each cable (B) for cracks and frayed connections**

tery. Hydrogen gas, which is highly flammable, is always present in the battery cells, so keep lighted tobacco and all other flames and sparks away from it. The electrolyte inside the battery is actually dilute sulfuric acid, which will cause injury if splashed on your skin or in your eyes. It will also ruin clothes and painted surfaces. When removing the battery cables, always detach the negative cable first and hook it up last!

1  Battery maintenance is an important procedure which will help ensure that you are not stranded because of a dead battery. Several tools are required for this procedure **(see illustration)**.
2  Before servicing the battery, always turn the engine and all accessories off and disconnect the cable from the negative terminal of the battery. The battery is located under the hood on four-cylinder 3-Series models and on 1996 Z3 models. On six-cylinder models and on 1997 and later Z3 models, the battery is located in the trunk **(see illustration)**. **Caution:** If the radio in your vehicle is equipped with an anti-theft system, make sure you have the correct activation code

**21.4  Remove the cell caps to check the water level in the battery - if the level is low, add distilled water only**

**21.8a  If corrosion (white, fluffy deposits) is evident, remove the cables from the battery terminals and clean the terminals . . .**

before disconnecting the battery.
3  The battery may either be a no-maintenance or a low-maintenance type. If it is a no-maintenance type, it will not be possible to remove the cell caps and check the level of electrolyte.
4  Remove the caps and check the electrolyte level in each of the battery cells **(see illustration)**. It must be above the plates. There's usually a split-ring indicator in each cell to indicate the correct level. If the level is low, add distilled water only, then install the cell caps. **Caution:** Overfilling the cells may cause electrolyte to spill over during periods of heavy charging, causing corrosion and damage to nearby components.
5  If the positive terminal and cable clamp on your vehicle's battery is equipped with a rubber protector, make sure that it's not torn or damaged. It should completely cover the terminal.
6  The external condition of the battery should be checked periodically. Look for damage such as a cracked case.
7  Check the tightness of the battery cable clamps to ensure good electrical connections and inspect the entire length of each cable, looking for cracked or abraded insulation and frayed conductors **(see illustration)**.
8  If corrosion (visible as white, fluffy

**21.8b ... and the cable clamps**

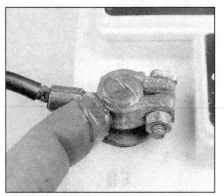

**21.8c Battery corrosion can be kept to a minimum by applying a thin film of petroleum jelly to the clamps and terminals after they are reconnected**

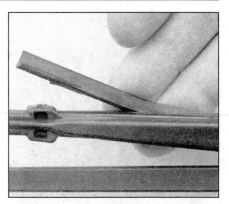

**22.2 Check the condition of the wiper blades; if they are cracked or show any signs of deterioration, or if the glass swept area is smeared, replace them**

deposits) is evident, remove the cables from the terminals, clean them with a battery brush and reinstall them **(see illustrations)**. Corrosion can be kept to a minimum by installing specially treated washers available at auto parts stores or by applying a layer of petroleum jelly or grease to the terminals and cable clamps after they are assembled **(see illustration)**.

9   Make sure that the battery carrier is in good condition and that the hold-down clamp bolt is tight. If the battery is removed (see Chapter 5 for the removal and installation procedure), make sure that no parts remain in the bottom of the carrier when it's reinstalled. When reinstalling the hold-down clamp, don't overtighten the bolt.

10   Corrosion on the carrier, battery case and surrounding areas can be removed with a solution of water and baking soda. Apply the mixture with a small brush, let it work, then rinse it off with plenty of clean water.

11   Any metal parts of the vehicle damaged by corrosion should be coated with a zinc-based primer, then painted.

12   Additional information on the battery and jump starting can be found in Chapter 5 and the front of this manual.

## Charging

**Note:** *The manufacturer recommends the battery be removed from the vehicle for charging because the gas which escapes during this procedure can damage the paint or interior, depending on the model. Fast charging with the battery cables connected can result in damage to the electrical system.*

13   Remove all of the cell caps (if equipped) and cover the holes with a clean cloth to prevent spattering electrolyte. Disconnect the negative battery cable and hook the battery charger leads to the battery posts (positive to positive, negative to negative), then plug in the charger. Make sure it is set at 12-volts if it has a selector switch. **Caution:** *If the radio in your vehicle is equipped with an anti-theft system, make sure you have the correct activation code before disconnecting the battery.*

14   If you're using a charger with a rate higher than two amps, check the battery regularly during charging to make sure it doesn't overheat. If you're using a trickle charger, you can safely let the battery charge overnight after you've checked it regularly for the first couple of hours.

15   If the battery has removable cell caps, measure the specific gravity with a hydrometer every hour during the last few hours of the charging cycle. Hydrometers are available inexpensively from auto parts stores - follow the instructions that come with the hydrometer. Consider the battery charged when there's no change in the specific gravity reading for two hours and the electrolyte in the cells is gassing (bubbling) freely. The specific gravity reading from each cell should be very close to the others. If not, the battery probably has a bad cell(s).

16   Some batteries with sealed tops have built-in hydrometers on the top that indicate the state of charge by the color displayed in the hydrometer window. Normally, a bright-colored hydrometer indicates a full charge and a dark hydrometer indicates the battery still needs charging. Check the battery manufacturer's instructions to be sure you know what the colors mean.

17   If the battery has a sealed top and no built-in hydrometer, you can hook up a voltmeter across the battery terminals to check

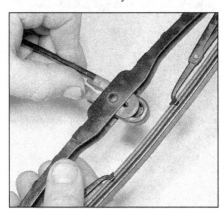

**22.5 To remove a wiper blade, pull the arm fully away from the windshield until it locks, swivel the blade 90-degrees, press the locking tab with your fingers and slide the blade out of the arm's hooked end**

the charge. A fully charged battery should read 12.6-volts or higher.

18   Further information on the battery and jump starting can be found in Chapter 5 and at the front of this manual.

## 22   Wiper blade check and replacement

1   Road film can build up on the wiper blades and affect their efficiency, so they should be washed regularly with a mild detergent solution.

### Check

*Refer to illustration 22.2*

2   The wiper and blade assembly should be inspected periodically. Even if you don't use your wipers, the sun and elements will dry out the rubber portions, causing them to crack and break apart **(see illustration)**. If inspection reveals hardened or cracked rubber, replace the wiper blades. If inspection reveals nothing unusual, wet the windshield, turn the wipers on, allow them to cycle several times, then shut them off. An uneven

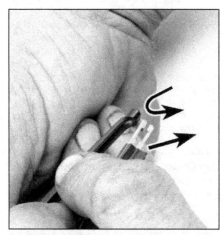

**22.6 Detach the end of the wiper element from the end of the frame, then slide the element out**

# Chapter 1 Tune-up and routine maintenance

wiper pattern across the glass or streaks over clean glass indicate that the blades should be replaced.

3 The operation of the wiper mechanism can loosen the fasteners, so they should be checked and tightened, as necessary, at the same time the wiper blades are checked (see Chapter 12 for further information regarding the wiper mechanism).

## Wiper blade replacement

*Refer to illustrations 22.5 and 22.6*

4 Pull the wiper/blade assembly away from the glass.
5 Press the retaining lever and slide the blade assembly down and out of the wiper arm **(see illustration)**.
6 Detach the end of the element from the wiper frame, then slide the element out of the frame **(see illustration)**.
7 Compare the new element with the old for length, design, etc.
8 Slide the new element into place and insert the end in the wiper frame to lock it in place.
9 Reinstall the blade assembly on the arm, wet the glass and check for proper operation.

# Inspection II

## 23 Spark plug replacement

### General

1 The correct functioning of the spark plugs is vital for the correct running and efficiency of the engine. It is essential that the plugs installed are appropriate for the engine (the suitable type is specified at the beginning of this Chapter). If this type is used, and the engine is in good condition, the spark plugs should not need attention between scheduled replacement intervals. Spark plug cleaning is rarely necessary, and should not be attempted unless specialized equipment is available, as damage can easily be caused to the firing ends.

### Four-cylinder engine

*Refer to illustrations 23.3, 23.4, 23.6a, 23.6b, 23.11, 23.12 and 23.14*

2 The spark plugs are located under a cover in the center of the cylinder head.
3 Working at the top of the valve cover, twist the two fasteners, and remove the spark plug cover from the center of the valve cover **(see illustration)**.
4 If the marks on the spark plug wires cannot be seen, mark the leads 1 to 4, corresponding to the cylinder the lead serves (No

**23.3 Removing the spark plug cover - six-cylinder engine**

1 cylinder is at the timing chain end of the engine). Using the plastic tool provided (clipped into the end of the spark plug wire plastic housing), pull the spark plug wires from the spark plugs **(see illustration)**.
5 It is advisable to remove any dirt from the spark plug recesses, using a clean brush, vacuum cleaner or compressed air before removing the plugs, to prevent dirt dropping into the cylinders.
6 Unscrew the plugs using a spark plug

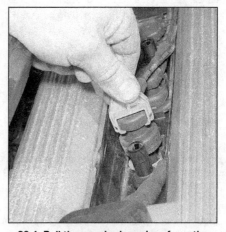

**23.4 Pull the spark plug wires from the spark plugs using the tool provided - six-cylinder engine**

socket and extension bar **(see illustrations)**. Keep the socket aligned with the spark plug - if it is forcibly moved to one side, the ceramic insulator may be broken off.
7 Examination of the spark plugs will give a good indication of the condition of the engine. If the insulator nose of the spark plug is clean and white, with no deposits, this is indicative of a weak mixture or too hot a plug (a hot plug transfers heat away from the elec-

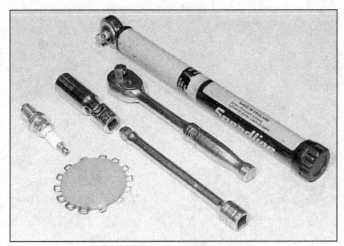

**23.6a Tools required for spark plug removal, gap adjustment and installation**

**23.6b Unscrew the spark plugs using a spark plug socket**

**23.11 Measuring the spark plug gap with a wire gauge**

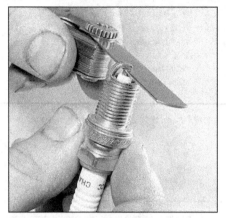

**23.12 Measuring the spark plug gap with a feeler blade**

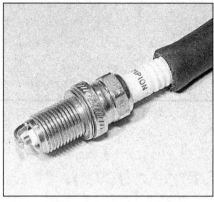

**23.14 A length of rubber hose will save time and prevent damaged threads when installing the spark plugs**

trode slowly, a cold plug transfers heat away quickly).

8  If the tip and insulator nose are covered with hard black-looking deposits, then this is indicative that the mixture is too rich. Should the plug be black and oily, then it is likely that the engine is fairly worn, as well as the mixture being too rich.

9  If the insulator nose is covered with light tan to grayish-brown deposits, then the mixture is correct, and it is likely that the engine is in good condition.

10  When buying new spark plugs, it is important to obtain the correct plugs for your specific engine (see Specifications).

11  The spark plug electrode gap is of considerable importance as, if it is too large or too small, the size of the spark and its efficiency will be seriously impaired. The gap should be set to the value given in the Specifications at the beginning of this Chapter **(see illustration)**.

12  To set the gap, measure it with a feeler gauge, then bend the outer plug electrode until the correct gap is achieved **(see illustration)**. The center electrode should never be bent, as this may crack the insulator and cause plug failure, if nothing worse. If using feeler gauges, the gap is correct when the appropriate-size blade is a firm sliding fit.

13  Special spark plug electrode gap adjusting tools are available from most auto parts stores, or from some spark plug manufacturers.

14  Before installing the spark plugs, check that the threaded connector sleeves (on top of the plug) are tight, and that the plug exterior surfaces and threads are clean. It is very often difficult to insert spark plugs into their holes without cross-threading them. To avoid this possibility, push a short length of hose over the end of the spark plug **(see illustration)**.

15  Remove the rubber hose (if used), and tighten the plug to the specified torque (see Specifications) using the spark plug socket and a torque wrench. Install the remaining plugs in the same way.

16  Connect the spark plug wires in their correct order, and clip the lead removal tool into position in its holder.

17  Install the spark plug cover, and secure with the fasteners.

### Six-cylinder engine

18  The spark plugs are located under the ignition coils in the center of the cylinder head.

19  Remove the ignition coils (see Chapter 5B).

20  It is advisable to remove any dirt from the spark plug recesses, using a vacuum cleaner or compressed air before removing the plugs, to prevent dirt dropping into the cylinders.

21  Unscrew the plugs using a spark plug wrench and extension bar **(see illustration 23.6b)**. Keep the socket aligned with the spark plug - if it is forcibly moved to one side, the ceramic insulator may be broken off.

22  Proceed as described in Steps 7 to 15.

23  Install the ignition coils (see Chapter 5B).

## 24  Air filter element replacement

### Four-cylinder engine

*Refer to illustrations 24.2 and 24.3*

1  The air cleaner assembly is located at the front left-hand corner of the engine compartment.

2  Release the four securing clips, and lift off the air cleaner cover **(see illustration)**.

3  Lift out the filter element **(see illustration)**.

**24.2 Releasing an air cleaner cover securing clip - four-cylinder engine**

**24.3 Lifting out the air filter element - four-cylinder engine**

# Chapter 1 Tune-up and routine maintenance

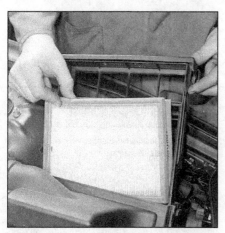

24.8 Lifting out the air cleaner element - six-cylinder engine

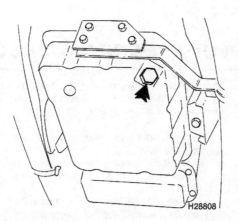

25.3 Automatic transmission fluid drain plug (arrow)

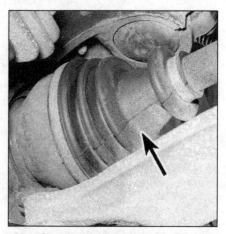

27.1 Check the condition of the CV joint boots (arrow)

4 Wipe out the air cleaner housing and the cover.
5 Lay the new filter element in position, then install the cover and secure with the clips.

## Six-cylinder engine

*Refer to illustration 24.8*

6 The air cleaner assembly is located at the front left-hand corner of the engine compartment.
7 Depress the securing clips, and slide the filter element tray up from the housing.
8 Lift out the filter element **(see illustration)**.
9 Wipe out the air cleaner housing and the tray.
10 Lay the new filter element in position, then slide the tray into the housing until it locks in position.

## 25 Automatic transmission fluid replacement

*Refer to illustration 25.3*
**Note:** *A new drain plug sealing ring will be required on installation.*

1 The transmission fluid should be drained with the transmission at operating temperature. If the vehicle has just been driven at least 20 miles, the transmission can be considered warm.
2 Immediately after driving the vehicle, park it on a level surface, apply the parking brake. If desired, jack up the vehicle and support it on axle stands to improve access, but make sure that the vehicle is level.
3 Working under the vehicle, loosen the transmission fluid pan drain plug about half a turn **(see illustration)**. Position a draining container under the drain plug, then remove the plug completely. If possible, try to keep the plug pressed into the fluid pan while unscrewing it by hand the last couple of turns.
4 Recover the sealing ring from the drain plug.

5 Unscrew the fluid pan bolts and carefully detach the pan. Clean the pan with solvent and remove all traces of gasket material from the mating surfaces on the pan and transmission.
6 Remove the screws and detach the oil screen. Clean the screen with solvent then reinstall it (or replace it with a new one, if necessary), tightening the screws to the torque listed in this Chapter's Specifications.
7 Reinstall the fluid pan, using a new gasket. Tighten the bolts to the torque listed in this Chapter's Specifications.
8 Install the drain plug, using a new sealing ring, and tighten it to the specified torque.
9 With reference to Section 5, fill the transmission with the specified quantity of the correct type of fluid (see Specifications).
10 Check the fluid level (see Section 5).
11 Note that it may be necessary to add or drain off a little fluid once the new fluid has reached the proper checking temperature.

## 26 Differential oil replacement

1 Park the vehicle on level ground.
2 Locate the filler/level plug in the center of the differential rear cover. Unscrew the plug and recover the sealing washer.
3 Place a suitable container beneath the differential, then unscrew the drain plug from the base of the rear cover and allow the oil to drain. Recover the sealing washer.
4 Inspect the sealing washers for signs of damage and replace if necessary.
5 When the oil has finished draining, install the drain plug and sealing washer and tighten it to the specified torque.
6 Refill the differential through the filler/level plug hole with the specified type of oil up to the base of the filler/level plug hole. Install the filler/level plug and take the car on a short journey so that the new oil is distributed fully around the final drive components.
7 On return, park on level ground and

allow the vehicle to stand for a few minutes. Unscrew the filler/level plug again. The oil level should reach the lower edge of the filler/level hole. To ensure that a true level is established, wait until the initial trickle has stopped, then add oil as necessary until a trickle of new oil can be seen emerging. The level will be correct when the flow ceases; use only good-quality oil of the specified type.
8 When the level is correct, install the filler/level plug and sealing washer and tighten it to the specified torque.

## 27 Driveaxle boot check

*Refer to illustration 27.1*

With the vehicle raised and securely supported on stands, slowly rotate the rear wheel. Inspect the condition of the constant velocity (CV) joint rubber boots, squeezing the boots to open out the folds **(see illustration)**. Check for signs of cracking, splits or deterioration of the rubber, which may allow the grease to escape, and lead to water and grit entry into the joint. Also check the security and condition of the retaining clips. Repeat these checks on the inner CV joints. If any damage or deterioration is found, the boots should be replaced (see Chapter 8).
At the same time, check the general condition of the CV joints themselves by first holding the driveaxle and attempting to rotate the wheel. Repeat this check by holding the inner joint and attempting to rotate the driveaxle. Any appreciable movement indicates wear in the joints, wear in the driveaxle splines, or a loose driveaxle retaining nut.

## 28 Parking brake shoe check

Referring to Chapter 9, remove the rear brake discs and inspect the parking brake shoes for signs of wear or contamination. Replace the shoes if necessary.

# Additional work to be carried out every second Inspection II

## 29 Manual transmission oil replacement

*Refer to illustrations 29.3 and 29.6*

**Note:** *New transmission oil drain plug and oil filler/level plug sealing rings may be required on installation.*

1   The transmission oil should be drained with the transmission at normal operating temperature. If the vehicle has just been driven at least 20 miles, the transmission can be considered warm.
2   Immediately after driving the vehicle, park it on a level surface, apply the parking brake. If desired, jack up the vehicle and support on axle stands to improve access, but make sure that the vehicle is level.
3   Working under the vehicle, loosen the transmission oil drain plug about half a turn **(see illustration)**. Position a draining container under the drain plug, then remove the plug completely. If possible, try to keep the plug pressed into the transmission while unscrewing it by hand the last couple of turns.
4   Where applicable, recover the sealing ring from the drain plug.
5   Install the drain plug, using a new sealing ring where applicable, and tighten to the specified torque.
6   Unscrew the oil filler/level plug from the side of the transmission, and recover the sealing ring, where applicable **(see illustration)**.
7   Fill the transmission through the filler/level plug hole with the specified quantity and type of oil (see Specifications), until the oil overflows from the filler/level plug hole.
8   Install the filler/level plug, using a new sealing ring where applicable, and tighten to the specified torque.
9   Where applicable, lower the vehicle to the ground.

## 30 Fuel filter replacement

**Warning:** *Gasoline is extremely flammable, so take extra precautions when you work on any part of the fuel system. Don't smoke or allow open flames or bare light bulbs near the work area, and don't work in a garage where a gas-type appliance (such as a water heater or clothes dryer) is present. Since gasoline is carcinogenic, wear fuel-resistant gloves when there's a possibility of being exposed to fuel, and, if you spill any fuel on your skin, rinse it off immediately with soap and water. Mop up any spills immediately and do not store fuel-soaked rags where they could ignite. The fuel system is under constant pressure, so, if any fuel lines are to be disconnected, the fuel pressure in the system must be relieved first (see Chapter 4 for more information). When you perform any kind of work on the fuel system, wear safety glasses and have a Class B type fire extinguisher on hand.*

### Four-cylinder engine

*Refer to illustrations 30.2 and 30.7*

1   Depressurize the fuel system (see Chapter 4A).
2   The fuel filter is located under the vehicle **(see illustration)**.
3   Jack up the vehicle and support it on axle stands.
4   Where applicable, remove the securing clips or nuts and withdraw the fuel filter cover.
5   If possible, clamp the fuel feed and return hoses to minimize fuel loss when the hoses are disconnected.
6   Place a container under the filter to catch escaping fuel, then loosen the hose clamps, and disconnect the fuel hoses from the filter.

29.3 Manual transmission oil drain plug (arrow)

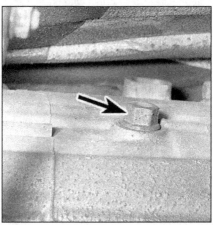

29.6 Manual transmission oil filler/level plug (arrow)

30.2 Fuel filter location - four-cylinder engine

30.7 Unscrewing the fuel filter clamp bolt

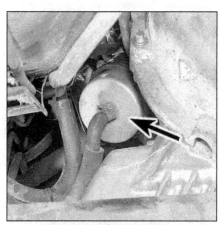

30.10 Fuel filter location (arrow) viewed from underneath the vehicle - six-cylinder engine

# Chapter 1  Tune-up and routine maintenance

7 Loosen the clamp nut or bolt until the filter can be slid from its mounting clamp (see illustration).
8 Installation is a reversal of removal, but make sure that the flow direction arrow on the filter points in the direction of fuel flow (i.e., towards the engine), and on completion, pressurize the fuel system with reference to Chapter 4A.

## Six-cylinder engine

*Refer to illustration 30.10*

9 Depressurize the fuel system (see Chapter 4B).
10 The fuel filter is located on a bracket bolted to the left-hand engine mounting (see illustration).
11 Working in the engine compartment, to improve access, remove the air ducting connecting the air mass meter to the throttle body, with reference to Chapter 4B if necessary.
12 Locate the fuel filter outlet hose, which connects to the fuel pipe under the intake manifold. Clamp the hose to minimize fuel spillage when the hose is disconnected, then loosen the hose clamp and disconnect the hose from the pipe.
13 Jack up the vehicle and support on axle stands.
14 Working under the vehicle, locate the fuel filter inlet hose, which connects to the fuel supply pipe under the left-hand side of the vehicle floor. As before, clamp the hose, then disconnect the hose from the pipe.
15 Again working under the vehicle, loosen the filter clamp bolt or nut, and slide the filter down from under the vehicle.
16 Disconnect the fuel hoses from the old filter, and connect them to the new filter.
17 Installation is a reversal of removal, but make sure that the flow direction arrow on the filter points in the direction of fuel flow (i.e., towards the engine), and on completion, pressurize the fuel system with reference to Chapter 4B.

## 31  Clutch check

This check is specified by BMW to check the clutch friction disc for wear. The check involves the use of a special tool, which fits into the clutch slave cylinder aperture in the transmission bellhousing, once the slave cylinder has been unbolted.
It is unlikely that the clutch wear will be significant unless the vehicle has covered a high mileage, or the clutch has been abused.
If in doubt as to the condition of the clutch friction disc, have the check carried out by a BMW dealer.

## 32  Brake fluid replacement

**Warning:** *Brake fluid can harm your eyes and damage painted surfaces, so use extreme caution when handling and pouring it. Do not use fluid that has been standing open for some time, as it absorbs moisture from the air. Excess moisture can allow the fluid to boil during periods of heavy braking, causing a dangerous loss of braking effectiveness.*

1 The procedure is similar to that for the bleeding of the hydraulic system as described in Chapter 9, except that the brake fluid reservoir should be emptied by siphoning, using a clean poultry baster or similar tool before starting, and allowance should be made for the old fluid to be expelled when bleeding a section of the circuit. **Warning:** *If a baster is used, never again use it for food preparation.*
2 Working as described in Chapter 9, open the first bleed screw in the sequence, and pump the brake pedal gently until nearly all the old fluid has been emptied from the master cylinder reservoir. **Note:** *Old brake fluid is invariably much darker in color than the new, making it easy to distinguish the two.*
3 Top-up to the "MAX" level with new fluid, and continue pumping until only the new fluid remains in the reservoir, and new fluid can be seen emerging from the bleed screw. Tighten the screw, and top the reservoir level up to the "MAX" level line.
4 Work through all the remaining bleed screws in the sequence until new fluid can be seen at all of them. Be careful to keep the master cylinder reservoir topped-up to above the "MIN" level at all times, or air may enter the system and greatly increase the length of the task.
5 When the operation is complete, check that all bleed screws are securely tightened, and that their dust caps are installed. Wash off all traces of spilled fluid, and recheck the master cylinder reservoir fluid level.
6 Check the operation of the brakes before taking the car on the road.

## 33  Coolant replacement

### Cooling system draining

**Warning:** *Wait until the engine is cold before starting this procedure. Do not allow antifreeze to come in contact with your skin, or with the painted surfaces of the vehicle. Rinse off spills immediately with plenty of water. Never leave antifreeze lying around in an open container, or in a puddle in the driveway or on the garage floor. Children and pets are attracted by its sweet smell, but antifreeze can be fatal if ingested.*

1 With the engine completely cold, cover the expansion tank cap with a wad of rag, and slowly turn the cap counterclockwise to relieve the pressure in the cooling system (a hissing sound will normally be heard). Wait until any pressure remaining in the system is released, then continue to turn the cap until it can be removed.
2 On models where expansion tank is built into the radiator, unscrew the bleed screw from the top of the expansion tank.
3 Where necessary, remove the retaining screws and remove the undercover from beneath the radiator.
4 Position a suitable container beneath the drain plug on the base of the radiator. Unscrew the drain plug and allow the coolant to drain into the container.
5 To fully drain the system, also unscrew the coolant drain plug from the right-hand side of the cylinder block and allow the remainder of the coolant to drain into the container.
6 If the coolant has been drained for a reason other than replacement, then provided it is clean and less than two years old, it can be re-used, though this is not recommended.
7 Once all the coolant has drained, install the bleed screw to the radiator. Install a new sealing washer to the block drain plug and tighten it to the specified torque.

### Cooling system flushing

8 If coolant replacement has been neglected, or if the antifreeze mixture has become diluted, then in time, the cooling system may gradually lose efficiency, as the coolant passages become restricted due to rust, scale deposits, and other sediment. The cooling system efficiency can be restored by flushing the system clean.
9 The radiator should be flushed independently of the engine, to avoid unnecessary contamination.

### Radiator flushing

10 To flush the radiator disconnect the top and bottom hoses and any other relevant hoses from the radiator, with reference to Chapter 3.
11 Insert a garden hose into the radiator top inlet. Direct a flow of clean water through the radiator, and continue flushing until clean water emerges from the radiator bottom outlet.
12 If after a reasonable period, the water still does not run clear, the radiator can be flushed with a good proprietary cooling system cleaning agent. It is important that their manufacturer's instructions are followed carefully. If the contamination is particularly bad, insert the hose in the radiator bottom outlet, and reverse-flush the radiator.

### Engine flushing

13 To flush the engine, remove the thermostat (see Chapter 3), then temporarily install the thermostat cover.
14 With the top and bottom hoses disconnected from the radiator, insert a garden hose into the radiator top hose. Direct a clean flow of water through the engine, and continue flushing until clean water emerges from the radiator bottom hose.
15 On completion of flushing, install the thermostat and reconnect the hoses with reference to Chapter 3.

### Cooling system filling

*Refer to illustrations 33.17 and 33.21*

16 Before attempting to fill the cooling system, make sure that all hoses and clips are in

**33.17 Where the expansion tank is an integral part of the radiator, unscrew the bleed screw from the top of the tank**

**33.21 Slowly fill the expansion tank until coolant free from air bubbles emerges from the bleed hole (arrow) then install the bleed screw**

good condition, and that the clips are tight and the radiator and cylinder block drain plugs are securely tightened. Note that an antifreeze mixture must be used all year round, to prevent corrosion of the engine components (see following sub-Section).
17  On models where expansion tank is built into the radiator, unscrew the bleed screw from the top of the expansion tank **(see illustration)**. On four-cylinder models also loosen the bleed screw which is situated on the top of the thermostat housing.
18  Remove the expansion tank filler cap and turn the heater temperature control knob to the maximum heat position. Fill the system by slowly pouring the coolant into the expansion tank to prevent airlocks from forming.
19  If the coolant is being replaced, begin by pouring in a couple of liters of water, followed by the correct quantity of antifreeze, then top-up with more water.
20  On four-cylinder models, as soon as coolant free from air bubbles emerges from the thermostat housing screw, tighten the screw securely.
21  Where the expansion tank is an integral part of the radiator, as coolant free from the air bubbles emerges from the radiator bleed hole, securely tighten the bleed screw **(see illustration)**.
22  Once the level in the expansion tank starts to rise, squeeze the radiator top and bottom hoses to help expel any trapped air in the system. Once all the air is expelled, top-up the coolant level to the "MAX" mark and install the expansion tank cap.
23  Start the engine and run it until it reaches normal operating temperature, then stop the engine and allow it to cool.
24  Check for leaks, particularly around disturbed components. Check the coolant level in the expansion tank, and top-up if necessary. Note that the system must be cold before an accurate level is indicated in the expansion tank. If the expansion tank cap is removed while the engine is still warm, cover the cap with a thick cloth, and unscrew the cap slowly to gradually relieve the system pressure (a hissing sound will normally be heard). Wait until any pressure remaining in the system is released, then continue to turn the cap until it can be removed.

### Antifreeze mixture

25  The antifreeze should always be replaced at the specified intervals. This is necessary not only to maintain the antifreeze properties, but also to prevent corrosion which would otherwise occur as the corrosion inhibitors become progressively less effective.
26  Always use an ethylene-glycol based antifreeze which is suitable for use in mixed-metal cooling systems. The quantity of antifreeze and levels of protection are indicated in the Specifications.
27  Before adding antifreeze, the cooling system should be completely drained, preferably flushed, and all hoses checked for condition and security.
28  After filling with antifreeze, a label should be attached to the expansion tank, stating the type and concentration of antifreeze used, and the date installed. Any subsequent topping-up should be made with the same type and concentration of antifreeze.

# Every 50,000 miles

### 34  Oxygen sensor replacement

Refer to Chapter 6 for the oxygen sensor replacement procedure.

# Every three years

### 35  Airbag system inspection

Refer to Chapter 12 and check the operation of the airbag warning light, the condition of all electrical connectors related to the system, and the mounting and appearance of all components. **Warning:** *Never use electrical test equipment on any airbag system component or related wiring harnesses.*

# Chapter 2 Part A
# Four-cylinder engines

## Contents

| | Section | | Section |
|---|---|---|---|
| Camshafts - removal and installation | 10 | Oil pan - removal and installation | 12 |
| Compression test - description and interpretation | 2 | Oil pump - removal, inspection and installation | 13 |
| Crankshaft pilot bearing - replacement | 16 | Oil seals - replacement | 14 |
| Crankshaft vibration damper/pulley and pulley hub - removal and installation | 5 | Timing chain covers - removal and installation | 6 |
| Cylinder head - removal and installation | 11 | Timing chain housing - removal and installation | 9 |
| Engine/transmission mounts - inspection and replacement | 17 | Timing chain sprockets and tensioner - removal and installation | 8 |
| Flywheel/driveplate - removal and installation | 15 | Timing chains - removal, inspection and installation | 7 |
| General information | 1 | Top Dead Center (TDC) for No. 1 piston - locating | 3 |
| | | Valve cover - removal and installation | 4 |

## Specifications

### General
Engine code
  1992 through 1995 ................................ M42
  1996 on ................................................. M44
Displacement
  M42 ...................................................... 109.6 cubic inches (1.8 liters)
  M44 ...................................................... 115.6 cubic inches (1.9 liters)
Direction of engine rotation ........................ Clockwise (viewed from front of vehicle)
No. 1 cylinder location ............................... Timing chain end of engine
Firing order ............................................... 1-3-4-2
Compression ratio ..................................... 10.0 : 1
Minimum compression pressure ................ 142 to 156 psi

### Camshafts
Endplay ..................................................... 0.0025 to 0.0059 inch
Oil clearance
  M42 engine ............................................ 0.0008 to 0.0024 inch
  M44 engine ............................................ 0.0016 to 0.0032 inch

### Lubrication system
Minimum oil pressure at idle speed ............ 18 psi
Oil pressure relief valve spring free-length ... 3.31 inches
Oil pump rotor clearances:
  Outer rotor-to-pump body ...................... 0.005 to 0.008 inch
  Outer rotor-to-inner rotor ....................... 0.005 to 0.008 inch
  Rotor endplay ........................................ 0.008 to 0.0035 inch

### Torque specifications
**Ft-lbs** (unless otherwise indicated)

Main bearing cap bolts
  Step 1 ................................................... 15
  Step 2 ................................................... Tighten an additional 50-degrees
Cylinder head bolts*
  Step 1 ................................................... 22
  Step 2 ................................................... Tighten an additional 90-degrees
  Step 3 ................................................... Tighten an additional 90-degrees

## Chapter 2 Part A  Four-cylinder engines

**Torque specifications**     **Ft-lbs** (unless otherwise indicated)

| | |
|---|---|
| Valve cover bolts | 79 in-lbs |
| Upper and lower timing chain cover bolts | 79 in-lbs |
| Crankshaft rear oil seal housing bolts | 79 in-lbs |
| Flywheel bolts* | 89 |
| Crankshaft vibration damper/pulley-to-hub bolts | 16 |
| Crankshaft pulley hub/sprocket bolt* | 244 |
| Connecting rod bearing cap bolts* | |
|    Step 1 | 15 |
|    Step 2 | Tighten an additional 70-degrees |
| Camshaft bearing cap nuts | 11 |
| Camshaft sprocket bolts | 11 |
| Timing chain tensioner cover plug | 18 |
| Oil pump cover | 79 in-lbs |
| Front subframe bolts* | 77 |

*Use new bolts*

## 1  General information

### How to use this Chapter

This Part of Chapter 2 describes the repair procedures that can reasonably be carried out on the engine while it remains in the vehicle. If the engine has been removed from the vehicle and is being disassembled as described in Part C, any preliminary dismantling procedures can be ignored.

Note that, while it may be possible physically to overhaul items such as the piston/connecting rod assemblies while the engine is in the car, such tasks are not usually carried out as separate operations. Usually, several additional procedures are required (not to mention the cleaning of components and oil passages); for this reason, all such tasks are classed as major overhaul procedures, and are described in Part C of this Chapter.

Part C describes the removal of the engine/transmission from the vehicle, and the full overhaul procedures that can then be carried out.

### Engine description

The four-cylinder engine is a double overhead camshaft design, mounted in-line, with the transmission bolted to the rear. A double timing chain drives the double overhead camshafts. On M42 engines, self-adjusting hydraulic lifters (or followers) are fitted between the camshafts and the valves. M44 engines are equipped with hydraulic pedestals and roller rockers instead of hydraulic lifters. Each camshaft is supported by bearings incorporated in bearing castings fitted to the cylinder head.

The crankshaft is supported in five main bearings of the usual shell-type. Endplay is controlled by thrust bearing shells on No. 4 main bearing.

The pistons are selected to be of matching weight, and incorporate fully floating wrist pins retained by circlips.

The rotor-type oil pump is located at the front of the engine, and is driven directly by the crankshaft.

### Repair operations possible with the engine in the vehicle

The following operations can be carried out without having to remove the engine from the vehicle:

a) Removal and installation of the cylinder head.
b) Removal and installation of the timing chain and sprockets.
c) Removal and installation of the camshafts.
d) Removal and installation of the oil pan.
e) Removal and installation of the connecting rod bearings, connecting rods, and pistons*.
f) Removal and installation of the oil pump.
g) Replacement of the engine/transmission mounts.
h) Removal and installation of the flywheel/driveplate.

* Although it is possible to remove these components with the engine in place, for reasons of access and cleanliness it is recommended that the engine be removed.

## 2  Compression test - description and interpretation

1  When engine performance is down, or if misfiring occurs which cannot be attributed to the ignition or fuel systems, a compression test can provide diagnostic clues as to the engine's condition. If the test is performed regularly, it can give warning of trouble before any other symptoms become apparent.

2  The engine must be fully warmed-up to normal operating temperature, the battery must be fully charged, and all the spark plugs must be removed (Chapter 1). The aid of an assistant will also be required.

3  Disable the ignition and fuel injection systems by removing the DME master relay, and the fuel pump relay, located in the main fuse box in the engine compartment (see Chapter 12).

4  Fit a compression tester to the No. 1 cylinder spark plug hole - the type of tester which screws into the plug thread is preferred.

5  Have the assistant hold the throttle wide open, and crank the engine on the starter motor. After one or two revolutions, the compression pressure should build up to a maximum figure, and then stabilize. Record the highest reading obtained.

6  Repeat the test on the remaining cylinders, recording the pressure in each.

7  All cylinders should produce very similar pressures; a difference of more than 20 psi between any two cylinders indicates a fault. Note that the compression should build up quickly in a healthy engine; low compression on the first stroke, followed by gradually increasing pressure on successive strokes, indicates worn piston rings. A low compression reading on the first stroke, which does not build up during successive strokes, indicates leaking valves or a blown head gasket (a cracked head could also be the cause). Deposits on the undersides of the valve heads can also cause low compression.

8  The recommended values for compression pressures are given in this Chapter's Specifications.

9  If the pressure in any cylinder is low, carry out the following test to isolate the cause. Introduce a teaspoonful of clean oil into that cylinder through its spark plug hole, and repeat the test.

10  If the addition of oil temporarily improves the compression pressure, this indicates that bore or piston wear is responsible for the pressure loss. No improvement suggests that leaking or burnt valves, or a blown head gasket, may be to blame.

11  A low reading from two adjacent cylinders is almost certainly due to the head gasket having blown between them; the presence of coolant in the engine oil will confirm this.

# Chapter 2 Part A  Four-cylinder engines 2A-3

3.4 Timing arrows on the camshaft sprockets positioned with No. 1 piston at TDC - M42 engine

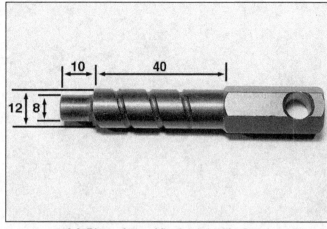

3.6 Dimensions of flywheel "locking" tool
*All dimensions in mm*

12 If one cylinder is about 20 percent lower than the others and the engine has a slightly rough idle, a worn camshaft lobe could be the cause.
13 If the compression reading is unusually high, the combustion chambers are probably coated with carbon deposits. If this is the case, the cylinder head should be removed and decarbonized.
14 On completion of the test, install the spark plugs (see Chapter 1) and reconnect the fuel pump relay and the DME master relay.

## 3 Top Dead Center (TDC) for No. 1 piston - locating

*Refer to illustrations 3.4, 3.6, 3.7, 3.10a and 3.10b*
**Note:** *To lock the engine in the TDC position, and to check the position of the camshafts, special tools will be required. These tools can easily be improvised - see text.*
1 Top Dead Center (TDC) is the highest point in the cylinder that each piston reaches as it travels up and down when the crankshaft turns. Each piston reaches TDC at the end of the compression stroke and again at the end of the exhaust stroke, but TDC generally refers to piston position on the compression stroke. No. 1 piston is at the timing belt/chain end of the engine.
2 Positioning No. 1 piston at TDC is an essential part of many procedures, such as timing chain removal and camshaft removal.
3 Remove the valve cover (see Section 4).
4 Using a wrench or socket on the crankshaft pulley bolt (if desired, remove the viscous cooling fan and shroud, as described in Chapter 3, to improve access), turn the crankshaft clockwise until the timing arrows on the camshaft sprockets are pointing vertically upwards, and the front cam lobes on the exhaust and intake camshafts are facing each other **(see illustration)**.
5 Pull the blanking plug from the timing hole in the left-hand rear corner flange of the cylinder block.
6 To "lock" the crankshaft in position, a special tool will now be required. BMW tool No. 11 2 300 can be used, but an alternative can be made up by machining a length of steel rod to the dimensions shown **(see illustration)**.
7 Insert the rod through the timing hole. If necessary, turn the crankshaft slightly until the rod enters the TDC hole in the flywheel **(see illustration)**.

3.7 Flywheel locking tool engaged with TDC hole in flywheel

8 The crankshaft is now "locked" in position with No. 1 piston at TDC.
9 Note also that the square flanges on the rear of the camshafts should be positioned with the sides of the flanges exactly at right-angles to the top surface of the cylinder head (this can be checked using a set-square), and the side of the flange with holes drilled into it uppermost.
10 For some operations it is necessary to

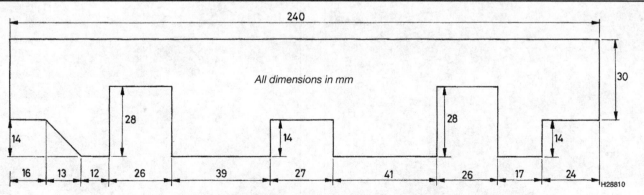

3.10a Make up a template from sheet metal to the dimensions shown

**3.10b Camshaft locking template in position on the cylinder head**

**4.4a Recover the heat shield from the edge of the valve cover**

lock the camshafts in position with No. 1 piston at TDC. This can be done by making up a template from sheet metal to the dimensions shown - when the camshafts are correctly positioned, the template will fit exactly over the flanges at the rear of the camshafts, and rest on the upper surface of the cylinder head **(see illustrations)**.

11  Do not attempt to turn the engine with the flywheel or camshaft locked in position, as engine damage may result. If the engine is to be left in the "locked" state for a long period of time, it is a good idea to place suitable warning notices inside the vehicle, and in the engine compartment. This will reduce the possibility of the engine being cranked on the starter motor.

## 4  Valve cover - removal and installation

**Note:** *New gaskets and/or seals may be required on installation - see text.*

### Removal

*Refer to illustrations 4.4a, 4.4b, 4.4c, 4.5 and 4.6*

1  Open the hood, then raise the hood to its fully open position (see Chapter 11).
2  Remove the ignition coil from the right strut tower (see Chapter 5B).
3  Loosen the securing screws and remove the spark plug cover from the center of the valve cover. Pull the connectors from the spark plugs.
4  Loosen the bolts securing the spark plug wire bracket to the edge of the valve cover and recover the heat shield, then lift the complete spark plug wire housing/ducting assembly from the valve cover **(see illustrations)**.
5  Disconnect the breather hose from the valve cover **(see illustration)**.
6  Unscrew the securing bolts and lift the valve cover from the engine **(see illustration)**. Recover the gaskets (note that there are separate gaskets at the center of the cover for the spark plug holes).

### Installation

*Refer to illustration 4.9*

7  Thoroughly clean the gasket faces of the valve cover and the engine.

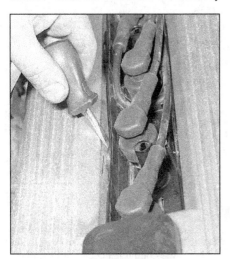

**4.4b Release the securing clips . . .**

**4.4c . . . and lift the complete spark plug wire housing/ducting assembly from the valve cover**

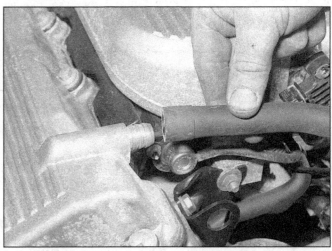

**4.5 Disconnect the breather hose from the valve cover**

# Chapter 2 Part A Four-cylinder engines

4.6 Lifting off the valve cover

4.9 Lay the gaskets in position on the valve cover

8   Check the condition of the sealing rubbers on the cover securing bolts, and replace if necessary. Ensure that the washers and rubber seals are correctly fitted to the securing bolts.
9   Examine the cover gaskets, and replace if necessary, then lay the gaskets in position on the valve cover **(see illustration)**.
10  Position the cover on the cylinder head, ensuring that the lugs on the gasket engage with the corresponding cut-outs in the rear of the cylinder head.
11  Install the cover securing bolts and tighten to the specified torque.
12  The remainder of the installation procedure is a reversal of removal, ensuring that the spark plug wires are correctly reconnected.

## 5   Crankshaft vibration damper/pulley and pulley hub - removal and installation

### Removal

*Refer to illustration 5.5*
**Note:** *If the pulley hub is removed, a new securing bolt will be required on installation, and a torque wrench capable of providing 244 ft-lbs of torque will be required.*
1   To improve access, remove the viscous cooling fan and fan shroud assembly (see Chapter 3).
2   Remove the drivebelt (see Chapter 1).
3   Unscrew the securing bolts, and remove the vibration damper/pulley from the hub. If necessary, counterhold the hub using a socket or wrench on the hub securing bolt.
4   To remove the hub, the securing bolt must be unscrewed. **Warning:** *The crankshaft pulley hub securing bolt is very tight. A tool will be required to counterhold the hub as the bolt is unscrewed. Do not attempt the job using inferior or poorly improvised tools, as injury or damage may result.*
5   Make up a tool to hold the pulley hub. A suitable tool can be fabricated using two lengths of steel bar, joined by a large pivot bolt. Bolt the holding tool to the pulley hub using the pulley-to-hub bolts **(see illustration)**.
6   Using a socket and a long breaker bar, loosen the pulley hub bolt. Note that the bolt is very tight.
7   Unscrew the pulley hub bolt, and remove the washer. Discard the bolt, a new one must be used on installation.
8   Withdraw the hub from the end of the crankshaft. If the hub is tight, use a puller to draw it off.
9   Recover the Woodruff key from the end of the crankshaft if it is loose.

### Installation

10  If the pulley hub has been removed, it is advisable to take the opportunity to replace the oil seal in the lower timing chain cover, with reference to Section 6.
11  If the pulley hub has been removed, proceed as follows, otherwise proceed to paragraph 15.
12  Where applicable, install the Woodruff key to the end of the crankshaft, then align the groove in the pulley hub with the key, and slide the hub onto the end of the crankshaft.
13  Install the washer, noting that the shoulder on the washer must face the hub, and fit a new hub securing bolt.
14  Bolt the holding tool to the pulley hub, as during removal, then tighten the hub bolt to the specified torque. Take care to avoid injury and/or damage.
15  Where applicable, unbolt the holding tool, and install the vibration damper/pulley, ensuring that the locating dowel on the hub engages with the corresponding hole in the damper/pulley.
16  Install the damper/pulley securing bolts, and tighten to the specified torque. Again, counterhold the pulley if necessary when tightening the bolts.
17  Install the drivebelt (see Chapter 1).
18  Install the viscous cooling fan and shroud (see Chapter 3).

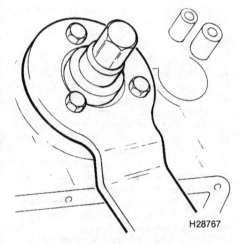

5.5 BMW special tool used to hold the crankshaft pulley hub

## 6   Timing chain covers - removal and installation

### Upper cover

**Note:** *New gaskets will be required on installation, and RTV sealant will be required.*

### Removal

*Refer to illustration 6.6*
1   Drain the cooling system (see Chapter 1).
2   If desired, to improve access on M42 engines, remove the viscous cooling fan and shroud (see Chapter 3).
3   Remove the valve cover (see Section 4).
4   Remove the camshaft position sensor from the cover (see Chapter 4A).
5   On M42 engines, remove the thermostat housing (see Chapter 3). On January 1997 and later M44 engines, remove the secondary air check valve (see Chapter 6).

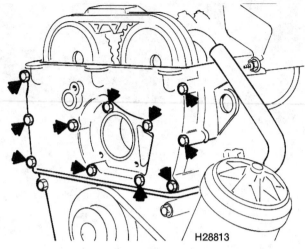

6.6 Upper timing chain cover securing bolt locations (arrows)

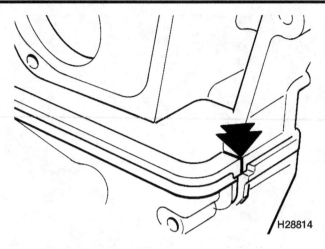

6.9 Apply a little RTV sealant to the joints between the lower gasket, timing chain housing and lower cover (arrow)

6  Unscrew the securing bolts, and withdraw the cover from the engine (see illustration). Recover the gaskets.

### Installation

*Refer to illustrations 6.9 and 6.11*

7  Commence installation by thoroughly cleaning the mating faces of the upper cover, lower cover and timing chain housing.
8  Position new gaskets on the timing chain housing, and on the top of the lower cover, using a little grease to hold the gaskets in place if necessary.
9  Apply a little RTV sealant to the joints between the lower gasket, timing chain housing, and lower cover (see illustration).
10  Offer the upper cover into position, then fit the two middle outer securing bolts - do not tighten the bolts at this stage.
11  Press down on the upper cover, for example using a screwdriver engaged with the camshaft sprocket (see illustration), until the top surfaces of the upper cover are exactly aligned with the top surface of the cylinder head. Hold the upper cover in this position, then tighten the two securing bolts to the specified torque.
12  Install the remaining cover securing bolts, and tighten to the specified torque.
13  Install the thermostat (see Chapter 3).
14  Install the camshaft position sensor (see Chapter 4A).
15  Install the valve cover (see Section 4).
16  Where applicable, install the viscous cooling fan and shroud (see Chapter 3).
17  Refill the cooling system (see Chapter 1).

### *Lower cover*

**Note:** *A new cover gasket and a new crankshaft oil seal will be required on installation.*

### Removal

*Refer to illustration 6.25*

18  Remove the upper timing chain cover as described previously in this Section.

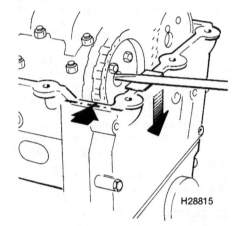

6.11 Press down on the cover using a screwdriver engaged with the camshaft sprocket

19  If not already done, remove the viscous cooling fan and shroud (see Chapter 3).
20  Remove the drivebelt (see Chapter 1).
21  The coolant pump pulley must now be removed. Counterhold the pulley by wrapping an old drivebelt around it and clamping tightly, then unscrew the securing bolts and withdraw the pulley.
22  Unbolt the crankshaft position sensor from its mounting bracket (see Chapter 4A), and move it to one side. Note that it is preferable to completely remove the sensor to avoid the possibility of damage.
23  Where applicable, release any wiring/ducting from the cover, and move clear of the working area.
24  Remove the crankshaft vibration damper/pulley and pulley hub (see Section 5).
25  Unscrew the securing bolts, noting their locations, and remove the lower timing chain cover (see illustration). Recover the gasket.

### Installation

26  Commence installation by thoroughly cleaning the mating faces of the lower cover,

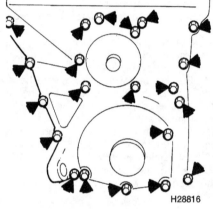

6.25 Lower timing chain cover bolt locations (arrows)

timing chain housing and upper cover.
27  Note the fitted depth of the crankshaft front oil seal in the cover, then lever out the seal.
28  Clean the oil seal housing, then fit a new oil seal, using a large socket or tube to drive the seal into position to the previously noted depth.
29  Check that the cover positioning dowels are in position in the timing chain housing.
30  Place a new gasket in position on the timing chain housing, using a little grease to hold the gasket in position if necessary.
31  Offer the lower cover into position, and install the securing bolts to their original locations. Tighten the bolts to the specified torque.
32  Install the crankshaft vibration damper/pulley and pulley hub (see Section 5).
33  Make sure that any wiring/ducting released during removal is correctly routed and repositioned during installation.
34  Install the crankshaft position sensor.
35  Install the coolant pump pulley and tighten the securing bolts.

# Chapter 2 Part A  Four-cylinder engines

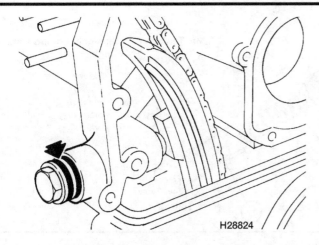

7.3 Unscrew the chain tensioner plug (arrow)

7.5 Upper chain guide securing bolts (arrows)

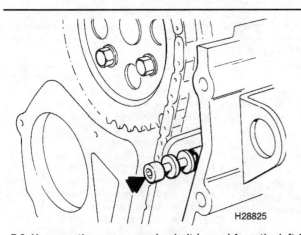

7.6 Unscrew the upper securing bolt (arrow) from the left-hand chain guide

7.7a Unscrew the camshaft sprocket securing bolts (arrows)

36  Install the drivebelt (see Chapter 1).
37  Install the upper timing chain cover as described previously in this Section.
38  Install the viscous cooling fan and shroud (see Chapter 3).

## 7  Timing chains - removal, inspection and installation

**Note:** *A new timing chain tensioner cover plug seal will be required on installation.*

### Removal

*Refer to illustrations 7.3, 7.5, 7.6, 7.7a, 7.7b and 7.10*

1  Position No. 1 piston at TDC, and lock the flywheel in position (see Section 3).
2  Remove the upper and lower timing chain covers (see Section 6).
3  Unscrew the chain tensioner plug from the right-hand side of the engine. Recover the sealing ring **(see illustration)**.
4  Withdraw the timing chain tensioner assembly from its housing.
5  Unscrew the securing bolts, and with-draw the upper chain guide from the cylinder head **(see illustration)**.
6  Unscrew the upper securing bolt from the left-hand chain guide **(see illustration)**.
7  On M44 engines, note the position of the sensor wheel on the intake camshaft. Unscrew the bolts securing the chain sprockets to the camshafts. Take care not to move the camshafts - if necessary, the camshafts can be counterheld using a 27 mm wrench on the flats provided between No. 5 and 6 cam lobes **(see illustrations)**.
8  Withdraw the sprockets from the camshafts, and disengage them from the chain. Note which way the sprockets are fitted to ensure correct installation.
9  Note the routing of the chain in relation to the sprockets, tensioner rail and the chain guides.
10  Unscrew the securing bolts, and remove

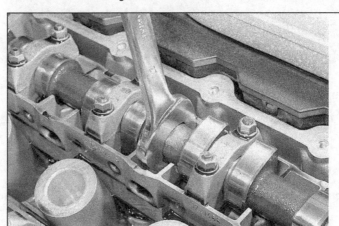

7.7b If necessary counterhold the camshafts using a wrench on the flats provided

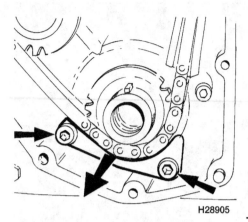

**7.10 Unscrew the securing bolts (arrows) and remove the lower chain guide**

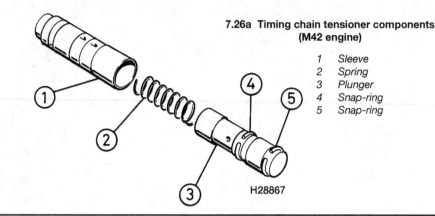

**7.26a Timing chain tensioner components (M42 engine)**

1. Sleeve
2. Spring
3. Plunger
4. Snap-ring
5. Snap-ring

the lower chain guide **(see illustration)**.

11 Manipulate the tensioner rail as necessary to enable the chain to be unhooked from the idler sprocket and crankshaft sprocket and lifted from the engine. **Caution:** *Once the timing chain has been removed, do not turn the crankshaft or the camshafts, as there is a danger of the valves hitting the pistons.*

12 If desired, the tensioner rail can now be removed after removing the clip from the lower pivot.

13 Similarly, the left-hand chain guide can be removed after unscrewing the remaining bolts.

## Inspection

14 The chain should be renewed if the sprockets are worn or if the chain is worn (indicated by excessive lateral play between the links, and excessive noise in operation). It is wise to replace the chain in any case if the engine is disassembled for overhaul. Note that the rollers on a very badly worn chain may be slightly grooved. To avoid future problems, if there is any doubt at all about the condition of the chain, replace it.

15 Examine the teeth on the sprockets for wear. Each tooth forms an inverted "V". If worn, the side of each tooth under tension will be slightly concave in shape when compared with the other side of the tooth (i.e., the teeth will have a hooked appearance). If the teeth appear worn, the sprockets must be renewed. Also check the chain guide and tensioner rail contact surfaces for wear, and replace any worn components as necessary.

## Installation

*Refer to illustrations 7.26a, 7.26b, 7.26c, 7.26d and 7.29*

16 Ensure that No. 1 piston is still positioned at TDC, with the crankshaft locked in position. Check the position of the camshafts using the template.

17 Where applicable, install the tensioner rail and the left-hand chain guide. Install the two lower chain guide securing bolts, but do not tighten them at this stage. Do not fit the upper securing bolt at this stage.

18 Engage the chain with the crankshaft sprocket, then install the lower chain guide and tighten the securing bolts.

19 Lay the chain in position around the left-hand chain guide and the tensioner rail, ensuring that the chain is routed as noted before removal.

20 Manipulate the camshaft sprockets until the timing arrows on the sprockets are pointing vertically upwards, then engage the chain with the sprockets.

21 Fit the sprockets to the camshafts, ensuring that the sprockets are installed the correct way as noted before removal. On M44 engines, install the sensor wheel on the intake cam sprocket with the arrow on the sensor wheel pointing up, and then install the sprocket securing bolts.

22 Tighten the sprocket securing bolts to the specified torque - if necessary, the camshafts can be counterheld using a 27 mm wrench on the flats provided between No. 5 and 6 cam lobes.

23 Install the upper securing bolt to the left-hand chain guide, but do not tighten it at this stage.

24 Install the upper chain guide and tighten the securing bolts.

25 The timing chain tensioner must now be fitted. **Note:** *The M44-style tensioner can be retrofitted to an M42 engine.* Before fitting, check that the tensioner plunger is retracted. If a new tensioner is being fitted, it should be

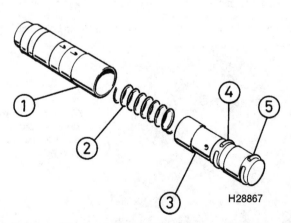

**7.26b Compress the tensioner until the first snap-ring (1) engages with the groove in the sleeve . . .**

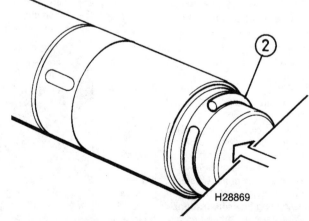

**7.26c . . . then compress the tensioner further until the second snap-ring (2) is engaged positively (M42 engine)**

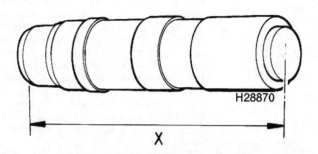

**7.26d Measure the overall length of the tensioner (M42 engine)**
Dimension X = 2.70 inches (68.5 mm)

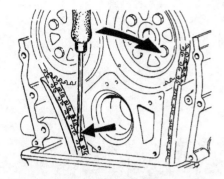

**7.29 Lever the timing chain and tensioner rail until the tensioner plunger is released (M42 engine)**

supplied with the plunger already in the retracted position.

26  On M42 engines, retract the tensioner as follows. **Caution:** *It is essential that the following procedure is carried out to ensure that the tensioner plunger is retracted. If the plunger is not fully retracted, it can lock in the extended position, causing the tensioner or timing chain to break, resulting in expensive engine damage.*

a) *Hold the tensioner upright, then knock the bottom end of the tensioner sleeve sharply on a solid surface such as a vise. This should cause the plunger to jump out of the end of the tensioner sleeve. Lift out the plunger and the spring.*
b) *Note the locations of the two circlips on the plunger* **(see illustration).**
c) *Thoroughly clean the components.*
d) *Ensure that the circlips are correctly located in their respective grooves on the plunger.*
e) *Slide the spring into the sleeve, and engage the plunger with the end of the spring.*
f) *Clamp the assembly in a vise, with the plunger resting in the sleeve so that both circlips are still visible.*
g) *Tighten the vise to compress the plunger into the sleeve, until the first snap-ring engages with the groove in the sleeve* **(see illustration).**
h) *Tighten the vise further to compress the plunger into the sleeve until the second snap-ring is heard to engage positively* **(see illustration).** *Do not push the plunger too far into the sleeve, or the snap-ring will be released, unlocking the plunger.*
i) *Loosen the vise - the plunger should stay retracted in the sleeve.*
j) *If the plunger comes out of the sleeve as the vise is loosened, or if the overall length of the tensioner assembly is greater than specified, then the procedure in paragraphs a) to i) must be repeated* **(see illustration).**

27  On M44 engines, place the tensioner between the jaws of a bench vise. Make sure the bench vise jaws are lined with soft jaw pads to protect the tensioner. Carefully squeeze the tensioner piston to expel the oil from the tensioner. Compress the piston to the second circlip (the one near the end). Repeat this procedure twice to ensure that all old oil has been removed from the tensioner.

28  Install the tensioner assembly, ensuring that it is fitted with the plunger against the tensioner rail, then install the tensioner plug using a new sealing ring. Tighten the plug to the specified torque.

29  On M42 engines, using a screwdriver, lever the timing chain and tensioner rail against the tensioner until the tensioner plunger is released from the sleeve to tension the chain **(see illustration)**. (This step is unnecessary on M44 engines, or if you're retrofitting an M44-style tensioner to an M42 engine.)

30  Once the chain is under tension, the left-hand chain guide securing bolts can be tightened. Using feeler gauges, position the guide to give an equal clearance between each side of the guide and the chain, then tighten the securing bolts.

31  Install the timing chain covers (see Section 6).

32  Remove the flywheel locking tool and the camshaft positioning template, and install the valve cover as described in Section 4. **Caution:** *When the engine is first started, it must be run at a speed of 3500 rpm for 20 seconds as soon as it starts - this is to ensure that the tensioner is primed with oil. A loud rattling sound will be heard until the tensioner is primed - do not be alarmed by the noise!*

### 8   Timing chain sprockets and tensioner - removal and installation

#### Camshaft sprockets and tensioner

1  Removal and installation are described as part of the timing chain removal procedure in Section 7.

#### Crankshaft sprocket

2  Remove the timing chain (see Section 7).
3  Slide the sprocket from the crankshaft, and recover the Woodruff key if it is loose.
4  Installation is a reversal of removal, but install the timing chain (see Section 7).

#### Idler sprocket

5  Remove the timing chain (see Section 7).
6  Unbolt the sprocket from the timing chain housing.
7  Installation is a reversal of removal, noting the following points.

a) *Ensure that the sprocket is installed the correct way - the greater projecting boss on the sprocket should be facing the timing chain housing.*
b) *Ensure that the washer is in place under the securing bolt.*
c) *Install the timing chain (see Section 7).*

### 9   Timing chain housing - removal and installation

**Note:** *A new timing chain housing gasket will be required on installation, and a new oil filter housing gasket and seal may be required.*

#### Removal

1  Remove the timing chain, tensioner rail and chain guides (see Section 7).
2  Remove the timing chain crankshaft and idler sprockets (see Section 8).
3  Remove the cylinder head (see Section 11).
4  Remove the oil pan (see Section 12).
5  Remove the alternator (see Chapter 5).
6  Disconnect the wiring from the oil pressure warning light switch.
7  If desired, unbolt the oil filter housing from the side of the timing chain housing. Be prepared for oil spillage, and recover the gasket and O-ring.
8  Unscrew the securing bolts, noting their locations, and withdraw the timing chain

# 2A-10  Chapter 2 Part A  Four-cylinder engines

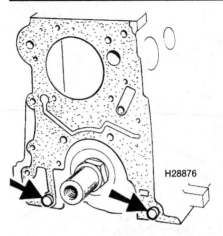

**9.10 Ensure that the locating dowels (arrows) are in place in the cylinder block**

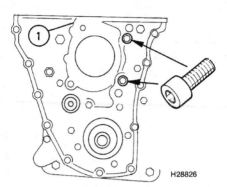

**9.11a Ensure that the socket-head bolts are installed in their correct locations, and note the upper right-hand bolt hole (1) is not used**

housing from the front of the cylinder block. Recover the gasket.

### Installation

*Refer to illustrations 9.10, 9.11a and 9.11b*

9  Thoroughly clean the mating faces of the timing chain housing and the cylinder block.
10  Ensure that the locating dowels are in place in the cylinder block, then lay a new gasket in position on the cylinder block **(see illustration)**.
11  Install the timing chain housing, and tighten the securing bolts to the specified torque. Ensure that the socket-head bolts are fitted to their correct locations, and note that the upper right-hand bolt hole is not used. Also ensure that the chain oil spray jet faces the idler sprocket location **(see illustrations)**.
12  Where applicable, install the oil filter housing using a new gasket and a new O-ring. Lubricate the O-ring with a little grease before fitting, and tighten the securing bolts.
13  Reconnect the oil pressure warning light switch wiring.
14  Install the alternator (see Chapter 5).
15  Install the oil pan (see Section 12).
16  Install the cylinder head (see Section 11).

17  Install the timing chain crankshaft and idler sprockets (see Section 8).
18  Install the chain guides, tensioner rail and timing chain (see Section 7).

## 10  Camshafts - removal and installation

**Caution:** *BMW special tool 11 3 260 is needed for this procedure on M42 engines and BMW special tool 11 5 130 is required on M44 engines. Because of the high degree of precision needed to manufacture these tools, they would be extremely difficult to fabricate. Do not attempt to remove and install the camshafts on either engine without the aid of the requisite tool, or expensive damage to the camshafts and/or bearings may result.*

### Removal

*Refer to illustration 10.12*

**Note:** *The following procedure applies to either camshaft.*

1  Remove the valve cover and upper timing chain cover (see Sections 4 and 6, respectively).
2  Position No. 1 piston at TDC, and lock the flywheel in position (see Section 3).
3  On M44 engines, rotate the engine until a pair of camshaft lobes is pointing up (the intake and exhaust lobes for the number one cylinder, for example). Using BMW special tool 11 5 130, compress each valve/valve spring assembly far enough to remove the rocker arms. After the first two rocker arms have been removed, rotate the engine again until the next pair of cam lobes is pointing up. Again, depress the valve with the special BMW tool and remove the rocker arms. And so on. Repeat this step until all rocker arms have been removed.
4  Unscrew the timing chain tensioner plug from the right-hand side of the engine. Recover the sealing ring. Withdraw the timing chain tensioner assembly from its housing.
5  Unscrew the securing bolts and withdraw the upper chain guide from the cylinder head.
6  Unscrew the upper securing bolts from the left-hand chain guide.
7  Unscrew the bolts securing the chain sprockets to the camshafts. Take care not to move the camshafts - if necessary, the camshafts can be counterheld using a 27 mm wrench on the flats provided between No. 5 and 6 cam lobes.
8  Withdraw the sprockets from the camshafts, and disengage them from the chain. Note which way the sprockets face to ensure correct installation.
9  Ensure that tension is kept on the timing chain - tie the chain up or support it using wire, to prevent it from dropping into the lower timing chain cover. **Caution:** *To avoid any possibility of piston-to-valve contact when installing the camshafts, it is necessary to ensure that none of the pistons are at TDC. Before proceeding further, remove the locking rod from the timing hole in the cylinder block, then turn the crankshaft approximately 90-degrees clockwise using a wrench or socket on the crankshaft pulley hub bolt.*
10  Remove the template from the camshafts.
11  Unscrew the spark plugs from the cylinder head.
12  Check the camshaft bearing caps for

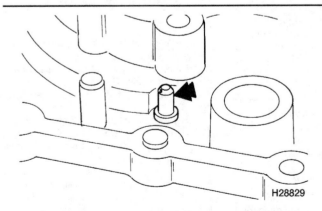

**9.11b Ensure that the oil spray jet (arrow) faces the idler sprocket location**

**10.12 Camshaft bearing cap identification marks**

# Chapter 2 Part A  Four-cylinder engines

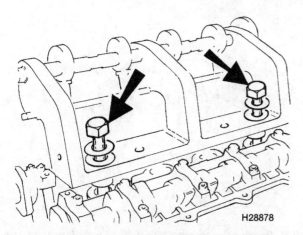

10.13 Fit BMW special tool 11 3 260 to the cylinder head by screwing the bolts (arrows) into the spark plug holes (M42 engine)

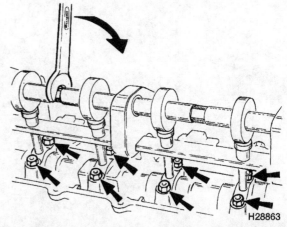

10.14 Apply pressure to the bearing caps, and unscrew the bearing cap nuts (arrows) (M42 engine)

identification marks. The caps are numbered from the timing chain end of the engine, and the marks can normally be read from the exhaust side of the engine. The exhaust camshaft bearing caps are marked "A1" to "A5", and the intake camshaft caps are marked "E1" to "E5" **(see illustration)**.

## M42 engine

*Refer to illustrations 10.13 and 10.14*

13  Mount BMW special tool 11 3 260 on the cylinder head by screwing the mounting bolts into the spark plug holes. Position the tool so that the plungers are located over the relevant camshaft bearing caps (i.e., intake or exhaust camshaft) **(see illustration)**.
14  Apply pressure to the camshaft bearing caps by turning the eccentric shaft on the tools using a wrench **(see illustration)**.
15  Unscrew the camshaft bearing cap nuts.
**Caution:** *Do not attempt to unscrew the camshaft bearing cap nuts without the special tools in place, as damage to the camshaft and/or bearings may result.*
16  Release the pressure on the special tool shaft, then unbolt the tool from the cylinder head.
17  Lift off the bearing caps, keeping them in order, then lift out the camshaft.
18  Repeat Steps 13 through 17 for the other camshaft.

## M44 engine

19  Loosen the camshaft bearing cap bolts gradually and evenly, remove the cam bearing caps and remove the camshafts.

## M42 engine

20  The camshaft bearing casting can now be lifted from the cylinder head. This should be done very slowly, as the cam followers will be released as the casting is lifted off - if the casting is lifted off awkwardly, the cam followers may fall out. Do not allow the cam followers to fall out and get mixed up, as they must be reinstalled in their original locations.
21  With the bearing casting removed, lift the cam followers from the cylinder head.

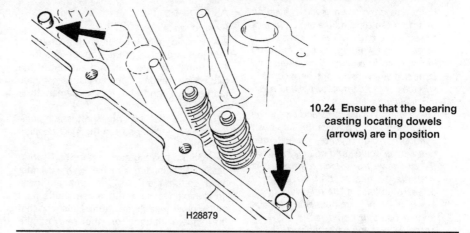

10.24 Ensure that the bearing casting locating dowels (arrows) are in position

Identify the followers for location, and store them upright in a container of clean engine oil to prevent the oil from draining from inside the followers. Do not forget to mark the cam followers "Intake" and "Exhaust". Store each cam follower in a labeled plastic cup filled with oil.

## M44 engine

22  Pull each hydraulic valve adjuster straight out of the cylinder head.

## Inspection

23  Clean all the components, including the bearing surfaces in the bearing castings and bearing caps. Examine the components carefully for wear and damage. In particular, check the bearing and cam lobe surfaces of the camshaft(s) for scoring and pitting. Examine the surfaces of the cam followers for signs wear or damage. Replace components as necessary.

## Installation

### M42 engine

*Refer to illustrations 10.24 and 10.27*

24  If the camshaft lower bearing castings have been removed, check that the mating faces of the bearing castings and the cylinder head are clean, and check that the bearing casting locating dowels are in position on the studs at No. 2 and 5 bearing locations **(see illustration)**.
25  The bearing casting(s) and cam followers must now be installed. The simplest method of installing these components is to retain the cam followers in the bearing casting, and install the components as an assembly.
26  Oil the bearing casting contact surfaces of the cam followers (avoid allowing oil onto the top faces of the followers at this stage), then fit each follower to its original location in the bearing casting. Once all the followers have been fitted, they must be retained in the bearing casting, so that they do not fall out as the assembly is installed to the cylinder head **(see illustrations in Chapter 2B, Section 10)**. To retain the cam followers in the bearing castings, proceed as follows:

a) *Apply a small amount of silicone or a similar adhesive compound to the top of each cam follower. The adhesive should protrude beyond the lower bearing surface of the casting. Do not use excessive adhesive, as there is a risk of contaminating the oil passages in the bearing castings.*

**10.27 Camshaft bearing casting identification mark**

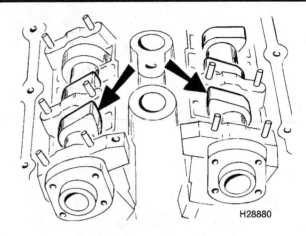

**10.31 Position the camshafts to that the tips of the front cam lobes (arrows) face one another**

b) *Press a wooden dowel (such as a length of broom handle) onto the top surface of the adhesive pads so that the pads stick to the dowel, holding the cam followers in the bearing casting. Ensure that no adhesive is pressed out between the surfaces of the cam followers and bearing casting.*

c) *Leave the adhesive pads and the dowel in position until the assembly has been installed to the cylinder head, then remove the dowel, and carefully remove the adhesive from each follower. It is essential to ensure that no trace of adhesive is left on the followers or on any of the engine components - serious engine damage could result if the oil passages become contaminated.*

27  With the cam followers retained in the bearing casting, install the casting to the cylinder head. Note that the exhaust side casting is marked "A" and the intake side casting is marked "E". When the castings are installed, the marks should face each other at the timing chain end of the cylinder head **(see illustration)**. **Caution:** *The cam followers expand when not subjected to load by the camshafts, and therefore require some time before they can be compressed. If the camshaft installation operation is carried out rapidly, there is a possibility that the "closed" valves will be forced open by the expanded cam followers, resulting in piston-to-valve contact. To minimize this possibility, after installing the camshaft(s) observe the following delays before turning the crankshaft back to the TDC position:*

| Temperature | Delay |
|---|---|
| Room temperature (68°F) | 4 minutes |
| 50°F to 68°F | 11 minutes |
| 32°F to 50°F | 30 minutes |

### M44 engine

28  Make sure that the bore for each hydraulic valve adjuster is neither scored nor worn, then lubricate each valve adjuster and push it straight into the cylinder head.

### All engines

*Refer to illustration 10.31*

29  First identify the camshafts to ensure that they are fitted in the correct locations. The camshafts are stamped in front of the rear square flanges. The exhaust camshaft is marked "A" and the intake camshaft is marked "E".

30  Ensure that the crankshaft is still positioned 90° clockwise from the TDC position (see Step 9).

31  Oil the bearing surfaces in the bearing casting. Position the camshaft on the cylinder head, so that the tips of the front cam lobes on the exhaust and intake camshafts face one another **(see illustration)**. Note also that the square flanges on the rear of the camshaft should be positioned with the sides of the flanges exactly at right-angles to the top surface of the cylinder head (this can be checked using a set-square), and the side of the flange with holes drilled into it uppermost.

32  Place the bearing caps in position, noting that the caps carry identification marks. The exhaust camshaft caps are marked "A1" to "A5", and the intake camshaft caps are marked "E1" to "E5". Place the bearing caps in their original locations as noted before removal.

33  On M42 engines, reassemble BMW special tool 11 3 260, and install it to the cylinder head as during removal. **Caution:** *Again, do not attempt to install the camshafts without the aid of the special tools.*

34  Apply pressure to the relevant bearing caps on M42 engines by turning the eccentric shaft on the tools using a wrench.

35  With pressure applied to the bearing caps on M42 engines, install the bearing cap securing nuts, and tighten them as far as possible by hand.

36  Tighten the bearing cap nuts to the specified torque, working progressively in a diagonal sequence.

37  Once the bearing cap nuts have been tightened on M42 engines, unbolt the tools used to apply pressure to the bearing caps.

38  Repeat this procedure to install the other camshaft.

39  Install the spark plugs.

40  Install the template used to check the position of the camshafts. If necessary, turn the camshaft(s) slightly using a wrench on the flats provided until the template can be fitted. **Caution:** *Note the precaution given in the **Caution** in paragraph 27 before proceeding.*

41  Turn the crankshaft back 90-degrees counterclockwise to the TDC position, then re-engage the locking rod with the flywheel to lock the crankshaft in position.

42  Manipulate the camshaft sprockets until the timing arrows on the sprockets are pointing vertically upwards, then engage the chain with the sprockets.

43  Place the sprockets on the camshafts, ensuring that the sprockets are installed facing the correct direction as noted before removal, then install the sprocket securing bolts.

44  Tighten the sprocket securing bolts to the specified torque - if necessary, the camshafts can be counterheld using a 27 mm wrench on the flats provided between No. 5 and 6 cam lobes.

45  Install and tighten the left-hand chain guide upper securing bolt.

46  Install the upper chain guide and tighten the securing bolts.

47  Install the timing chain tensioner (see Steps 25 through 29 in Section 7).

48  Remove the template used to lock the camshafts, and remove the locking tool from the flywheel.

49  On M44 engines, install the rocker arms. Rotate the engine until a pair of camshaft lobes is pointing up (the intake and exhaust lobes for the number one cylinder, for example). Using BMW special tool 11 5 130, compress each valve/valve spring assembly far enough to install the rocker arms. After the first two rocker arms have been installed, rotate the engine again until the next pair of cam lobes is pointing up. Again, depress the valve with the special BMW tool and install the rocker arms. And so on. Repeat this step

# Chapter 2 Part A  Four-cylinder engines

11.11 Unscrewing an upper chain guide securing bolt

11.19 Removing a cylinder head bolt

until all rocker arms have been installed.
50  Install the upper timing chain cover and the valve cover (see Sections 6 and 4).
51  The remainder of installation is the reverse of removal. **Caution:** *As described in the Caution in paragraph 27, the cam followers expand when not subjected to load by the camshafts. To minimize the possibility of piston-to-valve contact, after installing the camshaft(s), observe the following delays before cranking the engine:*

| Temperature | Delay |
|---|---|
| Room temperature (68°F) | 10 minutes |
| 50°F to 68°F | 30 minutes |
| 32°F to 50°F | 75 minutes |

**Caution:** *When the engine is first started, it must be run at a speed of 3500 rpm for 20 seconds as soon as it starts - this is to ensure that the tensioner is primed with oil.*

## 11  Cylinder head - removal and installation

**Warning:** *Wait until the engine is completely cool before beginning this procedure.*
**Note:** *New cylinder head bolts and a new cylinder head gasket will be required on installation.*

### Removal

*Refer to illustrations 11.11, 11.19 and 11.21*

1  Depressurize the fuel system (see Chapter 4A), then disconnect the battery negative lead. **Caution:** *If the radio in your vehicle is equipped with an anti-theft system, make sure you have the correct activation code before disconnecting the battery.*
2  Drain the cooling system (see Chapter 1).
3  Remove the air cleaner assembly and the airflow meter (see Chapter 4A).
4  Remove the upper and lower sections of the intake manifold (see Chapter 4A).
5  Remove the exhaust manifold (see Chapter 4A).
6  Disconnect the coolant hose from the cylinder head.

7  Remove the upper timing chain cover (see Section 6)
8  Position No. 1 piston at TDC, and lock the flywheel in position (see Section 3).
9  Unscrew the timing chain tensioner plug from the right-hand side of the engine. Recover the sealing ring.
10  Withdraw the timing chain tensioner assembly from its housing.
11  Unscrew the securing bolts, and withdraw the upper chain guide from the cylinder head **(see illustration)**.
12  Unscrew the upper securing bolt from the left-hand chain guide.
13  Unscrew the bolts securing the chain sprockets to the camshafts. Take care not to move the camshafts - if necessary, the camshafts can be counterheld using a 27 mm wrench on the flats provided between No. 5 and 6 cam lobes.
14  Withdraw the sprockets from the camshafts, and disengage them from the chain. Note which way the sprockets face to ensure correct installation.
15  Ensure that tension is kept on the timing chain - tie the chain up or support it using wire, to prevent it from dropping into the lower timing chain cover.
16  If not already done, disconnect the wiring plugs from the coolant temperature sensors located in the left-hand side of the cylinder head. **Caution:** *To avoid any possibility of piston-to-valve contact when installing the cylinder head, it is necessary to ensure that none of the pistons are at TDC. Before proceeding further, remove the locking rod from the timing hole in the cylinder block, then turn the crankshaft approximately 90-degrees clockwise using a wrench or socket on the crankshaft pulley hub bolt.*
17  Make a final check to ensure that all relevant hoses and wires have been disconnected to allow cylinder head removal.
18  Progressively loosen the cylinder head bolts, working in a spiral pattern from the outside of the head inwards.
19  Remove the cylinder head bolts **(see illustration)**. Where applicable recover the washers under the cylinder head bolts. Note that the cylinder head originally fitted in production has captive washers.
20  Release the cylinder head from the cylinder block and locating dowels by rocking it. Do not pry between the mating faces of the cylinder head and block, as this may damage the gasket faces.
21  Ideally, an assistant will now be required to help lift the cylinder head from the block - take care as the cylinder head is heavy **(see illustration)**.
22  Recover the cylinder head gasket.

### Inspection

23  Refer to Chapter 2C for details of cylinder head dismantling and reassembly.
24  The mating faces of the cylinder head and block must be perfectly clean before installing the head. Use a scraper to remove

11.21 Lifting off the cylinder head

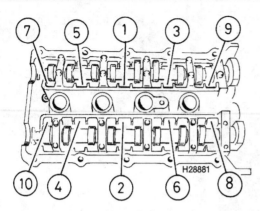

11.35a Cylinder head bolt tightening sequence

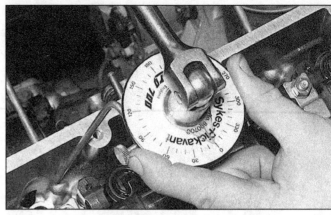

11.35b Angle-tightening a cylinder head bolt

all traces of gasket and carbon, and also clean the tops of the pistons. Take particular care with the aluminum cylinder head, as the soft metal is easily damaged. Also make sure that debris is not allowed to enter the oil and water passages. Using adhesive tape and paper, seal the water, oil and bolt holes in the cylinder block. To prevent carbon entering the gap between the pistons and bores, smear a little grease in the gap. After cleaning each piston, rotate the crankshaft so that the piston moves down the bore, then wipe out the grease and carbon with a cloth rag.

25   Check the block and head for nicks, deep scratches and other damage. If very slight, they may be removed from the cylinder block carefully with a file. More serious damage may be repaired by machining, but this is a specialist job.

26   If warpage of the cylinder head is suspected, use a straight-edge to check it for distortion, with reference to Chapter 2C.

27   Clean out the bolt holes in the block using a pipe cleaner or thin rag and a screwdriver. Make sure that all oil and water is removed, otherwise there is a possibility of the block being cracked by hydraulic pressure when the bolts are tightened.

28   Examine the bolt threads and the threads in the cylinder block for damage. If necessary, use the correct size tap to chase out the threads in the block.

## Installation

Refer to illustrations 11.35a and 11.35b

**Caution:** *If the camshafts have been removed from the cylinder head, note the Cautions given in Section 10, regarding expanded cam followers. Additionally, to minimize the possibility of piston-to-valve contact, after installing the camshaft(s) observe the following delays before installing the cylinder head.*

| Temperature | Delay |
|---|---|
| Room temperature (68°F) | 4 minutes |
| 50°F to 68°F | 11 minutes |
| 32°F to 50°F | 30 minutes |

29   Ensure that the mating faces of the cylinder block and head are spotlessly clean, that the cylinder head bolt threads are clean and dry, and that they screw in and out of their locations.

30   On M42 engines, make sure that the oil feed check valve and spacer are installed in the block.

31   Check that the cylinder head locating dowels are correctly positioned in the cylinder block. **Caution:** *To avoid any possibility of piston-to-valve contact when installing the cylinder head, it is necessary to ensure that none of the pistons are at TDC. Before proceeding further, if not already done, turn the crankshaft to position No. 1 piston at TDC (check that the locking rod can be engaged with the flywheel, then remove the locking rod and turn the crankshaft approximately 90-degrees clockwise using a wrench or socket on the crankshaft pulley hub bolt.*

32   Fit a new cylinder head gasket to the block, locating it over the dowels. Make sure that it is the correct way up. Note that 0.012-inch (0.3 mm) thicker-than-standard gaskets are available for use if the cylinder head has been machined (see Chapter 2C).

33   If not already done, fit the template to the cylinder head to ensure that the camshaft is correctly positioned (No. 1 piston at TDC) - see Section 3.

34   Lower the cylinder head into position. Ensure that the cylinder head engages with the locating dowels. Fit the new cylinder head bolts, complete with new washers where necessary, and tighten the bolts as far as possible by hand. Ensure that the washers are correctly seated in their locations in the cylinder head. **Note:** *Do not install washers on any bolts which are fitted to locations where there are already captive washers in the cylinder head. If a new cylinder head is installed (without captive washers), ensure that new washers are used on all the bolts.*

35   Tighten the bolts in the sequence shown, and in the steps given in this Chapter's Specifications - i.e., tighten all bolts in sequence to the Step 1 torque, then tighten all bolts in sequence to the Step 2 angle-torque, and so on **(see illustrations)**.

36   Turn the crankshaft back 90-degrees counterclockwise to the TDC position, then re-engage the locking rod with the flywheel to lock the crankshaft in position.

37   Manipulate the camshaft sprockets until the timing arrows on the sprockets are pointing vertically upwards, then engage the chain with the sprockets.

38   Install the sprockets on the camshafts, ensuring that the sprockets face the proper direction as noted before removal, then install the sprocket securing bolts.

39   Tighten the sprocket securing bolts to the specified torque - if necessary, the camshafts can be counterheld using a 27 mm wrench on the flats provided between No. 5 and 6 cam lobes.

40   Install and tighten the left-hand chain guide upper securing bolt.

41   Install the upper chain guide and tighten the securing bolts.

42   Install the timing chain tensioner (see Steps 25 through 29 in Section 7).

43   Further installation is a reversal of removal, bearing in mind the following points.

a) Install the upper timing chain cover (see Section 6).
b) On M44 engines, replace the seal for the oil supply tube **(see illustration 11.35a)**.
c) Install the valve cover (see Section 4).
d) Install the exhaust manifold (see Chapter 4A).
e) Install the lower and upper sections of the intake manifold (see Chapter 4A).
f) Install the air cleaner assembly and the air mass meter (see Chapter 4A).
g) On completion, refill the cooling system (see Chapter 1), and prime the fuel system (see Chapter 4A).

**Caution:** *When the engine is first started, it must be run at a speed of 3500 rpm for 20 seconds as soon as it starts - this is to ensure that the tensioner is primed with oil.*

## 12   Oil pan - removal and installation

**Note:** *A new gasket and sealant will be required on installation.*

### Removal

Refer to illustrations 12.13a, 12.13b, 12.16a and 12.16b

1   Open the hood, then raise the hood to its fully open position (see Chapter 11).

# Chapter 2  Part A  Four-cylinder engines

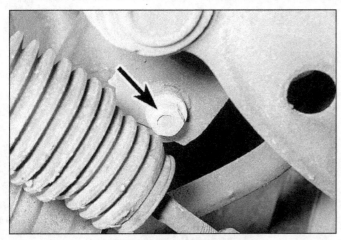

12.13a  Unscrew the front . . .

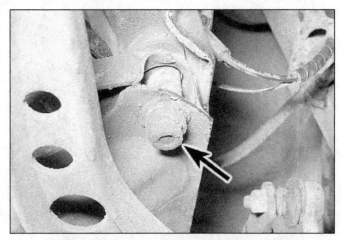

12.13b  . . . and rear subframe securing bolts (arrow)

2   Drain the engine oil (see Chapter 1).
3   Apply the parking brake, then jack up the front of the vehicle and support securely on axle stands. Remove any splash shields that are in the way.
4   Disconnect the vacuum hose from the brake booster.
5   Unscrew the bolt securing the dipstick tube to the intake manifold, then pull the lower end of the tube from the oil pan, and withdraw the dipstick tube assembly.
6   Position a hoist and lifting tackle over the engine compartment, and connect the lifting tackle to the front engine lifting bracket.
7   Disconnect the steering column shaft from the steering rack (see Chapter 10).
8   Unbolt the ground lead(s) from the engine mounting bracket(s).
9   Working under the vehicle, unscrew the nuts securing the left- and right-hand engine mountings to the brackets on the subframe.
10  Raise the lifting tackle to lift the engine approximately 1/4-inch.
11  Again working under the vehicle, unscrew the bolts securing the suspension lower arms to the body.
12  Support the center of the subframe, using a jack and a block of wood.
13  Unscrew the subframe securing bolts, then lower the subframe slightly using the jack **(see illustrations)**. **Caution:** *Do not remove the steering gear from the subframe.*
14  Where applicable, unclip the fuel pipes and/or the automatic transmission fluid cooler pipes from the brackets on the oil pan, and move the pipes clear of the working area.
15  Remove the drivebelt from the power steering pump, unbolt the power steering pump bracket and detach the pump and bracket from the engine (see Chapter 10). Hang the pump and bracket with a piece of wire to protect the power steering hoses.
16  Working under the vehicle, progressively unscrew and remove all the oil pan securing bolts. Note that the three lower transmission-to-engine bolts must be removed, as they screw into the oil pan **(see illustrations)**.
17  Lower the oil pan, and manipulate it out towards the rear of the vehicle. If necessary, lower the subframe further, using the jack, to give sufficient clearance. Similarly, if necessary, unbolt the oil pick-up pipe to ease oil pan removal. Recover the oil pan gasket, and discard it.

## Installation

18  Commence installation by thoroughly cleaning the mating faces of the oil pan and cylinder block.
19  Lightly coat the areas where the crankshaft rear oil seal housing and the timing chain housing join the cylinder block with a little non-hardening gasket sealant.
20  Place a new gasket in position on the oil pan flange. If necessary, apply more sealant (sparingly) to hold the gasket in place.
21  Install the oil pan to the cylinder block, ensuring that the gasket stays in place, and where applicable, install the oil pick-up pipe using a new gasket.
22  Install the oil pan securing bolts, tightening them finger-tight only at this stage.
23  Progressively tighten the oil pan-to-cylinder block securing bolts to the specified torque.
24  Tighten the oil pan-to-transmission and the transmission-to-engine bolts to the specified torque.
25  Install the power steering pump and bracket and install the power steering pump drivebelt (see Chapter 10).
26  Check the condition of the dipstick seal-

12.16a  Four of the oil pan securing bolts (arrows)

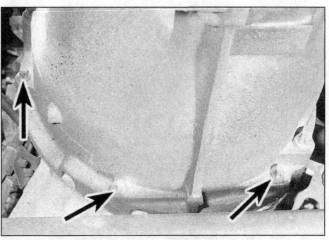

12.16b  The three lower transmission/transmission-to-engine bolts (arrows) must be unscrewed to remove the oil pan

**13.4a Unscrew the securing bolts . . .**

**13.4b . . . and remove the oil pump cover**

ing ring (at the oil pan end of the tube) and replace if necessary. Install the dipstick tube and tighten the bracket securing bolt.
27  On models with automatic transmission, secure the fluid cooler lines to the oil pan. Similarly, where applicable, clip the fuel pipes into position.
28  Raise the subframe using the jack, then install the securing bolts and tighten to the specified torque.
29  Install the bolts securing the suspension lower arms to the body, ensuring that the washers are in place, and tighten the bolts to the specified torque.
30  Lower the engine until the mountings are resting on the subframe, ensuring that the lugs on the engine mountings engage with the corresponding holes in the subframe. Install the engine mounting nuts and tighten them to the specified torque.
31  Disconnect and withdraw the engine lifting tackle and hoist.
32  Reconnect the steering shaft to the steering rack (see Chapter 10).
33  Further installation is a reversal of removal, but on completion refill the engine with oil (see Chapter 1).

## 13  Oil pump - removal, inspection and installation

### Removal

1  The oil pump is integral with the timing chain housing.
2  Remove the crankshaft vibration damper/pulley and pulley hub (see Section 5).
3  Remove the timing chain housing (see Section 9).

### Inspection

*Refer to illustrations 13.4a, 13.4b, 13.5a, 13.5b, 13.7, 13.11a, 13.11b, 13.11c, 13.11d and 13.11e*

4  Unscrew the oil pump cover from the rear of the timing chain housing to expose the oil pump rotors **(see illustrations)**.
5  Check the rotors for identification marks, and if necessary mark the rotors to ensure that they are installed in their original positions (mark the top faces of both rotors to ensure that they are installed the correct way up). Remove the rotors from the housing **(see illustrations)**.
6  Clean the housing and the rotors thoroughly, then install the rotors to the housing, ensuring that they are positioned as noted before removal.
7  Using feeler gauges, measure the clearance between the oil pump body and the outer rotor. Using the feeler gauges and a straight edge, measure the clearance (endplay) between each of the rotors and the oil pump cover mating face **(see illustration)**.

**13.5a Remove the inner . . .**

**13.5b . . . and outer rotors from the oil pump**

**13.7 Measuring the clearance between the oil pump body and the outer rotor**

# Chapter 2 Part A  Four-cylinder engines

**13.11a  Remove the pressure relief valve snap-ring . . .**

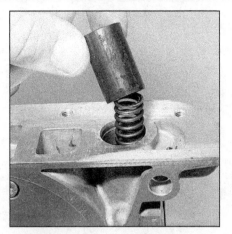

**13.11b  . . . and remove the sleeve . . .**

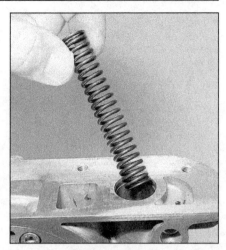

**13.11c  . . . spring . . .**

8   If the clearances are not as given in the Specifications, consult a BMW dealer regarding the availability of spare parts. The rotors should always be renewed as a matched pair. It may be necessary to replace the complete rotor/housing assembly as a unit.

9   If the clearances are within the tolerances given, remove the rotors, then pour a little engine oil into the housing. Install the rotors and turn them to lubricate all the contact surfaces.

10   Install the oil pump cover plate, and tighten the securing bolts to the specified torque.

11   To check the oil pressure relief valve, extract the snap-ring and remove the sleeve, spring and piston. Check that the free-length of the spring is as given in the Specifications **(see illustrations)**.

12   Reassemble the pressure relief valve using a reversal of the dismantling procedure.

### Installation

13   Install the timing chain housing (see Section 9).

14   Install the crankshaft vibration damper/pulley and pulley hub (see Section 5).

## 14   Oil seals - replacement

### Crankshaft front oil seal

1   Oil seal replacement is described as part of the lower timing chain cover removal and installation procedure in Section 6.

### Crankshaft rear oil seal

2   Proceed as described for six-cylinder engines in Chapter 2B, Section 14.

### Camshaft oil seals

3   No camshaft oil seals are fitted. Sealing is provided by the valve cover gasket and the timing chain cover gaskets.

## 15   Flywheel/driveplate - removal and installation

The procedure is as described for six-cylinder engines in Chapter 2B, Section 15.

## 16   Crankshaft pilot bearing - replacement

The procedure is as described for six-cylinder engines in Chapter 2B, Section 16.

## 17   Engine/transmission mounts - inspection and replacement

The procedure is as described for six-cylinder engines in Chapter 2B, Section 17.

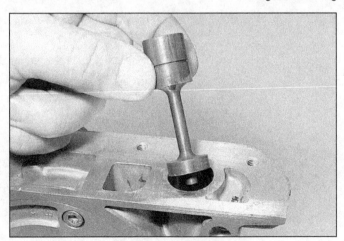

**13.11d  . . . and piston**

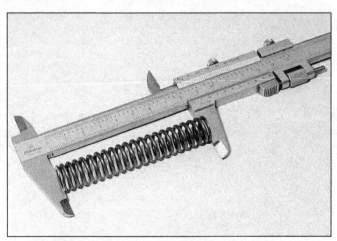

**13.11e  Check the length of the pressure relief valve spring**

## Notes

# Chapter 2 Part B
# Six-cylinder engines

## Contents

| | Section | | Section |
|---|---|---|---|
| Camshafts and followers - removal, inspection and installation.... | 10 | Oil pan - removal and installation................................................. | 12 |
| Compression test - description and interpretation ........................ | 2 | Oil pump and drive chain - removal, inspection and installation ... | 13 |
| Crankshaft pilot bearing - replacement........................................... | 16 | Oil seals - replacement ................................................................ | 14 |
| Crankshaft vibration damper/pulley and pulley hub - removal and installation................................................................... | 5 | Timing chain covers - removal and installation............................. | 6 |
| Cylinder head - removal and installation......................................... | 11 | Timing chain sprockets and tensioners - removal, inspection and installation......................................................................... | 8 |
| Engine oil and filter replacement ..................................... See Chapter 1 | | Timing chains - removal, inspection and installation .................... | 7 |
| Engine oil level check ....................................................... See Chapter 1 | | Top Dead Center (TDC) for No. 1 piston - locating....................... | 3 |
| Engine/transmission mounts - inspection and replacement............ | 17 | Valve cover - removal and installation .......................................... | 4 |
| Flywheel/driveplate - removal and installation ................................ | 15 | Variable valve timing system (VANOS) components - removal, inspection and installation ......................................................... | 9 |
| General information .......................................................................... | 1 | | |

## Specifications

### General

Engine code
  1995 and earlier................................................................. M50 B25
  1996 on............................................................................... M52 B28
Displacement
  1995 and earlier................................................................. 152.2 cubic inches (2.5 liters)
  1996 on............................................................................... 170.3 cubic inches (2.8 liters)
Direction of engine rotation ...................................................... Clockwise (viewed from front of vehicle)
No. 1 cylinder location .............................................................. Timing chain end
Firing order ............................................................................... 1-5-3-6-2-4
Compression ratio
  1992 ................................................................................... 10.0:1
  1993 thru 1995 .................................................................. 10.5:1
  1996 on ............................................................................. 10.2:1
Minimum compression pressure ............................................... 142 to 156 psi

### Camshafts
Endplay...................................................................................... 0.0008 to 0.0021 inch
Oil clearance.............................................................................. 0.006 to 0.013 inch

### Lubrication system
Minimum oil pressure at idle speed........................................... 7 psi
Oil pressure relief valve spring free-length ............................... 3.31 inches
Oil pump rotor clearance
  Outer rotor-to-pump body .................................................. 0.004 to 0.007 inch
  Outer rotor-to-inner rotor ................................................... 0.004 to 0.007 inch
  Rotor endplay..................................................................... 0.002 to 0.003 inch

## Torque specifications

Ft-lbs (unless otherwise indicated)

Main bearing cap bolts*
  1995 and earlier
    Step 1.................................................................................................. 15
    Step 2.................................................................................................. Tighten an additional 50-degrees
  1996 on
    Step 1.................................................................................................. 15
    Step 2.................................................................................................. Tighten an additional 70-degrees

Cylinder head bolts*
  1995 and earlier
    Step 1.................................................................................................. 22
    Step 2.................................................................................................. Tighten an additional 90-degrees
    Step 3.................................................................................................. Tighten an additional 90-degrees
  1996 on
    Step 1.................................................................................................. 30
    Step 2.................................................................................................. Tighten an additional 90-degrees
    Step 3.................................................................................................. Tighten an additional 90-degrees

Valve cover bolts ........................................................................................ 86 in-lbs
Upper and lower timing chain cover nuts and bolts
  M6 nuts/bolts ......................................................................................... 79 in-lbs
  M8 nuts/bolts ......................................................................................... 16
  M10 nuts/bolts ....................................................................................... 35
Crankshaft rear oil seal housing bolts
  M6 bolts ................................................................................................. 79 in-lbs
  M8 bolts ................................................................................................. 16
Flywheel bolts*........................................................................................... 89
Crankshaft vibration damper/pulley-to-hub bolts ..................................... 16
Crankshaft pulley hub bolt* ....................................................................... 300
Connecting rod bearing cap bolts*
  Step 1..................................................................................................... 15
  Step 2..................................................................................................... Tighten an additional 70-degrees
Camshaft bearing cap nuts ....................................................................... 11
Camshaft sprocket nuts/bolts
  Step 1..................................................................................................... 4
  Step 2..................................................................................................... 16
Primary timing chain tensioner cover plug ................................................ 28
VANOS solenoid valve............................................................................... 22
Oil feed pipe to VANOS adjustment unit .................................................. 24
VANOS oil feed pipe to oil filter housing .................................................. 37
Oil pump bolts
  M6 bolts ................................................................................................. 79 in-lbs
  M8 bolts ................................................................................................. 16
Oil pump cover .......................................................................................... 79 in-lbs
Oil pump sprocket nut ............................................................................... 18
Oil filter housing-to-cylinder block bolts................................................... 16
Front subframe bolts*................................................................................ 77

*Use new bolts

## 1 General information

### How to use this Chapter

This Part of Chapter 2 describes the repair procedures that can reasonably be carried out on the engine while it remains in the vehicle. If the engine has been removed from the vehicle and is being disassembled in the procedure outlined in Part C, any preliminary dismantling procedures can be ignored.

Note that, while it may be possible physically to overhaul items such as the piston/connecting rod assemblies while the engine is in the car, such tasks are not usually carried out as separate operations. Usually, several additional procedures are required (not to mention the cleaning of components and oil passages); for this reason, all such tasks are classed as major overhaul procedures, and are described in Part C of this Chapter.

Part C describes the removal of the engine/transmission from the car, and the full overhaul procedures that can then be carried out.

### Engine description

#### General

The six-cylinder engine is a double overhead camshaft design, mounted in-line, with the transmission bolted to the rear end.

A timing chain drives the exhaust camshaft; the intake camshaft is driven by a second chain from the end of the exhaust camshaft. Hydraulic cam followers are fitted between the camshafts and the valves. Each camshaft is supported by seven bearings incorporated in bearing castings fitted to the cylinder head.

The crankshaft runs in seven main bearings of the usual shell-type. Endplay is controlled by thrust bearing shells on No. 6 main bearing.

The pistons are selected to be of matching weight, and incorporate fully floating wrist pins retained by circlips.

The oil pump is chain-driven from the front of the crankshaft.

### VANOS variable camshaft timing control system

On 1993 and later models, a modified engine was introduced with a variable camshaft timing control system, known as

VANOS. The VANOS system uses data supplied by the DME engine management system (see Chapter 4B), to adjust the timing of the intake camshaft via a hydraulic control system (using engine oil as the hydraulic fluid). The camshaft timing is varied according to engine speed, retarding the timing (opening the intake valves later) at low and high engine speeds to improve low-speed driveability and maximum power respectively. At medium engine speeds, the camshaft timing is advanced (opening the intake valves earlier) to increase mid-range torque and to improve exhaust emissions.

## Repair operations possible with the engine in the vehicle

The following operations can be carried out without having to remove the engine from the vehicle:

a) Removal and installation of the cylinder head.
b) Removal and installation of the timing chain and sprockets.
c) Removal and installation of the camshafts.
d) Removal and installation of the oil pan.
e) Removal and installation of the connecting rod bearings, connecting rods, and pistons*.
f) Removal and installation of the oil pump.
g) Replacement of the engine/transmission mountings.
h) Removal and installation of the flywheel/driveplate.

* Although it is possible to remove these components with the engine in place, for reasons of access and cleanliness it is recommended that the engine be removed.

## 2 Compression test - description and interpretation

1  When engine performance is down, or if misfiring occurs which cannot be attributed to the ignition or fuel systems, a compression test can provide diagnostic clues as to the engine's condition. If the test is performed regularly, it can give warning of trouble before any other symptoms become apparent.
2  The engine must be fully warmed-up to normal operating temperature, the battery must be fully charged, and all the spark plugs must be removed (Chapter 1). The aid of an assistant will also be required.
3  Disable the ignition and fuel injection systems by removing the DME master relay, and the fuel pump relay, located in the main fuse box in the engine compartment (see Chapter 12).
4  Fit a compression tester to the No. 1 cylinder spark plug hole - the type of tester which screws into the plug thread is to be preferred.
5  Have the assistant hold the throttle wide open, and crank the engine on the starter motor. After one or two revolutions, the compression pressure should build up to a maximum figure, and then stabilize. Record the highest reading obtained.
6  Repeat the test on the remaining cylinders, recording the pressure in each.
7  All cylinders should produce very similar pressures; a difference of more than 20 psi between any two cylinders indicates a fault. Note that the compression should build up quickly in a healthy engine; low compression on the first stroke, followed by gradually increasing pressure on successive strokes, indicates worn piston rings. A low compression reading on the first stroke, which does not build up during successive strokes, indicates leaking valves or a blown head gasket (a cracked head could also be the cause). Deposits on the undersides of the valve heads can also cause low compression.
8  BMW recommended values for compression pressures are given in the Specifications.
9  If the pressure in any cylinder is low, carry out the following test to isolate the cause. Introduce a teaspoonful of clean oil into that cylinder through its spark plug hole, and repeat the test.
10  If the addition of oil temporarily improves the compression pressure, this indicates that bore or piston wear is responsible for the pressure loss. No improvement suggests that leaking or burnt valves, or a blown head gasket, may be to blame.
11  A low reading from two adjacent cylinders is almost certainly due to the head gasket having blown between them; the presence of coolant in the engine oil will confirm this.
12  If one cylinder is about 20 percent lower than the others and the engine has a slightly rough idle, a worn camshaft lobe could be the cause.
13  If the compression reading is unusually high, the combustion chambers are probably coated with carbon deposits. If this is the case, the cylinder head should be removed and decarbonized.
14  On completion of the test, install the spark plugs (see Chapter 1) and reconnect the fuel pump relay and the DME master relay.

## 3 Top Dead Center (TDC) for No. 1 piston - locating

*Refer to illustrations 3.4a, 3.4b, 3.5a, 3.5b, 3.6a, 3.6b, 3.8, 3.9a and 3.9b*

**Note:** *To lock the engine in the TDC position, and to check the position of the camshafts, special tools will be required. These tools can easily be improvised - see text.*

1  Top Dead Center (TDC) is the highest point in the cylinder that each piston reaches as it travels up and down when the crankshaft turns. Each piston reaches TDC at the end of the compression stroke and again at the end of the exhaust stroke, but TDC generally refers to piston position on the compression stroke. No. 1 piston is at the timing chain end of the engine.
2  Positioning No. 1 piston at TDC is an essential part of many procedures, such as timing chain removal and camshaft removal.
3  Remove the valve cover (see Section 4).
4  Unclip the plastic cover from the intake camshaft **(see illustrations)**.
5  Using a socket or wrench on the

3.4a  Release the securing clips . . .

3.4b  . . . and remove the cover from the intake camshaft

3.5a With No. 1 piston at TDC the tips of the front cam lobes should face one another..

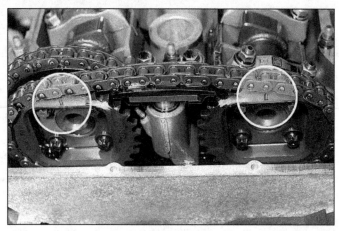

3.5b ... and the arrows on the camshaft sprockets should point vertically upwards

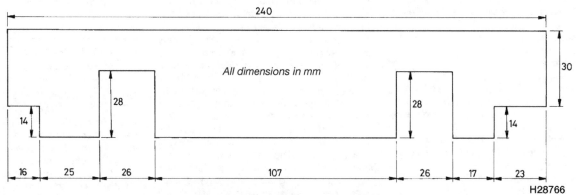

3.6a Make up a template from metal sheet to the dimensions shown

crankshaft pulley bolt, turn the engine clockwise until the tips of the front cam lobes on the exhaust and intake camshafts face one another. When the camshafts are correctly positioned, the arrows on the camshaft sprockets will point vertically upwards (note that there may be arrows at the top and bottom of the sprockets) **(see illustrations)**. Note also that the square flanges on the rear of the camshafts should be positioned with the sides of the flanges exactly at right-angles to the top surface of the cylinder head (this can be checked using a set-square), and the side of the flange with holes drilled into it uppermost.

6  A more accurate check on the camshaft positions can be made by making up a template from metal sheet to the dimensions shown - when the camshafts are correctly positioned, the template will fit exactly over the camshaft flanges, and rest on the upper surface of the cylinder head **(see illustrations)**. Note that it will be necessary to unbolt the rear camshaft cover studs from the cylinder head to enable the template to be fitted.

7  Pull the blanking plug from the timing hole in the left-hand rear corner flange of the cylinder block (access is greatly improved if the starter motor is removed - see Chapter 5).

8  To "lock" the crankshaft in position, a special tool will now be required. BMW tool No. 11 2 300 can be used, but an alternative can be made up by machining a length of

3.6b Template in place on upper surface of cylinder head with No. 1 piston at TDC

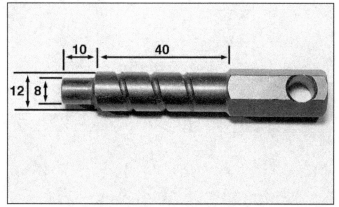

3.8 Dimensions of the flywheel "locking" tool

*All dimensions in mm*

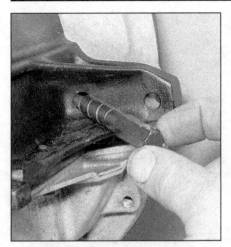

3.9a Insert the rod through the timing hole . . .

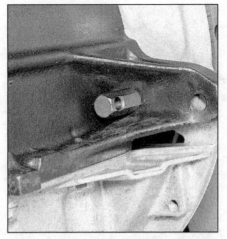

3.9b . . . until it enters the TDC hole in the flywheel - viewed with engine removed for clarity

4.4a Prying a securing nut cover plate from the fuel rail cover

steel rod to the dimensions shown **(see illustration)**.

9  Insert the rod through the timing hole. If necessary, turn the crankshaft slightly until the rod enters the TDC hole in the flywheel **(see illustrations)**.

10  The crankshaft is now "locked" in position with No. 1 piston at TDC. **Warning:** *If, for any reason, it is necessary to turn either or both of the camshafts with No. 1 piston positioned at TDC, and either of the timing chain tensioners slackened or removed (or the timing chains removed), the following precaution must be observed. Before turning the camshaft(s), the crankshaft must be turned approximately 30-degrees clockwise away from the TDC position (remove the locking rod from the TDC hole in the flywheel to do this) to prevent the possibility of piston-to-valve contact.*

11  **Do not** attempt to turn the engine with the flywheel or camshaft(s) locked in position, as engine damage may result. If the engine is to be left in the "locked" state for a long period of time, it is a good idea to place suitable warning notices inside the vehicle, and in the engine compartment. This will reduce the possibility of the engine being cranked on the starter motor.

## 4  Valve cover - removal and installation

*Refer to illustrations 4.4a, 4.4b, 4.4c, 4.5, 4.6, 4.7, 4.9, 4.10 and 4.11*

### Removal

**Note:** *New gaskets and/or seals may be required on installation - see text.*

1  Open the hood, then raise the hood to its fully open position (see Chapter 11).
2  Where necessary, to allow sufficient clearance for the valve cover to be removed, remove the heater/ventilation intake air ducting from the rear of the engine compartment as follows.

a) Lift the grille from the top of the ducting (on certain Coupe models, it will be necessary to remove the windshield wiper arms, then remove the plastic securing screws and lift off the complete scuttle grille assembly).
b) Working through the top of the ducting, remove the screws securing the cable ducting to the air ducting and move the cable ducting clear.
c) Unscrew the nuts and/or screw(s) securing the air ducting to the firewall (where applicable, bend back the heat shielding for access).
d) Remove the air ducting by pulling upwards.
e) Move the previously removed cable ducting clear of the valve cover.

3  Remove the engine oil filler cap.
4  Remove the plastic covers from the fuel rail and the top of the valve cover. To remove the covers, pry out the cover plates and unscrew the two securing nuts. To remove the cover from the cylinder head, lift and pull the cover forwards, then manipulate the cover over the oil filler neck **(see illustrations)**.
5  Unbolt the ground lead from the left-hand corner of the upper timing chain cover, and where applicable, unbolt the ground strap from the rear of the valve cover **(see illustration)**.

4.4b Unscrewing a cylinder head plastic cover securing nut

4.4c Removing the cylinder head plastic cover

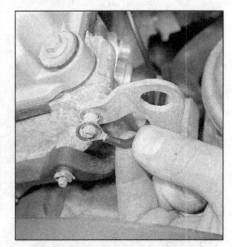

4.5 Disconnect the ground lead from the upper timing chain cover

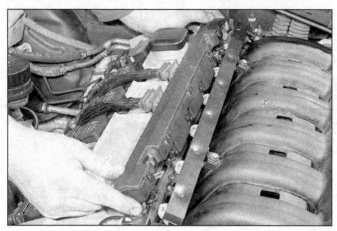

4.6 Pulling the wiring ducting from the fuel injectors

4.7 Disconnecting a coil wiring plug

6   Remove the two nuts securing the engine wiring ducting to the fuel rail, then pull the ducting up to release the wiring plugs from the fuel injectors **(see illustration)**.

7   Unclip the wiring connectors from the ignition coils. Recover the rubber seals if they are loose **(see illustration)**.

8   Release the wiring from the clips on the valve cover, then move the complete ducting/wiring assembly to one side, clear of the valve cover.

9   Unscrew the ignition coil securing nuts, then carefully pull the coils, complete with connectors, from the spark plugs **(see illustration)**. Note the locations of the ground leads and the coil wiring brackets.

10   Release the securing clip and disconnect the breather hose from the side of the valve cover **(see illustration)**.

11   Unscrew the securing bolts and the two studs (which also secure the coil ground leads) and lift off the valve cover. Note the locations

of all washers, seals and gaskets, and recover any which are loose **(see illustration)**.

## Installation

*Refer to illustration 4.13*

12   Commence installation by checking the condition of all seals and gaskets. Replace any which are deteriorated or damaged.

13   Clean the gasket/sealing faces of the cylinder head and the valve cover, then lay the main (outer) gasket and the two spark plug hole (center) gaskets in position on the cylinder head **(see illustration)**.

14   Lay the valve cover in position, taking care not to disturb the gaskets. Check that the tabs on the rear of the main gasket are correctly positioned in the cut-outs in the rear of the cylinder head.

15   Install the valve cover bolts and studs, ensuring that the seals are positioned as noted during removal, then tighten the bolts progressively to the specified torque.

16   Further installation is a reversal of the removal procedure, bearing in mind the following points.

a) Check that the ignition coil ground leads are correctly positioned as noted before removal.
b) Tighten the coil securing nuts to the specified torque.

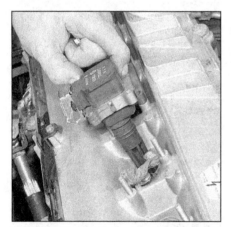

4.9 Removing an ignition coil

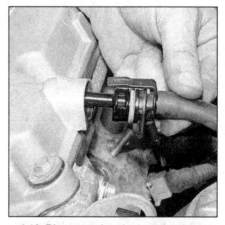

4.10 Disconnecting the breather hose from the valve cover

4.11 Lifting off the valve cover

4.13 Lay the gaskets in position on the cylinder head

# Chapter 2 Part B  Six-cylinder engines

5.4 Removing the vibration damper/pulley from the crankshaft

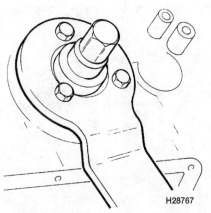

5.6 BMW special tool used to hold crankshaft pulley hub

5.8 Unscrew the pulley hub bolts and remove the washer . . .

c) *Check that the rubber seals are in place when reconnecting the spark plug wire plugs to the coils.*

## 5  Crankshaft vibration damper/pulley and pulley hub - removal and installation

*Refer to illustrations 5.4, 5.6, 5.8 and 5.9*

### Removal

**Note:** *If the pulley hub is removed, a new securing bolt will be required on installation, and a torque wrench capable of providing 303 ft-lbs of torque will be required.*

1  To improve access, unscrew the securing bolts and/or nuts, and remove the alternator air ducting from the front of the vehicle.
2  Again to improve access, remove the viscous cooling fan and fan shroud assembly (see Chapter 3).
3  Remove the drivebelt (see Chapter 1).
4  Unscrew the securing bolts, and remove the vibration damper/pulley from the hub **(see illustration)**. If necessary, counterhold the hub using a socket or wrench on the hub securing bolt.
5  To remove the hub, the securing bolt must be unscrewed. **Warning:** *The crankshaft pulley hub securing bolt is very tight. A tool will be required to counterhold the hub as the bolt is unscrewed. Do not attempt the job using inferior or poorly improvised tools, as injury or damage may result.*
6  Make up a tool to hold the pulley hub. A suitable tool can be fabricated using two lengths of steel bar, joined by a large pivot bolt. Bolt the holding tool to the pulley hub using the pulley-to-hub bolts **(see illustration)**.
7  Using a socket and a long breaker bar, loosen the pulley hub bolt. Note that the bolt is very tight.
8  Unscrew the pulley hub bolt, and remove the washer **(see illustration)**. Discard the bolt, a new one must be used on installation.
9  Withdraw the hub from the end of the crankshaft **(see illustration)**. If the hub is tight, use a puller to draw it off.
10  Recover the Woodruff key from the end of the crankshaft if it is loose.

### Installation

11  If the pulley hub has been removed, it is advisable to take the opportunity to replace the oil seal in the lower timing chain cover, with reference to Section 6.
12  If the pulley hub has been removed, proceed as follows, otherwise proceed to paragraph 16.
13  Where applicable, install the Woodruff key to the end of the crankshaft, then align the groove in the pulley hub with the key, and slide the hub onto the end of the crankshaft.
14  Install the washer, noting that the shoulder on the washer must face the hub, and fit a new hub securing bolt.
15  Bolt the holding tool to the pulley hub, as during removal, then tighten the hub bolt to the specified torque. Take care to avoid injury and/or damage.
16  Where applicable, unbolt the holding tool, and install the vibration damper/pulley, ensuring that the locating dowel on the hub engages with the corresponding hole in the damper/pulley.
17  Install the damper/pulley securing bolts, and tighten to the specified torque. Again, counterhold the pulley if necessary when tightening the bolts.
18  Install the drivebelt (see Chapter 1).
19  Install the viscous cooling fan and shroud (see Chapter 3).
20  Install the alternator air ducting.

## 6  Timing chain covers - removal and installation

*Refer to illustrations 6.2a, 6.2b, 6.3 and 6.4*

### Upper cover

#### 1992 models

**Note:** *A new gasket will be required on installation.*

5.9 . . . then withdraw the hub

### Removal

1  Remove the valve cover (see Section 4).
2  Release the securing clips, and remove the wiring ducting from the front of the upper timing chain cover **(see illustrations)**.

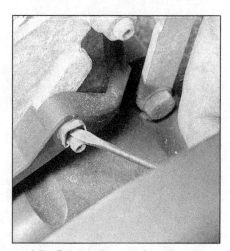

6.2a Release the securing clips . . .

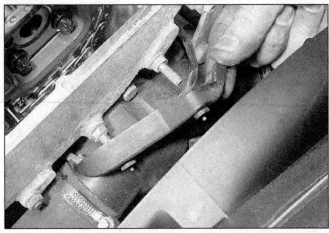

6.2b ... and remove the wiring ducting

6.3 Unbolt the engine lifting bracket

6.4 Removing the upper timing chain cover

6.13a Pull the cover from the idler pulley (arrow) ...

3  Unbolt the engine lifting bracket from the front left-hand corner of the cylinder head **(see illustration)**.
4  Unscrew the securing nuts and remove the upper timing chain cover from the front of the cylinder head **(see illustration)**. Recover the gasket.

6.13b ... then unscrew the bolt and remove the pulley

### Installation

5  Installation is a reversal of removal, bearing in mind the following points.
  a) Make sure that the dowel sleeves are in position on the top cover securing studs.
  b) Use a new cover gasket.
  c) Install the valve cover (see Section 4).

### 1993 and later models

6  On 1993 and later models, the upper timing chain cover is integral with the VANOS adjustment unit. Removal and installation of the VANOS adjustment unit is described in Section 9.

## Lower cover

**Note:** *New lower timing cover gaskets and a new crankshaft front oil seal will be required on installation. RTV sealant will be required to coat the cylinder head/cylinder block joint - see text.*

### Removal

*Refer to illustrations 6.13a, 6.13b, 6.13c, 6.13d, 6.16, 6.18, 6.19a and 6.19b*

7  Drain the cooling system (see Chapter 3).
8  Remove the oil pan (see Section 12).
9  Remove the valve cover (see Section 4).
10  Disconnect the two coolant hoses from the thermostat housing, and disconnect the coolant hose from the rear of the timing chain cover, behind the oil filter assembly.
11  Unbolt the thermostat housing from the front of the engine, and recover the gasket. Lift out the thermostat.
12  Unscrew the securing bolt, and withdraw the crankshaft position sensor from the front of the engine. Move the sensor to one side, clear of the working area.
13  Remove the drivebelt tensioner as follows **(see illustrations)**.
  a) Pull the cover from the idler pulley, then unscrew the pulley securing bolt, and remove the pulley.
  b) Pull the cover from the upper securing bolt, then unscrew the three securing bolts, and remove the tensioner assembly, complete with the hydraulic tensioner strut. Note that the tensioner strut is filled with oil, and must therefore be stored upright to avoid the oil draining.

14  The coolant pump pulley must now be removed. Counterhold the pulley by wrapping an old drivebelt around it and clamping tightly, then unscrew the securing bolts and

6.13c Pull the cover from the upper securing bolt

6.13d Removing the drivebelt tensioner

6.16 Driving a locating dowel from the lower timing chain cover

6.18 Remove the lower timing chain cover-to-cylinder head bolts. Note that this bolt secures the secondary timing chain guide - viewed with secondary timing chain removed for clarity

withdraw the pulley.
15  Remove the crankshaft damper/pulley and pulley hub (see Section 5).
16  Working at the top of the timing chain cover, drive out the two cover dowels. Drive out the dowels towards the rear of the engine, using a pin-punch (less than 3/16- inch diameter) **(see illustration)**.
17  On models with VANOS, it is now necessary to remove the VANOS adjustment unit (see Section 9), for access to the lower timing chain cover-to-cylinder head bolts.
18  Unscrew the three lower timing chain cover-to-cylinder head bolts, and lift the bolts from the cylinder head. Note that one of the bolts also secures the secondary timing chain guide **(see illustration)**.
19  Unscrew the lower timing chain cover-to-cylinder block bolts, then withdraw the cover from the front of the engine **(see illustrations)**. Recover the gaskets.

6.19a Unscrew the securing bolts (arrows)...

6.19b ... and remove the lower timing chain cover - viewed with cylinder head removed

# Chapter 2 Part B  Six-cylinder engines

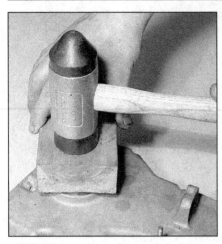

6.22  Fitting a new oil seal to the lower timing chain cover

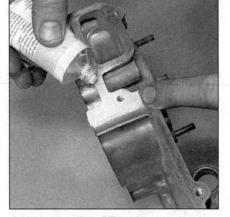

6.25  Apply a little RTV sealant to the area where the cylinder head gasket contacts the lower timing chain cover

7.3a  Using a wrench . . .

## Installation

*Refer to illustrations 6.22 and 6.25*

20  Commence installation by levering out the oil seal from the timing chain cover.
21  Thoroughly clean the mating faces of the cover, cylinder block and cylinder head.
22  Fit a new oil seal to the timing chain cover, using a large socket or tube, or a block of wood to drive the seal into position **(see illustration)**.
23  Drive the cover dowels into position in the top of the cover so that they protrude from the rear (cylinder block mating) face of the cover by approximately 1/8-inch.
24  Position new gaskets on the cover, and hold them in position using a little grease.
25  Apply a little RTV sealant to the cylinder head/cylinder block joint at the two points where the timing chain cover contacts the cylinder head gasket **(see illustration)**.
26  Offer the cover into position, ensuring that the gaskets stay in place. Make sure that the dowels engage with the cylinder block, and fit the cover securing bolts. Tighten the bolts finger-tight only at this stage.

27  Drive in the cover dowels until they are flush with the outer face of the cover.
28  Progressively tighten the cover securing bolts to the specified torque (do not forget the three cover-to-cylinder head bolts).
29  Where applicable, install the VANOS adjustment unit (see Section 9).
30  Install the crankshaft damper/pulley hub and damper/pulley (see Section 5).
31  The remainder of the installation procedure is a reversal of removal, bearing in mind the following points.

a) *Ensure that the drivebelt hydraulic tensioner strut is fitted correctly. The "TOP/OBEN" arrow must point upwards.*
b) *Install the drivebelt with reference to Chapter 1.*
c) *Install the thermostat and housing with reference to Chapter 3.*
d) *Install the valve cover (see Section 4).*
e) *Install the oil pan (see Section 12).*
f) *On completion, refill the cooling system and check the coolant level, (see Chapter 1 and "Weekly Checks" respectively).*

## 7  Timing chains - removal, inspection and installation

*Refer to illustrations 7.3a, 7.3b, 7.5, 7.6, 7.7 and 7.8*

### Secondary (exhaust-to-intake camshaft) chain

#### 1992 models

**Removal**

1  Position No. 1 piston at TDC, and lock the flywheel in position, (see Section 3).
2  Remove the upper timing chain cover, (see Section 6).
3  Unscrew the primary timing chain tensioner plunger cover plug from the right-hand side of the engine. Recover the sealing ring **(see illustrations)**. **Warning:** *The chain tensioner plunger has a strong spring. Take care when unscrewing the cover plug.*
4  Recover the spring and withdraw the tensioner plunger.
5  Press the secondary timing chain tensioner pad down, and lock it in position using

7.3b  . . . unscrew the primary timing chain tensioner assembly

7.5  Lock the secondary chain tensioner pad down using a length of rod (arrow)

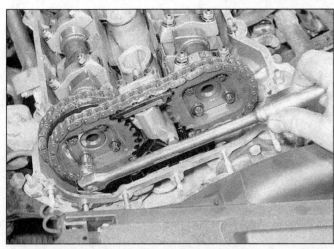

**7.6 Unscrew the bolts securing the secondary chain sprockets to the camshafts**

**7.7 Withdraw the sprockets complete with the secondary timing chain**

a tool made up from a length of welding rod or similar material. Insert the tool through the holes in the top of the tensioner to hold the tensioner plunger down **(see illustration)**.

6   Unscrew the bolts securing the chain sprockets to the camshafts **(see illustration)**. Take care not to move the camshafts - if necessary, the camshafts can be counterheld using a 24 mm wrench on the flats provided between No. 10 and 11 cam lobes.

7   Withdraw the sprockets, complete with the secondary chain, from the front of the camshafts **(see illustration)**.

8   Recover the camshaft position sensor plate from the front of the intake camshaft **(see illustration)**.

### Inspection

9   The chain should be replaced if the sprockets are worn or if the chain is worn (indicated by excessive lateral play between the links, and excessive noise in operation). It is wise to replace the chain in any case if the engine is disassembled for overhaul. Note that the rollers on a very badly worn chain may be slightly grooved. To avoid future problems, if there is any doubt at all about the condition of the chain, replace it.

10   Examine the teeth on the sprockets for wear. Each tooth forms an inverted "V". If worn, the side of each tooth under tension will be slightly concave in shape when compared with the other side of the tooth (i.e., the teeth will have a hooked appearance). If the teeth appear worn, the sprockets must be replaced. Also check the chain guide and tensioner contact surfaces for wear, and replace any worn components as necessary.

### Installation

*Refer to illustration 7.14*

11   Ensure that No. 1 piston is still positioned at TDC, with the crankshaft locked in position. Check the position of the camshafts using the template described in Section 3.

12   Where applicable, fit the camshaft position sensor plate to the intake camshaft.

13   Lay the chain over the sprockets, ensuring that the arrows on the front faces of the sprockets are pointing upwards (note that some sprockets have two arrows, which must point vertically up and down), then position the sprockets on the camshafts.

14   Install the sprocket securing bolts and tighten them finger-tight only at this stage **(see illustration)**.

15   Install the primary timing chain tensioner plunger, ensuring that the guide lugs engage with the tensioner rail.

16   Fit the tensioner spring, then fit the cover plug, using a new seal, and tighten the plug to the specified torque.

17   Remove the tool locking the secondary timing chain tensioner in position.

18   Again, check the position of the camshafts using the template, then tighten the intake camshaft sprocket securing bolts to the specified torque in the two stages given in the Specifications.

19   Similarly, tighten the exhaust camshaft sprocket bolts to the specified torque.

20   Withdraw the flywheel locking rod from the timing hole in the cylinder block, and where applicable remove the template from the camshafts.

21   Rotate the engine through two complete revolutions clockwise (note that the engine is easier to turn with the spark plugs removed),

**7.8 Recover the camshaft position sensor plate from the inlet camshaft**

**7.14 Install the sprocket securing bolts. Note that the arrows on the sprockets must be positioned vertically**

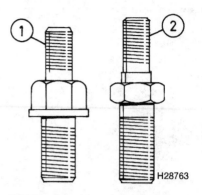

7.31 Do not mix up the valve cover studs (1) with the camshaft position sensor plate securing studs (2)

7.40a Lift out the plunger and spring . . .

7.40b . . . then unscrew the securing bolts . . .

7.40c . . . and withdraw the secondary chain tensioner

7.41a Unscrew the securing bolts . . .

then install the locking rod to the timing hole in the cylinder block, ensuring that the tool engages with the flywheel, and check that the template can be refitted to the camshafts without turning the camshafts (the timing arrows on the camshaft sprockets should be pointing vertically upwards). If not, the secondary timing chain and/or sprockets have been incorrectly installed.

22  Remove the camshaft template from the cylinder head, and remove the locking rod from the flywheel (install the blanking plug to the timing hole).

23  Install the upper timing chain cover (see Section 6).

### 1993 and later models

#### Removal

24  Remove the VANOS adjustment unit (see Section 9).

25  Remove the exhaust camshaft sprocket securing bolts (the bolts should already have been loosened), and withdraw the thrust plate from the sprocket.

26  Unscrew the intake camshaft sprocket securing nuts, and withdraw the thrust plate from the front of the sprocket.

27  Withdraw the sprockets, complete with the secondary chain, from the front of the camshafts.

28  If desired, the camshaft position sensor plate can be removed from the front of the intake camshaft as follows.
 a) Unscrew the three securing studs from the camshaft flange.
 b) Withdraw the thrust plate.
 c) Withdraw the sensor plate.

#### Inspection

29  Refer to paragraphs 9 and 10.

#### Installation

*Refer to illustration 7.31*

30  Ensure that No. 1 piston is still positioned at TDC, with the crankshaft locked in position. Check the position of the camshafts using the template.

31  Where applicable, install the camshaft position sensor plate to the intake camshaft. Ensure that the thrust plate is refitted, and make sure that the correct studs are used to secure the sensor plate and thrust plate. Tighten the studs to the specified torque.
**Caution:** *It is possible to mix up the valve cover studs and the camshaft position sensor plate securing studs. The camshaft position sensor plate studs are longer, and have a narrower "hexagon" section* **(see illustration)**.

32  Lay the chain over the sprockets, noting that when the sprockets are refitted, the securing bolt holes/studs on the camshafts must be centered in the elongated holes in the sprockets. Note that the intake camshaft sprocket fits with the flat side facing the VANOS adjustment unit, and the raised collar facing the camshaft.

33  Fit the sprockets to the camshafts, ensuring that the securing bolts holes/studs are aligned in the center of the elongated sprocket holes.

34  Fit the thrust plate to the intake camshaft sprocket, then fit the sprocket securing nuts, and tighten the nuts to the specified torque.

35  Fit the thrust plate to the exhaust camshaft sprocket, then fit the sprocket securing bolts. Tighten the bolts finger-tight only at this stage.

36  Install the VANOS adjustment unit (see Section 9).

### Primary (crankshaft-to-exhaust camshaft) chain

#### 1992 models

#### Removal

*Refer to illustrations 7.40a, 7.40b, 7.40c, 7.41a, 7.41b, 7.42, 7.44, 7.45 and 7.46*

37  Remove the secondary timing chain as described previously in this Section.

38  Remove the tool locking the secondary

# Chapter 2 Part B Six-cylinder engines

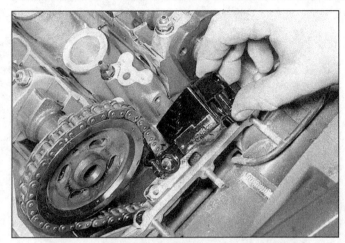

7.41b ... and withdraw the secondary chain guide

7.42 Withdraw the primary timing chain sprocket complete with the chain

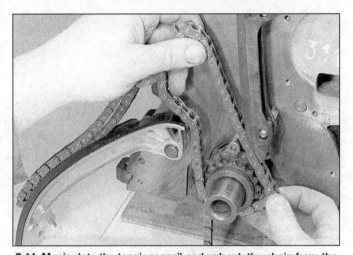

7.44 Manipulate the tensioner rail and unhook the chain from the crankshaft sprocket - viewed with engine removed

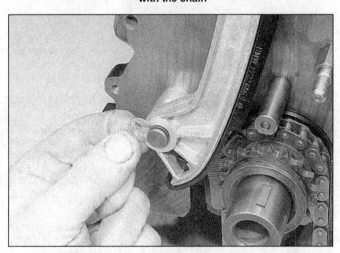

7.45 Remove the clip from the lower pivot to remove the tensioner rail - viewed with engine removed

timing chain tensioner in position.
39  Remove the lower timing chain cover (see Section 6).
40  Remove the tool locking the secondary chain tensioner plunger in position, then lift out the plunger and spring, unscrew the securing bolts, and withdraw the secondary chain tensioner from the cylinder head (see illustrations).
41  Unscrew the securing bolts and withdraw the secondary chain guide (see illustrations).
42  Withdraw the primary timing chain sprocket from the exhaust camshaft, complete with the chain (see illustration). Remove the sprocket. Note which way the sprocket faces to ensure correct installation.
43  Note the routing of the chain in relation to the tensioner rail and the chain guide.
44  Manipulate the tensioner rail as necessary to enable the chain to be unhooked from the crankshaft sprocket and lifted from the engine (see illustration). **Warning:** *Once the primary timing chain has been removed, do not turn the crankshaft or the camshafts, as there is a danger of the valves hitting the pistons.*

45  If desired, the tensioner rail can now be removed after removing the clip from the lower pivot (see illustration).
46  Similarly, the chain guide can be removed after releasing the upper and lower retaining clips. Take care when releasing the retaining clips, as the clips are easily broken (see illustration).

### Inspection
47  Refer to paragraphs 9 and 10.

### Installation
*Refer to illustration 7.51*
48  Ensure that No. 1 piston is still positioned at TDC, with the crankshaft locked in position. Check the position of the camshafts using the template.
49  Commence installation by engaging the chain with the crankshaft sprocket.
50  Where applicable, install the chain guide and the tensioner rail, ensuring that the chain is correctly routed in relation to the guide and tensioner rail, as noted before removal. Take care when installing the chain guide, as the clips are easily broken.
51  Manipulate the exhaust camshaft primary chain sprocket until the timing arrow on the sprocket is pointing vertically upwards, then engage the chain with the sprocket. Fit the sprocket to the exhaust camshaft, aligning the sprocket so that the tapped holes in the camshaft flange are positioned at the left-hand ends of the elongated slots in the

7.46 Release the retaining clips to remove the chain guide

7.51 The tapped holes in the camshaft flange should be positioned at the left-hand ends of the elongated slots in the sprocket

8.12a Withdraw the secondary chain tensioner plunger...

sprocket **(see illustration)**. Ensure that the sprocket is facing the correct way as noted before removal.
52  Install the secondary chain guide and the secondary chain tensioner. Note that the tensioner plunger fits with the cut-out in the plunger positioned on the right-hand side of the engine.
53  Install the secondary timing chain as described previously in this Section, but do not install the valve cover until the lower timing chain cover has been installed.
54  Install the lower timing chain cover (see Section 6).

### 1993 and later models

**Note:** *BMW special tool No. 11 3 390 or a suitable equivalent will be required to carry out this operation.*

### Removal

55  Remove the secondary timing chain as described previously in this Section.
56  Unscrew the timing chain tensioner plunger cover plug from the right-hand side of the engine. Recover the sealing ring. **Warning:** *The chain tensioner plunger has a strong spring. Take care when unscrewing the cover plug.*
57  Recover the spring and withdraw the tensioner plunger.
58  Follow the procedure outlined in steps 38 through 46 of this Section.

### Inspection

59  Refer to paragraphs 9 and 10.

### Installation

60  Follow the procedure outlined in steps 48 through 52 of this Section.
61  Fit special tool No. 11 3 390 into the tensioner aperture (see Section 9), then turn the adjuster screw on the tool until the end of the screw just touches the tensioning rail. Note that the exhaust camshaft sprocket should now have moved counterclockwise so that the tapped holes in the camshaft flange are centered in the elongated holes in the sprocket.
62  Install the secondary timing chain as described previously in this Section.

## 8  Timing chain sprockets and tensioners - removal and installation

### Camshaft sprockets

1  Removal, inspection and installation of the sprockets is described as part of the secondary timing chain removal and installation procedure in Section 7.

### Crankshaft sprocket

#### Removal

2  The sprocket is combined with the oil pump drive sprocket. On some engines, the sprocket may be a press-fit on the end of the crankshaft.
3  Remove the primary timing chain (see Section 7).
4  Slide the sprocket from the front of the crankshaft. If the sprocket is a press-fit, use a three-legged puller to pull the sprocket from the crankshaft. Protect the threaded bore in the front of the crankshaft by installing the pulley hub bolt, or by using a metal spacer between the puller and the end of the crankshaft. Note which way round the sprocket faces to ensure correct installation.
5  Once the sprocket has been removed, recover the Woodruff key from the slot in the crankshaft if it is loose.

#### Inspection

6  Inspection is described along with the timing chain inspection procedure in Section 7.

#### Installation

7  Where applicable, install the Woodruff key to the slot in the crankshaft.
8  Slide the sprocket into position on the crankshaft. Ensure that the sprocket is fitted the correct way round as noted before removal. If a press-fit sprocket is to be installed, before installing, the sprocket must be heated to a temperature of 300-degrees F. Do not exceed this temperature, as damage

8.12b ...spring...

to the sprocket may result.
9  Once the sprocket has been heated to the specified temperature, align the slot in the sprocket with the Woodruff key, then tap the sprocket into position using a socket or metal tube. **Warning:** *When the sprocket is heated, take precautions against burns - the metal will stay hot for some time.*
10  Install the primary timing chain (see Section 7).

### Secondary chain tensioner

#### Removal

*Refer to illustrations 8.12a, 8.12b, 8.12c and 8.13*

11  Remove the secondary timing chain (see Section 7).
12  Remove the tool locking the secondary timing chain tensioner in position, then withdraw the plunger, spring and plunger housing **(see illustrations)**.
13  Unscrew the securing bolts and withdraw the chain tensioner housing from the cylinder head **(see illustration)**.

8.12c ... and plunger housing

8.13 Withdrawing the secondary timing chain tensioner housing

### Inspection

14 Inspect the tensioner, and replace if necessary. Check the plunger and the plunger housing for wear and damage. Inspect the chain contact face of the plunger slipper for wear, and check the condition of the spring. Replace any components which are worn or damaged.

15 When installing the plunger to the tensioner, note that the cut-out in the plunger should be positioned on the right-hand side of the engine when the assembly is refitted.

### Installation

16 Install the chain tensioner and tighten the securing bolts to the specified torque.

17 Install the tool to lock the tensioner in position.

18 Install the secondary timing chain (see Section 7).

### Primary chain tensioner

19 Removal and installation is described as part of the primary timing chain removal procedure in Section 7.

## 9  Variable valve timing system (VANOS) components - removal, inspection and installation

*Refer to illustrations 9.4, 9.10 and 9.13*

### VANOS adjustment unit

**Note:** *Before the installation procedure can be completed, the operation of the VANOS system and the intake camshaft position adjustment must be checked by a BMW dealer or other qualified repair shop. BMW special tool No. 11 3 390 or a suitable equivalent will be required to carry out this operation. A new VANOS unit gasket will be required on installation.*

### Removal

1 Unscrew the securing bolts and/or nuts, and remove the alternator air ducting from the front of the vehicle.

2 Remove the viscous cooling fan and fan cowl assembly (see Chapter 3).

3 Remove the valve cover (see Section 4).

4 Unscrew the union bolt, and disconnect the oil feed pipe from the front of the VANOS adjustment unit **(see illustration)**. Recover the sealing rings.

5 Disconnect the solenoid valve wiring connector. The connector is clipped to the wiring harness located behind the oil filter housing.

6 Unscrew the securing nut and bolt, and remove the engine lifting bracket from the front of the engine.

7 Release the securing clips, and remove the wiring ducting from the front of the VANOS adjustment unit.

8 Unclip the plastic cover from the intake camshaft.

9 Position No. 1 piston at TDC, and lock the flywheel in position, and check the position of the camshafts using the template, (see Section 3).

10 Unscrew the two cover plugs from the front of the VANOS adjustment unit to expose the lower exhaust camshaft sprocket securing bolts **(see illustration)**. Recover the sealing rings.

11 Fully loosen the four exhaust camshaft sprocket securing bolts. The two lower bolts

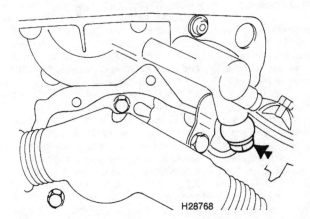

9.4 VANOS adjustment unit oil feed pipe union bolt (arrow)

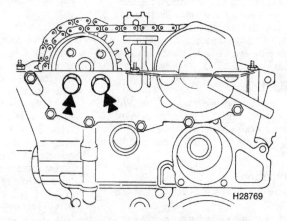

9.10 Unscrew the two cover plugs (arrows) to expose the camshaft sprocket securing bolts

# Chapter 2 Part B Six-cylinder engines

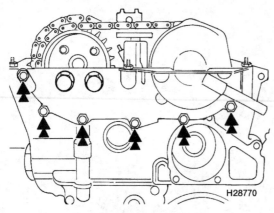

9.13 Unscrew the securing nuts (arrows) and remove the VANOS adjustment unit

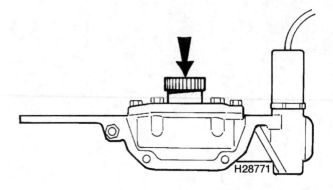

9.15 Press the splined shaft into the VANOS adjustment unit as far as the stop

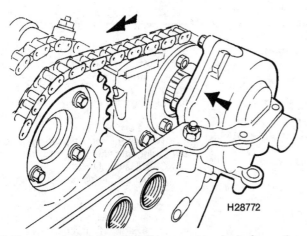

9.21 The sprockets and chain will turn counterclockwise as the VANOS adjustment unit is fitted

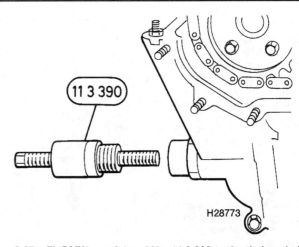

9.27a Fit BMW special tool No. 11 3 390 to the timing chain tensioner aperture . . .

are reached through the holes in the front of the VANOS adjustment unit.

12 Press the secondary timing chain tensioner pad down, and lock it in position using a tool made up from a length of welding rod or similar material. Insert the tool through the holes in the top of the tensioner to hold the tensioner plunger down **(see illustration 7.5)**.

13 Unscrew the securing nuts and remove the VANOS adjustment unit from the front of the engine **(see illustration)**. Recover the gasket.

### Inspection

14 To test the operation of the VANOS adjustment unit, special equipment is required. Testing must therefore be entrusted to a BMW dealer or other qualified repair shop.

### Installation

*Refer to illustrations 9.15, 9.21, 9.27a and 9.27b*

15 Commence installation by pressing the splined shaft into the VANOS adjustment unit until the shaft reaches the stop in the housing **(see illustration)**.

16 Make sure that the dowel sleeves are in position on the top VANOS adjustment unit securing studs in the cylinder head.

17 Fit a new gasket over the studs on the cylinder head, and apply a little sealant to the corners of the joint surfaces between the cylinder head and the VANOS adjustment unit.

18 Turn the intake camshaft sprocket clockwise as far as possible by hand (the camshafts should be locked in position by fitting the template described in Section 3, so turn the sprocket until it reaches the clockwise stop).

19 Offer the VANOS adjustment unit into position and, if necessary, rotate the splined shaft on the VANOS adjustment unit slightly until the internal splines on the VANOS adjustment unit shaft engage with the splines on the camshaft. Do not turn the camshaft or the camshaft sprocket.

20 It is now necessary to engage the VANOS adjustment unit shaft outer splines with the internal splines on the camshaft sprocket. Turn the camshaft sprocket slowly by hand counterclockwise until the VANOS adjustment unit shaft splines mesh with the sprocket. **Warning:** *It is essential to ensure that that the FIRST suitable spline meshes when the sprocket is turned back counterclockwise from its clockwise stop.*

21 Push the VANOS adjustment unit fully onto the cylinder head studs, noting that the camshaft sprockets and chain will turn counterclockwise slightly as the VANOS adjustment unit is pushed into position (this is due to the helical sprocket splines). As the unit is pushed into position, guide the sprockets and chain counterclockwise as necessary by hand **(see illustration)**.

22 Install and tighten the VANOS adjustment unit securing nuts.

23 Remove the tool locking the secondary timing chain tensioner in position.

24 Unscrew the primary timing chain tensioner plunger cover plug from the right-hand side of the engine **(see illustrations 7.3a and 7.3b)**. Recover the sealing ring. **Warning:** *The chain tensioner plunger has a strong spring. Take care when unscrewing the cover plug.*

25 Recover the spring and withdraw the tensioner plunger.

26 Fit special tool No. 11 3 390 into the tensioner aperture, then turn the adjuster screw

# Chapter 2 Part B Six-cylinder engines

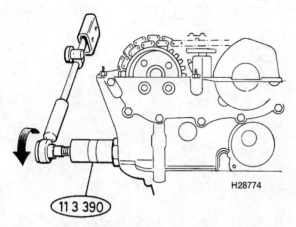

9.27b ... then apply the specified torque (see text) to the tool

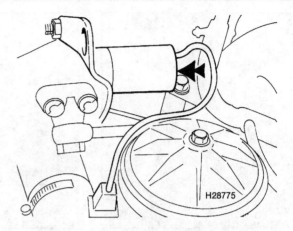

9.41 VANOS solenoid valve location (arrow)

on the tool until the end of the screw just touches the tensioner rail.
27  Using a torque wrench, apply a torque of 12 in-lbs to the adjusting screw on the special tool **(see illustrations)**.
28  Tighten the exhaust camshaft sprocket securing bolts to the specified torque.
29  Remove the template from the camshafts, then withdraw the locking rod from the timing hole in the cylinder block.
30  Rotate the engine through two complete revolutions clockwise, then install the locking rod to the timing hole in the cylinder block, ensuring that the tool engages with the flywheel.
31  Install the template to check the position of the camshafts. If the template cannot be fitted with the flywheel locked in position, the VANOS adjustment unit has been incorrectly installed.
32  Unscrew the special tool (No. 11 3 390) from the tensioner aperture.
33  Install the primary timing chain tensioner plunger, ensuring that the guide lugs engage with the tensioner rail.
34  Fit the tensioner spring, then fit the cover plug, using a new seal, and tighten the plug to the specified torque.
35  Install the camshaft sprocket securing bolt cover plugs to the front of the VANOS adjustment unit, using new sealing rings. Tighten the plugs to the specified torque.
36  It is now necessary to check the operation of the VANOS system and to adjust the intake camshaft position. Special tools are required to do this, and the operation must be entrusted to a BMW dealer or other qualified repair shop.
37  Once the VANOS system has been checked and adjusted, remove the camshaft template and flywheel locking tool. The remainder of the installation procedure is a reversal of removal, bearing in mind the following points.

a) Use new sealing rings when reconnecting the oil feed pipe to the VANOS adjustment unit.
b) Install the valve cover with reference to Section 4.

c) Install the viscous cooling fan and cowl assembly (see Chapter 3).

## VANOS solenoid valve

**Note:** *A new sealing ring will be required on installation.*

### Removal

*Refer to illustration 9.41*

38  Ensure that the ignition is switched off.
39  Unscrew the securing bolts and/or nuts, and remove the alternator air ducting from the front of the vehicle.
40  Disconnect the solenoid valve wiring connector, which is clipped to the engine wiring harness behind the oil filter assembly.
41  Using an open-ended wrench, unscrew the solenoid valve and recover the seal **(see illustration)**.

### Inspection

42  Check that the solenoid plunger can be pulled freely back and forth by hand. If not, the solenoid must be renewed.
43  Similarly, check that the hydraulic piston in the VANOS adjustment unit can be moved easily. If it is difficult to move the hydraulic piston, the complete VANOS adjustment unit must be renewed.

### Installation

44  Installation is a reversal of removal, but use a new sealing ring.

## 10  Camshafts and followers - removal, inspection and installation

*Refer to illustrations 10.4a, 10.4b, 10.7, 10.12, 10.13, 10.14 and 10.19*

**Warning:** *BMW special tools 11 3 260 and 11 3 270 will be required for this operation. These tools are extremely difficult to improvise due to their rugged construction and the need for accurate manufacture. Do not attempt to remove and install the camshafts without the aid of the special tools, as expensive damage to the camshafts and/or bearings may result.*

### 1992 models

#### Removal

1  Open the hood, then raise the hood to its fully open position (see Chapter 11).
2  Unscrew the securing bolts and/or nuts, and remove the alternator air ducting from the front of the vehicle.
3  Remove the secondary timing chain (see Section 7).
4  Trace the wiring back from the camshaft position sensor, then disconnect the sensor connector. Unscrew the securing bolt, and remove the sensor from the cylinder head **(see illustrations)**.

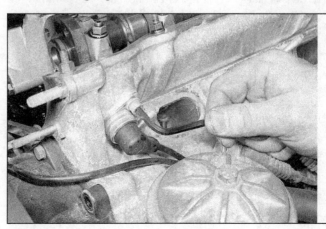

10.4a Unscrew the securing bolt . . .

10.4b ... and withdraw the camshaft position sensor

10.7 Withdraw the secondary chain guide

10.12 Unscrew the four camshaft cover securing studs

10.13 BMW special tools 11 3 260 and 11 3 270 fitted to cylinder head

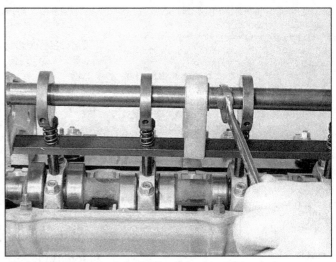

10.14 Turn the eccentric shaft using a wrench to apply pressure to the bearing caps

5   Remove the tool locking the secondary timing chain tensioner in position. Withdraw the plunger and spring.
6   Unscrew the securing bolts, and withdraw the secondary chain tensioner from the cylinder head.
7   Unscrew the securing bolts and withdraw the secondary chain guide **(see illustration)**.
8   Withdraw the primary timing chain sprocket from the exhaust camshaft, complete with the chain. Remove the sprocket. **Caution:** *Keep tension on the chain, and tie up the end of the chain using wire or string to prevent it from dropping into the lower timing chain cover and/or disengaging from the crankshaft sprocket.* **Warning:** *To avoid any possibility of piston-to-valve contact when installing the camshaft(s), it is necessary to ensure that none of the pistons are at TDC. Before proceeding further, remove the locking rod from the timing hole in the cylinder block, then turn the crankshaft approximately 30-degrees clockwise using a wrench or socket on the crankshaft pulley hub bolt.*

9   Remove the template from the camshafts.
10  Unscrew the spark plugs from the cylinder head.
11  Check the camshaft bearing caps for identification marks. The caps are numbered from the timing chain end of the engine, and the marks can normally be read from the exhaust side of the engine. The exhaust camshaft bearing caps are marked "A1" to "A7", and the intake camshaft caps are marked "E1" to "E7".
12  Unscrew the four camshaft cover securing studs from the center of the cylinder head **(see illustration)**.
13  Assemble BMW special tools 11 3 260 and 11 3 270, and mount the tools on the cylinder head by screwing the mounting bolts into the spark plug holes. Position the tools so that the plungers are located over the relevant camshaft bearing caps (i.e., intake or exhaust camshaft) **(see illustration)**.
14  Apply pressure to the camshaft bearing caps by turning the eccentric shaft on the tools using a wrench **(see illustration)**.

15  Unscrew the camshaft bearing cap nuts. **Warning:** *Do not attempt to unscrew the camshaft bearing cap nuts without the special tools in place, as damage to the camshaft and/or bearings may result.*

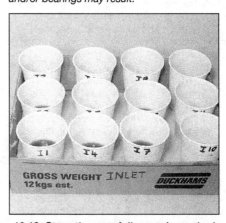

10.19 Store the cam followers in marked cups of clean engine oil

# Chapter 2 Part B Six-cylinder engines

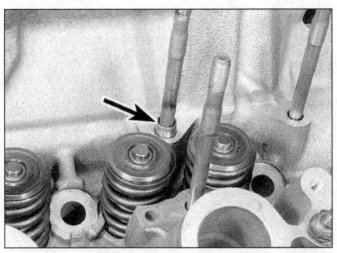

10.22 Bearing casting locating dowel (arrow) on cylinder head stud at No. 2 bearing location

10.26a Apply a small amount of silicone or a similar adhesive compound to the top of each can follower. The adhesive should protrude beyond the lower bearing surface of the casting. Do not use excessive amounts of adhesive, as there is a risk of contaminating the oil passages in the bearing castings

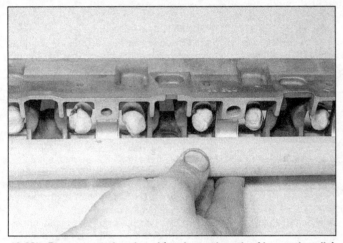

10.26b Press a wooden dowel (such as a length of broom handle) onto the top surface of the adhesive pads so that the pads stick to the dowel, holding the cam followers in the bearing casting. Ensure that no adhesive is pressed out between the surfaces of the cam followers and bearing casting

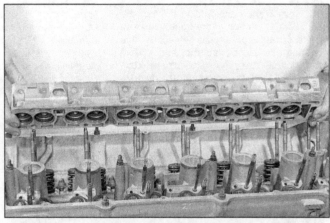

10.26c Leave the adhesive pads and the dowel in position until the assembly has been refitted to the cylinder head, then remove the dowel, and carefully remove the adhesive from each follower. It is *essential* to ensure that no trace of adhesive is left on the followers or on any of the engine components - serious engine damage could result if the oil passages become contaminated

16  Release the pressure on the special tool shaft, then unbolt the tools from the cylinder head.
17  Lift off the bearing caps, keeping them in order, then lift out the camshaft.
18  The camshaft bearing casting can now be lifted from the cylinder head. This should be done very slowly, as the cam followers will be released as the casting is lifted off - if the casting is lifted off awkwardly, the cam followers may fall out. Do not allow the cam followers to fall out and get mixed up, as they must be fitted to their original locations.
19  With the bearing casting removed, lift the cam followers from the cylinder head. Identify the followers for location, and store them upright in a container of clean engine oil to prevent the oil from draining from inside the followers **(see illustration)**. Do not forget to mark the cam followers "Intake" and "Exhaust".
20  Repeat the procedure on the remaining camshaft.

### Inspection
21  Clean all the components, including the bearing surfaces in the bearing castings and bearing caps. Examine the components carefully for wear and damage. In particular, check the bearing and cam lobe surfaces of the camshaft(s) for scoring and pitting. Examine the surfaces of the cam followers for signs wear or damage. Replace components as necessary.

### Installation
*Refer to illustrations 10.22, 10.26a, 10.26b and 10.26c*
22  If the camshaft lower bearing castings have been removed, check that the mating faces of the bearing castings and the cylinder head are clean, and check that the bearing casting locating dowels are in position on the studs at No. 2 and 7 bearing locations **(see illustration)**.
23  The bearing casting(s) and cam followers must now be refitted.

24  The simplest method of installing these components is to retain the cam followers in the bearing casting, and install the components as an assembly.
25  Oil the bearing casting contact surfaces of the cam followers (avoid allowing oil onto the top faces of the followers at this stage), then fit each follower to its original location in the bearing casting.
26  Once all the followers have been fitted, they must be retained in the bearing casting, so that they do not fall out as the assembly is refitted to the cylinder head **(see illustrations)**.
27  With the cam followers retained in the bearing casting, install the casting to the cylinder head. Note that the exhaust side casting is marked "A" and the intake side casting is marked "E". When the castings are refitted, the marks should face each other at the timing chain end of the cylinder head.
**Warning:** *The cam followers expand when not subjected to load by the camshafts, and*

therefore require some time before they can be compressed. If the camshaft installation operation is carried out rapidly, there is a possibility that the "closed" valves will be forced open by the expanded cam followers, resulting in piston-to-valve contact. To minimize this possibility, after installing the camshaft(s) observe the following delays before turning the crankshaft back to the TDC position:

| Temperature | Delay |
|---|---|
| Room temperature (68°F) | 4 minutes |
| 50°F to 68°F | 11 minutes |
| 32°F to 50°F | 30 minutes |

28  First identify the camshafts to ensure that they are fitted in the correct locations. The intake camshaft has a small cut-out in the top of the front flange, and the exhaust camshaft has a plain front flange.

29  Ensure that the crankshaft is still positioned 30-degrees clockwise from the TDC position (see **Warning** at the end of paragraph 8).

30  Position the camshaft on the cylinder head, so that the tips of the front cam lobes on the exhaust and intake camshafts face one another. Note also that the square flanges on the rear of the camshaft should be positioned with the sides of the flanges exactly at right-angles to the top surface of the cylinder head (this can be checked using a set-square), and the side of the flange with holes drilled into it uppermost.

31  Place the bearing caps in position, noting that the caps carry identification marks. The exhaust camshaft caps are marked "A1" to "A7", and the intake camshaft caps are marked "E1" to "E7". Place the bearing caps in their original locations as noted before removal.

32  Re-assemble BMW special tools 11 3 260 and 11 3 270, and install them to the cylinder head as during removal. **Warning:** *Again, do not attempt to install the camshafts without the aid of the special tools.*

33  Apply pressure to the relevant bearing caps by turning the eccentric shaft on the tools using a wrench.

34  With pressure applied to the bearing caps, install the bearing cap securing nuts, and tighten them as far as possible by hand.

35  Tighten the bearing cap nuts to the specified torque, working progressively in a diagonal sequence.

36  Once the bearing cap nuts have been tightened, unbolt the tools used to apply pressure to the bearing caps.

37  Repeat the procedure on the remaining camshaft.

38  Install the spark plugs, and install the camshaft cover securing studs to the cylinder head.

39  Install the template used to check the position of the camshafts. If necessary, turn the camshaft(s) slightly using a wrench on the flats provided until the template can be fitted. **Warning:** *Note the precaution given in the warning at the end of paragraph 27 before proceeding.*

40  Turn the crankshaft back 30-degrees counterclockwise to the TDC position, then re-engage the locking rod with the flywheel to lock the crankshaft in position.

41  Manipulate the exhaust camshaft primary chain sprocket until the timing arrow on the sprocket is pointing vertically upwards, then engage the chain with the sprocket.

42  Fit the sprocket to the exhaust camshaft, aligning the sprocket so that the tapped holes in the camshaft flange are positioned at the left-hand ends of the elongated slots in the sprocket.

43  Install the secondary chain guide and tighten the securing bolts.

44  Install the secondary chain tensioner and tighten the securing bolts.

45  Temporarily install the tool to lock the secondary chain tensioner in position, then install the secondary timing chain (see Section 7).

46  Install the camshaft position sensor and reconnect the wiring plug.

47  Install the alternator air ducting, then lower the hood. **Warning:** *Follow the procedure outlined in the warning at the end of paragraph 27, the cam followers expand when not subjected to load by the camshafts To minimize the possibility of piston-to-valve contact, after installing the camshaft(s), observe the following delays before cranking the engine:*

| Temperature | Delay |
|---|---|
| Room temperature (68°F) | 10 minutes |
| 50°F to 68°F | 30 minutes |
| 32°F to 50°F | 75 minutes |

## 1993 and later models

**Note:** *BMW special tool No. 11 3 390 or a suitable equivalent will be required to carry out this operation.*

### Removal

48  Remove the secondary timing chain, and then remove the camshaft position sensor plate, (see Section 7).

49  Unscrew the primary timing chain tensioner plunger cover plug from the right-hand side of the engine. Recover the sealing ring. **Warning:** *The chain tensioner plunger has a strong spring. Take care when unscrewing the cover plug.*

50  Recover the spring and withdraw the tensioner plunger.

51  Follow the procedure outlined in steps 5 through 8 of this Section. **Caution:** *Keep tension on the primary timing chain, and tie up the end of the chain using wire or string to prevent it from dropping into the lower timing chain cover and/or disengaging from the crankshaft sprocket.* **Warning:** *To avoid any possibility of piston-to-valve contact when installing the camshaft(s), it is necessary to ensure that none of the pistons are at TDC. Before proceeding further, remove the locking rod from the timing hole in the cylinder block, then turn the crankshaft approximately 30-degrees clockwise using a wrench or socket on the crankshaft pulley hub bolt.*

52  Follow the procedure outlined in steps 9 through 13.

53  Unscrew the securing nuts, and remove the No. 1 (timing chain end) intake camshaft bearing cap. Note that the bearing cap is located on dowels.

54  Apply pressure to the camshaft bearing caps by turning the eccentric shaft on the special tool.

55  Follow the procedure outlined in steps 15 through 20.

### Inspection

56  Follow the procedure outlined in step 21.

57  If necessary, the splined VANOS shaft on the front of the intake camshaft can be renewed. To do this, proceed as follows.

  a) *Clamp the camshaft carefully in a soft-jawed vise.*
  b) *Unscrew the splined shaft using a hexagon key.*
  c) *Fit the new splined shaft and tighten to the specified torque.*

### Installation

58  If the camshaft lower bearing castings have been removed, check that the mating faces of the bearing castings and the cylinder head are clean, and check that the bearing casting locating dowels are in position on the studs at No. 2 and 7 bearing locations.

59  Follow the procedure outlined in steps 23 through 27.

60  If both camshafts have been removed, first identify the camshafts to ensure that they are fitted in the correct locations. The intake camshaft has a triangular-shaped front flange, and the splined VANOS shaft screwed into the front (timing chain) end.

61  Ensure that the crankshaft is still positioned 30-degrees clockwise from the TDC position (see **Warning** at the end of paragraph 51).

62  Follow the procedure outlined in steps 30 through 44.

63  Fit special tool No. 11 3 390 into the primary timing chain tensioner aperture, then turn the adjuster screw on the tool until the end of the screw just touches the tensioning rail **(see illustration 9.27a)**. Note that the exhaust camshaft sprocket should now have moved counterclockwise so that the tapped holes in the camshaft flange are centered in the elongated holes in the sprocket.

64  Temporarily install the tool to lock the secondary chain tensioner in position, then install the camshaft position sensor plate and the secondary timing chain (see Section 7). **Warning:** *Follow the procedure outlined in the warning at the end of paragraph 27, the cam followers expand when not subjected to load by the camshafts To minimize the possibility of piston-to-valve contact, after installing the camshaft(s), observe the following delays before cranking the engine:*

| Temperature | Delay |
|---|---|
| Room temperature (68°F) | 10 minutes |
| 50°F to 68°F | 30 minutes |
| 32°F to 50°F | 75 minutes |

# Chapter 2 Part B Six-cylinder engines

11.10 Disconnect the two coolant hoses (arrows) from the thermostat housing

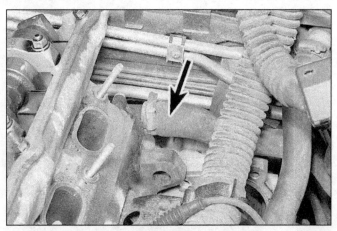

11.11 Disconnect the coolant hose (arrow) from the rear left-hand corner of the cylinder head

## 11 Cylinder head - removal and installation

*Refer to illustrations 11.10, 11.11, 11.15, 11.16 and 11.18*

**Warning:** *Wait until the engine is completely cool before beginning this procedure.*

### 1992 models

**Note:** *New cylinder head bolts and a new cylinder head gasket will be required on installation.*

### Removal

1 Drain the cooling system (see Chapter 1).
2 Remove the intake and exhaust manifolds (see Chapter 4B).
3 Remove the secondary timing chain (see Section 7).
4 Trace the wiring back from the camshaft position sensor, then disconnect the sensor connector. Unscrew the securing bolt, and remove the sensor from the cylinder head.
5 Remove the tool locking the secondary timing chain tensioner in position. Lift out the tensioner pad and spring (note which way the tensioner pad fits to ensure correct installation).
6 Unscrew the securing bolts, and withdraw the secondary chain tensioner from the cylinder head.
7 Unscrew the securing bolts and withdraw the secondary chain guide.
8 Withdraw the primary timing chain sprocket from the exhaust camshaft, complete with the chain. Remove the sprocket. Note which way round the sprocket is fitted to ensure correct installation. **Caution:** *Keep tension on the chain, and tie up the end of the chain using wire or string to prevent it from dropping into the lower timing chain cover and/or disengaging from the crankshaft sprocket.* **Warning:** *To avoid any possibility of piston-to-valve contact when installing the cylinder head, it is necessary to ensure that none of the pistons are at TDC. Before proceeding further, remove the locking rod from the timing hole in the cylinder block, then turn the crankshaft approximately 30-degrees clockwise using a wrench or socket on the crankshaft pulley hub bolt.*
9 Unscrew the bolts securing the lower timing chain cover to the cylinder head (note that one of the bolts also secures the secondary timing chain tensioner).
10 Disconnect the two coolant hoses from the thermostat cover at the front of the cylinder head **(see illustration)**.
11 Disconnect the coolant hose from the rear left-hand corner of cylinder head **(see illustration)**.
12 Disconnect the remaining small coolant hose from the left-hand side of the cylinder head.
13 Disconnect wiring plugs from the temperature sensors located in the left-hand side of the cylinder head.
14 Unscrew the securing bolt and remove the crankshaft position sensor from the front of the engine. Trace the wiring back from the sensor, then disconnect the wiring plug and remove the sensor. **Caution:** *Mark the wiring plug for identification, as it is possible to mix up the camshaft sensor and crankshaft sensor wiring plugs.*
15 Progressively loosen the cylinder head bolts, working in a spiral pattern from the outside of the head inwards **(see illustration)**.
16 Remove the cylinder head bolts, and recover the washers **(see illustration)**. Note that some of the washers may be captive in the cylinder head, in which case they cannot

11.15 Loosen the cylinder head bolts, working in a spiral pattern . . .

11.16 . . . then withdraw the bolts

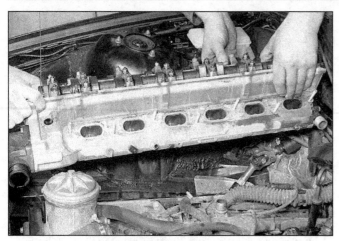

**11.18 Lifting off the cylinder head**

**11.28 Fit a new cylinder head gasket**

be withdrawn.

17 Release the cylinder head from the cylinder block and locating dowels by rocking it. Do not pry between the mating faces of the cylinder head and block, as this may damage the gasket faces.

18 Ideally, two assistants will now be required to help remove the cylinder head. Have one assistant hold the timing chain up, clear of the cylinder head, making sure that tension is kept on the chain. With the aid of another assistant, lift the cylinder head from the block - take care, as the cylinder head is heavy **(see illustration)**. As the cylinder head is removed, feed the timing chain through the aperture in the front of the cylinder head, and support it from the cylinder block using the wire.

19 Recover the cylinder head gasket.

### Inspection

20 Refer to Chapter 2C for details of cylinder head dismantling and reassembly.

21 The mating faces of the cylinder head and block must be perfectly clean before installing the head. Use a scraper to remove all traces of gasket and carbon, and also clean the tops of the pistons. Take particular care with the aluminum cylinder head, as the soft metal is easily damaged. Also make sure that debris is not allowed to enter the oil and water passages. Using adhesive tape and paper, seal the water, oil and bolt holes in the cylinder block. To prevent carbon entering the gap between the pistons and bores, smear a little grease in the gap. After cleaning each piston, rotate the crankshaft so that the piston moves down the bore, then wipe out the grease and carbon with a cloth rag.

22 Check the block and head for nicks, deep scratches and other damage. If very slight, they may be removed from the cylinder block carefully with a file. More serious damage may be repaired by machining, but this is a specialist job.

23 If warpage of the cylinder head is suspected, use a straight-edge to check it for distortion, with reference to Chapter 2C.

24 Clean out the bolt holes in the block using a pipe cleaner or thin rag and a screwdriver. Make sure that all oil and water is removed, otherwise there is a possibility of the block being cracked by hydraulic pressure when the bolts are tightened.

25 Examine the bolt threads and the threads in the cylinder block for damage. If necessary, use the correct size tap to chase out the threads in the block.

### Installation

*Refer to illustrations 11.28, 11.31, 11.32a and 11.32b*

**Warning:** *If the camshafts have been removed from the cylinder head, note the warnings given in Section 10, regarding expanded cam followers. Additionally, to minimize the possibility of piston-to-valve contact, after installing the camshaft(s) observe the following delays before installing the cylinder head.*

| Temperature | Delay |
|---|---|
| Room temperature (68°F) | 4 minutes |
| 50°F to 68°F | 11 minutes |
| 32°F to 50°F | 30 minutes |

26 Ensure that the mating faces of the cylinder block and head are spotlessly clean, that the cylinder head bolt threads are clean and dry, and that they screw in and out of their locations.

27 Check that the cylinder head locating dowels are correctly positioned in the cylinder block. **Warning:** *To avoid any possibility of piston-to-valve contact when installing the cylinder head, it is necessary to ensure that none of the pistons are at TDC. Before proceeding further, if not already done, turn the crankshaft to position No. 1 piston at TDC (check that the locking rod can be engaged with the flywheel, then remove the locking rod and turn the crankshaft approximately 30-degrees clockwise using a wrench or socket on the crankshaft pulley hub bolt.*

28 Fit a new cylinder head gasket to the block, locating it over the dowels. Make sure that it is the correct way up **(see illustration)**. Note that 0.012-inch (0.3 mm) thicker-than-standard gaskets are available for use if the cylinder head has been machined (see Chapter 2C).

**11.31 Fit the washers by guiding them into position using a length of welding rod or stiff wire**

29 If not already done, fit the template to the cylinder head to ensure that the camshafts are correctly positioned (No. 1 piston at TDC) - see Section 3.

30 Lower the cylinder head onto the block, engaging it over the dowels.

31 Fit the new cylinder head bolts, complete with new washers where necessary, and tighten the bolts as far as possible by hand. Ensure that the washers are correctly seated in their locations in the cylinder head **(see illustration)**. **Note:** *Do not fit washers to any bolts which are fitted to locations where there are already captive washers in the cylinder head. If a new cylinder head is fitted (without captive washers), ensure that new washers are fitted to all the bolts.*

32 Tighten the bolts in the order shown, and in the stages given in the Specifications - i.e., tighten all bolts in sequence to the Step 1 torque, then tighten all bolts in sequence to the Step 2 torque, and so on **(see illustrations)**.

33 Install and tighten the bolts securing the lower timing chain cover to the cylinder head.

34 Turn the crankshaft back 30-degrees counterclockwise to the TDC position, then

# Chapter 2 Part B Six-cylinder engines

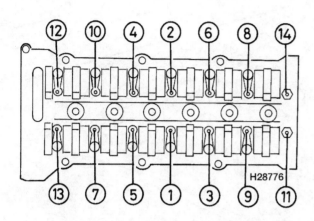

11.32a Cylinder head bolt tightening sequence

11.32b Tightening a cylinder head bolt using an angle-gauge

re-engage the locking rod with the flywheel to lock the crankshaft in position.

35  Manipulate the exhaust camshaft primary chain sprocket until the timing arrow on the sprocket is pointing vertically upwards (note that some sprockets have two arrows opposite each other), then engage the chain with the sprocket.

36  Fit the sprocket to the exhaust camshaft, aligning the sprocket so that the tapped holes in the camshaft flange are positioned at the left-hand ends of the elongated slots in the sprocket.

37  Install the secondary chain guide and tighten the securing bolts.

38  Install the secondary chain tensioner and tighten the securing bolts. Install the spring and tensioner pad, ensuring that the pad is fitted the correct way round (as noted before removal).

39  Temporarily install the tool to lock the secondary chain tensioner in position, then install the secondary timing chain (see Section 7).

40  Further installation is a reversal of removal, but install the intake and exhaust manifolds (see Chapter 4B), and on completion refill the cooling system (see Chapter 1).

## 1993 and later models

Note: *BMW special tool No. 11 3 390 or a suitable equivalent will be required to carry out this operation. New cylinder head bolts and a new cylinder head gasket will be required on installation.*

### Removal

41  Follow the procedure outlined in steps 1 through 4.

42  Unscrew the primary timing chain tensioner plunger cover plug from the right-hand side of the engine. Recover the sealing ring.

Warning: *The chain tensioner plunger has a strong spring. Take care when unscrewing the cover plug.*

43  Recover the spring and withdraw the tensioner plunger.

44  Follow the procedure outlined in steps 5 through 19.

### Inspection

45  Follow the procedure outlined in steps 20 through 25.

### Installation

46  Follow the procedure outlined in steps 28 through 38, noting the warning at the beginning of paragraph 28.

47  Fit special tool No. 11 3 390 into the primary timing chain tensioner aperture, then turn the adjuster screw on the tool until the end of the screw just touches the tensioning rail (see illustration 9.27a). Note that the exhaust camshaft sprocket should now have moved counterclockwise so that the tapped holes in the camshaft flange are centered in the elongated holes in the sprocket.

48  Follow the procedure outlined in steps 39 through 40.

## 12 Oil pan - removal and installation

*Refer to illustrations 12.20, 12.22, 12.23, 12.24 and 12.25*

### 1992 models

Note: *A new oil pan gasket and/or a new dipstick tube sealing ring may be required on installation, and suitable gasket sealant will be required.*

### Removal

1  Drain the engine oil (see Chapter 1).

2  Apply the parking brake, then jack up the front of the vehicle and support securely on axle stands.

3  Remove the exhaust system (see Chapter 4B).

4  Open the hood, then raise the hood to its fully open position (see Chapter 11).

5  Unscrew the securing bolts and/or nuts, and remove the alternator air ducting from the front of the vehicle.

6  Remove the air cleaner assembly and air mass meter, (see Chapter 4B).

7  Remove the heater/ventilation intake air ducting from the rear of the engine compartment as follows.

a) Lift the grille from the top of the ducting (on certain Coupe models, it will be necessary to remove the securing screws and lift off the complete scuttle grille assembly).

b) Working through the top of the ducting, remove the screws securing the cable ducting to the air ducting and move the cable ducting clear.

c) Unscrew the nuts and/or screw(s) securing the air ducting to the firewall (where applicable, bend back the heat shielding for access).

d) Remove the air ducting by pulling upwards.

e) Move the previously removed cable ducting clear of the valve cover.

8  Remove the viscous cooling fan and fan cowl assembly (see Chapter 3).

9  Release the radiator upper securing clips with reference to Chapter 3.

10  Unscrew the dipstick tube bracket securing bolt, and pull the dipstick tube from the cylinder block.

11  Unbolt the power steering fluid reservoir from the left-hand engine mounting bracket, and move the reservoir to one side, taking care not to strain the fluid hose.

12  Remove the drivebelt (see Chapter 1).

13  Unbolt the power steering pump support bracket from the oil pan, then unbolt the power steering pump, and move it to one side, clear of the engine, leaving the fluid lines connected. Ensure that the pump is adequately supported, and take care not to strain the fluid lines.

14  Similarly, on models with air conditioning, unbolt the air conditioning compressor from the engine and suspend it to one side, leaving the refrigerant lines connected.

Warning: *Do not disconnect the refrigerant lines - refer to Chapter 3 for precautions to be taken.*

15  Unscrew the nuts securing the left- and right-hand engine mounting brackets to the engine mountings. Loosen the nuts approximately four complete turns.

**12.20 The rear oil pan securing bolts (arrows) are accessible through the access slots in the bellhousing - viewed with transmission and flywheel removed for clarity**

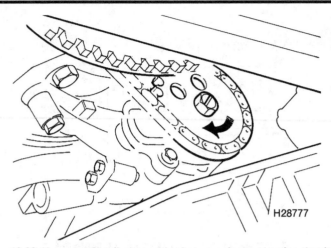

**12.22 Unscrew the oil pump sprocket securing nut, noting that it has a left-hand thread**

16 Unbolt the ground cable(s) from the engine mounting bracket(s).

17 Where applicable, unclip any pipes, hoses and/or wiring from the engine mounting brackets.

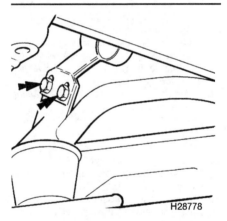

**12.23 Unbolt the oil pick-up pipe from the mounting bracket**

18 Connect an engine hoist and lifting tackle to the engine lifting bracket on the front of the cylinder head, and to the engine mounting brackets. Adjust the lifting tackle so that the engine is equally supported at all three points.

19 Slowly lift the engine as far as possible, continually checking that no pipes, hoses or wiring is being stretched or trapped.

20 Working under the vehicle, progressively unscrew and remove all the oil pan securing bolts. Note that the rear oil pan securing bolts are accessible through the access slots provided in the transmission bellhousing **(see illustration)**. Also note that the three lower transmission-to-engine bolts must be removed, as they screw into the oil pan.

21 Lower the oil pan as far as possible (with the engine in the car, the oil pan cannot be removed until the oil pump and pick-up tube have been removed).

22 The oil pump drive sprocket securing nut must now be unscrewed - note that the nut has a left-hand thread, i.e. it must be turned clockwise to loosen it **(see illustration)**. If necessary, prevent the crankshaft from turning as the nut is unscrewed using a wrench or sprocket on the crankshaft sprocket bolt. **Caution:** *The oil pump drive sprocket securing nut has a left-hand thread.*

23 Unbolt the oil pump pick-up tube from its mounting bracket **(see illustration)**.

24 Unscrew the securing bolts and lower the oil pump into the oil pan **(see illustration)**.

25 Slide the oil pan (with the oil pump) rearwards, and manipulate it out from under the vehicle **(see illustration)**.

26 Recover the oil pan gasket.

### Installation

*Refer to illustration 12.28*

27 Commence installation by thoroughly cleaning the mating faces of the oil pan and cylinder block. Check the condition of the gasket and replace if necessary.

28 Lightly coat the areas where the crankshaft rear oil seal housing and front timing chain cover join the cylinder block with a little gasket sealant **(see illustration)**.

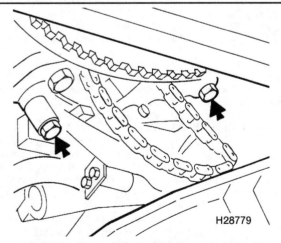

**12.24 Unscrew the oil pump securing bolts (arrows)**

**12.25 Removing the oil pan - viewed with engine removed for clarity**

## Chapter 2 Part B  Six-cylinder engines

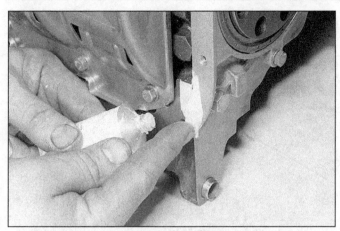

**12.28 Applying sealant to the crankshaft rear oil seal housing/cylinder block joint**

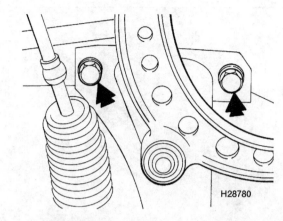

**12.49 Unscrew the subframe securing bolts (arrows)**

29  Place the gasket in position on the oil pan flange. If necessary, apply more sealant (sparingly) to hold the gasket in place.
30  Offer the oil pump into position, ensuring that it locates on the positioning dowels. Install and tighten the oil pump securing bolts.
31  Install and tighten the bolts securing the oil pump pick-up tube to its mounting bracket.
32  Install the oil pump drive sprocket, ensuring that it engages with the splines on the oil pump shaft. Make sure that the drive chain is correctly positioned on the crankshaft and oil pump sprockets.
33  Install the oil pump drive sprocket securing nut, and tighten it to the specified torque - note that the nut has a left-hand thread, i.e. it must be turned counterclockwise to tighten it. If necessary, prevent the crankshaft from turning as during removal.
**Caution:** *The oil pump drive sprocket securing nut has a left-hand thread.*
34  Offer the oil pan up to the cylinder block, ensuring that the gasket stays in place, and install the oil pan securing bolts, tightening them finger-tight only at this stage.
35  Progressively tighten the oil pan-to-cylinder block securing bolts to the specified torque.
36  Tighten the oil pan-to-transmission and the transmission-to-engine bolts to the specified torque.
37  Further installation is a reversal of the removal procedure, noting the following points.
a) When lowering the engine into position, make sure that no pipes, hoses and/or wiring are trapped.
b) Tighten the engine mounting nuts to the specified torque.
c) Install and tension the drivebelt (see Chapter 1).
d) When installing the dipstick tube, check the condition of the sealing ring (at the oil pan end of the tube), and replace if necessary.
e) Install the viscous cooling fan and cowl assembly (see Chapter 3).

f) Install the exhaust system with reference to Chapter 4B.
g) On completion, refill the engine with oil (see Chapter 1).

### 1993 and later models

**Note:** *A new oil pan gasket and/or a new dipstick tube sealing ring may be required on installation, and suitable gasket sealant will be required.*

#### Removal

*Refer to illustration 12.49*

38  On models with VANOS (variable camshaft timing control - see Section 1), the front suspension subframe must be lowered in order to remove the oil pan. Proceed as follows.
39  Follow the procedure outlined in steps 1 through 6 inclusive.
40  Set the steering wheel to the straight-ahead position, then loosen the clamp screw, and disconnect the steering intermediate shaft from the steering gear pinion. Refer to *Steering gear - removal and installation* in Chapter 10 for details.
41  Siphon the power steering fluid from the reservoir, or alternatively, unbolt the reservoir, and drain the fluid into a suitable container. Assuming that the fluid is clean and uncontaminated, save it in a sealed container for re-use.
42  Working under the vehicle, unscrew the unions and disconnect the power steering fluid lines from the steering gear. Be prepared for fluid spillage.
43  Where applicable, unclip any pipes, hoses and/or wiring from the engine mounting brackets.
44  Connect an engine hoist and lifting tackle to the engine lifting bracket on the front of the cylinder head, and to the engine mounting brackets. Adjust the lifting tackle so that the engine is equally supported at all three points. If necessary, remove the air cleaner assembly and air mass meter, (see Chapter 4B) to allow sufficient clearance for the lifting tackle to be connected to the left-hand engine mounting bracket.
45  Working under the vehicle, unscrew the

nuts securing the left- and right-hand engine mountings to the brackets on the subframe.
46  Raise the lifting tackle to lift the engine approximately 1/4-inch.
47  Again working under the vehicle, unscrew the bolts securing the suspension lower arms to the body.
48  Support the center of the subframe, using a jack and a block of wood.
49  Unscrew the subframe securing bolts, then lower the subframe slightly using the jack **(see illustration). Caution:** *Do not remove the steering gear from the subframe.*
50  On models with automatic transmission, release the fluid cooler lines from the oil pan.
51  Unscrew the dipstick tube bracket securing bolt, and pull the dipstick tube from the oil pan.
52  Progressively unscrew and remove all the oil pan securing bolts. Note that the rear oil pan securing bolts are accessible through the access slots provided in the transmission bellhousing. Also note that the three lower transmission to engine bolts must be removed, as they screw into the oil pan.
53  Lower the oil pan from the engine, and manipulate it out from under the vehicle. If necessary, lower the subframe further, using the jack, to give sufficient clearance.
54  Recover the oil pan gasket, and discard it.

#### Installation

55  Commence installation by thoroughly cleaning the mating faces of the oil pan and cylinder block. Check the condition of the gasket and replace if necessary.
56  Lightly coat the areas where the crankshaft rear oil seal housing and front timing chain cover join the cylinder block with a little gasket sealant.
57  Place a new gasket in position on the oil pan flange. If necessary, apply more sealant (sparingly) to hold the gasket in place.
58  Offer the oil pan up to the cylinder block, ensuring that the gasket stays in place, and install the oil pan securing bolts, tightening them finger-tight only at this stage.
59  Progressively tighten the oil pan-to-

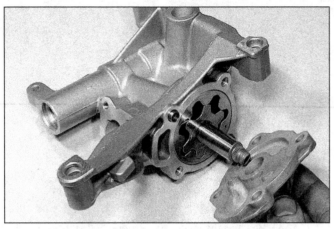

13.3 Removing the cover from the oil pump

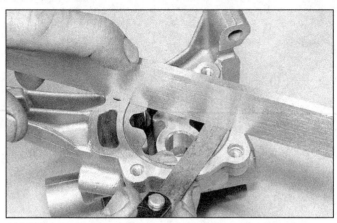

13.6 Measuring the clearance between the outer oil pump rotor and the pump cover mating face

cylinder block securing bolts to the specified torque.
60  Tighten the oil pan-to-transmission bell-housing bolts to the specified torque.
61  Check the condition of the dipstick sealing ring (at the oil pan end of the tube) and replace if necessary. Install the dipstick tube and tighten the bracket securing bolt.
62  On models with automatic transmission, secure the fluid cooler lines to the oil pan.
63  Raise the subframe using the jack, then install the securing bolts and tighten to the specified torque.
64  Install the bolts securing the suspension lower arms to the body, ensuring that the washers are in place, and tighten the bolts to the specified torque.
65  Lower the engine until the mountings are resting on the subframe, ensuring that the lugs on the engine mountings engage with the corresponding holes in the subframe. Install the engine mounting nuts and tighten them to the specified torque.
66  Disconnect and withdraw the engine lifting tackle and hoist.
67  Further installation is a reversal of removal, bearing in mind the following points.
   a) Top-up the fluid level in the power steering fluid reservoir, and bleed the system (see Chapter 10).
   b) Reconnect the steering intermediate shaft to the steering gear pinion, with reference to "Steering gear - removal and installation" in Chapter 10.
   c) On completion, refill the engine with oil (see Chapter 1).

### 13 Oil pump and drive chain - removal, inspection and installation

#### Oil pump

**Removal and installation**

1  Removal and installation of the oil pump is described in Section 12 as part of the oil pan removal and installation procedure. Note that the oil pick-up pipe must be fitted to the pump before the pump is refitted. With the oil pump removed, it can be inspected by following the procedure outlined in the following steps.

**Inspection**

*Refer to illustrations 13.3, 13.6, 13.7 and 13.8*
**Note:** *A new pick-up pipe gasket, a new relief valve spring cap O-ring and a new relief valve snap-ring will be required on installation.*

2  Unbolt the pick-up pipe from the pump. Recover the gasket.
3  Unbolt the cover from the front of the pump **(see illustration)**.
4  Withdraw the driveshaft/rotor and the outer rotor from the pump body.
5  Check the pump body, rotors and cover for any signs of scoring, wear or cracks. If any wear or damage is evident, fit new rotors or replace the complete pump, depending on the extent of the damage. Note that it is wise to replace the complete pump as a unit.
6  Install the rotors to the pump body, then using feeler gauges, measure the clearance between the outer rotor and the pump body. Using the feeler gauges and a straight edge, measure the clearance (endplay) between each of the rotors and the oil pump cover mating face **(see illustration)**. Compare the measurements with the values given in the Specifications, and if necessary replace any worn components, or replace the complete pump as a unit.
7  To remove the pressure relief valve components, press the valve into its housing slightly, using a metal tool, then extract the snap-ring from the top of the housing using snap-ring pliers **(see illustration)**. **Warning:** *The relief valve has a strong spring. Take care when removing the snap-ring.*
8  Withdraw the spring cap, spring and

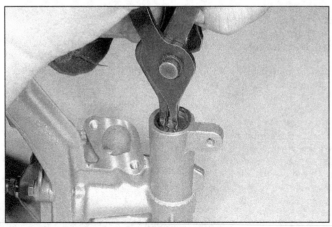

13.7 Extract the snap-ring . . .

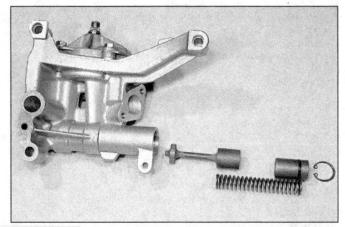

13.8 . . . and withdraw the oil pressure relief valve components

# Chapter 2 Part B Six-cylinder engines 2B-27

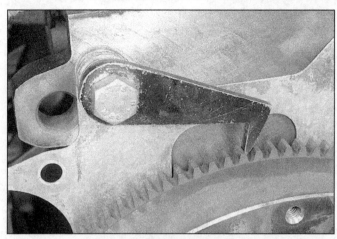

15.3 Toothed tool used to lock flywheel in position when unscrewing flywheel bolts

15.4 Withdrawing the flywheel from the crankshaft

piston from the relief valve housing **(see illustration)**.
9   Measure the free-length of the relief valve spring, and compare it with the value given in the Specifications. Replace the spring if the free-length is not as specified.
10  Fit a new O-ring seal to the top of the relief valve spring cap, then install the components to the housing using a reversal of the removal procedure. Take care not to damage the surface of the spring cap during fitting, and secure the components using a new O-ring.
11  Install the rotors to the pump body, then install the cover to the pump. Ensure that the locating dowels are in position in the pump cover. Install and tighten the cover securing bolts.
12  Thoroughly clean the pick-up pipe strainer, then install the pick-up pipe to the pump using a new gasket. Note that the tab on the gasket must face the pick-up pipe strainer.

## Oil pump drive chain

### Removal
13  Remove the primary timing chain (see Section 7).
14  Withdraw the chain from the crankshaft sprocket.

### Inspection
15  Proceed as described for the secondary timing chain in Section 7.

### Installation
16  Install the chain to the crankshaft sprocket, then install the primary timing chain (see Section 7).

## 14  Oil seals - replacement

### Crankshaft front oil seal
1   The procedure is described as part of the lower timing chain cover removal and installation procedure in Section 6.

### Crankshaft rear oil seal
**Note:** *A new oil seal housing gasket will be required on installation.*
2   Remove the flywheel/driveplate, (see Section 15).
3   Working at the bottom of the oil seal housing, unscrew the bolts securing the rear of the oil pan to the housing.
4   Unscrew the bolts securing the oil seal housing to the cylinder block.
5   If the housing is stuck to the oil pan gasket, run a sharp, thin blade between the housing and the oil pan gasket. Take care not to damage the oil pan gasket.
6   Withdraw the housing from the cylinder block. If the housing is stuck, tap it gently using a soft-faced mallet. Do not lever between the housing and the cylinder block, as this may damage the gasket surfaces.
7   Recover the gasket.
8   Thoroughly clean all traces of old gasket and sealant from the mating faces of the oil seal housing and the cylinder block. Again, take care not to damage the oil pan gasket. If the oil pan gasket has been damaged during removal, it is advisable to fit a new one with reference to Section 12.
9   Support the oil seal housing on blocks of wood, then drive out the seal from the rear of the housing using a hammer and drift.
10  Clean the seal mating surfaces in the housing.
11  Lightly grease the outer edge of the new oil seal, and carefully drive it into position in the housing, using either a large tube of the correct diameter, or a block of wood, to avoid damage to the seal.
12  Ensure that the locating dowels are in position in the rear of the cylinder block, then locate a new oil seal housing gasket over the dowels.
13  Lightly grease the inner lips of the oil seal, then carefully offer the housing to the cylinder block, sliding the oil seal over the crankshaft flange. Take care not to damage the oil seal lips.
14  Install the housing-to-cylinder block and the oil pan-to-housing bolts, and tighten them lightly by hand.
15  Tighten the housing-to-cylinder block bolts to the specified torque, then tighten the oil pan-to-housing bolts to the specified torque.
16  Install the flywheel/driveplate (see Section 15).

### Camshaft oil seals
17  No camshaft oil seals are fitted. Sealing is provided by the valve cover gasket and the timing chain cover gaskets.

## 15  Flywheel/driveplate - removal and installation

*Refer to illustrations 15.3 and 15.4*

### Removal
**Note:** *New flywheel/driveplate securing bolts will be required on installation, and thread-locking compound may be required.*
1   Remove the manual transmission (see Chapter 7A), or the automatic transmission (see Chapter 7B), as applicable.
2   On models with a manual transmission, remove the clutch (see Chapter 6).
3   In order to unscrew the securing bolts, the flywheel/driveplate must be locked in position. This can be done by bolting a toothed tool (engage the tooth with the starter ring gear) to the cylinder block using one of the engine-to-transmission bolts **(see illustration)**.
4   Progressively unscrew the securing bolts, then withdraw the flywheel/driveplate from the crankshaft **(see illustration)**. Note that the flywheel/driveplate locates on dowels. **Warning:** *Take care as the flywheel/driveplate is heavy!*
5   Recover the engine/transmission intermediate plate, noting its orientation.

### Installation
*Refer to illustrations 15.6 and 15.8*
6   Install the engine/transmission interme-

**15.6 Ensure that the engine/transmission intermediate plate is correctly located**

**15.8 Fitting a flywheel bolt**

diate plate, ensuring that it is correctly located on the dowel(s) **(see illustration)**.

7  Install the flywheel/driveplate to the end of the crankshaft, ensuring that the locating dowel engages.

8  Examine the threads of the new securing bolts. If the threads are not already coated with thread-locking compound, then apply suitable thread-locking compound to them, then install the bolts **(see illustration)**.

9  Tighten the bolts progressively in a diagonal sequence to the specified torque. Counterhold the flywheel/driveplate by reversing the tool used during removal.

10  Where applicable, install the clutch (see Chapter 6).

11  Install the manual transmission or the automatic transmission, as applicable, (see Chapter 7A or 7B respectively).

**16.1 Crankshaft pilot bearing (arrow)**

## 16 Crankshaft pilot bearing - replacement

*Refer to illustration 16.1*

1  On manual transmission models, a ball bearing assembly is fitted to the end of the crankshaft to support the end of the transmission input shaft **(see illustration)**.

2  To replace the bearing, proceed as follows.

3  Remove the clutch (see Chapter 6).

4  Using a slide hammer fitted with a suitable adapter, remove the bearing from the end of the crankshaft.

5  Thoroughly clean the bearing housing in the end of the crankshaft.

6  Tap the new bearing into position, up to the stop, using a tube or socket on the bearing outer race.

7  Install the clutch (see Chapter 6).

## 17 Engine/transmission mounts - inspection and replacement

### Inspection

1  Two engine mounts are used, one on either side of the engine.

2  If improved access is required, raise the front of the vehicle and support it securely on axle stands.

3  Check the mounting rubber to see if it is cracked, hardened or separated from the metal at any point. Replace the mount if any such damage or deterioration is evident.

4  Check that all the mounting fasteners are securely tightened.

5  Using a large screwdriver or a crowbar, check for wear in the mount by carefully levering against it to check for free play. Where this is not possible, enlist the aid of an assistant to move the engine/transmission back and forth, or from side to side, while you observe the mount. While some free-play is to be expected, even from new components, excessive wear should be obvious. If excessive free-play is found, check first that the fasteners are correctly secured, then replace any worn components as required.

### Replacement

6  Support the engine, either using a hoist and lifting tackle connected to the engine lifting brackets (refer to *Engine - removal and installation* in Part C of this Chapter), or by positioning a jack and interposed block of wood under the oil pan. Ensure that the engine is adequately supported before proceeding.

7  Unbolt the power steering fluid reservoir from the left-hand engine mounting bracket, and disconnect the ground lead(s) from the mounting bracket(s).

8  Unscrew the nuts securing the left- and right-hand engine mounting brackets to the mount rubbers, then unbolt the mounting brackets from the cylinder block, and remove the mounts.

9  Unscrew the nuts securing the mounts to the body, then withdraw the mounts. Recover the metal protector plates from the mounts if they are loose.

10  Installation is a reversal of removal, but ensure that the metal protector plates are in position on the mounts, and securely tighten all fixings.

# Chapter 2 Part C
# General engine overhaul procedures

## Contents

| | Section | | Section |
|---|---|---|---|
| Crankshaft - inspection | 13 | Engine overhaul - reassembly sequence | 15 |
| Crankshaft - installation and main bearing oil clearance check | 17 | Engine - removal and installation | 4 |
| Crankshaft - removal | 10 | Engine removal - methods and precautions | 3 |
| Cylinder block/crankcase - cleaning and inspection | 11 | General information | 1 |
| Cylinder head and valves - cleaning and inspection | 7 | Main and connecting rod bearings - inspection | 14 |
| Cylinder head - disassembly | 6 | Piston/connecting rod assembly - inspection | 12 |
| Cylinder head - reassembly | 8 | Piston/connecting rod assembly - installation and connecting rod bearing oil clearance check | 18 |
| Engine - initial start-up after overhaul | 19 | Piston/connecting rod assembly - removal | 9 |
| Engine overhaul - disassembly sequence | 5 | Piston rings - installation | 16 |
| Engine overhaul - general information | 2 | | |

## Specifications

### Cylinder head

| | |
|---|---|
| Maximum gasket face distortion (warpage) | 0.0012 inch |
| Maximum gasket face machining limit* | 0.010 to 0.014 inch |
| New cylinder head height | 5.508 to 5.516 inches |
| Minimum cylinder head height after machining | 5.494 inches |

*If 0.012 inch (0.30 mm) of metal or more is removed from the cylinder head gasket face, a 0.012 inch (0.30 mm) thicker cylinder head gasket must be used when installing the cylinder head.

### Valves

| | |
|---|---|
| Valve face angle | 45-degrees |
| Valve head diameter | |
|   Intake | 1.299 inches |
|   Exhaust | 1.201 inches |
| Valve stem diameter | |
|   1992 | 0.275 inch |
|   1993 and later | |
|     Intake | 0.235 inch |
|     Exhaust | 0.234 inch |
| Stem-to-guide clearance (measured at valve head with top of valve stem flush with guide) | 0.020 inch |

### Cylinder block

| | |
|---|---|
| Cylinder bore diameter | |
|   Standard | 3.3071 inches |
|   Intermediate | 3.3103 inches |
|   1st oversize | 3.3170 inches |
|   2nd oversize | 3.3268 inches |
| Maximum cylinder bore out-of-round | 0.0004 inch |
| Maximum cylinder bore taper | 0.0004 inch |

### Crankshaft and connecting rods

| | |
|---|---|
| Crankshaft endplay | 0.0031 to 0.0064 inch |
| Maximum run-out | 0.006 inch |
| Connecting rod side clearance | 0.0016 to 0.010 inch |
| Main bearing journal diameter | |
|   Standard | |
|     Yellow | 2.3616 to 2.3618 inches |
|     Green | 2.3613 to 2.3615 inches |
|     White | 2.3611 to 2.3613 inches |

## Crankshaft and connecting rods (continued)

| | |
|---|---|
| 1st undersize (0.001 inch) | |
|   Yellow | 2.3518 to 2.3520 inches |
|   Green | 2.3515 to 2.3517 inches |
|   White | 2.3512 to 2.3514 inches |
| 2nd undersize (0.023 inch) | |
|   Yellow | 2.3419 to 2.3421 inches |
|   Green | 2.3416 to 2.3418 inches |
|   White | 2.3414 to 2.3416 inches |
| Connecting rod bearing journal diameter | |
|   Four-cylinder engine | |
|     Standard | 1.7720 to 1.7726 inches |
|     1st undersize | 1.7622 to 1.7628 inches |
|     2nd undersize | 1.7523 to 1.7530 inches |
|   Six-cylinder engine | |
|     1992 | |
|       Standard | 1.7720 to 1.7727 inches |
|       1st undersize | 1.7622 to 1.7628 inches |
|       2nd undersize | 1.7523 to 1.7530 inches |
|     1993 and later | |
|       Standard | 1.7707 to 1.7634 inches |
|       1st undersize | 1.7608 to 1.7614 inches |
|       2nd undersize | 1.7509 to 1.7516 inches |
| Main bearing oil clearance | 0.0008 to 0.0023 inch |
| Connecting rod bearing oil clearance | 0.0008 to 0.0023 inch |

## Pistons and piston rings

| | |
|---|---|
| Piston diameter | |
|   Standard | 3.3063 inches |
|   Intermediate | 3.3094 inches |
|   1st oversize | 3.3161 inches |
|   2nd oversize | 3.3260 inches |
| Piston-to-cylinder bore clearance | 0.0004 to 0.0016 inch |
| Piston ring end gaps | |
|   Four-cylinder engine | |
|     Top compression ring | 0.008 to 0.016 inch |
|     Second compression ring | 0.008 to 0.016 inch |
|     Oil control ring | 0.008 to 0.018 inch |
|   Six-cylinder engine | |
|     Top compression ring | 0.008 to 0.016 inch |
|     Second compression ring | 0.008 to 0.016 inch |
|     Oil control ring | 0.008 to 0.018 inch |
| Piston ring-to-groove clearance | |
|   Four-cylinder engine | |
|     Top compression ring | 0.0008 to 0.0020 inch |
|     Second compression ring | 0.0008 to 0.0020 inch |
|     Oil control ring | 0.0008 to 0.0022 inch |
|   Six-cylinder engine | |
|     1992 | |
|       Top compression ring | 0.0008 to 0.0020 inch |
|       Second compression ring | 0.0008 to 0.0020 inch |
|       Oil control ring | 0.0008 to 0.0022 inch |
|     1993 and later | |
|       Top compression ring | 0.0008 to 0.0023 inch |
|       Second compression ring | 0.0008 to 0.0025 inch |
|       Oil control ring | 0.0008 to 0.0020 inch |

## Torque specifications*

| | Ft-lbs |
|---|---|
| Connecting rod bearing cap bolts | |
|   Step 1 | 15 |
|   Step 2 | Tighten an additional 70-degrees |
| Main bearing cap bolts | |
|   1995 and earlier | |
|     Step 1 | 15 |
|     Step 2 | Tighten an additional 50-degrees |
|   1996 on | |
|     Step 1 | 15 |
|     Step 2 | Tighten an additional 70-degrees |

*Refer to Chapter 2A or 2B for additional specifications

# Chapter 2 Part C  General engine overhaul procedures

## 1  General information

Included in this Part of Chapter 2 are details of removing the engine/transmission from the car and general overhaul procedures for the cylinder head, cylinder block/crankcase and all other engine internal components.

The information given ranges from advice concerning preparation for an overhaul and the purchase of replacement parts, to detailed step-by-step procedures covering removal, inspection, renovation and installation of engine internal components.

After Section 5, all instructions are based on the assumption that the engine has been removed from the car. For information concerning in-car engine repair, as well as the removal and installation of those external components necessary for full overhaul, refer to Part A or B of this Chapter, as applicable and to Section 5. Ignore any preliminary disassembly operations described in Parts A or B that are no longer relevant once the engine has been removed from the car.

Apart from torque wrench settings, which are given at the beginning of Parts A and B, all specifications relating to engine overhaul are at the beginning of this Part of Chapter 2.

## 2  Engine overhaul - general information

*Refer to illustration 2.4*

1  It is not always easy to determine when, or if, an engine should be completely overhauled, as a number of factors must be considered.

2  High mileage is not necessarily an indication that an overhaul is needed, while low mileage does not preclude the need for an overhaul. Frequency of servicing is probably the most important consideration. An engine which has had regular and frequent oil and filter changes, as well as other required maintenance, should give many thousands of miles of reliable service. Conversely, a neglected engine may require an overhaul very early in its life.

3  Excessive oil consumption is an indication that piston rings, valve seals and/or valve guides are in need of attention. Make sure that oil leaks are not responsible before deciding that the rings and/or guides are worn. Perform a compression test, (see Chapter 2A or2B, as applicable), to determine the likely cause of the problem.

4  Check the oil pressure with a gauge fitted in place of the oil pressure switch **(see illustration)**, and compare it with that specified. If it is extremely low, the main and connecting rod bearings, and/or the oil pump, are probably worn out.

5  Loss of power, rough running, knocking or metallic engine noises, excessive valve gear noise, and high fuel consumption may also point to the need for an overhaul, especially if they are all present at the same time. If a complete service does not remedy the situation, major mechanical work is the only solution.

6  A full engine overhaul involves restoring all internal parts to the specification of a new engine. During a complete overhaul, the pistons and the piston rings are renewed, and the cylinder bores are reconditioned. New main and connecting rod bearings are generally fitted; if necessary, the crankshaft may be reground, to compensate for wear in the journals. The valves are also serviced as well, since they are usually in less-than-perfect condition at this point. Always pay careful attention to the condition of the oil pump when overhauling the engine, and replace it if there is any doubt as to its serviceability. The end result should be an as-new engine that will give many trouble-free miles.

7  Critical cooling system components such as the hoses, thermostat and water pump should be renewed when an engine is overhauled. The radiator should be checked carefully, to ensure that it is not clogged or leaking. Also, it is a good idea to replace the oil pump whenever the engine is overhauled.

8  Before beginning the engine overhaul, read through the entire procedure, to familiarize yourself with the scope and requirements of the job. Overhauling an engine is not difficult if you follow carefully all of the instructions, have the necessary tools and equipment, and pay close attention to all specifications. It can, however, be time-consuming. Plan on the car being off the road for a minimum of two weeks, especially if parts must be taken to an automotive machine shop for repair or reconditioning. Check on the availability of parts and make sure that any necessary special tools and equipment are obtained in advance. Most work can be done with typical hand tools, although a number of precision measuring tools are required for inspecting parts to determine if they must be renewed. Often the machine shop will handle the inspection of parts and offer advice concerning reconditioning and replacement.

9  Always wait until the engine has been completely disassembled, and until all components (especially the cylinder block/crankcase and the crankshaft) have been inspected, before deciding what service and repair operations must be performed by an automotive machine shop. The condition of these components will be the major factor to consider when determining whether to overhaul the original engine, or to buy a reconditioned unit. Do not, therefore, purchase parts or have overhaul work done on other components until they have been thoroughly inspected. As a general rule, time is the primary cost of an overhaul, so it does not pay to fit worn or sub-standard parts.

10  As a final note, to ensure maximum life and minimum trouble from a reconditioned engine, everything must be assembled with care, in a spotlessly clean environment.

## 3  Engine removal - methods and precautions

1  If you have decided that the engine must be removed for overhaul or major repair work, several preliminary steps should be taken.

2  Locating a suitable place to work is extremely important. Adequate work space, along with storage space for the car, will be needed. If a workshop or garage is not available, at the very least, a flat, level, clean work surface is required.

3  Cleaning the engine compartment and engine/transmission before beginning the removal procedure will help keep tools clean and organized.

4  An engine hoist or A-frame will also be necessary. Make sure the equipment is rated in excess of the weight of the engine. Safety is of primary importance, considering the potential hazards involved in lifting the engine out of the car.

5  If this is the first time you have removed an engine, an assistant should ideally be available. Advice and aid from someone more experienced would also be helpful. There are many instances when one person cannot simultaneously perform all of the operations required when lifting the engine out of the vehicle.

6  Plan the operation ahead of time. Before starting work, arrange for the hire of or obtain all of the tools and equipment you will need. Some of the equipment necessary to perform engine/transmission removal and installation safely and with relative ease (in addition to an engine hoist) is as follows: a heavy duty floor jack, complete sets of wrenches and sockets (see *"Tools and Working Facilities"*), wooden blocks, and plenty of rags and cleaning solvent for mopping up spilled oil, coolant and fuel. If the hoist must be rented, make sure that you arrange for it in advance, and perform all of the operations possible without it beforehand. This will save you money and time.

2.4  Remove the oil pressure switch and install an oil pressure test gauge in its place

7 Plan for the car to be out of use for quite a while. A machine shop will be required to perform some of the work which the do-it-yourselfer cannot accomplish without special equipment. These places often have a busy schedule, so it would be a good idea to consult them before removing the engine, in order to accurately estimate the amount of time required to rebuild or repair components that may need work.
8 Always be extremely careful when removing and installing the engine/transmission. Serious injury can result from careless actions. Plan ahead and take your time, and a job of this nature, although major, can be accomplished successfully.
9 On all models, then engine is removed by first removing the transmission, then lifting the engine out from above the vehicle.

## 4 Engine - removal and installation

### Four-cylinder engine

*Refer to illustration 4.22*
*Note: This is an involved operation. Read through the procedure thoroughly before starting work, and ensure that adequate lifting tackle and/or jacking/support equipment is available. Make notes during disassembly to ensure that all wiring/hoses and brackets are correctly repositioned and routed on installation.*

### Removal

1 Depressurize the fuel system (see Chapter 4A), then disconnect the battery negative lead. **Caution:** *If the radio in your vehicle is equipped with an anti-theft system, make sure you have the correct activation code before disconnecting the battery.*
2 Drain the cooling system and engine oil (see Chapter 1).
3 Remove the manual transmission (see Chapter 7A) or the automatic transmission (see Chapter 7B), as applicable.
4 Unless a hoist is available which is capable of lifting the engine out over the front of the vehicle with the vehicle raised, it will now be necessary to remove the axle stands and lower the vehicle to the ground. Ensure that the engine is adequately supported during the lowering procedure.
5 To improve access and working room, temporarily support the engine from underneath the oil pan, using a floor jack and interposed block of wood, then disconnect and withdraw the hoist and lifting tackle used to support the engine during transmission/transmission removal. **Warning:** *Ensure that the engine is securely and safely supported by the jack before disconnecting the lifting tackle.*
6 Remove the radiator (see Chapter 3).
7 Remove the ignition coil(s) (see Chapter 5B).
8 Remove the air cleaner/airflow meter assembly (see Chapter 4A).
9 If not already done, remove the heater/ventilation intake air ducting from the rear of the engine compartment as follows.

a) Lift the grille from the top of the ducting (on certain coupe models, it will be necessary to remove the securing screws and lift off the complete cowl grille assembly).
b) Working through the top of the ducting, remove the screws securing the cable ducting to the air ducting and move the cable ducting clear.
c) Unscrew the nuts and/or screw(s) securing the air ducting to the firewall (where applicable, bend back the heat shielding for access).
d) Remove the air ducting by pulling upwards.
e) Move the previously removed cable ducting clear of the valve cover.

10 Remove the drivebelt with reference to Chapter 1.
11 Unbolt the power steering pump (see Chapter 10), and move it to one side, leaving the fluid lines connected.
12 Similarly, where applicable, unbolt the air conditioning compressor from the engine, and support it clear of the working area, (see Chapter 3). **Warning:** *Do not disconnect the refrigerant lines - refer to Chapter 3 for precautions to be taken.*
13 Unbolt the power steering reservoir, and move the reservoir to one side, leaving the fluid lines connected.
14 Unbolt the ground lead(s) from the engine mounting bracket(s).
15 Remove the upper and lower sections of the intake manifold (see Chapter 4A).
16 Disconnect the heater coolant hoses from the heater pipe and the heater valve on the engine compartment firewall, then disconnect the hoses from the engine, and remove the coolant hose assembly. Note the hose routing to aid installation.
17 Disconnect the wiring from the following components:

a) Alternator.
b) Temperature sensors in left-hand side of cylinder head.
c) Idle speed control valve (mounted on the left-hand side of the engine).
d) Oil pressure warning light switch (located in the rear of the oil filter housing).
e) Camshaft and crankshaft position sensors and knock sensors, as applicable (the wiring plugs are attached to a bracket on the left-hand side of the engine - make sure that the connectors are marked to ensure correct installation).

18 Check that all relevant wiring has been disconnected from the engine to enable engine removal.
19 Unscrew the securing bolts, and release any clips (noting their locations) and release the wiring harness/wiring ducting assembly from the engine. Note the routing of all wiring, and note the locations of the brackets to ensure correct installation (note that the oxygen sensor and back-up light switch wiring also forms part of the engine harness). Move

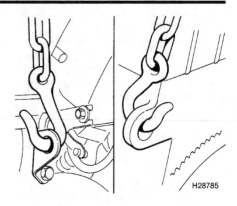

**4.22 Engine lifting eye and lifting bracket locations - four-cylinder engine**

the assembly to one side, clear of the engine.
20 Make a final check to ensure that all relevant hoses, pipes and wiring have been disconnected from the engine and moved clear to allow the engine to be lifted out.
21 Install the front intake manifold support bracket (which incorporates the engine lifting bracket - the bracket was removed during the manifold removal procedure) to the cylinder head, and tighten the securing bolts.
22 Reposition the lifting tackle and hoist to support the engine both from the lifting eye at the rear left-hand corner of the cylinder block, and from the lifting bracket at the front of the cylinder head **(see illustration)**. Raise the hoist to just take the weight of the engine.
23 Unscrew the nuts securing the left- and right-hand engine mounting brackets to the mounting rubbers, then unbolt the mounting brackets from the cylinder block, and remove the mountings.
24 With the aid of an assistant, raise the hoist, and lift the engine from the engine compartment.

### Installation

25 Installation is a reversal of removal, bearing in mind the following points.

a) Tighten all fasteners to the specified torque.
b) Ensure that all wiring, hoses and brackets are positioned and routed as noted before removal.
c) Install the drivebelt with reference to Chapter 1.
d) Install the lower and upper sections of the intake manifold (see Chapter 4A).
e) Install the radiator (see Chapter 3).
f) Install the manual transmission or automatic transmission (see Chapter 7A or 7B respectively).
g) On completion, refill the engine with oil, and refill the cooling system (see Chapter 1).

### Six-cylinder engine

*Refer to illustrations 4.32a, 4.32b, 4.33, 4.38 and 4.39*
*Note: This is an involved operation. Read through the procedure thoroughly before*

# Chapter 2 Part C  General engine overhaul procedures

4.32a Disconnecting the coolant hose from the coolant pump housing - six-cylinder engine

4.32b Removing the coolant hose assembly - six-cylinder engine

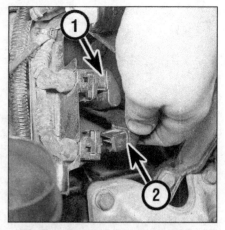

4.33 Disconnecting the camshaft (1) and crankshaft (2) position sensor wiring plugs - six-cylinder engine

starting work, and ensure that adequate lifting tackle and/or jacking/support equipment is available. Make notes during disassembly to ensure that all wiring/hoses and brackets are correctly repositioned and routed on installation.

## Removal

26  Follow the procedure outlined in steps 1 through 7.
27  Remove the engine oil filler cap.
28  Remove the plastic cover from the top of the valve cover. To remove the cover, pry out the cover plates and unscrew the two securing nuts, then lift and pull the cover forwards. Manipulate the cover over the oil filler neck.
29  Unbolt the ground lead from the left-hand corner of the upper timing chain cover, and where applicable, unbolt the ground strap from the rear of the valve cover.
30  Release the locking clips, and disconnect the wiring connectors from the ignition coils. Recover the rubber seals if they are loose.
31  Release the spark plug wires from the clips on the valve cover, then move the complete ducting/spark plug wire assembly to

one side, clear of the engine.
32  Disconnect the coolant hose from the rear of the coolant pump housing, then disconnect the hose from the heater matrix pipe on the engine compartment firewall, and remove the coolant hose assembly. Note the hose routing to aid installation **(see illustrations)**.
33  Disconnect the wiring from the following components **(see illustration)**:

a) Alternator.
b) Temperature sensors in left-hand side of cylinder head.
c) Idle speed control valve (mounted on a bracket on the left-hand side of the engine).
d) Oil pressure warning light switch (located in the rear of the oil filter housing).
e) Camshaft and crankshaft position sensors and knock sensors, as applicable (the wiring plugs are attached to a bracket on the left-hand side of the engine - make sure that the connectors are marked to ensure correct installation).
f) VANOS solenoid valve (where applicable).

34  Check that all relevant wiring has been

disconnected from the engine to enable engine removal.
35  Unscrew the securing bolts, and release any clips (noting their locations) and release the wiring harness/wiring ducting assembly, complete with the idle speed control valve and mounting bracket, from the engine. Note the routing of all wiring, and note the locations of the brackets to ensure correct installation (note that the oxygen sensor and back-up light switch wiring also forms part of the engine harness). Move the assembly to one side, clear of the engine.
36  Follow the procedure outlined in steps 9 through 13.
37  Unbolt the intake manifold support brackets from the engine, noting their locations to ensure correct installation.
38  Unbolt the fuel filter bracket/dipstick tube assembly from the engine, then pull the dipstick tube from the cylinder block. Separate the dipstick tube and the fuel filter bracket, then remove the dipstick tube. Move the fuel filter/bracket assembly to one side, clear of the engine **(see illustration)**.
39  Follow the procedure outlined in steps 20 through 24 **(see illustration)**.

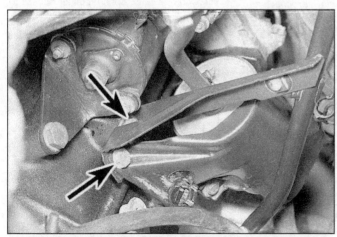

4.38 Fuel filter bracket/dipstick tube assembly securing bolts (arrows) - six-cylinder engine

4.39 Lifting a six-cylinder engine from the vehicle

6.3a Compressing the valve springs using a spring compressor tool

6.3b Removing a valve stem oil seal

## Installation

40 Installation is a reversal of removal, bearing in mind the following points.
  a) Tighten all fasteners to the specified torque.
  b) Ensure that all wiring, hoses and brackets are positioned and routed as noted before removal.
  c) Install the drivebelt with reference to Chapter 1.
  d) Install the radiator (see Chapter 3).
  e) Install the manual transmission or automatic transmission (see Chapter 7A or 7B respectively).
  f) On completion, refill the engine with oil, and refill the cooling system (see Chapter 1).

## 5 Engine overhaul - disassembly sequence

1 It is much easier to disassemble and work on the engine if it is mounted on a portable engine stand. These stands can often be rented from a rental yard. Before the engine is mounted on a stand, the flywheel/driveplate should be removed, so that the stand bolts can be tightened into the end of the cylinder block/crankcase.
2 If a stand is not available, it is possible to disassemble the engine with it blocked up on a sturdy workbench, or on the floor. Be extra-careful not to tip or drop the engine when working without a stand.
3 If you are going to obtain a reconditioned engine, all the external components must be removed first, to be transferred to the replacement engine (just as they will if you are doing a complete engine overhaul yourself). These components include the following:
  a) Ancillary unit mounting brackets (oil filter, starter, alternator, power steering pump, etc.)
  b) Thermostat and housing (see Chapter 3).
  c) Dipstick tube.
  d) All electrical switches and sensors.
  e) Intake and exhaust manifolds - where applicable (see Chapter 4).
  f) Ignition coils and spark plugs - as applicable (see Chapter 4).
  g) Flywheel/driveplate (see Chapter 2A or 2B).

Note: When removing the external components from the engine, pay close attention to details that may be helpful or important during installation. Note the fitted position of gaskets, seals, spacers, pins, washers, bolts, and other small items.

4 If you are obtaining a "short block" (which consists of the engine cylinder block/crankcase, crankshaft, pistons and connecting rods all assembled), then the cylinder head, oil pan, oil pump, and timing chain will have to be removed also.
5 If you are planning a complete overhaul, the engine can be disassembled, and the internal components removed, in the order given below, referring to Part A or B of this Chapter unless otherwise stated.
  a) Intake and exhaust manifolds - where applicable (see Chapter 4).
  b) Timing belt or chains, sprockets and tensioner(s).
  c) Cylinder head.
  d) Flywheel/driveplate.
  e) Oil pan.
  f) Oil pump.
  g) Piston/connecting rod assemblies (Section 9).
  h) Crankshaft (Section 10).
6 Before beginning the disassembly and overhaul procedures, make sure that you have all of the correct tools necessary. Refer to "Tools and working facilities" for further information.

## 6 Cylinder head - disassembly

Refer to illustrations 6.3a, 6.3b and 6.6
Note: New and reconditioned cylinder heads are available from the manufacturer, and from engine overhaul specialists. Be aware that some specialist tools are required for the disassembly and inspection procedures, and new components may not be readily available. It may therefore be more practical and economical for the home mechanic to purchase a reconditioned head, rather than disassemble, inspect and recondition the original head. A valve spring compressor tool will be required for this operation.

### Four-cylinder engine

1 Remove the cylinder head (see Chapter 2A).
2 Remove the camshafts, cam followers and camshaft bearing castings (see Chapter 2A).
3 Using a valve spring compressor, compress the spring(s) on each valve in turn until the keepers can be removed. Release the compressor, and lift off the spring retainer, springs and spring seats. Using a pair of pliers, carefully extract the valve stem oil seal from the top of the guide (see illustrations).
4 If, when the valve spring compressor is screwed down, the spring retainer refuses to

6.6 Place each valve and its associated components in a labeled cellophane bag

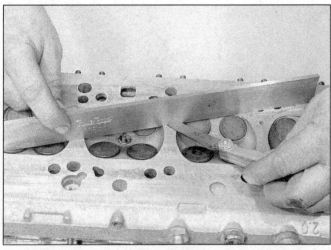

**7.6 Checking the cylinder head gasket face for distortion**

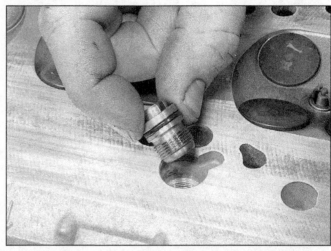

**7.10 Fit a new O-ring to the cylinder head oil pressure check valve**

free and expose the keepers, gently tap the top of the tool, directly over the retainer, with a light hammer. This will free the retainer.
5   Withdraw the valve through the combustion chamber.
6   It is essential that each valve is stored together with its keepers, retainer, springs, and spring seats. The valves should also be kept in their correct sequence, unless they are so badly worn that they are to be renewed. If they are going to be kept and used again, place each valve assembly in a labeled cellophane bag or similar small container **(see illustration)**. Note that No. 1 valve is nearest to the timing chain end of the engine.

### Six-cylinder engine

7   Remove the cylinder head (see Chapter 2B).
8   Remove the camshafts, cam followers and camshaft bearing castings (see Chapter 2B).
9   Follow the procedure outlined in steps 3 through 6.

## 7   Cylinder head and valves - cleaning and inspection

1   Thorough cleaning of the cylinder head and valve components, followed by a detailed inspection, will enable you to decide how much valve service work must be carried out during the engine overhaul. **Note:** *If the engine has been severely overheated, it is best to assume that the cylinder head is warped - check carefully for signs of this.*

### Cleaning

2   Scrape away all traces of old gasket material from the cylinder head.
3   Scrape away the carbon from the combustion chambers and ports, then wash the cylinder head thoroughly with a suitable solvent.
4   Scrape off any heavy carbon deposits that may have formed on the valves, then use a power-operated wire brush to remove deposits from the valve heads and stems.

### Inspection

*Refer to illustrations 7.6, 7.10, 7.14 and 7.17*
**Note:** *Be sure to perform all the following inspection procedures before concluding that the services of a machine shop or engine overhaul specialist are required. Make a list of all items that require attention.*

#### Cylinder head

5   Inspect the head very carefully for cracks, evidence of coolant leakage, and other damage. If cracks are found, a new cylinder head should be obtained.
6   Use a straight-edge and feeler gauge to check that the cylinder head gasket surface is not distorted **(see illustration)**. If it is, it may be possible to have it machined, provided that the cylinder head is not reduced to less than the specified height. **Note:** *If 0.012 inch (0.3 mm) is machined off the cylinder head, a 0.012 inch (0.3 mm) thicker cylinder head gasket must be fitted when the engine is reassembled. This is necessary in order to maintain the correct dimensions between the valve heads, valve guides and cylinder head gasket face.*
7   Examine the valve seats in each of the combustion chambers. If they are severely pitted, cracked, or burned, they will need to be renewed or re-cut by an engine overhaul specialist. If they are only slightly pitted, this can be removed by grinding-in the valve heads and seats with fine valve-grinding compound, as described later in this Section.
8   Check the valve guides for wear by inserting the relevant valve, and checking for side-to-side motion of the valve. A very small amount of movement is acceptable. If the movement seems excessive, remove the valve. Measure the valve stem diameter (see later in this Section), and replace the valve if it is worn. If the valve stem is not worn, the wear must be in the valve guide, and the guide must be reamed, and corresponding oversize (stem) valves fitted. The reaming of valve guides is best carried out by a BMW dealer or engine overhaul specialist, who will have the necessary tools available.
9   If reaming the valve guides, the valve seats should be re-cut or re-ground only *after* the guides have been reamed.
10   On six-cylinder engines, unscrew the oil pressure check valve from the bottom of the cylinder head. Check that the valve can be blown through from bottom-to-top, but not from top-to-bottom. Thoroughly clean the valve and fit a new O-ring, then install the valve to the cylinder head and tighten securely **(see illustration)**.
11   Examine the bearing surfaces in the cylinder head or bearing castings (as applicable) and the bearing caps for signs of wear or damage.
12   Check the camshaft bearing casting mating faces on the cylinder head for distortion. Use a straight-edge and feeler gauge to check that the cylinder head faces are not distorted. If the distortion is outside the specified limit, the cylinder head and bearing castings must be renewed.

#### Valves

**Warning:** *The exhaust valves may be filled with sodium to improve their heat transfer. Sodium is a highly reactive metal, which will ignite or explode spontaneously on contact with water (including water vapor in the air). These valves must NOT be disposed of as ordinary scrap. Seek advice from a BMW dealer or your local authority when disposing of the valves.*
13   Examine the head of each valve for pitting, burning, cracks, and general wear. Check the valve stem for scoring and wear ridges. Rotate the valve, and check for any obvious indication that it is bent. Look for pits or excessive wear on the tip of each valve stem. Replace any valve that shows any such signs of wear or damage.
14   If the valve appears satisfactory at this stage, measure the valve stem diameter at

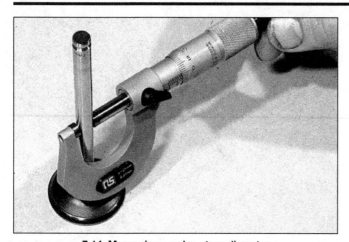

7.14 Measuring a valve stem diameter

7.17 Grinding-in a valve

several points using a micrometer **(see illustration)**. Any significant difference in the readings obtained indicates wear of the valve stem. Should any of these conditions be apparent, the valve(s) must be renewed.

15 If the valves are in satisfactory condition, they should be ground (lapped) into their respective seats, to ensure a smooth, gas-tight seal. If the seat is only lightly pitted, or if it has been re-cut, fine grinding compound *only* should be used to produce the required finish. Coarse valve-grinding compound should *not* be used, unless a seat is badly burned or deeply pitted. If this is the case, the cylinder head and valves should be inspected by an expert, to decide whether seat re-cutting, or even the replacement of the valve or seat insert (where possible) is required.

16 Valve grinding is carried out as follows. Place the cylinder head upside-down on a bench.

17 Smear a trace of (the appropriate grade of) valve-grinding compound on the seat face, and press a suction grinding tool onto the valve head **(see illustration)**. With a semi-rotary action, grind the valve head to its seat, lifting the valve occasionally to redistribute the grinding compound. A light spring placed under the valve head will greatly ease this operation.

18 If coarse grinding compound is being used, work only until a dull, matte even surface is produced on both the valve seat and the valve, then wipe off the used compound, and repeat the process with fine compound. When a smooth unbroken ring of light gray matte finish is produced on both the valve and seat, the grinding operation is complete. *Do not* grind-in the valves any further than absolutely necessary, or the seat will be prematurely sunk into the cylinder head.

19 When all the valves have been ground-in, carefully wash off *all* traces of grinding compound using a suitable solvent, before reassembling the cylinder head.

**Valve components**

20 Examine the valve springs for signs of damage and discoloration. No minimum free length is specified by BMW, so the only way of judging valve spring wear is by comparison with a new component.

21 Stand each spring on a flat surface, and check it for squareness. If any of the springs are damaged, distorted or have lost their tension, obtain a complete new set of springs. It is normal to replace the valve springs as a matter of course if a major overhaul is being carried out.

22 Replace the valve stem oil seals regardless of their apparent condition.

**Cam followers/valve lifters**

23 Examine the contact surfaces for wear or scoring. If excessive wear is evident, the component(s) should be renewed.

## 8 Cylinder head - reassembly

**Note:** *New valve stem oil seals should be fitted, and a valve spring compressor tool will be required for this operation.*

### Four-cylinder engine

1 Follow the procedure outlined in steps 4 through 9.
2 Install the camshaft bearing castings, the cam followers and the camshafts (see Chapter 2A).
3 Install the cylinder head (see Chapter 2A).

### Six-cylinder engine

*Refer to illustrations 8.4, 8.5a, 8.5b, 8.6a, 8.6b, 8.7a, 8.7b, 8.7c and 8.8*

4 Lubricate the stems of the valves, and insert the valves into their original locations **(see illustration)**. If new valves are being fitted, insert them into the locations to which they have been ground.
5 Working on the first valve, dip the new

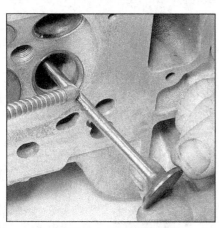

8.4 Lubricate the valve stem

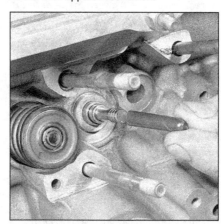

8.5a Fit the protective sleeve to the valve stem . . .

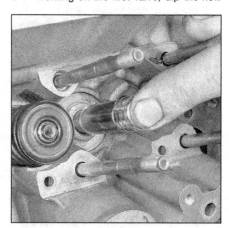

8.5b . . . then fit the oil seal using a deep socket

# Chapter 2 Part C  General engine overhaul procedures

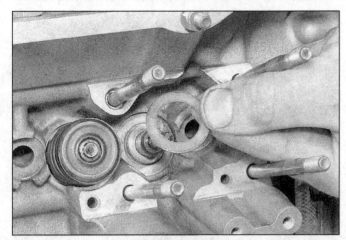

8.6a  Fit the outer ...

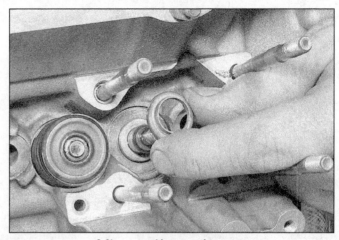

8.6b  ... and inner spring seats

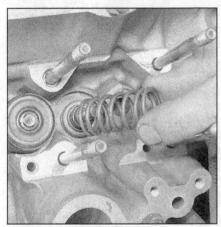

8.7a  Fit the inner ...

8.7b  ... and outer valve springs ...

**illustrations)**.
8  Compress the valve spring(s), and locate the keepers in the recess in the valve stem **(see illustration)**. Release the compressor, then repeat the procedure on the remaining valves.
9  With all the valves installed, support the cylinder head on blocks of wood and, using a hammer and block of wood, tap the end of each valve stem to settle the components.
10  Install the camshaft bearing castings, cam followers and camshafts (see Chapter 2B).
11  Install the cylinder head (see Chapter 2B).

## 9  Piston/connecting rod assembly - removal

*Refer to illustrations 9.3, 9.5 and 9.7*

1  On four-cylinder engines, remove the cylinder head and oil pan (see Chapter 2A).
2  On six-cylinder engines, remove the cylinder head, oil pan and oil pump (see Chapter 2B).
3  Where applicable, unbolt the oil baffle

valve stem seal in fresh engine oil. New seals are normally supplied with protective sleeves which should be fitted to the tops of the valve stems to prevent the keeper grooves from damaging the oil seals. If no sleeves are supplied, wind a little thin tape around the top of the valve stems to protect the seals. Carefully locate the seal over the valve and onto the guide. Take care not to damage the seal as it is passed over the valve stem. Use a suitable socket or metal tube to press the seal firmly onto the guide **(see illustrations)**.
6  Install the spring seat(s) **(see illustrations)**.
7  Locate the valve spring(s) on top of the seat(s), then install the spring retainer **(see**

8.7c  ... followed by the spring retainer

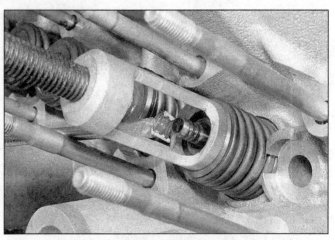

8.8  Use a little dab of grease to hold the keepers in position on the valve stem while the spring compressor is released

# Chapter 2 Part C  General engine overhaul procedures

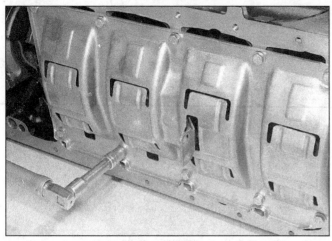

9.3 Unbolt the oil baffle plate from the cylinder block - six-cylinder engine

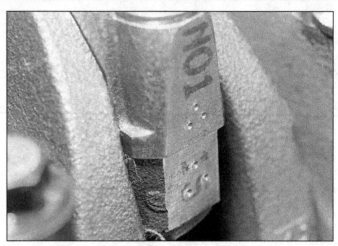

9.5 Connecting rod bearing cap marks

9.7 Removing a connecting rod bearing cap

10.5 Remove the oil seal carrier from the rear of the cylinder block

from the bottom of the cylinder block **(see illustration)**.

4   If there is a pronounced wear ridge at the top of any bore, it may be necessary to remove it with a scraper or ridge reamer, to avoid piston damage during removal. Such a ridge indicates excessive wear of the cylinder bore.

5   Check the connecting rods and bearing caps for identification marks. Both rods and caps should be marked with the cylinder number on the exhaust manifold side of the engine. Note that No. 1 cylinder is at the timing chain end of the engine. If no marks are present, using a hammer and center-punch, paint or similar, mark each connecting rod and connecting rod bearing cap with its respective cylinder number on the flat machined surface provided - ensure that the marks are made on the exhaust manifold side of the connecting rods **(see illustration)**.

6   Turn the crankshaft to bring pistons 1 and 4 (four-cylinder engine), or 1 and 6 (six-cylinder engine), as applicable, to BDC (bottom dead center).

7   Unscrew the bolts from No. 1 piston connecting rod bearing cap. Take off the cap, and recover the bottom half bearing shell **(see illustration)**. If the bearing shells are to be re-used, tape the cap and the shell together.

8   Using a hammer handle, push the piston up through the bore, and remove it from the top of the cylinder block. Recover the bearing shell, and tape it to the connecting rod for safe-keeping.

9   Loosely install the connecting rod cap to the connecting rod, and secure with the bolts - this will help to keep the components in their correct order.

10   Remove No. 4 piston assembly (four-cylinder engine), or No. 6 piston assembly (six-cylinder engine), as applicable, in the same way.

11   Turn the crankshaft as necessary to bring the remaining pistons to BDC, and remove them in the same way.

## 10  Crankshaft - removal

*Refer to illustrations 10.5, 10.8, 10.9 and 10.13*

1   On four-cylinder engines, remove the oil pan, the timing chain housing, and the flywheel/driveplate (see Chapter 2A).

2   On six-cylinder engines, remove the oil pan, the primary timing chain, and the flywheel/driveplate, (see Chapter 2B).

3   Remove the pistons and connecting rods, (see Section 9). If no work is to be done on the pistons and connecting rods, there is no need to remove the cylinder head, or to push the pistons out of the cylinder bores. The pistons should just be pushed far enough up the bores so that they are positioned clear of the crankshaft journals. **Warning:** *If the pistons are pushed up the bores, and the cylinder head is still fitted, take care not to force the pistons into the open valves.*

4   Check the crankshaft endplay (see Section 13), then proceed as follows.

5   Loosen and remove the retaining bolts, and remove the oil seal carrier from the rear (flywheel/driveplate) end of the cylinder block, along with its gasket **(see illustration)**.

6   On six-cylinder engines, if not already done, remove the oil pump drive chain and, if necessary, remove the crankshaft sprocket with reference to Part B of this Chapter.

7   On four-cylinder engines, the main bearing caps should be numbered 1 to 5 from the timing chain end of the engine. Caps No. 1

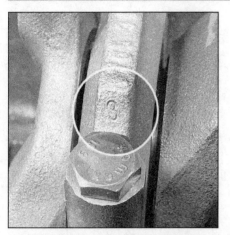

**10.8 Main bearing cap identification number - six-cylinder engine**

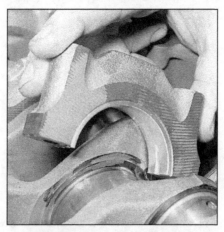

**10.9 Lifting off a main bearing cap**

**10.13 Lifting an upper main bearing shell from the cylinder block**

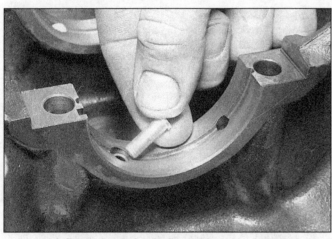

**11.2 Remove the piston oil spray jet tubes from the main bearing locations**

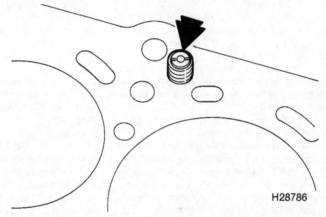

**11.3a Oil pressure check valve location in cylinder block - early four-cylinder engines**

to 3 are numbered, cap No. 4 has shoulders machined into the end faces, and cap No. 5 is unmarked. If the bearing caps are not marked, mark them accordingly using a center-punch.

8 On six-cylinder engines, the main bearing caps should be numbered 1 to 7 on the exhaust side of the engine, starting from the timing chain end of the engine **(see illustration)**. If not, mark them accordingly using a center-punch.

9 Loosen and remove the main bearing cap retaining bolts, and lift off each bearing cap **(see illustration)**. Recover the lower bearing shells, and tape them to their respective caps for safe-keeping.

10 On four-cylinder engines, note that the lower thrust bearing shell, which controls crankshaft endplay is fitted to No. 4 main bearing cap.

11 On six-cylinder engines, note that the lower thrust bearing shell, which controls crankshaft endplay is fitted to No. 6 main bearing cap. Also note the oil pick-up tube support bracket, which is secured by the No. 5 main bearing cap bolts.

12 Lift out the crankshaft. Take care as the crankshaft is heavy.

13 Recover the upper bearing shells from the cylinder block **(see illustration)**, and tape them to their respective caps for safe-keeping. Again, note the location of the upper thrust bearing shell.

## 11 Cylinder block/crankcase - cleaning and inspection

*Refer to illustrations 11.2, 11.3a, 11.3b, 11.9 and 11.16*

### Cleaning

1 Remove all external components and electrical switches/sensors from the block. For complete cleaning, the core plugs should ideally be removed. To remove the core plugs, carefully drive one side of the core plug into the block with a hammer and punch. Grasp the protruding edge with a large pair of adjustable pliers and pull the core plug from the block.

2 Where applicable, pull the piston oil jet spray tubes from the bearing locations in the cylinder block. The tubes are fitted to No. 2 to 5 bearing locations on four-cylinder engines, and No. 2 to 7 bearing locations on six-cylinder engines **(see illustration)**.

3 On four-cylinder engines, where applicable, remove the oil pressure check valve from the top face of the cylinder block. A screw-in type check valve may be fitted, or on later engines, a calibrated jet may be fitted, with a rubber-lined spacer sleeve above

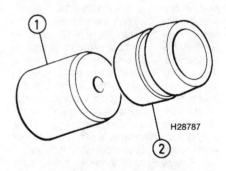

**11.3b Oil pressure calibrated jet components - later four-cylinder engines**

1  Jet   2  Spacing sleeve

## Chapter 2 Part C General engine overhaul procedures

11.9 Cleaning a cylinder block threaded hole using a tap

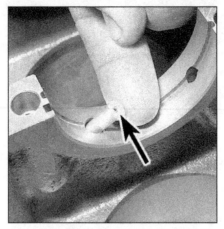

11.16 Clean the holes (arrow) in the oil spray tubes

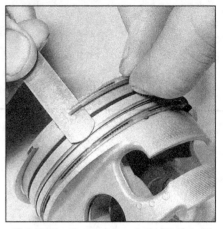

12.2 Removing a piston ring with the aid of a feeler gauge

(see illustrations).

4 Scrape all traces of gasket from the cylinder block/crankcase, taking care not to damage the gasket/sealing surfaces.

5 Remove all oil gallery plugs (where fitted). The plugs are usually very tight - they may have to be drilled out, and the holes re-tapped. Use new plugs when the engine is reassembled.

6 If any of the castings are extremely dirty, all should be steam-cleaned.

7 After the castings are returned, clean all oil holes and oil galleries one more time. Flush all internal passages with warm water until the water runs clear. Dry thoroughly, and apply a light film of oil to all mating surfaces, to prevent rusting. Also oil the cylinder bores. If you have access to compressed air, use it to speed up the drying process, and to blow out all the oil holes and galleries. **Warning:** *Wear eye protection when using compressed air!*

8 If the castings are not very dirty, you can do an adequate cleaning job with hot, soapy water and a stiff brush. Take plenty of time, and do a thorough job. Regardless of the cleaning method used, be sure to clean all oil holes and galleries very thoroughly, and to dry all components well. Protect the cylinder bores as described above, to prevent rusting.

9 All threaded holes must be clean, to ensure accurate torque readings during reassembly. To clean the threads, run the correct-size tap into each of the holes to remove rust, corrosion, thread sealant or sludge, and to restore damaged threads **(see illustration)**. If possible, use compressed air to clear the holes of debris produced by this operation. **Note:** *A good alternative is to inject aerosol-applied water-dispersing lubricant into each hole, using the long spout usually supplied.* **Warning:** *Wear eye protection when cleaning out these holes in this way!*

10 Ensure that all threaded holes in the cylinder block are dry.

11 After coating the mating surfaces of the new core plugs with suitable sealant, fit them to the cylinder block. Make sure that they are driven in straight and seated correctly, or leakage could result. A large socket with an outside diameter which will just fit into the core plug can be used to drive core plugs into position.

12 Apply suitable sealant to the new oil gallery plugs, and insert them into the holes in the block. Tighten them securely.

13 Where applicable, thoroughly clean the oil pressure check valve/calibrated jet (see Paragraph 3), then install the components as follows.

14 If a screw-in type check valve is fitted, check that the valve can be blown through from bottom-to-top, but not from top-to-bottom. Thoroughly clean the valve, and where applicable, fit a new O-ring, then install the valve and tighten securely.

15 If a calibrated jet is fitted, install the jet, ensuring that it is fitted the correct way up, with the stepped collar at the bottom, then install the spacer sleeve. **Warning:** *If the calibrated jet is fitted incorrectly, it may starve the oil supply to the cylinder head.*

16 Where applicable, thoroughly clean the piston oil spray tubes which fit in the bearing locations in the cylinder block, then install the tubes **(see illustration)**.

17 If the engine is not going to be reassembled right away, cover it with a large plastic bag to keep it clean; protect all mating surfaces and the cylinder bores as described above, to prevent rusting.

### Inspection

18 Visually check the castings for cracks and corrosion. Look for stripped threads in the threaded holes. If there has been any history of internal water leakage, it may be worthwhile having an engine overhaul specialist check the cylinder block/crankcase with special equipment. If defects are found, have them repaired if possible, or replace the assembly.

19 Check each cylinder bore for scuffing and scoring. Check for signs of a wear ridge at the top of the cylinder, indicating that the bore is excessively worn.

20 If the necessary measuring equipment is available, measure the bore diameter of each cylinder at the top (just under the wear ridge), center, and bottom of the cylinder bore, parallel to the crankshaft axis.

21 Next, measure the bore diameter at the same three locations, at right-angles to the crankshaft axis. Compare the results with the figures given in the Specifications. If there is any doubt about the condition of the cylinder bores, seek the advice of a BMW dealer or suitable engine reconditioning specialist.

22 If the cylinder bore wear exceeds the permitted tolerances, or if the cylinder walls are badly scored or scuffed, then the cylinders will have to be rebored by a suitably-qualified specialist, and new oversize pistons will have to be fitted. A BMW dealer or automotive machine shop will normally be able to supply suitable oversize pistons when carrying out the reboring work.

## 12 Piston/connecting rod assembly - inspection

*Refer to illustrations 12.2, 12.13a, 12.13b, 12.17a and 12.17b*

1 Before the inspection process can begin, the piston/connecting rod assemblies must be cleaned, and the original piston rings removed from the pistons.

2 Carefully expand the old rings over the top of the pistons. The use of two or three old feeler gauges will be helpful in preventing the rings dropping into empty grooves **(see illustration)**. Be careful not to scratch the piston with the ends of the ring. The rings are brittle, and will snap if they are spread too far. They are also very sharp - protect your hands and fingers. Note that the third ring incorporates an expander. Always remove the rings from the top of the piston. Keep each set of rings with its piston if the old rings are to be re-used. Note which way up each ring is fitted.

3 Scrape away all traces of carbon from the top of the piston. A hand-held wire brush (or a piece of fine emery cloth) can be used, once the majority of the deposits have been scraped away.

# Chapter 2 Part C  General engine overhaul procedures

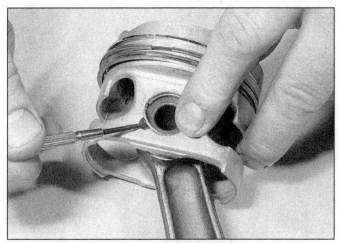

**12.13a  Prying out the circlips . . .**

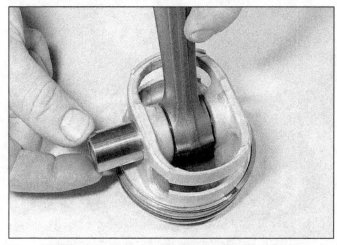

**12.13b  . . . to remove the piston pins from the pistons**

4  Remove the carbon from the ring grooves in the piston, using an old ring. Break the ring in half to do this (be careful not to cut your fingers - piston rings are sharp). Be careful to remove only the carbon deposits - do not remove any metal, and do not nick or scratch the sides of the ring grooves.

5  Once the deposits have been removed, clean the piston/connecting rod assembly with a suitable solvent, and dry thoroughly. Make sure that the oil return holes in the ring grooves are clear.

6  If the pistons and cylinder bores are not damaged or worn excessively, and if the cylinder block does not need to be rebored, the original pistons can be refitted. Measure the piston diameters, and check that they are within limits for the corresponding bore diameters. If the piston-to-bore clearance is excessive, the block will have to be rebored, and new pistons and rings fitted. Normal piston wear shows up as even vertical wear on the piston thrust surfaces, and slight looseness of the top ring in its groove. New piston rings should always be used when the engine is reassembled.

7  Carefully inspect each piston for cracks around the skirt, around the piston pin holes, and at the piston ring "lands" (between the ring grooves).

8  Look for scoring and scuffing on the piston skirt, holes in the piston crown, and burned areas at the edge of the crown. If the skirt is scored or scuffed, the engine may have been suffering from overheating, and/or abnormal combustion which caused excessively high operating temperatures. The cooling and lubrication systems should be checked thoroughly. Scorch marks on the sides of the pistons show that blow-by has occurred. A hole in the piston crown, or burned areas at the edge of the piston crown, indicates that abnormal combustion (pre-ignition, knocking, or detonation) has been occurring. If any of the above problems exist, the causes must be investigated and corrected, or the damage will occur again. The causes may include incorrect ignition timing, intake air leaks, or incorrect air/fuel mixture.

9  Corrosion of the piston, in the form of pitting, indicates that coolant has been leaking into the combustion chamber and/or the crankcase. Again, the cause must be corrected, or the problem may persist in the rebuilt engine.

10  New pistons can be purchased from a BMW dealer.

11  Examine each connecting rod carefully for signs of damage, such as cracks around the bearing cap. Check that the rod is not bent or distorted. Damage is highly unlikely, unless the engine has been seized or badly overheated. Detailed checking of the connecting rod assembly can only be carried out by a BMW dealer or engine repair specialist with the necessary equipment.

12  The piston pins are of the floating type, secured in position by two circlips. The pistons and connecting rods can be separated as follows.

13  Using a small flat-bladed screwdriver, pry out the circlips, and push out the piston pin **(see illustrations)**. Hand pressure should be sufficient to remove the pin. Identify the piston and rod to ensure correct reassembly. Discard the circlips - new ones *must* be used on installation.

14  Examine the piston pin and connecting rod bushing for signs of wear or damage. It should be possible to push the piston pin through the connecting rod bushing by hand, without noticeable play. Wear can be cured by renewing both the pin and bushing. Bushing replacement, however, is a specialist job - press facilities are required, and the new bushing must be reamed accurately.

15  The connecting rods themselves should not be in need of replacement, unless seizure or some other major mechanical failure has occurred. Check the alignment of the connecting rods visually, and if the rods are not straight, take them to an engine overhaul specialist for a more detailed check.

16  Examine all components, and obtain any new parts from your BMW dealer. If new pistons are purchased, they will be supplied complete with piston pins and circlips. Circlips can also be purchased individually.

17  Position the piston in relation to the connecting rod, so that when the assembly is refitted to the engine, the identifying cylinder numbers on the connecting rod and cap are positioned on the exhaust manifold side of the engine, and the installation direction arrow on the piston crown points towards the timing chain end of the engine **(see illustrations)**.

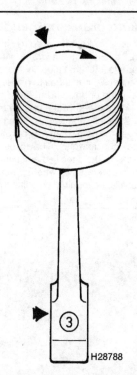

**12.17a  The cylinder number markings should be positioned on the exhaust manifold side of the engines, and the arrow on the piston crown should point towards the timing chain end of the engine**

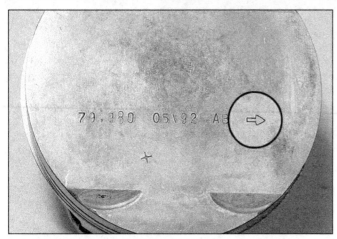

12.17b Installation direction arrow on six-cylinder engine piston crown

13.2 Measuring crankshaft endplay using a dial gauge

18  Apply a smear of clean engine oil to the piston pin. Slide it into the piston and through the connecting rod bushing. Check that the piston pivots freely on the rod, then secure the piston pin in position with two new circlips. Ensure that each snap-ring is correctly located in its groove in the piston.

## 13  Crankshaft - inspection

*Refer to illustrations 13.2 and 13.3*

### Checking crankshaft endplay

1  If the crankshaft endplay is to be checked, this must be done when the crankshaft is still installed in the cylinder block/crankcase, but is free to move.
2  Check the endplay using a dial gauge in contact with the end of the crankshaft. Push the crankshaft fully one way, and then zero the gauge. Push the crankshaft fully the other way, and check the endplay. The result can be compared with the specified amount, and will give an indication as to whether new thrust bearing shells are required **(see illustration)**.

3  If a dial gauge is not available, feeler gauges can be used. First push the crankshaft fully towards the flywheel end of the engine, then use feeler gauges to measure the gap between the web of No. 4 connecting rod journal and the thrust bearing shell on four-cylinder engines, or between No. 6 connecting rod journal and the thrust bearing shell on six-cylinder engines **(see illustration)**.

### Inspection

*Refer to illustration 13.10*

4  Clean the crankshaft using a suitable solvent, and dry it, preferably with compressed air if available. Be sure to clean the oil holes with a pipe cleaner or similar probe, to ensure that they are not obstructed. **Warning:** *Wear eye protection when using compressed air!*
5  Check the main and connecting rod bearing journals for uneven wear, scoring, pitting and cracking.
6  Connecting rod bearing wear is accompanied by distinct metallic knocking when the engine is running (particularly noticeable when the engine is pulling from low speed)

and some loss of oil pressure.
7  Main bearing wear is accompanied by severe engine vibration and rumble - getting progressively worse as engine speed increases - and again by loss of oil pressure.
8  Check the bearing journal for roughness by running a finger lightly over the bearing surface. Any roughness (which will be accompanied by obvious bearing wear) indicates that the crankshaft requires regrinding (where possible) or replacement.
9  If the crankshaft has been reground, check for burrs around the crankshaft oil holes (the holes are usually chamfered, so burrs should not be a problem unless regrinding has been carried out carelessly). Remove any burrs with a fine file or scraper, and thoroughly clean the oil holes as described previously.
10  Using a micrometer, measure the diameter of the main and connecting rod bearing journals, and compare the results with the Specifications **(see illustration)**. By measuring the diameter at a number of points around each journal's circumference, you will be able to determine whether or not the journal is out-of-round. Take the measurement at

13.3 Measuring crankshaft endplay using feeler gauges - six-cylinder engine shown

13.10 Measuring a crankshaft main bearing journal

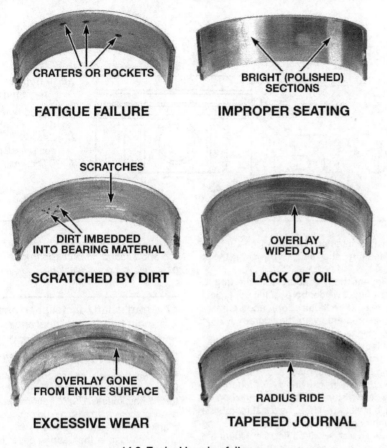

**14.2 Typical bearing failures**

each end of the journal, near the webs, to determine if the journal is tapered.

11  Check the oil seal contact surfaces at each end of the crankshaft for wear and damage. If the seal has worn a deep groove in the surface of the crankshaft, consult an engine overhaul specialist; repair may be possible, but otherwise a new crankshaft will be required.

12  If the crankshaft journals have not already been reground, it may be possible to have the crankshaft reconditioned, and to fit oversize shells (see Section 17). If no oversize shells are available and the crankshaft has worn beyond the specified limits, it will have to be renewed. Consult your BMW dealer or engine specialist for further information on parts availability.

## 14  Main and connecting rod bearings - inspection

*Refer to illustration 14.2*

1  Even though the main and connecting rod bearings should be renewed during the engine overhaul, the old bearings should be retained for close examination, as they may reveal valuable information about the condition of the engine. The bearing shells are graded by thickness, the grade of each shell being indicated by the color code marked on it.

2  Bearing failure can occur due to lack of lubrication, the presence of dirt or other foreign particles, overloading the engine, or corrosion **(see illustration)**. Regardless of the cause of bearing failure, the cause must be corrected (where applicable) before the engine is reassembled, to prevent it from happening again.

3  When examining the bearing shells, remove them from the cylinder block/crankcase, the connecting rods and the connecting rod bearing caps. Lay them out on a clean surface in the same general position as their location in the engine. This will enable you to match any bearing problems with the corresponding crankshaft journal. *Do not* touch any shell's bearing surface with your fingers while checking it, or the delicate surface may be scratched.

4  Dirt and other foreign matter gets into the engine in a variety of ways. It may be left in the engine during assembly, or it may pass through filters or the crankcase ventilation system. It may get into the oil, and from there into the bearings. Metal chips from machining operations and normal engine wear are often present. Abrasives are sometimes left in engine components after reconditioning, especially when parts are not thoroughly cleaned using the proper cleaning methods. Whatever the source, these foreign objects often end up embedded in the soft bearing material, and are easily recognized. Large particles will not embed in the bearing, and will score or gouge the bearing and journal. The best prevention for this cause of bearing failure is to clean all parts thoroughly, and keep everything spotlessly clean during engine assembly. Frequent and regular engine oil and filter changes are also recommended.

5  Lack of lubrication (or lubrication breakdown) has a number of interrelated causes. Excessive heat (which thins the oil), overloading (which squeezes the oil from the bearing face) and oil leakage (from excessive bearing clearances, worn oil pump or high engine speeds) all contribute to lubrication breakdown. Blocked oil passages, which usually are the result of misaligned oil holes in a bearing shell, will also oil-starve a bearing, and destroy it. When lack of lubrication is the cause of bearing failure, the bearing material is wiped or extruded from the steel backing of the bearing. Temperatures may increase to the point where the steel backing turns blue from overheating.

6  Driving habits can have a definite effect on bearing life. Full-throttle, low-speed operation (laboring the engine) puts very high loads on bearings, tending to squeeze out the oil film. These loads cause the bearings to flex, which produces fine cracks in the bearing face (fatigue failure). Eventually, the bearing material will loosen in pieces, and tear away from the steel backing.

7  Short-distance driving leads to corrosion of bearings, because insufficient engine heat is produced to drive off the condensed water and corrosive gases. These products collect in the engine oil, forming acid and sludge. As the oil is carried to the engine bearings, the acid attacks and corrodes the bearing material.

8  Incorrect bearing installation during engine assembly will lead to bearing failure as well. Tight-fitting bearings leave insufficient bearing oil clearance, and will result in oil starvation. Dirt or foreign particles trapped behind a bearing shell result in high spots on the bearing, which lead to failure.

9  *Do not* touch any shell's bearing surface with your fingers during reassembly; there is a risk of scratching the delicate surface, or of depositing particles of dirt on it.

10  As mentioned at the beginning of this Section, the bearing shells should be renewed as a matter of course during engine overhaul; to do otherwise is false economy. Refer to Section 17 for details of bearing shell selection.

## 15  Engine overhaul - reassembly sequence

1  Before reassembly begins, ensure that all new parts have been obtained, and that all necessary tools are available. Read through

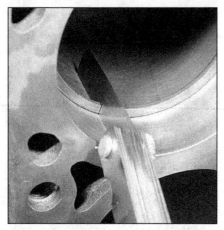

16.5 Measuring a piston ring end gap

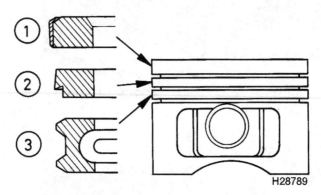

16.10 Typical piston ring fitting

1 Top compression ring
2 Second compression ring
3 Oil control ring

the entire procedure to familiarize yourself with the work involved, and to ensure that all items necessary for reassembly of the engine are at hand. In addition to all normal tools and materials, thread-locking compound will be needed. A suitable tube of liquid sealant will also be required for the joint faces that are fitted without gaskets.

2 In order to save time and avoid problems, engine reassembly can be carried out in the following order, referring to Part A or B of this Chapter unless otherwise stated:

a) Crankshaft (Section 10).
b) Piston/connecting rod assemblies (Section 9).
c) Oil pump.
d) Oil pump and oil pan.
e) Flywheel/driveplate.
f) Cylinder head.
g) Timing chain, tensioner and sprockets.
h) Engine external components.

3 At this stage, all engine components should be absolutely clean and dry, with all faults repaired. The components should be laid out (or in individual containers) on a completely clean work surface.

## 16 Piston rings - installation

*Refer to illustrations 16.5 and 16.10*

1 Before fitting new piston rings, the ring end gaps must be checked as follows.
2 Lay out the piston/connecting rod assemblies and the new piston ring sets, so that the ring sets will be matched with the same piston and cylinder during the end gap measurement and subsequent engine reassembly.
3 Insert the top ring into the first cylinder, and push it down the bore using the top of the piston. This will ensure that the ring remains square with the cylinder walls. Position the ring near the bottom of the cylinder bore, at the lower limit of ring travel. Note that the top and second compression rings are different. The second ring is easily identified by the step on its lower surface, and by the fact that its outer face is tapered.
4 Measure the end gap using feeler gauges.
5 Repeat the procedure with the ring at the top of the cylinder bore, at the upper limit of its travel **(see illustration)**, and compare the measurements with the figures given in the Specifications.
6 If the gap is too small (unlikely if genuine BMW parts are used), it must be enlarged, or the ring ends may contact each other during engine operation, causing serious damage. Ideally, new piston rings providing the correct end gap should be fitted. As a last resort, the end gap can be increased by filing the ring ends very carefully with a fine file. Mount the file in a vise equipped with soft jaws, slip the ring over the file with the ends contacting the file face, and slowly move the ring to remove material from the ends. Take care, as piston rings are sharp, and are easily broken.
7 With new piston rings, it is unlikely that the end gap will be too large. If the gaps are too large, check that you have the correct rings for your engine and for the particular cylinder bore size.
8 Repeat the checking procedure for each ring in the first cylinder, and then for the rings in the remaining cylinders. Remember to keep rings, pistons and cylinders matched up.
9 Once the ring end gaps have been checked and if necessary corrected, the rings can be fitted to the pistons.
10 Fit the piston rings using the same technique as for removal. Fit the bottom (oil control) ring first, and work up. When fitting the oil control ring, where applicable, first insert the expander, then fit the ring with its gap positioned 180° from the expander gap. Ensure that the second compression ring is fitted the correct way up, with its identification mark (either a dot of paint or the word "TOP" stamped on the ring surface) at the top, and the stepped surface at the bottom **(see illustration)**. Arrange the gaps of the top and second compression rings 120° either side of the oil control ring gap, but make sure that none of the rings gaps are positioned over the piston pin hole. **Note:** *Always follow any instructions supplied with the new piston ring sets - different manufacturers may specify different procedures. Do not mix up the top and second compression rings, as they have different cross-sections.*

## 17 Crankshaft - installation and main bearing oil clearance check

### Selection of bearing shells

1 Bearing shells are color-coded, and the upper and lower bearing shells may be of different thickness. BMW recommend that bearing shells matching the color code on the crankshaft should be fitted to the bearing shell locations in the bearing caps. Bearing shells matching the color codes on the inside of the crankcase are fitted to the upper bearing shell locations in the crankcase. If the color code markings in the crankcase have worn off, fit upper bearing shells with the same color code as the crankshaft. Three different bearing shell thickness are available, color-coded yellow, green and white. Consult a BMW dealer for further details.

### Main bearing oil clearance check

*Refer to illustrations 17.4a, 17.4b, 17.9 and 17.12*

2 The oil clearance check can be carried out using the original bearing shells. However, it is preferable to use a new set, since the results obtained will be more conclusive.
3 Clean the backs of the bearing shells, and the bearing locations in both the cylinder block/crankcase and the main bearing caps.
4 Press the bearing shells into their locations, ensuring that the tab on each shell engages in the notch in the cylinder block/crankcase or bearing cap. Take care not to touch any shell's bearing surface with your fingers. Note that the upper bearing shells have an oil groove running along the full length of the bearing surface, whereas the lower shells have a short, tapered oil groove at each end. If the original bearing shells are being used for the check, ensure that they are refitted in their original locations. The thrust bearing shells fit in No. 4 bearing location on four-cylinder engines, or No. 6 bear-

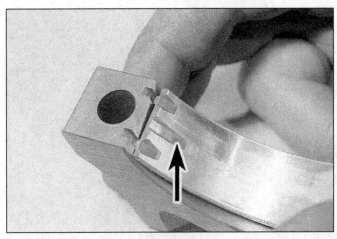

17.4a The lower bearing shells have short tapered oil grooves (arrow)

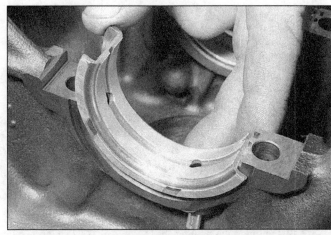

17.4b The thrust bearing shells fit in No. 6 bearing location on six-cylinder engines

ing location on six-cylinder engines **(see illustrations)**.

5  The clearance can be checked in either of two ways.

6  One method (which will be difficult to achieve without a range of internal micrometers or internal/external expanding calipers) is to install the main bearing caps to the cylinder block/crankcase, with bearing shells in place. With the original cap retaining bolts tightened to the specified torque, measure the internal diameter of each assembled pair of bearing shells. If the diameter of each corresponding crankshaft journal is measured and then subtracted from the bearing internal diameter, the result will be the main bearing oil clearance.

7  The second (and more accurate) method is to use a product known as "Plastigage". This consists of a fine thread of perfectly-round plastic, which is compressed between the bearing shell and the journal. When the shell is removed, the plastic is deformed, and can be measured with a special card gauge supplied with the kit. The oil clearance is determined from this gauge. The procedure for using Plastigage is as follows.

8  With the main bearing upper shells in place, carefully lay the crankshaft in position. Do not use any lubricant; the crankshaft journals and bearing shells must be perfectly clean and dry.

9  Cut several lengths of the appropriate-size Plastigage (they should be slightly shorter than the width of the main bearings), and place one length on each crankshaft journal axis **(see illustration)**.

10  With the main bearing lower shells in position, install the main bearing caps. Starting with the center main bearing and working outwards, tighten the original main bearing cap bolts progressively to their specified torque, in the two stages given in the Specifications.

Take care not to disturb the Plastigage, and *do not* rotate the crankshaft at any time during this operation.

11  Remove the main bearing cap bolts and carefully lift off the caps, keeping then in order. Again, take great care not to disturb the Plastigage or rotate the crankshaft. If any of the bearing caps are difficult to remove, free them by carefully tapping them with a soft-faced mallet.

12  Compare the width of the crushed Plastigage on each journal to the scale printed on the Plastigage envelope, to obtain the main bearing oil clearance **(see illustration)**. Compare the clearance measured with that given in the Specifications at the beginning of this Chapter.

13  If the clearance is significantly different from that expected, the bearing shells may be the wrong size (or excessively worn, if the original shells are being re-used). Before deciding that different-size shells are required, make sure that no dirt or oil was trapped between the bearing shells and the caps or block when the clearance was measured. If the Plastigage was wider at one end than at the other, the crankshaft journal may be tapered.

14  If the clearance is not as specified, use the reading obtained, along with the shell thickness quoted in the Specifications, to calculate the necessary grade of bearing shells required. When calculating the bearing clearance required, bear in mind that it is always better to have the oil clearance towards the lower end of the specified range, to allow for wear in use.

17.9 Plastigage in place on a crankshaft main bearing journal

17.12 Measure the width of the deformed Plastigage using the scale on the card

# 2C-18  Chapter 2 Part C  General engine overhaul procedures

**17.19  Lubricate the bearing shells**

**17.22a  Lightly oil the threads of the main bearing cap bolts**

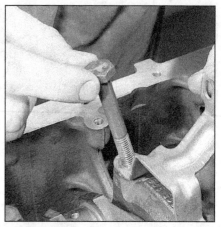

**17.22b  Ensure that the oil pick-up tube support bracket is in position on No. 5 main bearing cap bolts - six-cylinder engines**

15  Where necessary, obtain the required grades of bearing shell, and repeat the oil clearance checking procedure as described above.

16  On completion, carefully scrape away all traces of the Plastigage material from the crankshaft and bearing shells. Use your fingernail, or a wooden or plastic scraper which is unlikely to score the bearing surfaces.

## Final crankshaft installation

*Refer to illustrations 17.19, 17.22a, 17.22b, 17.23a and 17.23b*

**Note:** *New main bearing cap bolts must be used when finally installing the crankshaft.*

17  Carefully lift the crankshaft out of the cylinder block once more.

18  Where applicable, ensure that the oil spray jets are fitted to the bearing locations in the cylinder block.

19  Place the bearing shells in their locations as described earlier. If new shells are being fitted, ensure that all traces of protective grease are cleaned off using solvent. Wipe dry the shells and connecting rods with a lint-free cloth. Liberally lubricate each bearing shell in the cylinder block/crankcase and cap with clean engine oil **(see illustration)**.

20  Lower the crankshaft into position so that No. 1 and 4 cylinder connecting rod journals (four-cylinder engines), or No. 1 and 6 cylinder connecting rod journals (six-cylinder engines), as applicable, will be at BDC, ready for fitting No. 1 piston. Check the crankshaft endplay (see Section 13).

21  Lubricate the lower bearing shells in the main bearing caps with clean engine oil. Make sure that the locating lugs on the shells engage with the corresponding recesses in the caps.

22  Fit the main bearing caps to their correct locations, ensuring that they are fitted the correct way (the bearing shell tab recesses in the block and caps must be on the same side). Thoroughly clean the new main bearing cap bolts, and lightly oil the threads, then insert the bolts, tightening them only loosely at this stage. On six-cylinder engines, ensure that the oil pick-up tube support bracket is correctly in position on the No. 5 main bearing cap bolts **(see illustrations)**.

23  Tighten the main bearing cap bolts to the specified torque, in the two steps given in the Specifications **(see illustrations)**.

24  Check that the crankshaft rotates freely.

25  Fit a new crankshaft rear oil seal to the oil seal carrier, then install the oil seal carrier using a new gasket, (see Chapter 2A or 2B, as applicable).

26  Where applicable, on six-cylinder engines, install the crankshaft sprocket and the oil pump drive chain, (see Chapter 2B).

27  Install the piston/connecting rod assemblies (see Section 18).

28  On four-cylinder engines, install the flywheel/driveplate, the timing chain housing, and the oil pan, (see Chapter 2A).

29  On six-cylinder engines, install the flywheel/driveplate, the primary timing chain, and the oil pan, (see Chapter 2B).

## 18  Piston/connecting rod assembly - installation and connecting rod bearing oil clearance check

### Selection of bearing shells

1  There are a number of sizes of connecting rod bearing shell produced by BMW; a standard size for use with the standard crankshaft, and oversize for use once the

**17.23a  Tighten the main bearing cap bolts to the specified torque . . .**

**17.23b  . . . then through the specified angle**

# Chapter 2 Part C  General engine overhaul procedures

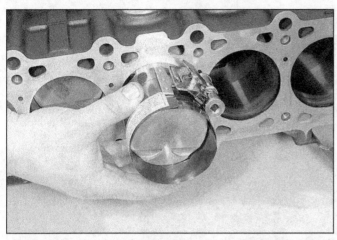

**18.18a Insert the piston/connecting rod assembly into the top of the cylinder bore . . .**

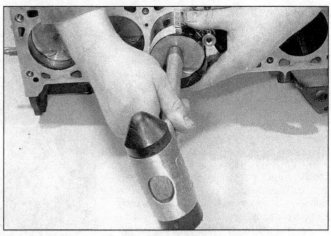

**18.18b . . . then tap the assembly into the cylinder**

crankshaft journals have been reground.

2  Consult your BMW dealer for the latest information on parts availability. To be safe, always quote the diameter of the crankshaft connecting rod journals when ordering bearing shells.

3  Prior to installing the piston/connecting rod assemblies, it is recommended that the connecting rod bearing oil clearance is checked as follows.

### Connecting rod bearing oil clearance check

4  Clean the backs of the bearing shells, and the bearing locations in both the connecting rod and bearing cap.

5  Press the bearing shells into their locations, ensuring that the tab on each shell engages in the notch in the connecting rod and cap. Take care not to touch any shell's bearing surface with your fingers. If the original bearing shells are being used for the check, ensure that they are refitted in their original locations. The clearance can be checked in either of two ways.

6  One method is to install the connecting rod bearing cap to the connecting rod, using the marks made or noted on removal to ensure that they are fitted the correct way around, with the bearing shells in place. With the original cap retaining bolts correctly tightened, use an internal micrometer or vernier caliper to measure the internal diameter of each assembled pair of bearing shells. If the diameter of each corresponding crankshaft journal is measured and then subtracted from the bearing internal diameter, the result will be the connecting rod bearing oil clearance.

7  The second, and more accurate method is to use Plastigage (see Section 17).

8  Ensure that the bearing shells are correctly fitted. Place a strand of Plastigage on each (cleaned) connecting rod journal.

9  Install the (clean) piston/connecting rod assemblies to the crankshaft, and install the connecting rod bearing caps, using the marks made or noted on removal to ensure that they are fitted the correct way around.

10  Install the original bearing cap bolts, and tighten the bolts to the specified torque in the two steps given in the Specifications. Take care not to disturb the Plastigage, nor rotate the connecting rod during the tightening sequence.

11  Disassemble the assemblies without rotating the connecting rods. Use the scale printed on the Plastigage envelope to obtain the connecting rod bearing oil clearance.

12  If the clearance is significantly different from that expected, the bearing shells may be the wrong size (or excessively worn, if the original shells are being re-used). Make sure that no dirt or oil was trapped between the bearing shells and the caps or block when the clearance was measured. If the Plastigage was wider at one end than at the other, the crankshaft journal may be tapered.

13  On completion, carefully scrape away all traces of the Plastigage material from the crankshaft and bearing shells. Use your fingernail, or some other object which is unlikely to score the bearing surfaces.

### Final piston/connecting rod installation

*Refer to illustrations 18.18a, 18.18b, 18.19 and 18.20*

**Note:** *New connecting rod cap bolts must be used when finally installing the piston/connecting rod assemblies. A piston ring compressor tool will be required for this operation.*

14  Note that the following procedure assumes that the main bearing caps are in place (see Section 17).

15  Ensure that the bearing shells are correctly fitted as described earlier. If new shells are being fitted, ensure that all traces of the protective grease are cleaned off using solvent. Wipe dry the shells and connecting rods with a lint-free cloth.

16  Lubricate the cylinder bores, the pistons, and piston rings, then lay out each piston/connecting rod assembly in its respective position.

17  Start with assembly No. 1. Make sure

**18.19 Lubricate the lower connecting rod bearing shells before fitting the caps**

that the piston rings are still spaced (see Section 16), then clamp them in position with a piston ring compressor.

18  Insert the piston/connecting rod assembly into the top of cylinder No. 1. Ensure that the arrow on the piston crown points towards the timing chain end of the engine, and that the identifying marks on the connecting rods and caps are positioned as noted before removal. Using a block of wood or hammer handle against the piston crown, tap the assembly into the cylinder until the piston crown is flush with the top of the cylinder **(see illustrations)**.

19  Ensure that the bearing shell is still correctly installed. Liberally lubricate the connecting rod journal and both bearing shells **(see illustration)**. Taking care not to mark the cylinder bores, pull the piston/connecting rod assembly down the bore and onto the connecting rod journal. Install the connecting rod bearing cap. Note that the bearing shell locating tabs must abut each other.

20  Fit new bearing cap securing bolts, then tighten the bolts evenly and progressively to the Step 1 torque setting. Once both bolts have been tightened to the Step 1 setting,

angle-tighten them through the specified Step 2 angle, using a socket and extension bar. It is recommended that an angle-measuring gauge is used during this stage of the tightening, to ensure accuracy **(see illustration)**. If a gauge is not available, use a dab of white paint to make alignment marks between the bolt and bearing cap prior to tightening; the marks can then be used to check that the bolt has been rotated sufficiently during tightening.

21  Once the bearing cap bolts have been correctly tightened, rotate the crankshaft. Check that it turns freely; some stiffness is to be expected if new components have been fitted, but there should be no signs of binding or tight spots.

22  Install the remaining piston/connecting rod assemblies in the same way.

23  Where applicable, install the oil baffle to the bottom of the cylinder block.

24  On four-cylinder engines, install the cylinder head and oil pan (see Chapter 2A).

25  On six-cylinder engines, install the cylinder head, oil pump and oil pan (see Chapter 2B).

18.20 Using an angle-measuring gauge to tighten the connecting rod bearing cap bolts

## 19  Engine - initial start-up after overhaul

**Warning:** *If the camshafts have been removed, observe the recommended delays between installing the camshafts and starting the engine - refer to the relevant camshaft removal and installation procedure in Chapter 2A or 2B, as applicable for details.*

1  With the engine refitted in the vehicle, double-check the engine oil and coolant levels. Make a final check that everything has been reconnected, and that there are no tools or rags left in the engine compartment.

2  Disable the ignition and fuel injection systems by removing the DME master relay, and the fuel pump relay, located in the main relay box (see Chapter 12), then turn the engine on the starter motor until the oil pressure warning light goes out.

3  Install the relays (and ensure that the fuel pump fuse is fitted), and switch on the ignition to prime the fuel system.

4  Start the engine, noting that this may take a little longer than usual, due to the fuel system components having been disturbed.

**Caution:** *On four-cylinder engines, if the timing chain tensioner has been removed (see Chapter 2A), the engine must be run at 3500 rpm for 20 seconds as soon as it starts - this is to ensure that the tensioner is primed with oil.*

5  While the engine is idling, check for fuel, water and oil leaks. Don't be alarmed if there are some odd smells and smoke from parts getting hot and burning off oil deposits.

6  Assuming all is well, keep the engine idling until hot water is felt circulating through the top hose, then switch off the engine.

7  After a few minutes, recheck the oil and coolant levels (see Chapter 1), and top-up as necessary.

8  If new pistons, rings or crankshaft bearings have been fitted, the engine must be treated as new, and run-in for the first 500 miles. *Do not* operate the engine at full-throttle, or allow it to labor at low engine speeds in any gear. It is recommended that the oil and filter are changed at the end of this period.

# Chapter 3
# Cooling, heating and ventilation systems

## Contents

| | Section |
|---|---|
| Air conditioning compressor drivebelt - check and replacement | See Chapter 1 |
| Air conditioning system components - removal and installation | 12 |
| Air conditioning system - general information and precautions | 11 |
| Antifreeze mixture | See Chapter 1 |
| Coolant level check | See Weekly checks |
| Coolant pump - removal and installation | 7 |
| Cooling fan - removal and installation | 4 |
| Cooling system - draining | See Chapter 1 |
| Cooling system electrical switches - testing, removal and installation | 6 |
| Cooling system - filling | See Chapter 1 |
| Cooling system - flushing | See Chapter 1 |
| Cooling system hoses - disconnection and replacement | 2 |
| General information and precautions | 1 |
| Heater/ventilation components (Z3 models) - removal and installation | 10 |
| Heater/ventilation components (3-Series models) - removal and installation | 9 |
| Heating and ventilation system - general information | 8 |
| Radiator - removal, inspection and installation | 5 |
| System flushing | See Chapter 1 |
| Thermostat - removal, testing and installation | 3 |

## Specifications

### General
Expansion tank cap opening pressure .................................................. 12.6 to 15.4 psi

### Thermostat
Opening temperatures
   Four-cylinder engine ........................................................................ 176-degrees F
   Six-cylinder engine ........................................................................... 190-degrees F

### Torque specifications      Ft-lbs (unless otherwise indicated)
Thermostat cover bolts ........................................................................... 79 in-lbs
Cooling fan viscous coupling to coolant pump ........................................ 22
Coolant pump nuts/bolts
   M6 nuts/bolts ................................................................................... 79 in-lbs
   M8 nuts/bolts ................................................................................... 16

## 1 General information and precautions

### General information

The cooling system is of pressurized type, comprising a pump, an aluminum crossflow radiator, cooling fan, and a thermostat. The system functions as follows. Cold coolant from the radiator passes through the hose to the coolant pump where it is pumped around the cylinder block and head passages. After cooling the cylinder bores, combustion surfaces and valve seats, the coolant reaches the underside of the thermostat, which is initially closed. The coolant passes through the heater and is returned through the cylinder block to the coolant pump.

When the engine is cold the coolant circulates only through the cylinder block, cylinder head, expansion tank and heater. When the coolant reaches a predetermined temperature, the thermostat opens and the coolant passes through to the radiator. As the coolant circulates through the radiator it is cooled by the inrush of air when the car is in forward motion. Airflow is supplemented by the action of the cooling fan. Upon reaching the radiator, the coolant is now cooled and the cycle is repeated.

On M42 four-cylinder and all six-cylinder engines, the cooling fan is driven via a viscous coupling. The viscous coupling varies the fan speed, according to engine temperature. At low temperatures, the coupling provides very little resistance between the fan and pump pulley so only a slight amount of drive is transmitted to the cooling fan. As the temperature of the coupling increases, so does its internal resistance therefore increasing drive to the cooling fan.

1996 and later 3-Series models powered by an M44 four-cylinder engine coupled to a manual transmission, and all Z3 models powered by the M44 engine, use a two-speed electric fan mounted behind the radiator. The fan is activated by a thermostatic fan switch when the coolant temperature exceeds the low-speed or high-speed threshold.

Six-cylinder models are also equipped with a two-speed electric auxiliary cooling fan, which is mounted behind the bumper and grille, and in front of the A/C condenser. The auxiliary fan is primarily intended to help cool the radiator and condenser when the air conditioning system is running, but it is also switched on when coolant temperature is high. Two types of auxiliary fan have been used on six-cylinder 3-Series models: Type 1 (up to September 1992), and Type 2 (after September 1992). All six-cylinder Z3 models use the second type.

Refer to Section 11 for information on the air conditioning system.

## Precautions

**Warning 1:** *Do not attempt to remove the expansion tank filler cap or disturb any part of the cooling system while the engine is hot, as there is a high risk of scalding. If the expansion tank filler cap must be removed before the engine and radiator have fully cooled (even though this is not recommended) the pressure in the cooling system must first be relieved. Cover the cap with a thick layer of cloth, to avoid scalding, and slowly unscrew the filler cap until a hissing sound can be heard. When the hissing has stopped, indicating that the pressure has reduced, slowly unscrew the filler cap until it can be removed; if more hissing sounds are heard, wait until they have stopped before unscrewing the cap completely. At all times keep well away from the filler cap opening.*

**Warning 2:** *Do not allow antifreeze to come into contact with skin or painted surfaces of the vehicle. Rinse off spills immediately with plenty of water. Never leave antifreeze lying around in an open container or in a puddle in the driveway or on the garage floor. Children and pets are attracted by its sweet smell. Antifreeze can be fatal if ingested.*

**Warning 3:** *Refer to Section 11 for precautions to be observed when working on models equipped with air conditioning.*

3.4 Remove the thermostat housing . . .

## 2 Cooling system hoses - disconnection and replacement

**Warning:** *Refer to the **Warnings** given in Section 1 of this Chapter before proceeding.*

1 If the checks described in Chapter 1 reveal a faulty hose, it must be renewed as follows.

2 First drain the cooling system (see Chapter 1). If the coolant is not due for replacement, it may be re-used if it is collected in a clean container.

3 To disconnect a hose, release its retaining clamps, then move them along the hose, clear of the relevant inlet/outlet union. Carefully work the hose free. While the hoses can be removed with relative ease when new, or when hot, do not attempt to disconnect any part of the system while it is still hot.

4 Note that the radiator inlet and outlet unions are fragile; do not use excessive force when attempting to remove the hoses. If a hose proves to be difficult to remove, try to release it by rotating the hose ends before attempting to free it. If all else fails, cut the hose with a sharp knife, then slit it so that it can be peeled off in two pieces. Although this may prove expensive if the hose is otherwise undamaged, it is preferable to buying a new radiator.

5 When fitting a hose, first slide the clamps onto the hose, then work the hose into position. If spring-type clamps were originally fitted, it is a good idea to replace them with screw type clamps when installing the hose. If the hose is stiff, use a little soapy water as a lubricant, or soften the hose by soaking it in hot water.

6 Work the hose into position, checking that it is correctly routed, then slide each clamp along the hose until it passes over the flared end of the relevant inlet/outlet union, before securing it in position with the retaining clamp.

7 Refill the cooling system with reference to Chapter 1.

8 Check thoroughly for leaks as soon as possible after disturbing any part of the cooling system.

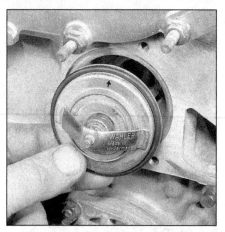

3.5 . . . and lift out the thermostat and sealing ring

## 3 Thermostat - removal, testing and installation

**Warning:** *Wait until the engine is completely cool before beginning this procedure.*

### Removal

*Refer to illustrations 3.4 and 3.5*

**Note:** *A new thermostat sealing ring and (where fitted) housing gasket/seal will be required on installation.*

1 Disconnect the battery negative lead.
**Caution:** *If the radio in your vehicle is equipped with an anti-theft system, make sure you have the correct activation code before disconnecting the battery.*

2 Drain the cooling system (see Chapter 1). To improve access to the thermostat housing, remove the cooling fan and coupling (see Section 4).

3 Loosen the retaining clamp(s) and disconnect the coolant hose(s) from the thermostat housing on the front of the cylinder head/timing chain cover (as applicable).

4 Loosen and remove the retaining screws and remove the thermostat housing **(see illustration)**. Recover the housing gasket/seal (where fitted). On six-cylinder engines, it will be necessary to unbolt the engine lifting bracket to allow the housing to be removed.

5 Lift the thermostat out from the cylinder head and recover its sealing ring **(see illustration)**.

### Testing

6 A rough test of the thermostat may be made by suspending it with a piece of string in a container full of water. Heat the water to bring it to a boil - the thermostat must open by the time the water boils. If not, replace it.

7 If a thermometer is available, the precise opening temperature of the thermostat may be determined, and compared with the figures given in the Specifications. The opening temperature is also marked on the thermostat.

8 A thermostat which fails to close as the water cools must also be renewed.

# Chapter 3 Cooling, heating and ventilation systems

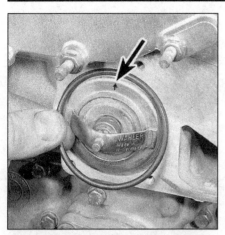

3.9a Ensure the thermostat is fitted with its bleed hole/arrow marking (arrow) uppermost and fit the new sealing ring

3.9b Where necessary, fit a new gasket/seal to the thermostat housing

4.2 On six-cylinder models unclip the alternator cooling duct and remove it from the engine compartment

## Installation

*Refer to illustrations 3.9a and 3.9b*

9  Installation is a reversal of removal, bearing in mind the following points.
   a) *Fit the thermostat and new sealing ring to the head/timing cover. Note that the thermostat should be fitted with its bleed hole/arrow marking uppermost* **(see illustration)**.
   b) *Fit the new gasket/seal (where necessary) and tighten the thermostat cover bolts to the specified torque setting* **(see illustration)**.
   c) *Where necessary, install the cooling fan (see Section 5).*
   d) *On completion refill the cooling system (see Chapter 1).*

## 4  Cooling fan - removal and installation

### Mechanical fan (M42 four-cylinder and all six-cylinder engines)

#### Removal

*Refer to illustrations 4.2, 4.4a and 4.4b*

**Note:** *A special 32 mm narrow open-ended wrench will be required to remove the fan and viscous coupling assembly.*

1  Undo the retaining screws and remove the plastic cover from above the radiator.
2  On six-cylinder 3-Series models, disconnect the auxiliary cooling ducts from the shroud and remove them from the engine compartment **(see illustration)**.
3  Release the fan shroud upper retaining clips by pulling out their center pins then lift the shroud upwards and out of position. **Note:** *It may be necessary to remove the fan to enable the shroud to be removed.*
4  Using the special open-ended wrench unscrew the fan clutch from the coolant pump and remove the cooling fan **(see illus-**

4.4a Loosen the cooling fan coupling using the special open-ended wrench . . .

4.4b . . . and remove the cooling fan

**trations)**. **Note:** *The fan clutch has a left-hand thread.*
5  If necessary, loosen and remove the retaining bolts and separate the cooling fan from the clutch noting which way around the fan is fitted.

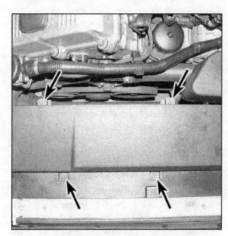

4.9 To detach the cover from the radiator, remove these fasteners (arrows)

#### Installation

6  Where necessary, install the fan to the fan clutch and securely tighten its retaining bolts. Make sure the fan is fitted the correct way around. **Note:** *If the fan is installed facing the wrong way, the efficiency of the cooling system will be significantly reduced.*
7  Install the fan and shroud. Screw the fan onto the coolant pump and tighten it to the specified torque. Engage the fan shroud with the lugs on the radiator and secure it in position with the retaining clips.
8  Reconnect the auxiliary cooling ducts (where fitted) to the shroud then install the plastic cover to the hood crossmember and securely tighten its retaining screws.

### Electric fan (3-Series models with M44 engine and manual transmission; Z3 models with M44 engine)

#### Removal

*Refer to illustrations 4.9, 4.10 and 4.11*

9  Remove the cover from the top of the radiator **(see illustration)**.

## 3-4  Chapter 3  Cooling, heating and ventilation systems

4.10 Before removing the fan assembly, unplug the electrical connector (arrow)

4.11 To detach the shroud from the radiator, remove the bolt (arrow) from each upper corner (right bolt shown)

4.13 When installing the fan shroud, make sure that the locator pin at each lower corner is fully seated into its corresponding alignment notch (arrow) (right locator pin and alignment notch shown)

10  Unplug the electrical connector from the fan switch at the bottom left corner of the radiator **(see illustration)**.
11  Remove the retaining screws from the left and right sides of the fan shroud **(see illustration)**.
12  Lift the fan and shroud assembly out of the engine compartment.

### Installation
*Refer to illustration 4.13*
13  Installation is the reverse of removal. Make sure that the tabs on the fan housing are aligned with the retaining tabs on the radiator **(see illustration)**.

## Auxiliary cooling fan (all 3-Series models and six-cylinder Z3 models)

### Type 1 (3-Series models manufactured through September 1992)
14  Remove the seven fasteners from the front radiator air shroud and remove the shroud.
15  Remove the three fan assembly mounting screws.
16  Tilt the fan to the front and unplug the electrical connector.
17  Lift the fan up and remove it from the engine compartment.
18  Installation is the reverse of removal.

### Type 2 (3-Series models manufactured after September 1992; all six-cylinder Z3 models)
19  Remove the front bumper assembly (see Chapter 11).
20  On 3-Series models, remove the front grille.
21  Unplug the electrical connector from the fan.
22  Remove the four fan assembly mounting screws.
23  Remove the fan assembly through the bumper opening.
24  Installation is the reverse of removal.

## 5  Radiator - removal, inspection and installation

**Warning:** *Wait until the engine is completely cool before beginning this procedure.*
**Note:** *If leakage is the reason for wanting to remove the radiator, bear in mind that minor leaks can be often be cured using a radiator sealant with the radiator installed.*

### Removal
*Refer to illustrations 5.9a, 5.9b and 5.9c*
1  Disconnect the battery negative lead.
**Caution:** *If the radio in your vehicle is equipped with an anti-theft system, make sure you have the correct activation code before disconnecting the battery.*
2  Drain the cooling system (see Chapter 1).
3  Remove the cooling fan shroud (see steps 1 through 3 in Section 4).
4  Loosen the retaining clamp and disconnect the upper and lower hoses from the radiator.
5  On models equipped with air conditioning, disconnect the wiring connector from the auxiliary cooling fan temperature switch which is screwed into the end of the radiator.
6  On models where the expansion tank is incorporated in the radiator, disconnect the wiring connector from the coolant level sensor which is fitted to the base of the expansion tank.
7  Where necessary, loosen the retaining clamp and disconnect the expansion tank hose from the base of the radiator.
8  On models with automatic transmission where the oil cooler is incorporated into the side on the radiator, loosen the union nuts and disconnect the cooler pipes from the end of the radiator. Be prepared for some fluid loss and plug the pipe ends to minimize fluid loss and prevent entry of dirt into the hydraulic system. Discard the sealing rings fitted to the pipe unions; new ones should be used on installation.
9  On all models, using a flat-bladed

5.9a Insert a flat-bladed screwdriver into the upper retaining clip . . .

screwdriver, release the upper retaining clamps and lift the radiator upwards and out of position **(see illustrations)**. Recover the mounting rubbers from the front of the body.

### Inspection
10  If the radiator has been removed due to suspected blockage, reverse flush it (see Chapter 1). Clean dirt and debris from the radiator fins, using a soft brush. Be careful, as the fins are easily damaged, and are sharp.
11  If necessary, a radiator specialist can perform a "flow test" on the radiator, to establish whether an internal blockage exists.
12  A leaking radiator must be referred to a specialist for permanent repair. Do not attempt to weld or solder a leaking radiator, as damage may result.
13  In an emergency, minor leaks from the radiator can be cured by using a suitable radiator sealant in accordance with the manufacturers instructions with the radiator installed.

# Chapter 3 Cooling, heating and ventilation systems 3-5

5.9b ... then release the clip by levering the screwdriver carefully in the direction of the arrow

5.9c Lifting the radiator out of position

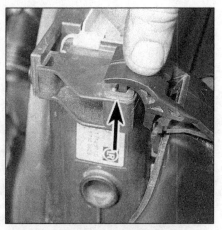

5.17 On installation, ensure the retaining clip is correctly engaged with the radiator (arrow) and clip it in position

14 If the radiator is to be sent for repair or renewed, remove the cooling fan switch/level sensor (where fitted).
15 Inspect the radiator lower mounting rubbers for signs of damage or deterioration and replace if necessary.
16 On models where the expansion tank is incorporated into the radiator, if necessary, undo the retaining screws and remove the tank retaining plate from the top of the radiator. Remove the tank and recover its upper and lower sealing rings. Fit new sealing rings to the tank and locate it in the radiator. Install the retaining plate and securely tighten its retaining screws.

## Installation

*Refer to illustration 5.17*

17 Installation is the reverse of removal, noting the following points.
 a) Ensure that the lower mounting rubbers are correctly located in the body then lower the radiator into position, engage it with the mountings and secure it in position with the retaining clamps **(see illustration)**.
 b) On automatic transmission models, fit new sealing rings to the oil cooler pipe unions and securely tighten the union nuts. On completion check the transmission fluid level (see Chapter 1).
 c) Ensure that the fan cowl is correctly located with the lugs on the radiator and secure it in position with the clamps.
 d) Reconnect the hoses and securely tighten the retaining clamps.
 e) On completion, reconnect the battery and refill the cooling system (see Chapter 1).

## 6 Cooling system electrical switches - testing, removal and installation

**Warning:** *The engine must be completely cool before removing any of the following switches/sensors.*

### Auxiliary electric cooling fan thermostatic switch - models with air conditioning

#### Testing

1 Testing of the air conditioning system should be entrusted to a BMW dealer or other qualified repair shop.

#### Removal

2 The switch is located in the right-hand side of the radiator. The engine and radiator should be cold before removing the switch.
3 Disconnect the battery negative lead. **Caution:** *If the radio in your vehicle is equipped with an anti-theft system, make sure you have the correct activation code before disconnecting the battery.*
4 Either drain the cooling system to below the level of the switch (see Chapter 1), or have ready a suitable plug which can be used to plug the switch aperture in the radiator while the switch is removed. If a plug is used, take great care not to damage the radiator, and do not use anything which will allow foreign matter to enter the radiator.
5 Disconnect the wiring plug from the switch.
6 Carefully unscrew the switch from the radiator and recover the sealing washer.

#### Installation

7 Installation is a reversal of removal using a new sealing washer. On completion, refill the cooling system (see Chapter 1).
8 Start the engine and run it until it reaches normal operating temperature, then continue to run the engine and check that the cooling fan cuts in and functions correctly.

### Coolant temperature gauge sensor

#### Testing

9 The coolant temperature gauge, mounted in the instrument panel, is fed with a stabilized voltage supply from the instrument panel feed (through the ignition switch and a fuse), and its ground is controlled by the sensor.
10 The sensor unit is screwed into the left-hand side of the cylinder head. There are two sensors screwed into the head; the temperature gauge sensor is the rear sensor of the two, the front one being the fuel injection system temperature sensor. To improve access to the sensor, on four-cylinder models disconnect the coolant hose and free the cable channel from the manifold (see Chapter 4A), and on six-cylinder engines remove the alternator cooling duct (see Chapter 4B).
11 The sensor contains a thermistor, which consists of an electronic component whose electrical resistance decreases at a predetermined rate as its temperature rises. When the coolant is cold, the sensor resistance is high, current flow through the gauge is reduced, and the gauge needle points towards the "cold" end of the scale. If the sensor is faulty, it must be renewed.
12 If the gauge develops a fault, first check the other instruments; if they do not work at all, check the instrument panel electrical feed. If the readings are erratic, there may be a fault in the instrument panel assembly. If the fault lies in the temperature gauge alone, check it as follows.
13 If the gauge needle remains at the "cold" end of the scale, disconnect the wiring connector from the sensor unit, and ground the temperature gauge wire (see "Wiring diagrams" for details) to the cylinder head. If the needle then deflects when the ignition is switched on, the sensor unit is proved faulty, and should be renewed. If the needle still does not move, remove the instrument panel (Chapter 12) and check the continuity of the wiring between the sensor unit and the gauge, and the feed to the gauge unit. If continuity is shown, and the fault still exists, then the gauge is faulty and should be renewed.
14 If the gauge needle remains at the "hot" end of the scale, disconnect the sensor wire. If the needle then returns to the "cold" end of the scale when the ignition is switched on, the sensor unit is proved faulty and should be renewed. If the needle still does not move, check the remainder of the circuit as described previously.

# Chapter 3 Cooling, heating and ventilation systems

7.6 If the coolant pump is a tight fit, draw the pump out of position using two jacking bolts (arrows)

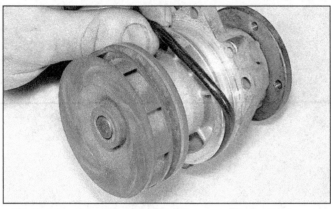

7.7 Recover the sealing ring from the rear of the coolant pump

## Removal

15  Either partially drain the cooling system to just below the level of the sensor (see Chapter 1), or have ready a suitable plug which can be used to plug the sensor aperture while it is removed. If a plug is used, take great care not to damage the sensor unit aperture, and do not use anything which will allow foreign matter to enter the cooling system.
16  Disconnect the battery negative lead.
**Caution:** *If the radio in your vehicle is equipped with an anti-theft system, make sure you have the correct activation code before disconnecting the battery.*
17  Disconnect the wiring from the sensor (see paragraph 10).
18  Unscrew the sensor unit from the cylinder head and recover its sealing washer.

## Installation

19  Fit a new sealing washer to the sensor unit and fit the it to the head, tightening it securely.
20  Reconnect the wiring connector then refill the cooling system (see Chapter 1).

## Fuel injection system coolant temperature sensor

### Testing

21  The sensor unit is screwed into the left-hand side of the cylinder head. There are two sensors screwed into the head; the fuel injection system temperature sensor is the front sensor of the two, the rear one being the temperature gauge sensor. Testing of these sensors should be entrusted to a BMW dealer or other qualified repair shop (see Chapter 4).

### Removal and installation

22  Refer to paragraphs 15 to 20.

## 7   Coolant pump - removal and installation

**Warning:** *Wait until the engine is completely cool before beginning this procedure.*

### Removal

*Refer to illustrations 7.6 and 7.7*

1  Drain the cooling system (see Chapter 1).

2  Remove the fan shroud, the cooling fan and, on engines with a mechanical fan, the clutch (see Section 4).
3  Detach the hoses from the thermostat housing (see Section 3).
4  Loosen the coolant pump pulley bolts then remove the drivebelt (see Chapter 1).
5  Unscrew the retaining bolts and remove the pulley from the pump, noting which way around it is fitted.
6  Loosen and remove the pump retaining bolts/nuts (as applicable) and withdraw the pump. If the pump is a tight fit, screw two M6 bolts into the jacking holes on either side of the pump and use the bolts to draw the pump out of position **(see illustration)**.
7  Recover the sealing ring from the rear of the pump **(see illustration)**.

### Installation

8  Fit a *new* sealing ring to the rear of the pump and lubricate it with a smear of grease to ease installation.
9  Locate the pump in position and install the retaining bolts/nuts. Tighten the bolts/nuts evenly and progressively to the specified torque, making sure the pump is drawn squarely into position.
10  Install the pulley to the pump, making sure it is the correct way around, and screw in its retaining bolts.
11  Install the drivebelt (see Chapter 1) then securely tighten the pulley bolts.
12  Reconnect the coolant hoses to the thermostat housing.
13  Install the cooling fan assembly (see Section 4).
14  Refill the cooling system (see in Chapter 1).

## 8   Heating and ventilation system - general information

1  The heating/ventilation system consists of a four-speed blower motor, face-level vents in the center and at each end of the dash, and air ducts to the front and rear footwells.
2  The control unit is located in the dash, and the controls operate flap valves to deflect and mix the air flowing through the various

parts of the heating/ventilation system. The flap valves are contained in the air distribution housing, which acts as a central distribution unit, passing air to the various ducts and vents.
3  Cold air enters the system through the grille at the rear of the engine compartment. A pollen filter is fitted to the inlet to filter out dust, spores and soot from the incoming air.
4  The airflow, which can be boosted by the blower, then flows through the various ducts, according to the settings of the controls. Stale air is expelled through ducts at the rear of the vehicle. If warm air is required, the cold air is passed through the heater core, which is heated by the engine coolant.
5  If necessary, the outside air supply can be closed off, allowing the air inside the vehicle to be recirculated. This can be useful to prevent unpleasant odors entering from outside the vehicle, but should only be used briefly, as the recirculated air inside the vehicle will soon deteriorate.
6  Certain models may be fitted with heated front seats. The heat is produced by electrically heated mats in the seat and backrest cushions (see Chapter 12). The temperature is regulated automatically by a thermostat, and cannot be adjusted.

## 9   Heater/ventilation components (3-Series models) - removal and installation

**Warning:** *On models equipped with airbags, always disable the airbag system before working in the vicinity of the impact sensors, steering column or instrument panel to avoid the possibility of accidental deployment of the airbag, which could cause personal injury (see Chapter 12).*

## Models without air conditioning

### Heater/ventilation control unit

*Refer to illustrations 9.4, 9.5a, 9.5b, 9.6 and 9.7*

1  Disconnect the battery negative lead.
**Caution:** *If the radio in your vehicle is equipped with an anti-theft system, make*

# Chapter 3  Cooling, heating and ventilation systems     3-7

9.4 Pull off the knobs from each of the heater controls

9.5a Remove the retaining screws (arrows) . . .

9.5b . . . and remove the heater control unit front panel

sure you have the correct activation code before disconnecting the battery.

2   Remove the clock/multi-information display, if applicable (see Chapter 12).

3   Unclip the storage compartment from the front of the center console and remove it, disconnecting the wiring from the cigarette lighter.

4   Pull off the control knobs from each of the heater controls **(see illustration)**.

5   Loosen and remove the retaining screws and unclip the front panel from the heater control unit **(see illustrations)**.

6   Free the control unit from the rear of the dash then maneuver it downwards and out through the center console aperture **(see illustration)**.

7   Disconnect the wiring connector(s) then unclip the control cables and release each cable from the control unit **(see illustration)**. Note each cables correct fitted location and routing; to avoid confusion on installing label each cable as it is disconnected.

8   Installation is reversal of removal. Ensure that the control cables are correctly routed and reconnected to the control panel, as noted before removal. Clip the outer cables in position and check the operation of each knob/lever before installing the storage compartment to the center console.

### Heater/ventilation control cables

9   Remove the heater/ventilation control unit from the dash as described above in paragraphs 1 to 6, detaching the relevant cable from the control unit.

10   Remove the glovebox (see Chapter 11).

11   Follow the run of the cable behind the dash, taking note of its routing, and disconnect the cable from the air distribution housing.

12   Fit the new cable by reversing the removal procedure, ensuring that it is correctly routed and free from kinks and obstructions. Check the operation of the control knob then install the control unit as described previously in this Section.

### Heater core

*Refer to illustrations 9.18, 9.19a, 9.19b, 9.22, 9.23, 9.24a and 9.24b*

**Warning:** *Wait until the engine is completely cool before beginning this procedure.*

13   Working in the engine compartment, remove the rubber seal from the top of the heating/ventilation system inlet and release the grille from the inlet. On models where the grille is an integral part of the windshield wiper motor cover panel, to improve access remove the wiper arms and remove the one-piece cover panel (see Chapter 12).

14   Remove the retaining screws and free the wiring harness duct from the inlet duct.

15   Loosen and remove the retaining screws and retaining plate and remove the inlet from the firewall. **Note:** *On six-cylinder engines it may be necessary to remove the injector and spark plug covers from the engine to enable the inlet to be removed.*

16   Unscrew the expansion tank cap (referring to the **Warning** in Section 1) to release any pressure present in the cooling system then securely install the cap.

17   Clamp both heater hoses as close to the firewall as possible to minimize coolant loss. Alternatively, drain the cooling system (see Chapter 1).

18   Unscrew the heater core pipe union nut

9.6 Maneuver the control unit downwards and out through the center console aperture

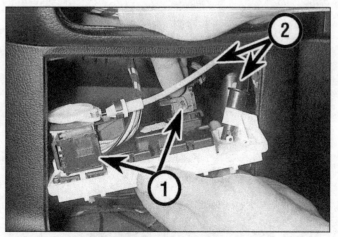

9.7 Release the retaining clips and disconnect the wiring connectors (1) and cables (2)

# 3-8  Chapter 3  Cooling, heating and ventilation systems

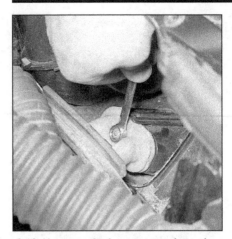

9.18 Unscrew the heater core pipe union retaining nut . . .

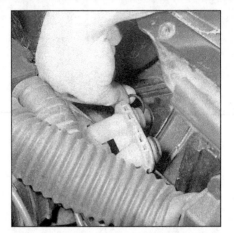

9.19a . . . then detach the pipes and recover the sealing rings

9.19b Loosen and remove the nut and washer (arrow) from the pipe stud

and disconnect the pipe union **(see illustration)**. Be prepared for coolant spillage as the nut is loosened and mop up any coolant.

19  Recover the sealing rings from the heater pipes and unscrew the nut and washer from the pipe retaining stud **(see illustrations)**.

20  Loosen and remove the driver's side lower dash panel retaining screws then unclip the panel and remove it from the vehicle.

21  Be prepared for coolant spillage and position a suitable container beneath the union on the end of the core. To be safe also cover the carpet with rags.

22  Unclip the wiring from the side of the coolant pipes then unscrew the bolt securing the coolant pipes to the heater core and allow the coolant to drain into the container **(see illustration)**.

23  Once the flow of coolant stops, remove the pipes from the heater core and recover the sealing rings **(see illustration)**.

24  Remove the retaining screws and slide the heater core out of position **(see illustrations)**. **Note:** *Keep the heater core unions uppermost as the core is removed to prevent coolant spillage. Mop up any spilled coolant immediately and wipe the affected area with a damp cloth to prevent staining.*

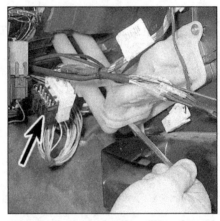

9.22 Unclip the wiring connectors (arrow) then unscrew the retaining bolt . . .

25  Installation is the reverse of removal, using new sealing rings. On completion, refill the cooling system (see Chapter 1).

## Heater blower motor

*Refer to illustrations 9.27, 9.28a, 9.28b, 9.28c and 9.28d*

26  Disconnect the battery negative cable.

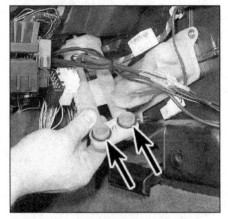

9.23 . . . and detach the coolant pipes from the end of the core. Recover the sealing rings (arrows)

**Caution:** *If the radio in your vehicle is equipped with an anti-theft system, make sure you have the correct activation code before disconnecting the battery. Remove the engine compartment heater inlet (see steps 13 through 15).*

27  Unclip the blower motor end covers and remove them from the housing **(see illustra-

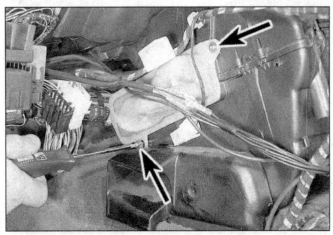

9.24a Remove the retaining screws (arrow) . . .

9.24b . . . and slide the heater core out of position

## Chapter 3 Cooling, heating and ventilation systems

9.27 Unclip the end covers and remove them from the motor housing

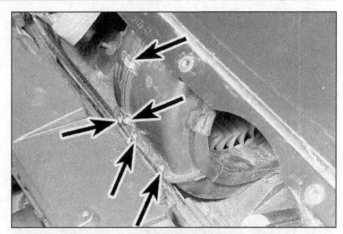

9.28a Remove the retaining clips (arrows) . . .

9.28b . . . and unclip the blower motor covers

9.28c Unhook the retaining clip . . .

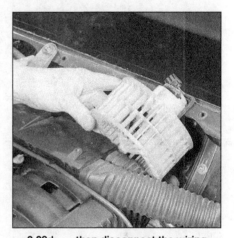

9.28d . . . then disconnect the wiring connector and maneuver the motor out of position

tion). Where necessary, also remove the pollen filter.

28  Release the retaining clips and lift off the blower motor covers. Disconnect the motor wiring then release the motor retaining clip and maneuver the motor out of position (see illustrations). Note: *On six-cylinder models,* *it may be necessary to remove the injector cover from the head to gain the required clearance necessary to remove the motor (see Chapter 4B).*

29  Installation is a reversal of the removal procedure making sure the motor is correctly clipped into the housing and the housing covers are securely installed.

### Heater blower motor resistor

*Refer to illustrations 9.31 and 9.32*

30  Remove the glovebox (see Chapter 11).
31  Disconnect the wiring connector from the resistor (see illustration).
32  Release the retaining clips and maneu-

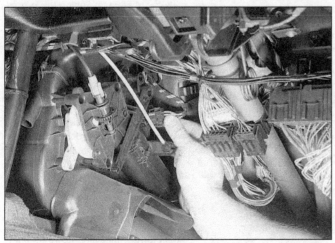

9.31 Disconnect the wiring connector . . .

9.32 . . . then release the retaining clips and remove the blower motor resistor from the housing

## Chapter 3 Cooling, heating and ventilation systems

10.4 To detach the climate control assembly from the center console, remove these two screws (arrows) from the radio recess

10.5 Push the climate control assembly into the center console, unplug the electrical connectors and then pull it out through the radio recess

ver the resistor out from the front of the air distribution housing (see illustration).

33 Installation is the reverse of removal.

### Heater coolant valve

**Warning:** *Wait until the engine is completely cool before beginning this procedure.*

34 The coolant valve is mounted onto the engine compartment firewall. Unscrew the expansion tank cap (referring to the **Warning** in Section 1) to release any pressure present in the cooling system then securely install the cap.

35 Clamp both heater hoses as close to the coolant valve as possible to minimize coolant loss.

36 Disconnect the valve wiring connector(s).

37 Loosen the retaining clips and disconnect the hoses from the valve then unclip the valve and remove it from the engine compartment.

38 Installation is the reverse of removal.

## Models equipped with air conditioning

**Note:** *The following information is only applicable to manually controlled air conditioning systems. At the time of writing no information was available on models with the automatic "Climatronic" system.*

### Heater control unit - models with manually adjusted controls

39 Refer to the information given in paragraphs 1 to 8.

### Heater control unit - models automatic air conditioning system

40 Remove the clock/multi-information display, if applicable (see Chapter 12).

41 Reach in behind the control unit and press it out from the dash.

42 Disconnect the wiring connectors and remove the control unit.

43 Installation is the reverse of removal.

### Heater core

**Warning:** *Wait until the engine is completely cool before beginning this procedure.*

44 Remove the center console (see Chapter 11).

45 Remove the heater control unit as described earlier.

46 Follow the procedure outlined in steps 13 through 19.

47 From inside the vehicle, be prepared for coolant spillage and position a suitable container beneath the union on the end of the heater core. To be safe also cover the carpet with rags.

48 Unscrew the bolts securing the coolant pipes to the heater core and allow the coolant to drain into the container.

49 Once the flow of coolant stops, disconnect the pipes from the heater core and recover the sealing rings.

50 Release the air conditioning system control unit from its retaining clips and disconnect its wiring connectors.

51 Release the wiring harness from its retaining clips and position it clear of the front of the air distribution housing.

52 Loosen and remove the retaining bolts and remove the cover from the air distribution housing.

53 Carefully unclip the linkage lever and release the air ducts from the housing then maneuver the heater core out of position.

**Note:** *Keep the heater core unions uppermost as the core is removed to prevent coolant spillage. Mop up any spilled coolant immediately and wipe the affected area with a damp cloth to prevent staining.*

54 Installation is the reverse of removal, using new sealing rings. On completion, refill the cooling system (see Chapter 1).

### Heater blower motor

55 Remove the heating/ventilation inlet (see steps 13 through 15).

56 Release the retaining clips and remove the cover from the top of the blower motor.

57 Disconnect the wiring from the blower motor.

58 Release the side retaining clips and remove the top section of the blower motor housing.

59 Release the retaining clips and free the fan from each side of the motor spindle.

60 Make identification marks on the fans, to ensure that they are correctly positioned on installation then lift the motor and fans separately out from the housing.

61 Installation is the reverse of removal. Install the motor and fans separately, making sure the fans are correctly positioned, and clip the fans onto the motor spindle, aligning the fan cutouts with the spindle locating pins.

### Heater blower motor resistor

62 Follow the procedure outlined in steps 30 through 33.

### Heater coolant valve

63 Follow the procedure outlined in steps 34 through 38

## 10 Heater/ventilation components (Z3 models) - removal and installation

1 Disconnect the cable from the negative battery terminal.

2 If you're replacing the heater control valve or the heater core, drain the cooling system. **Warning:** *Wait until the engine is completely cool before beginning this procedure.*

### Climate control assembly

*Refer to illustrations 10.4 and 10.5*

3 Remove the radio (see Chapter 12).

4 Remove the two climate control assembly retaining screws (see illustration) from the radio recess..

5 Push the climate control assembly into the console (see illustration), turn the assembly over, unplug the electrical connectors and remove the assembly through the radio cavity.

6 Installation is the reverse of removal.

### Heater control valve

*Refer to illustration 10.7*

7 Unplug the electrical connector from the heater control valve (see illustration).

8 Locate the heater hose clamps at the

# Chapter 3 Cooling, heating and ventilation systems 3-11

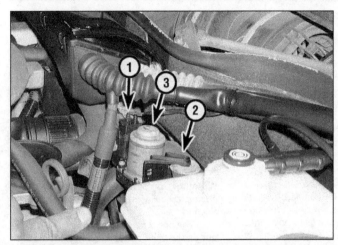

10.7 To remove the heater control valve, unplug the electrical leads (1), disengage the upper bracket (2) and the lower bracket (not visible in this photo) and remove the valve (3)

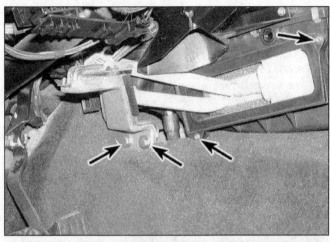

10.12 Remove the knee bolster bracket bolts (left arrows) and the bracket, and then remove the heater core retaining screws (right arrows)

heater control valve, which is located at the rear of the engine compartment, in front of the firewall. Loosen the clamps. Clearly label the heater hoses, then disconnect the hoses from the valve.

9  Detach the heater control valve from the lower mounting bracket, then detach it from the upper mounting bracket.

10  Installation is the reverse of removal. Refill the cooling system when you're done (see Chapter 1).

## Heater core

Refer to illustrations 10.12 and 10.14

11  Locate the heater hose clamps at the firewall, loosen the clamps and disconnect the heater hoses from the heater core pipes.

12  Remove the trim panel from the lower left part of the dash and remove the lower knee bolster (see Chapter 11). Remove the bracket for the knee bolster **(see illustration)**.

13  Remover the heater core retaining screws **(see illustration 10.8)**.

14  Squeeze the coolant pipe retaining clips **(see illustration)** together, then detach the coolant pipes from the heater core.

15  Pull the heater core sideways and remove it.

16  Installation is the reverse of removal.

17  Refill the cooling system when you're done (see Chapter 1).

## Blower motor

Refer to illustrations 10.21 and 10.22

18  Remove the rubber seal from the air intake grille in front of the windshield (see Chapter 12).

19  Remove the air intake grille retainers and remove the grille (see Chapter 12).

20  Remove the windshield wiper motor and linkage (see Chapter 12).

21  Disconnect the blower motor cover retaining straps **(see illustration)** and remove the cover.

22  Remove the fan cover and unplug the electrical connector **(see illustration)** from the blower motor resistor pack.

23  Remove the fan retaining strap **(see illustration 10.18)**. Remove the blower motor.

24  Installation is the reverse of removal.

10.14 To disconnect the heater core pipes from these fittings, squeeze the upper and lower tabs on the fittings (arrows) and pull firmly (there are O-rings inside the fittings, so the heater core pipes are a tight fit)

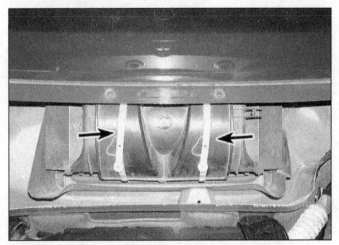

10.21 To remove the blower motor cover, disconnect these two straps (arrows)

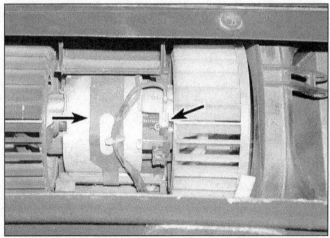

10.22 To remove the blower motor, unplug the electrical connector (right arrow) and remove the retaining strap (left arrow)

## 11 Air conditioning system - general information and precautions

### General information

1 The air conditioning system enables the temperature of incoming air to be lowered, and dehumidifies the air, which makes for rapid demisting and increased comfort.
2 The cooling side of the system works in the same way as a domestic refrigerator. Refrigerant gas is drawn into a belt-driven compressor and passes into a condenser mounted in front of the radiator, where it loses heat and becomes liquid. The liquid passes through an expansion valve to an evaporator, where it changes from liquid under high pressure to gas under low pressure. This change is accompanied by a drop in temperature, which cools the evaporator. The refrigerant returns to the compressor and the cycle begins again.
3 Air blown through the evaporator passes to the air distribution unit, where it is mixed with hot air blown through the heater core to achieve the desired temperature in the passenger compartment.
4 The heating side of the system works in the same way as on models without air conditioning (see Section 8).
5 The operation of the system is controlled electronically by the coolant temperature switch (see Section 6), and pressure switches which are screwed into the compressor high-pressure line. Any problems with the system should be referred to a BMW dealer or other qualified repair shop.

### Precautions

**Warning:** *The refrigerant is potentially dangerous and should only be handled by qualified persons. If it is splashed onto the skin it can cause frostbite. It is not itself poisonous, but in the presence of an open flame (including a cigarette) it forms a poisonous gas. Uncontrolled discharging of the refrigerant is dangerous and potentially damaging to the environment.*
**Caution:** *Do not operate the air conditioning system if it is known to be short of refrigerant, as this may damage the compressor.*
6 When an air conditioning system is fitted, it is necessary to observe special precautions whenever dealing with any part of the system, its associated components and any items which require disconnection of the system. If for any reason the system must be disconnected, entrust this task to your BMW dealer or an automotive air conditioning specialist.

## 12 Air conditioning system components - removal and installation

**Warning:** *Do not attempt to open the refrigerant circuit. Refer to the precautions given in Section 11.*

The only operation which can be carried out easily without discharging the refrigerant is replacement of the compressor drivebelt, which is covered in Chapter 1. All other operations must be referred to a BMW dealer or an automotive air conditioning specialist.

If necessary the compressor can be unbolted and moved aside, without disconnecting its flexible hoses, after removing the drivebelt.

# Chapter 4 Part A
# Fuel and exhaust systems - four-cylinder engines

## Contents

| | Section | | Section |
|---|---|---|---|
| Air cleaner assembly - removal and installation | 2 | Fuel injection system - general information | 6 |
| Air filter element replacement | See Chapter 1 | Fuel injection system - testing and adjustment | 10 |
| Engine management system check | See Chapter 1 | Fuel pump/fuel gauge sender unit - removal and installation | 8 |
| Exhaust system - general information, removal and installation | 14 | Fuel tank - removal and installation | 3 |
| Exhaust system and mounting check | See Chapter 1 | General information and precautions | 1 |
| Fuel filter replacement | See Chapter 1 | Manifolds - removal and installation | 13 |
| Fuel gauge sender unit - removal and installation | 9 | Throttle body - removal and installation | 11 |
| Fuel injection system components - removal and installation | 12 | Throttle cable - removal, installation and adjustment | 4 |
| Fuel injection system - depressurization and priming | 7 | Throttle pedal - removal and installation | 5 |

## Specifications

### Fuel system data
Fuel pump type.................................................. Electric, immersed in tank
Fuel pressure regulator rating............................. 43.5 ± 1 psi
Specified idle speed........................................... 850 ± 40 rpm (not adjustable - controlled by ECM)
Specified idle mixture CO content....................... 0.5 to 1.5% (not adjustable - controlled by ECM)

### Torque specifications                            Ft-lbs (unless otherwise indicated)
Intake manifold nuts
    M6 nuts...................................................... 84 in-lbs
    M7 nuts...................................................... 132 in-lbs
    M8 nuts...................................................... 16
Exhaust manifold nuts
    M6 nuts...................................................... 84 in-lbs
    M7 nuts...................................................... 15
    M8 nuts...................................................... 16
Crankshaft position sensor bolt........................... 60 in-lbs
Camshaft position sensor bolt............................. 60 in-lbs
Fuel rail-to-intake manifold bolts.......................... 84 in-lbs
Fuel tank mounting bolts..................................... 17
Fuel tank retaining strap bolts.............................. 72 in-lbs

# Chapter 4 Part A Fuel and exhaust systems - four-cylinder engines

2.2 Removing the air cleaner assembly

4.1a Withdraw the cover from the throttle linkage . . .

## 1 General information and precautions

### General information

The fuel supply system consists of a fuel tank (which is mounted under the rear of the vehicle, with an electric fuel pump immersed in it), a fuel filter, and fuel feed and return lines. The fuel pump supplies fuel to the fuel rail, which acts as a reservoir for the four fuel injectors which inject fuel into the intake ports. The fuel filter incorporated in the feed line from the pump to the fuel rail ensures that the fuel supplied to the injectors is clean.

Refer to Section 6 for further information on the operation of the fuel injection system, and to Section 14 for information on the exhaust system.

### Precautions

**Warning 1:** *Gasoline is extremely flammable, so take extra precautions when you work on any part of the fuel system. Don't smoke or allow open flames or bare light bulbs near the work area, and don't work in a garage where a gas-type appliance (such as a water heater or a clothes dryer) is present. Since gasoline is carcinogenic, wear latex gloves when there's a possibility of being exposed to fuel, and, if you spill any fuel on your skin, rinse it off immediately with soap and water. Mop up any spills immediately and do not store fuel-soaked rags where they could ignite. When you perform any kind of work on the fuel system, wear safety glasses and have a Class B type fire extinguisher on hand.*

**Warning 2:** *Residual pressure will remain in the fuel lines long after the vehicle was last used. When disconnecting any fuel line, first depressurize the fuel system (see Section 7).*

## 2 Air cleaner assembly - removal and installation

### Removal

*Refer to illustration 2.2*

1  Remove the airflow meter and air cleaner cover assembly (see Section 12).
2  Loosen the two nuts securing the air cleaner casing to the mounting bracket, then pull the casing upwards, and unclip the intake duct from the front body panel **(see illustration)**.

### Installation

3  Installation is a reversal of removal, but ensure that the lower mounting engages with the plastic lug on the body, and install the air-flow meter (see Section 12).

## 3 Fuel tank - removal and installation

Refer to Part B of this Chapter.

## 4 Throttle cable - removal, installation and adjustment

*Refer to illustrations 4.1a, 4.1b and 4.1c*

Proceed as described in Part B of this Chapter, but note that it may be necessary to remove the securing screw(s), and fold back or withdraw the cover from the top of the throttle linkage for access to the throttle cable **(see illustrations)**.

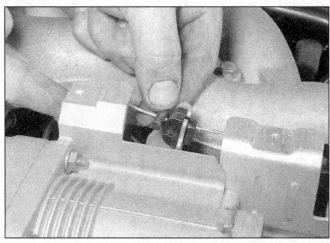

4.1b . . . for access to the throttle cable

4.1c Throttle cable adjuster screw (arrow)

# Chapter 4 Part A Fuel and exhaust systems - four-cylinder engines

## 5 Throttle pedal - removal and installation

Refer to Part B of this Chapter.

## 6 Fuel injection system - general information

1 An integrated engine management system known as DME (Digital Motor Electronics) is fitted to all models, and the system controls all fuel injection and ignition system functions using a central ECM (Engine Control Module). A Bosch Motronic engine management system is fitted to all models.
2 On most models, the system incorporates a closed-loop catalytic converter and an evaporative emission control system, and complies with the very latest emission control standards. Refer to Chapter 5B for information on the ignition side of the system; the fuel side of the system operates as follows.
3 The fuel pump (which is immersed in the fuel tank) supplies fuel from the tank to the fuel rail, via a filter. Fuel supply pressure is controlled by the pressure regulator in the fuel rail. When the optimum operating pressure of the fuel system is exceeded, the regulator allows excess fuel to return to the tank.
4 The electrical control system consists of the ECM, along with the following sensors:
  a) *Airflow meter - informs the ECM of the quantity of air entering the engine.*
  b) *Throttle position sensor - informs the ECM of the throttle position, and the rate of throttle opening/closing.*
  c) *Coolant temperature sensor - informs the ECM of engine temperature.*
  d) *Crankshaft position sensor - informs the ECM of the crankshaft position and speed of rotation.*
  e) *Camshaft position sensor - informs the ECM of the intake camshaft position.*
  f) *Oxygen sensor - informs the ECM of the oxygen content of the exhaust gases (explained in greater detail in Chapter 6).*

5 All the above signals are analyzed by the ECM which selects the fueling response appropriate to those values. The ECM controls the fuel injectors (varying the pulse width - the length of time the injectors are held open - to provide a richer or weaker mixture, as appropriate). The mixture is constantly varied by the ECM, to provide the best setting for cranking, starting (with either a hot or cold engine), warm-up, idle, cruising and acceleration.
6 The ECM also has full control over the engine idle speed, via an auxiliary air valve which bypasses the throttle valve. When the throttle valve is closed, the ECM controls the opening of the valve, which in turn regulates the amount of air entering the manifold, and so controls the idle speed.
7 The ECM also controls the exhaust and evaporative emission control systems, which are described in detail in Chapter 6.
8 The throttle body is coolant heated.
9 These engines are equipped with a Differential Air Intake System (DISA). Variable length intake tracts incorporated in the intake manifold are operated by a butterfly valve according to engine speed and load. This improves engine torque at low and medium engine speeds. The butterfly valve is operated by a vacuum actuator installed under the manifold.
10 If there is an abnormality in any of the readings obtained from the sensors, the ECM enters its back-up mode. In this event, it ignores the abnormal sensor signal and assumes a pre-programmed value which will allow the engine to continue running (albeit at reduced efficiency). If the ECM enters this back-up mode, the relevant fault code will be stored in the ECM memory.
11 If a fault is suspected, the vehicle should be taken to a BMW dealer at the earliest opportunity. A complete test of the engine management system can then be carried out, using a special electronic diagnostic test unit which is simply plugged into the system's diagnostic connector.

## 7 Fuel injection system - depressurization and priming

### Depressurization

1 Remove the fuel pump fuse from the fuse box. The fuse is located in the engine compartment fuse box, and the exact location is given on the fuse box cover (see Chapter 12).
2 Start the engine, and wait for it to stall. Switch off the ignition.
3 Remove the fuel filler cap.
4 The fuel system is now depressurized.
**Note:** *Place a rag around fuel lines before disconnecting, to prevent any residual fuel from spilling onto the engine.*
5 Disconnect the battery negative cable before working on any part of the fuel system. **Caution:** *If the stereo in your vehicle is equipped with an anti-theft system, make sure you have the correct activation code before disconnecting the battery.*

### Priming

6 Install the fuel pump fuse, then switch on the ignition and wait for a few seconds for the fuel pump to run, building up fuel pressure. Switch off the ignition unless the engine is to be started.

## 8 Fuel pump/fuel gauge sender unit - removal and installation

Refer to Part B of this Chapter.

## 9 Fuel gauge sender unit - removal and installation

Refer to Part B of this Chapter.

## 10 Fuel injection system - testing and adjustment

### Testing

1 If a fault appears in the fuel injection system, first ensure that all the system wiring connectors are securely connected and free of corrosion. Ensure that the fault is not due to poor maintenance; i.e., check that the air filter element is clean, the spark plugs are in good condition and correctly gapped, the cylinder compression pressures are correct, and that the engine breather hoses are clear and undamaged, referring to Chapters 1, 2 and 5 for further information.
2 If these checks fail to reveal the cause of the problem, the vehicle should be taken to a BMW dealer or other qualified repair shop for testing. A wiring block connector is incorporated in the engine management circuit, into which a special electronic diagnostic tester can be plugged. The connector is clipped to the right-hand suspension tower. The tester will locate the fault quickly and simply, alleviating the need to test all the system components individually, which is a time-consuming operation that also carries a risk of damaging the ECM.

### Adjustment

3 Experienced home mechanics with a considerable amount of skill and equipment (including a tachometer and an accurately calibrated exhaust gas analyzer) may be able to check the exhaust CO level and the idle speed. However, if these are found to be in need of adjustment, the car *must* be taken to a BMW dealer or other qualified repair shop for further testing.
4 To adjust the CO level or idle speed, special diagnostic equipment is required.

## 11 Throttle body - removal and installation

**Warning:** *Wait until the engine is completely cool before beginning this procedure.*

### Removal

*Refer to illustration 11.9*
1 Disconnect the battery negative cable.
**Caution:** *If the stereo in your vehicle is equipped with an anti-theft system, make sure you have the correct activation code before disconnecting the battery.*
2 Where applicable, remove the securing screw(s), and lift the cover from the throttle body.
3 Disconnect the throttle cable(s) from the throttle body and the cable support bracket.
4 Loosen the clamp screw securing the air intake ducting to the throttle body.
5 Remove the upper section of the air cleaner, complete with the airflow meter and the air intake ducting (see Section 12).
6 Disconnect the throttle position sensor wiring plug.

**11.9 Throttle body securing nuts and bolts (arrows)**

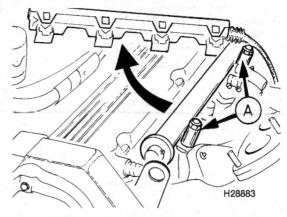

**12.4 Detach the wiring ducting from the fuel rail**

A Fuel rail securing bolts

7 Disconnect the vacuum hoses from the throttle body.
8 Disconnect the coolant hose(s) from the throttle body. Be prepared for coolant spillage, and clamp the open ends of the hoses to prevent further coolant loss.
9 Unscrew the four securing nuts/bolts, and withdraw the throttle body from the intake manifold (see illustration). Recover the throttle body heater which fits between the throttle body and the manifold, and recover the gaskets which fit either side of the heater.

## Installation

10 Installation is a reversal of removal, bearing in mind the following points.
  a) Use new gaskets on both sides of the throttle body heater.
  b) Reconnect and if necessary adjust the throttle cable (see Section 4).
  c) On completion, check and if necessary top-up the coolant level (see Chapter 1).

## 12 Fuel injection system components - removal and installation

### Engine Control Module (ECM)

1 Proceed as described in Part B of this Chapter, but note that for access to the ECM, the battery must be removed (see Chapter 5A).

### Fuel rail and injectors

**Warning:** Refer to the Warning in Section 1 before proceeding.
**Note:** New fuel injector O-rings should be used on installation.

#### Removal

Refer to illustration 12.4

2 Depressurize the fuel system (see Section 7), then disconnect the battery negative cable. **Caution:** If the stereo in your vehicle is equipped with an anti-theft system, make sure you have the correct activation code before disconnecting the battery.
3 Remove the upper section of the intake manifold (see Section 13).
4 Pull up the wiring ducting to release the wiring plugs from the fuel injectors (see illustration).
5 Loosen the hose clamps and disconnect the fuel feed and return hoses from the fuel rail. Be prepared for fuel spillage and take adequate fire precautions. Plug the open ends of the pipes and hoses to prevent dirt entry and further fuel spillage.
6 Disconnect the vacuum hose from the fuel pressure regulator.
7 Unscrew the two bolts securing the fuel rail to the intake manifold.
8 Carefully pull the fuel rail upwards to release the fuel injectors from the cylinder head, then withdraw the complete fuel rail/fuel injector assembly.
9 To remove a fuel injector from the fuel rail, proceed as follows.
  a) Pry off the metal securing clip, using a screwdriver.
  b) Pull the fuel injector from the fuel rail.

#### Installation

Refer to illustration 12.11

10 Before installation, it is wise to replace all the fuel injector O-rings as a matter of course.
11 Check that the plastic washer at the bottom of each injector is positioned above the lower O-ring (see illustration).
12 Lightly lubricate the fuel injector O-rings with a little petroleum jelly or clean engine oil.
13 Where applicable, install the fuel injectors to the fuel rail, ensuring that the securing clips are correctly fitted. Note that the injectors should be positioned so that the wiring sockets are uppermost when the assembly is reinstalled.
14 Slide the fuel rail/fuel injector assembly into position, ensuring that the injectors engage with their bores in the intake manifold.
15 Further installation is a reversal of removal, but install the upper section of the intake manifold (see Section 13), and pressurize the fuel system (install the fuel pump fuse and switch on the ignition) and check for leaks before starting the engine.

### Fuel pressure regulator

**Warning:** Refer to the **Warning** notes in Section 1 before proceeding.
**Note:** New O-rings may be required on installation.

#### Removal

Refer to illustrations 12.22 and 12.23
16 Depressurize the fuel system (see Section 7), then disconnect the battery negative cable. **Caution:** If the stereo in your vehicle is equipped with an anti-theft system, make sure you have the correct activation code before disconnecting the battery.
17 Pull the vacuum hose from the pressure regulator.
18 Where applicable, to improve access, disconnect the breather hose from the front of the valve cover.
19 Note the angle of the vacuum intake on the pressure regulator, so that it can be reinstalled in the same position.
20 Where applicable, to give sufficient clearance to remove the pressure regulator, unbolt the engine lifting bracket from the cylinder head.

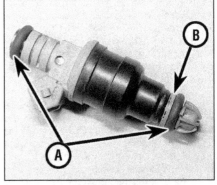

**12.11 Fuel injector O-rings (A) and plastic washer (B)**

# Chapter 4 Part A  Fuel and exhaust systems - four-cylinder engines

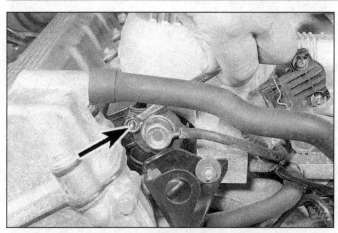

12.22  Unscrewing the pressure regulator clamping bracket nut (arrow)

12.23  Pull out the pressure regulator retaining ring locking clip (A)

21  On models where the pressure regulator is secured by a clamping ring, unscrew the clamp bolt and remove the clamping ring and lift off the thrustwasher.

22  On models where the regulator is secured by a clamping bracket retained by a nut, unscrew the securing nut and withdraw the clamping bracket **(see illustration)**.

23  On models where the regulator is secured by a retaining ring retained by a locking clip, pull out the locking clip and remove the retaining ring **(see illustration)**.

24  Twist and pull the regulator from the fuel rail. Note that it will be difficult to pull the regulator from the fuel rail due to the tight-fitting O-rings.

## Installation

*Refer to illustration 12.25*

25  Before installation, check the condition of the O-rings, and replace if necessary **(see illustration)**.

26  Installation is a reversal of removal, bearing in mind the following points.

  a)  Ensure that the regulator is pushed firmly into position in the end of the fuel rail.

  b)  Make sure that the regulator vacuum intake is positioned as noted before removal.

  c)  Where applicable, ensure that the thrustwasher is correctly positioned.

  d)  On models where the regulator is retained by a ring and locking clip, make sure that the cut-out in the retaining ring is aligned with the regulator vacuum intake (to avoid damage to the vacuum hose).

  e)  On completion, pressurize the fuel system (install the fuel pump fuse and switch on the ignition) and check for leaks before starting the engine.

## Airflow meter

*Note: A new airflow meter seal will be required on installation.*

### Removal

*Refer to illustrations 12.28a and 12.28b*

27  Disconnect the battery negative cable.

**Caution:** *If the stereo in your vehicle is equipped with an anti-theft system, make sure you have the correct activation code before disconnecting the battery.*

28  Turn the locking collar, and disconnect the wiring plug from the airflow meter **(see illustrations)**.

29  Remove the screw securing the carbon canister hose to the airflow meter.

30  Loosen the hose clamp and disconnect

12.25  Check the condition of the O-rings (arrows)

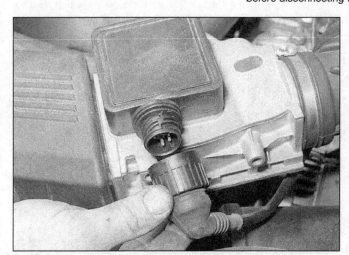

12.28a  Disconnecting the electrical connector from an air-volume type airflow meter, which is used on 1992 through 1995 M42 (1.8L) engines

12.28b  Electrical connector (arrow) on a hot-film (air mass) airflow meter, which is used on 1996 and later M44 (1.9L) engines

**12.36 Disconnecting the throttle position sensor electrical connector**

**12.42 Crankshaft position sensor electrical connector (arrow) - pre-1994 model**

the air ducting from the airflow meter.
31 Release the securing clips, and lift off the air cleaner cover, complete with the airflow meter.
32 Disconnect the breather hose(s) from the air ducting, and the airflow meter (as applicable), and release any wires or hoses from the clips on the air ducting.
33 Working inside the air cleaner cover, unscrew the four securing nuts, and withdraw the airflow meter from the air cleaner cover. Recover the seal.

### Installation

34 Installation is a reversal of removal, but use a new seal between the airflow meter and the air cleaner cover, and ensure that any wire/hoses are securely positioned in their clips.

## Throttle position sensor

*Note: A new O-ring may be required on installation.*

### Removal

*Refer to illustration 12.36*

35 Disconnect the battery negative cable.
**Caution:** *If the stereo in your vehicle is equipped with an anti-theft system, make sure you have the correct activation code before disconnecting the battery.*
36 Disconnect the wiring plug from the sensor **(see illustration)**.
37 Unscrew the two securing screws and withdraw the sensor from the throttle body. Where applicable, recover the O-ring.

### Installation

38 Installation is a reversal of removal, but where applicable, check the condition of the O-ring and replace if necessary, and ensure that the O-ring is correctly positioned.

## Coolant temperature sensor

39 The sensor is located in the left-hand side of the cylinder head. Refer to Chapter 3 for removal and installation details.

## Crankshaft position sensor

### Removal

*Refer to illustrations 12.42 and 12.44*

40 To improve access, remove the viscous cooling fan and fan shroud assembly (see Chapter 3).
41 Remove the upper section of the intake manifold (see Section 13), and release the wiring ducting from the manifold.
42 Trace the wiring back from the sensor to the connector, located on a bracket behind the oil filter. Note that there are two wiring connectors, and the crankshaft position sensor is the lower connector - if both connectors are to be disconnected, mark them to avoid confusion on installation **(see illustration)**.
43 Release the wiring from any clips and brackets on the engine.
44 Unscrew the sensor securing bolt, then withdraw the sensor from its mounting bracket **(see illustration)**.

### Installation

45 Installation is a reversal of removal, bearing in mind the following points.

**12.44 Crankshaft position sensor securing bolt (arrow)**

a) *Ensure that the sensor wiring is securely refitted to its securing clips to avoid the wiring from rubbing on the drivebelt.*
b) *Ensure that the sensor wiring connector is correctly reconnected as noted before removal.*
c) *Where applicable, install the upper section of the intake manifold (see Section 13).*
d) *Install the viscous cooling fan and fan shroud assembly (see Chapter 3).*

## Camshaft position sensor

### Removal

*Refer to illustration 12.51*

46 The sensor is located in the front of the upper timing chain cover.
47 Disconnect the battery negative cable.
**Caution:** *If the stereo in your vehicle is equipped with an anti-theft system, make sure you have the correct activation code before disconnecting the battery.*
48 On 1994 and later models, remove the upper section of the intake manifold (see Section 13), and release the wiring ducting from the manifold.

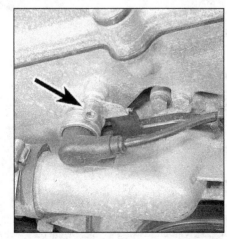

**12.51 Camshaft position sensor securing bolt (arrow)**

# Chapter 4 Part A  Fuel and exhaust systems - four-cylinder engines

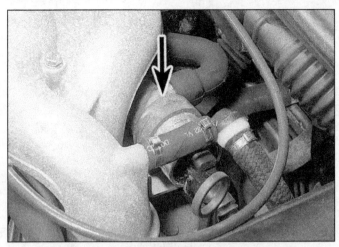

12.56 Idle speed control valve location (arrow)

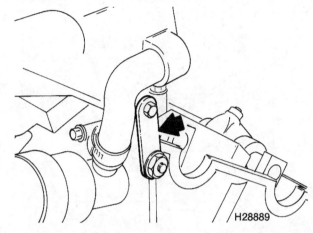

13.7 Intake manifold upper section rear support bracket (arrow)

49  Trace the wiring back from the sensor, and locate the wiring connector attached to the bracket behind the oil filter. Note that there are two connectors, and the camshaft position sensor connector is the top connector - if both connectors are to be disconnected, mark them to avoid confusion on installation. Disconnect the wiring connector.
50  Remove the screw, and/or release the clips securing the wiring ducting to the front of the engine, then release the sensor wiring from the ducting.
51  Unscrew the sensor securing bolt, and withdraw the sensor (see illustration). Recover the O-ring if it is loose.

### Installation
52  Installation is a reversal of removal, bearing in mind the following points.
a)  Make sure that the sensor O-ring is correctly seated.
b)  Ensure that the wiring connector is correctly reconnected as noted before removal.
c)  Where applicable, install the upper section of the intake manifold (see Section 13).

## Oxygen sensor
53  Refer to Chapter 6.

## Idle speed control valve

### Removal
*Refer to illustration 12.56*
54  Disconnect the battery negative cable.
**Caution:** *If the stereo in your vehicle is equipped with an anti-theft system, make sure you have the correct activation code before disconnecting the battery.*
55  Disconnect the wiring plug from the valve.
56  Pull the valve, complete with its rubber mounting, from the retaining bracket (see illustration).
57  Disconnect the hoses from the valve, noting their locations to ensure correct installation.
58  Withdraw the valve.

### Installation
59  Installation is a reversal of removal, but make sure that the hoses are correctly reconnected.

## 13  Manifolds - removal and installation

### Intake manifold upper section
**Warning:** *Wait until the engine is completely cool before beginning this procedure.*
**Note:** *New gaskets will be required on installation.*

### Removal
*Refer to illustrations 13.7, 13.9, 13.15 and 13.17*
1  Disconnect the battery negative cable.
**Caution:** *If the stereo in your vehicle is equipped with an anti-theft system, make sure you have the correct activation code before disconnecting the battery.*
2  Where necessary, to allow sufficient clearance for the manifold to be removed, remove the heater/ventilation intake air ducting from the rear of the engine compartment as follows.
a)  Lift the grille from the top of the ducting (on certain Coupe models, it will be necessary to remove the windshield wiper arms, then remove the plastic securing screws and lift off the complete scuttle grille assembly).
b)  Working through the top of the ducting, remove the screws securing the cable ducting to the air ducting and move the cable ducting clear.
c)  Unscrew the nuts and/or screw(s) securing the air ducting to the firewall (where applicable, bend back the heat shielding for access).
d)  Remove the air ducting by pulling upwards.
3  Move the previously removed cable ducting clear of the valve cover.

4  Where applicable, unscrew the bolt securing the charcoal canister hose to the airflow meter assembly.
5  Loosen the clamp screw securing the air intake ducting to the throttle body.
6  Remove the upper section of the air cleaner, complete with the airflow meter and the air intake ducting (see Section 12).
7  Unbolt the rear manifold support bracket (see illustration).
8  Disconnect the idle speed control valve hose from the rear of the manifold.
9  Loosen the nut securing the manifold to the front support bracket (see illustration).
10  Disconnect the throttle cable from the throttle linkage at the throttle body (see Section 4).
11  Disconnect the throttle position sensor wiring plug.
12  Disconnect the vacuum hoses from the throttle body.
13  Disconnect the vacuum hose from the fuel pressure regulator.
14  Disconnect the coolant hoses from the throttle body. Be prepared for coolant spillage, and clamp the open ends of the hoses to prevent further coolant loss.
15  Where applicable, unscrew the nut

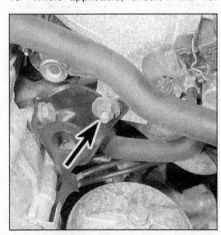

13.9 Loosen the nut (arrow) securing the manifold to the front support bracket

# Chapter 4 Part A  Fuel and exhaust systems - four-cylinder engines

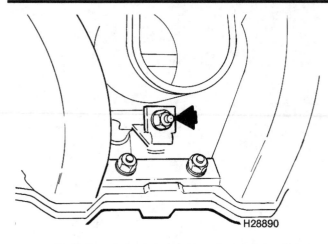

13.15 Throttle body heater securing nut (arrow)

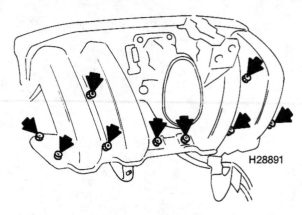

13.17 Intake manifold upper section securing nuts and bolts (arrows)

securing the throttle body heater **(see illustration)**.

16  Disconnect and release any remaining hoses and wiring looms from the manifold, noting their locations and routing to ensure correct installation.

17  Unscrew the securing nuts/bolts and lift off the upper section of the intake manifold **(see illustration)**. Recover the gasket.

### Installation

18  Installation is a reversal of removal, bearing in mind the following points.
   a) *Use new gaskets.*
   b) *Ensure that the upper section of the manifold engages securely on the dowels.*
   c) *Ensure that all hoses are connected to their correct locations as noted before removal.*
   d) *Reconnect and if necessary adjust the throttle cable (see Section 4).*
   e) *Where applicable, on completion check the coolant level (see Chapter 1).*

## Intake manifold lower section

**Warning:** *Wait until the engine is completely cool before beginning this procedure.*
**Note:** *A new gasket will be required on installation.*

### Removal

19  Depressurize the fuel system (see Section 7), then disconnect the battery negative cable. **Caution:** *If the stereo in your vehicle is equipped with an anti-theft system, make sure you have the correct activation code before disconnecting the battery.*

20  Remove the upper section of the manifold as described previously in this Section.

21  Pull up the wiring ducting to release the wiring plugs from the fuel injectors. Move the wiring ducting to one side, clear of the working area.

22  Loosen the hose clips, and disconnect the fuel feed and return hoses from the fuel rail. Be prepared for fuel spillage, and take adequate fire precautions. Plug the open ends of the pipes and hoses to prevent dirt entry and further fuel spillage.

23  Where applicable, unscrew the clamp bolt securing the fuel pipes to the manifold bracket.

24  Where applicable, unbolt the manifold support bracket.

25  Disconnect and release any remaining hoses and wiring looms from the manifold, noting their locations and routing to ensure correct installation.

26  Unscrew the securing nuts and withdraw the manifold from the cylinder head. Recover the gasket.

### Installation

27  Installation is a reversal of removal, bearing in mind the following points.
   a) *Fit the manifold using a new gasket, and tighten the securing nuts to the specified torque. Ensure that the manifold engages correctly with the locating dowels.*
   b) *On M44 engines, remove the old seals above and below the crankcase vent valve and discard them. Install new upper and lower crankcase vent valve seals in the lower intake manifold.*
   c) *Ensure that all wiring and hoses are correctly routed and reconnected as noted before removal.*
   d) *Install the upper section of the manifold as described previously in this Section.*
   e) *On completion, pressurize the fuel system (install the fuel pump fuse and switch on the ignition) and check for leaks before starting the engine.*

## Exhaust manifold

**Note:** *New gaskets and new manifold securing nuts will be required on installation.*

### Removal

28  To improve access, jack up the front of the vehicle and support securely on axle stands.

29  Working underneath the vehicle, unscrew the nuts securing the exhaust front section to the manifold.

30  Working at the transmission exhaust bracket, unscrew the bolt(s) securing the two exhaust mounting clamp halves together.

31  Loosen the bolt securing the clamp halves to the bracket on the transmission, then lower the exhaust downpipes down from the manifold studs. Recover the gasket(s).

32  Working in the engine compartment, unscrew the manifold securing nuts.

33  Withdraw the manifold from the studs and recover the gasket(s).

34  It is possible that some of the manifold studs may be unscrewed from the cylinder head when the manifold securing nuts are unscrewed. In this event, the studs should be screwed back into the cylinder head once the manifolds have been removed, using two manifold nuts locked together.

### Installation

35  Installation is a reversal of removal, but use new gaskets and new manifold securing nuts.

## 14  Exhaust system - general information, removal and installation

### General information

1  The original equipment exhaust system consists of two sections. The front section incorporates the catalytic converter and the oxygen sensor (where applicable) and the center expansion box. The rear section incorporates the rear muffler.

2  The system is suspended throughout its length by rubber mountings and a metal bracket.

### Complete system

**Note:** *New exhaust front section-to-manifold gaskets and securing nuts will be required on installation.*

# Chapter 4 Part A  Fuel and exhaust systems - four-cylinder engines

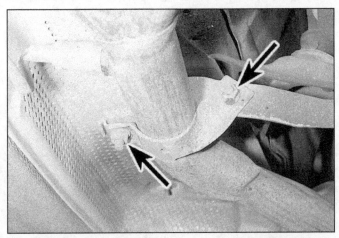

**14.6 Unscrew the exhaust clamp nuts and bolts (arrows)**

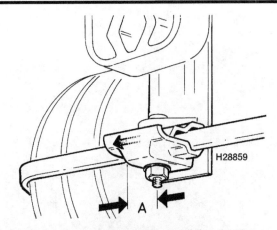

**14.10a Preload the exhaust system by sliding the clamps towards the rear of the muffler box**

*A = 5/8-inch*

## Removal

*Refer to illustration 14.6*

3  Jack up the vehicle and support securely on axle stands.
4  On models with a catalytic converter, turn the securing ring and disconnect the oxygen sensor electrical connector at the bracket under the vehicle.
5  Unscrew the securing nuts and disconnect the exhaust front sections from the manifold. Recover the gasket.
6  Unscrew the clamp bolt securing the two halves of the transmission exhaust mounting bracket together, then unscrew the clamp pivot bolt, and pivot the clamp halves away from the exhaust system **(see illustration)**.
7  Slide the rear exhaust mounting rubbers from the brackets on the exhaust system.
8  Working at the rear of the exhaust system, unscrew the nuts securing the rear muffler box mounting clamps.
9  Withdraw the complete exhaust system from under the vehicle.

## Installation

*Refer to illustrations 14.10a and 14.10b*

10  Installation is a reversal of removal, bearing in mind the following points **(see illustrations)**.

a) *Use new gaskets when reconnecting the exhaust front section to the manifold. Also use new nuts, and coat the threads of the new nuts with copper grease.*
b) *When installing the rear muffler box clamps, slide the clamps onto the muffler box and allow the system to hang so that the mountings are free from tension, then slide the clamps towards the rear of the muffler by 5/8-inch to give a preload, forcing the system towards the front of the vehicle.*

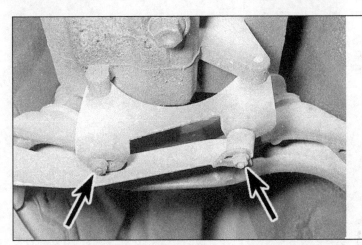

c) *Check the position of the tailpipes in relation to the cut-out in the rear valance, and if necessary adjust the exhaust mountings to give sufficient clearance between the system and the valance.*
d) *Once the mountings have been reconnected and tightened, loosen the two nuts and bolts securing the exhaust mounting bracket to the transmission bracket, and if necessary slide the bracket within the elongated holes to release any sideways tension on the system. Once the system is correctly positioned, tighten the nuts and bolts.*

## Front section

### Removal

11  If desired, the exhaust front section can be removed leaving the rear section in place.
12  Follow the procedure outlined in Steps 3 through 7.
13  Counterhold the bolts, and unscrew the clamp nuts securing the exhaust front section

**14.10b Loosen the nuts (arrows) and slide the bracket to release any tension on the exhaust system**

to the rear section, then slide the front section from the rear section and remove the front section from under the vehicle.

### Installation

14  Refer to Step 10.

## Rear section

### Removal

15  Counterhold the bolts, and unscrew the clamp nuts securing the exhaust front section to the rear section.
16  Working at the rear of the exhaust system, unscrew the nuts securing the rear muffler mounting clamps, then slide the rear section from the front section and withdraw the rear section from under the vehicle.

### Installation

17  Installation is a reversal of removal, but coat the threads of the clamp mounting nuts and bolts with a little copper grease before installing, and preload the rear mounting (see Step 10b).

**Notes**

# Chapter 4 Part B
# Fuel and exhaust systems - six-cylinder engines

## Contents

| | Section | | Section |
|---|---|---|---|
| Air cleaner assembly - removal and installation | 2 | Fuel injection system - general information | 6 |
| Air filter element replacement | See Chapter 1 | Fuel injection system - testing and adjustment | 10 |
| Engine management system check | See Chapter 1 | Fuel pump/fuel gauge sender unit - removal and installation | 8 |
| Exhaust system - general information, removal and installation | 14 | Fuel tank - removal and installation | 3 |
| | | General information and precautions | 1 |
| Exhaust system and mounting check | See Chapter 1 | Manifolds - removal and installation | 13 |
| Fuel filter replacement | See Chapter 1 | Throttle body - removal and installation | 11 |
| Fuel gauge sender unit - removal and installation | 9 | Throttle cable - removal, installation and adjustment | 4 |
| Fuel injection system components - removal and installation | 12 | Throttle pedal - removal and installation | 5 |
| Fuel injection system - depressurization and priming | 7 | | |

## Specifications

### Fuel system data
| | |
|---|---|
| Fuel pump type | Electric, immersed in tank |
| Fuel pressure regulator rating | 43.5 ± 1 psi |
| Specified idle speed | 700 ± 40 rpm (not adjustable - controlled by ECM) |
| Specified idle mixture CO content | 0.5 to 1.5% (not adjustable - controlled by ECM) |

### Torque specifications
Ft-lbs (unless otherwise indicated)

| | |
|---|---|
| Intake manifold nuts | |
|   M6 nuts | 84 in-lbs |
|   M7 nuts | 132 in-lbs |
|   M8 nuts | 16 |
| Exhaust manifold nuts | |
|   M6 nuts | 84 in-lbs |
|   M7 nuts | 15 |
|   M8 nuts | 16 |
| Crankshaft position sensor bolt | 60 in-lbs |
| Camshaft position sensor bolt | 60 in-lbs |
| Air temperature sensor to intake manifold | 120 in-lbs |
| Fuel rail-to-intake manifold bolts | 84 in-lbs |
| Fuel tank mounting bolts | 17 |
| Fuel tank retaining strap bolts | 72 in-lbs |

2.1a  Disconnect the air mass meter electrical connector

2.1b  Disconnect the air ducting from the throttle body

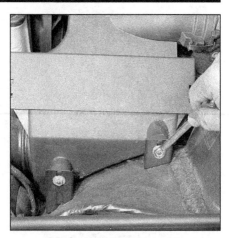

2.2a  Loosen the air cleaner securing nuts . . .

# 1  General information and precautions

## General information

The fuel supply system consists of a fuel tank (which is mounted under the rear of the vehicle, with an electric fuel pump immersed in it), a fuel filter, fuel feed and return lines. The fuel pump supplies fuel to the fuel rail, which acts as a reservoir for the six fuel injectors which inject fuel into the inlet tracts. The fuel filter incorporated in the feed line from the pump to the fuel rail ensures that the fuel supplied to the injectors is clean.

Refer to Section 6 for further information on the operation of the fuel injection system, and to Section 14 for information on the exhaust system.

## Precautions

**Warning 1:** *Gasoline is extremely flammable, so take extra precautions when you work on any part of the fuel system. Don't smoke or allow open flames or bare light bulbs near the work area, and don't work in a garage where a gas-type appliance (such as a water heater or a clothes dryer) is present. Since gasoline is carcinogenic, wear latex gloves when there is a possibility of being exposed to fuel, and, if you spill any fuel on your skin, rinse it off immediately with soap and water. Mop up any spills immediately and do not store fuel-soaked rags where they could ignite. When you perform any kind of work on the fuel system, wear safety glasses and have a Class B type fire extinguisher on hand.*

**Warning 2:** *Residual pressure will remain in the fuel lines long after the vehicle was last used. When disconnecting any fuel line, first depressurize the fuel system as described in Section 7.*

# 2  Air cleaner assembly - removal and installation

*Refer to illustrations 2.1a, 2.1b, 2.2a, 2.2b, 2.2c, 2.3 and 2.5*

## Removal

1  Remove the air mass meter, as described in Section 12. Alternatively, disconnect the electrical connector from the air mass meter, loosen the clip securing the air ducting to the throttle body, and leave the air mass meter attached to the air cleaner casing **(see illustrations)**.

2  Loosen the two nuts securing the air cleaner casing to the mounting bracket, then pull the casing upwards, and unclip the intake duct from the front body panel **(see illustrations)**. *Note: If the intake duct cannot be released from the front body panel, unbolt the alternator air ducting from the top of the front body panel, and reach down behind the front grille panel for access to the air intake duct securing clip.*

3  Where applicable, remove the securing bolt, and withdraw the coolant bypass valve from the air cleaner casing, leaving the coolant hoses connected **(see illustration)**.

4  If the assembly is being removed complete with the air ducting and air mass meter, disconnect the breather hoses from the bottom of the air ducting, then withdraw the assembly from the engine compartment.

## Installation

5  Installation is a reversal of removal, but ensure that the lower mounting engages with

2.2b  . . . then pull the casing upwards

2.2c  Reach behind the grille to depress the air intake duct securing clip (arrow) - viewed with headlight removed for clarity

2.3  Withdrawing the coolant bypass valve from the air cleaner casing

# Chapter 4 Part B  Fuel and exhaust systems - six-cylinder engines

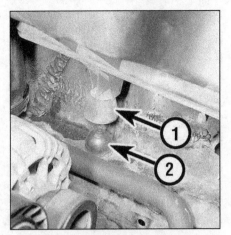

2.5 Ensure that the air cleaner lower mounting (1) engages with the lug (2) on the body

3.5 Disconnect the fuel hoses (arrows)

3.11a Left-hand fuel tank strap securing bolts (arrows)

the plastic lug on the body, and where applicable install the air mass meter with reference to Section 12 **(see illustration)**.

## 3  Fuel tank - removal and installation

*Refer to illustrations 3.5, 3.11a and 3.11b*

**Warning:** *Refer to the* **Warnings** *in Section 1 before proceeding.*

### Removal

**Note:** *New fuel hose clamps will be required on installation.*

1  Depressurize the fuel system (see Section 7), then disconnect the battery negative cable. **Caution:** *If the stereo in your vehicle is equipped with an anti-theft system, make sure you have the correct activation code before disconnecting the battery.*
2  Before removing the fuel pump/fuel gauge sender unit, all fuel should be drained from the fuel tank. Since a fuel tank drain plug is not provided, it is preferable to carry out the removal operation when the tank is nearly empty. Before proceeding, disconnect the battery negative cable and siphon or hand-pump the remaining fuel from the tank. **Caution:** *If the stereo in your vehicle is equipped with an anti-theft system, make sure you have the correct activation code before disconnecting the battery.*
3  Working under the rear seats, disconnect the electrical connectors and the fuel hoses from the fuel pump and fuel gauge senders, with reference to Sections 8 and 9.
4  Jack up the rear of the vehicle and support it securely on axle stands.
5  Working under the rear of the vehicle, disconnect the two fuel hoses from the pipes in front of the fuel tank **(see illustration)**. Be prepared for fuel spillage, and clamp or plug the open ends of the hoses and pipes to prevent dirt entry and further fuel spillage.
6  Remove the rear section of the exhaust system as described in Section 14.

7  Remove the driveshaft as described in Chapter 8.
8  Disconnect the parking brake cables from the parking brake lever as described in Chapter 9, then release the cables from the clips under the vehicle and pull them through the retaining clips at the rear of the fuel tank.
9  Loosen the hose clamp, and disconnect the fuel filler hose from the right-hand side of the fuel tank.
10  Support the fuel tank using a floor jack and block of wood.
11  Unscrew the fuel tank strap securing bolts, noting the locations of any washers to ensure correct installation, then pivot the strap down, and lower the tank sufficiently to disconnect the vent pipes from the top of the tank **(see illustrations)**.
12  Disconnect the vent pipes, and withdraw the tank from under the rear of the vehicle.

### Installation

13  Installation is a reversal of removal, bearing in mind the following points.
   a) Ensure that any washers are installed on the fuel tank strap bolts as noted before removal.
   b) Reconnect and adjust the parking brake cables as described in Chapter 9.

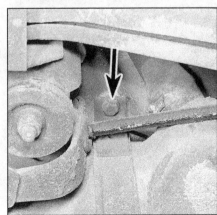

3.11b Right-hand fuel tank strap securing bolt (arrow)

   c) Install the driveshaft as described in Chapter 8.
   d) Use new hose clamps when reconnecting the fuel hoses.
   e) On completion, before starting the engine, check that the fuel tank is correctly grounded by measuring the electrical resistance between the tank filler pipe ground and the rear wheel hub - the reading should be approximately 0.6 ohms. Fill the tank with at least two gallons of fuel before attempting to start the engine.

## 4  Throttle cable - removal, installation and adjustment

*Refer to illustrations 4.1a, 4.1b and 4.4*

### Removal

1  Working in the engine compartment, depress the retaining tangs, and release the cable end fitting from the bracket on the end of the throttle lever. Pry the end fitting from the end of the cable **(see illustrations)**.
2  Pry the cable grommet from the bracket

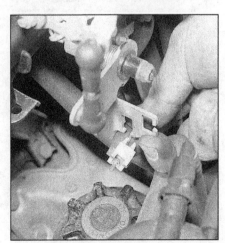

4.1a Release the cable end fitting from the bracket . . .

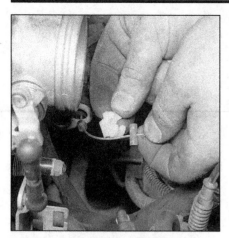

4.1b ... then pry the end fitting from the cable

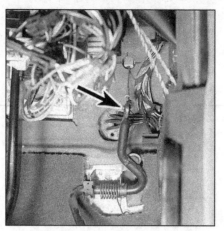

4.4 Pull the end of the throttle cable from the grommet (arrow)

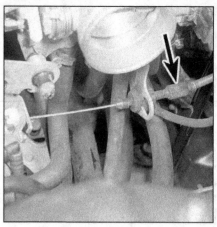

4.10 Throttle cable adjuster screw (arrow)

on the throttle body, and withdraw the cable through the bracket to free it from the throttle body.

3 Working inside the vehicle, remove the two securing screws, and withdraw the driver's side lower dash panel for access to the pedals.

4 Pull the end of the throttle cable from the grommet in the end of the throttle cable operating lever **(see illustration)**.

5 Working on the interior side of the engine compartment firewall, depress the retaining clip and push the cable guide from the firewall into the engine compartment.

6 Make a careful note of the routing of the cable, then release it from any clips and brackets, and withdraw the cable assembly from the engine compartment.

## Installation

7 Installation is a reversal of removal, bearing in mind the following points.
 a) Ensure that the cable is correctly routed as noted before removal.
 b) Check the condition of the grommet in the end of the throttle cable operating lever, and replace if necessary.
 c) On completion check the cable adjustment as described in the following Steps.

## Adjustment

*Refer to illustration 4.10*

### Models with manual transmission

8 Check that the accelerator pedal is in the rest position, against the pedal idle stop.

9 Check that the throttle lever on the throttle body is in the rest (idle) position.

10 Turn the cable adjuster screw at the throttle body bracket to eliminate free-play in the cable **(see illustration)**.

11 Have an assistant fully depress the throttle pedal, and check that, with the pedal fully depressed, there is still 0.020-inch of free-play at the throttle valve in the throttle body.

12 If necessary, turn the pedal full-throttle stop (screwed into the floor) to give the spec- ified free-play (see Step 11) at the throttle valve. On some models, it will be necessary to loosen a locknut before the full-throttle stop can be adjusted. Note that turning the full-throttle stop by 1.5 turns will adjust the throttle valve free-play by 0.020-inch.

13 Where applicable, tighten the full-throttle stop locknut on completion.

### Models with automatic transmission

14 At the time of writing, no information was available regarding throttle cable adjustment for models with automatic transmission. Consult a BMW dealer for advice.

## 5 Throttle pedal - removal and installation

*Refer to illustration 5.2*

### Pedal assembly

**Warning:** *Once the throttle pedal has been removed, it must be replaced. Removal will damage the pedal retaining clips, and if the original pedal is reinstalled, it could work loose, causing an accident.*

### Removal

1 Pry off the clip securing the top of the pedal assembly to the throttle cable operating lever.

2 Press down on the carpet under the pedal, then bend back the lower pedal retaining clips, and lever the pedal upwards to release it from the floor **(see illustration)**. Discard the pedal - a new pedal must be used on installation. Withdraw the pedal.

### Installation

3 Place the new pedal in position, engaging the top of the pedal with the throttle cable operating lever.

4 Push the pedal down to engage the lower retaining clips with the floor plate.
**Warning:** *Ensure that the retaining clips snap securely into place.*

5 Install the clip to secure the pedal to the throttle cable operating lever.

6 Check the throttle cable adjustment as described in Section 4.

### Throttle cable operating lever

### Removal

7 Working inside the vehicle, remove the two securing screws, and withdraw the driver's side lower dash panel for access to the pedals.

8 Pull the end of the throttle cable from the grommet in the end of the throttle cable operating lever.

9 Pry off the clip securing the top of the pedal assembly to the throttle cable operating lever.

10 Pry the locking clip from the end of the throttle cable operating lever pivot shaft, noting its orientation to ensure correct installation.

11 Disconnect the return spring from the lever, then slide the lever to the left against the pressure of the return spring on the shaft, to free the right-hand end of the pivot shaft.

12 Withdraw the lever assembly.

### Installation

13 Installation is a reversal of removal, bearing in mind the following points.
 a) Ensure that the return spring is correctly positioned on the lever.

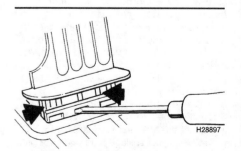

5.2 Releasing the throttle pedal from the floor

- b) Ensure that the locking clip is correctly installed on the end of the lever pivot shaft, as noted before removal.
- c) Ensure that the clip securing the pedal to the cable operating lever is correctly installed.
- d) Check the condition of the grommet in the end of the throttle cable operating lever, and replace if necessary.
- e) On completion, check the cable adjustment as described in Section 4.

## 6 Fuel injection system - general information

1 An integrated engine management system known as DME (Digital Motor Electronics) is installed on all models, and the system controls all fuel injection and ignition system functions using a central ECM (Engine Control Module). A Bosch Motronic or Siemens engine management system may be used, depending on model, but in most respects the systems are similar (sensors and actuators are identical) - the only significant difference is in the ECM.

2 On most models, the system incorporates a closed-loop catalytic converter and an evaporative emission control system, and complies with the very latest emission control standards. Refer to Chapter 5 Part B for information on the ignition side of the system; the fuel side of the system operates as follows.

3 The fuel pump (which is immersed in the fuel tank) supplies fuel from the tank to the fuel rail, via a filter. Fuel supply pressure is controlled by the pressure regulator in the fuel rail. When the optimum operating pressure of the fuel system is exceeded, the regulator allows excess fuel to return to the tank.

4 The electrical control system consists of the ECM, along with the following sensors:
- a) **Air mass meter** - informs the ECM of the mass of air entering the engine.
- b) **Throttle position sensor** - informs the ECM of the throttle position, and the rate of throttle opening/closing.
- c) **Coolant temperature sensor** - informs the ECM of engine temperature.
- d) **Inlet air temperature sensor** - informs the ECM of the temperature of the air passing through the intake manifold.
- e) **Crankshaft position sensor** - informs the ECM of the crankshaft position and speed of rotation.
- f) **Camshaft position sensor** - informs the ECM of the inlet camshaft position.
- g) **Oxygen sensor** - informs the ECM of the oxygen content of the exhaust gases (explained in greater detail in Chapter 6).

5 All the above signals are analyzed by the ECM which selects the fueling response appropriate to those values. The ECM controls the fuel injectors (varying the pulse width - the length of time the injectors are held open - to provide a richer or weaker mixture, as appropriate). The mixture is constantly varied by the ECM, to provide the best setting for cranking, starting (with either a hot or cold engine), warm-up, idle, cruising and acceleration.

6 The ECM also has full control over the engine idle speed, via an auxiliary air valve which bypasses the throttle valve. When the throttle valve is closed, the ECM controls the opening of the valve, which in turn regulates the amount of air entering the manifold, and so controls the idle speed.

7 The ECM also controls the exhaust and evaporative emission control systems, which are described in detail in Chapter 6.

8 The throttle body is coolant heated, and on certain models, a thermostatic coolant bypass valve in the air cleaner housing ensures that warm coolant reaches the throttle body quickly (before the thermostat opens) when the ambient air temperature is low.

9 If there is an abnormality in any of the readings obtained from the sensors, the ECM enters its back-up mode. In this event, it ignores the abnormal sensor signal and assumes a pre-programmed value which will allow the engine to continue running (albeit at reduced efficiency). If the ECM enters this back-up mode, the relevant fault code will be stored in the ECM memory.

10 If a fault is suspected, the vehicle should be taken to a BMW dealer at the earliest opportunity. A complete test of the engine management system can then be carried out, using a special electronic diagnostic test unit which is simply plugged into the system's diagnostic connector.

## 7 Fuel injection system - depressurization and priming

### Depressurization

1 Remove the fuel pump fuse from the fuse box. The fuse is located in the engine compartment fuse box, and the exact location is given on the fuse box cover.

2 Start the engine, and wait for it to stall. Switch off the ignition.

3 Remove the fuel filler cap.

4 The fuel system is now depressurized.
**Note:** *Place a rag around fuel line fittings before disconnecting them to prevent any residual fuel from spilling onto the engine.*

5 Disconnect the battery negative cable before working on any part of the fuel system. **Caution:** *If the stereo in your vehicle is equipped with an anti-theft system, make sure you have the correct activation code before disconnecting the battery.*

### Priming

6 Install the fuel pump fuse, then switch on the ignition and wait for a few seconds for the fuel pump to run, building up fuel pressure. Switch off the ignition unless the engine is to be started.

## 8 Fuel pump/fuel gauge sender unit - removal and installation

**Warning:** *Refer to the* **Warnings** *in Section 1 before proceeding.*
**Note:** *A new sealing ring and a new locking ring must be used on installation.*

### Removal

*Refer to illustration 8.5*

1 The fuel pump is integral with the right-hand fuel gauge sender, in the right-hand half of the fuel tank.

2 Disconnect the battery negative cable.
**Caution:** *If the stereo in your vehicle is equipped with an anti-theft system, make sure you have the correct activation code before disconnecting the battery.*

3 Before removing the fuel pump/fuel gauge sender unit, all fuel should be drained from the fuel tank. Since a fuel tank drain plug is not provided, it is preferable to carry out the removal operation when the tank is nearly empty. Before proceeding, disconnect the battery negative cable and siphon or hand-pump the remaining fuel from the tank.
**Caution:** *If the stereo in your vehicle is equipped with an anti-theft system, make sure you have the correct activation code before disconnecting the battery.*

4 Remove the rear seat cushion with reference to Chapter 11.

5 Lift up the floor insulation to expose the fuel pump/sender unit cover. Remove the securing screws and withdraw the cover from the floor **(see illustration)**.

6 Disconnect the two electrical connectors from the fuel pump/sender unit.

7 Loosen the hose clamps, then disconnect the fuel hoses from the fuel pump/sender unit. Mark the hoses to ensure that they are correctly reconnected. Be prepared for fuel spillage.

8 Unscrew the fuel/pump sender unit locking ring and remove it from the tank. This is best accomplished by using a large pair of pliers to push on two opposite raised ribs on the locking ring **(see illustration 9.3b)**. Alternatively, use two screwdrivers on the raised

**8.5 Withdraw the cover to expose the fuel pump/sender unit**

**9.3a** Disconnecting the electrical connector from the left-hand fuel gauge sender unit

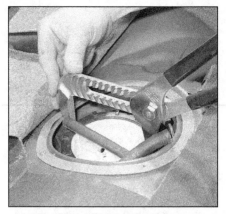

**9.3b** Using a large pair of pliers to unscrew the fuel gauge sender unit locking ring

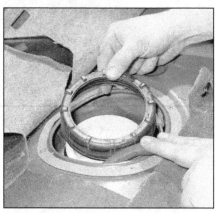

**9.3c** Remove the locking ring . . .

ribs, but take care not to damage the locking ring. Turn the ring counterclockwise until it can be unscrewed by hand.

9 Carefully lift the fuel pump/sender unit from the fuel tank, taking great care not to bend the sender unit float arm (gently push the float arm towards the unit if necessary). Recover the sealing ring.

## Installation

10 Before installation, check the condition of the fuel pick-up strainer, and clean if necessary.

11 Installation is a reversal of removal, bearing in mind the following points.

a) *Use a new sealing ring and a new locking ring.*
b) *To allow the unit to pass through the aperture in the fuel tank, press the float arm against the fuel pick-up strainer.*
c) *When the unit is reinstalled, the raised ribs on the unit must align with the corresponding mark on the fuel tank.*

## 9 Fuel gauge sender unit - removal and installation

*Refer to illustrations 9.3a, 9.3b, 9.3c, 9.3d and 9.3e*

**Warning:** *Refer to the* **Warnings** *in Section 1 before proceeding.*

1 Two fuel level sender units are installed, one in each (left- and right-hand) half of the fuel tank.

2 The right-hand sender unit is integral with the fuel pump, and removal and installation are described in Section 8.

3 Removal and installation of the left-hand sender unit is also as described in Section 8, noting the following differences **(see illustrations)**.

a) *There is only one electrical connector.*
b) *The fuel return and pick-up pipes are pressed against the bottom of the fuel tank. When installing the unit, check that the pipes are pressed firmly into position, and check that the float arm is still free to move.*

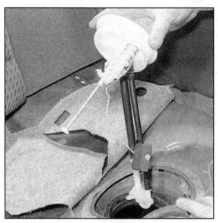

**9.3d** . . . then lift out the sender unit . . .

## 10 Fuel injection system - testing and adjustment

### Testing

1 If a fault appears in the fuel injection system, first ensure that all the system wiring connectors are securely connected and free of corrosion. Ensure that the fault is not due to poor maintenance; i.e., check that the air cleaner filter element is clean, the spark plugs are in good condition and correctly gapped, the cylinder compression pressures are correct, and that the engine breather hoses are clear and undamaged, referring to Chapters 1, 2 and 5 for further information.

2 If these checks fail to reveal the cause of the problem, the vehicle should be taken to a BMW dealer or other qualified repair shop for testing. A wiring block connector is incorporated in the engine management circuit, into which a special electronic diagnostic tester can be plugged. The connector is clipped to the right-hand suspension tower. The tester will locate the fault quickly and simply, alleviating the need to test all the system components individually, which is a time-consuming operation that also carries a risk of damaging the ECM.

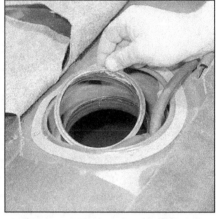

**9.3e** . . . and recover the sealing ring

### Adjustment

3 Experienced home mechanics with a considerable amount of skill and equipment (including a tachometer and an accurately calibrated exhaust gas analyzer) may be able to check the exhaust CO level and the idle speed. However, if these are found to be in need of adjustment, the car *must* be taken to a BMW dealer or other qualified repair shop for further testing.

4 To adjust the CO level or idle speed, special diagnostic equipment is required.

## 11 Throttle body - removal and installation

*Refer to illustrations 11.5 and 11.7*

**Warning:** *Wait until the engine is completely cool before beginning this procedure.*

**Note:** *A new sealing ring must be used on installation.*

### Removal

1 Remove the air mass meter as described in Section 12.

2 Disconnect the throttle cable from the throttle linkage, and move the cable to one side, with reference to Section 4.

# Chapter 4 Part B  Fuel and exhaust systems - six-cylinder engines

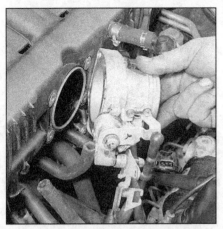

11.5  Removing the throttle body

11.7  Use a new sealing ring when installing the throttle body

12.3a  Remove the relay cover . . .

3  Loosen the hose clamp, and disconnect the coolant hoses from the bottom of the throttle body. Be prepared for coolant spillage, and plug or clamp the open ends of the hoses.
4  Disconnect the electrical connector from the throttle position sensor.
5  Unscrew the securing bolts, and remove the throttle body from the intake manifold (see illustration).
6  Recover the sealing ring.

## Installation

7  Installation is a reversal of removal, but use a new sealing ring, and install the air mass meter with reference to Section 12 (see illustration). Check the coolant level (see Chapter 1).

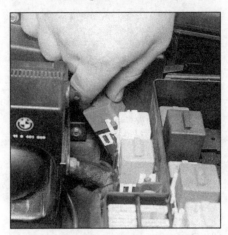

12.3b  . . . and withdraw the rearmost relay

12.4a  Unscrew the securing screws . . .

## 12  Fuel injection system components - removal and installation

### Engine Control Module (ECM)

Refer to illustrations 12.3a, 12.3b, 12.4a, 12.4b, 12.4c, 12.5 and 12.6

#### Removal

1  Disconnect the battery negative cable.

**Caution:** *If the stereo in your vehicle is equipped with an anti-theft system, make sure you have the correct activation code before disconnecting the battery.*

2  Where necessary, for improved access, unbolt the anti-theft alarm siren, and move it to one side, clear of the ECM housing.
3  Similarly, for improved access, unclip the cover from the engine compartment fuse box, then remove the relay cover and with-

draw the rearmost relay from its connector (see illustrations).
4  Unscrew the securing screws, and remove the ECM cover from the engine compartment firewall. Unclip the wiring harness from the cover as the cover is removed (see illustrations).
5  Release the locking clip, and pull the electrical connector from the ECM. Take great care not to damage the pins on the ECM or

12.4b  . . . then remove the ECM cover . . .

12.4c  . . . and unclip the wiring harness

12.5  Disconnect the electrical connector . . .

# Chapter 4 Part B Fuel and exhaust systems - six-cylinder engines

12.6 ... and remove the ECM

12.13 Disconnect the fuel feed hose from the fuel rail

12.16 Pulling the wiring ducting from the fuel injectors

the wiring connector **(see illustration)**.
6  Pull the ECM forwards and manipulate it from its housing **(see illustration)**.

## Installation

7  Installation is a reversal of removal.

## Fuel rail and injectors

*Refer to illustrations 12.13, 12.16, 12.17, 12.18, 12.19 and 12.21*

**Warning:** *Refer to the* **Warnings** *in Section 1 before proceeding.*

**Note:** *New fuel injector O-rings should be used on installation.*

### Removal

8  Depressurize the fuel system as described in Section 7, then disconnect the battery negative cable. **Caution:** *If the stereo in your vehicle is equipped with an anti-theft system, make sure you have the correct activation code before disconnecting the battery.*
9  Where necessary, to allow sufficient clearance for the fuel rail to be removed, remove the heater/ventilation inlet air ducting from the rear of the engine compartment as follows.

  a) *Lift the grille from the top of the ducting (on certain Coupe models, it will be necessary to remove the windshield wiper arms, then remove the plastic securing screws and lift off the complete cowl grille assembly).*
  b) *Working through the top of the ducting, remove the screws securing the cable ducting to the air ducting and move the cable ducting clear.*
  c) *Unscrew the nuts and/or screw(s) securing the air ducting to the firewall (where applicable, bend back the heat shielding for access).*
  d) *Remove the air ducting by pulling upwards.*
  e) *Move the previously removed cable ducting clear of the valve cover.*

10  Remove the engine oil filler cap.
11  Remove the plastic cover from the top of the valve cover. To remove the cover, pry out the cover plates and unscrew the two securing nuts, then lift and pull the cover forwards. Manipulate the cover over the oil filler neck.
12  Pry out the cover plates, then unscrew the bolts and remove plastic cover from the top of the fuel injectors.
13  Loosen the hose clamp, and disconnect the fuel feed hose from the front of the fuel rail **(see illustration)**. Be prepared for fuel spillage, and take adequate fire precautions. Plug the open end of the fuel pipe and hose to prevent dirt entry and further fuel spillage.
14  Similarly, disconnect the fuel return hose from the fuel pressure regulator at the rear of the fuel rail.
15  Disconnect the vacuum hose from the fuel pressure regulator.
16  Remove the two nuts securing the engine wiring ducting to the fuel rail, then pull the ducting up to release the electrical connectors from the fuel injectors **(see illustration)**.
17  Unscrew the two bolts securing the fuel rail to the intake manifold **(see illustration)**.
18  Carefully pull the fuel rail upwards to release the fuel injectors from the cylinder head, then withdraw the complete fuel rail/fuel injector assembly **(see illustration)**.
19  To remove a fuel injector from the fuel rail, proceed as follows.

  a) *Pry off the metal securing clip, using a screwdriver.*
  b) *Pull the fuel injector from the fuel rail* **(see illustration)**.

### Installation

20  Before installation, it is wise to replace all the fuel injector O-rings as a matter of course.

12.17 Unscrew the bolts securing the fuel rail

12.18 Withdraw the fuel rail/fuel injector assembly

# Chapter 4 Part B Fuel and exhaust systems - six-cylinder engines

12.19 Removing a fuel injector

21 Check that the plastic washer at the bottom of each injector is positioned above the lower O-ring **(see illustration)**.
22 Lightly lubricate the fuel injector O-rings with a little petroleum jelly or SAE 90 gear oil.
23 Where applicable, install the fuel injectors to the fuel rail, ensuring that the securing clips are correctly installed. Note that the injectors should be positioned so that the wiring sockets are uppermost when the assembly is reinstalled.
24 Slide the fuel rail/fuel injector assembly into position, ensuring that the injectors engage with their bores in the intake manifold.
25 Further installation is a reversal of removal, but pressurize the fuel system (install the fuel pump fuse and switch on the ignition) and check for leaks before starting the engine.

## Fuel pressure regulator

Refer to illustrations 12.27a, 12.27b, 12.27c and 12.28
**Warning:** Refer to the **Warnings** in Section 1 before proceeding.
**Note:** New O-rings may be required on installation.

12.27c ... and pull the regulator from the fuel rail

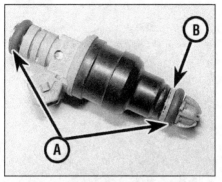

12.21 Fuel injector O-rings (A) and plastic washer (B)

### Removal

26 Remove the fuel rail/fuel injector assembly as described previously in this Section.
27 Unscrew the clamp bolt and remove the pressure regulator clamping ring, then lift off the thrust washer, and pull the pressure regulator from the fuel rail. Note that it will be difficult to pull the regulator from the fuel rail due to the tight-fitting O-rings **(see illustrations)**.

### Installation

28 Before installation, check the condition of the O-rings, and replace if necessary **(see illustration)**.
29 Installation is a reversal of removal, but ensure that the thrustwasher is correctly positioned, and install the fuel rail/fuel injector assembly as described previously in this Section.

## Air mass meter

Refer to illustrations 12.32, 12.36a and 12.36b
**Note:** New filter grilles and/or a new O-ring may be required on installation.

### Removal

30 Disconnect the battery negative cable.
**Caution:** If the stereo in your vehicle is equipped with an anti-theft system, make sure you have the correct activation code before disconnecting the battery.

12.28 Check the condition of the O-rings (arrows)

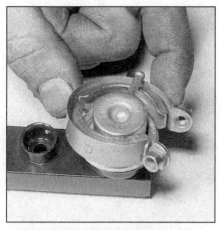

12.27a Remove the pressure regulator clamping ring . . .

12.27b . . . lift off the thrust washer . . .

31 Turn the locking collar, and disconnect the electrical connector from the air mass meter. Where applicable, release the wiring from any clips on the air ducting and air mass meter.
32 Loosen the hose clamp, and disconnect the air ducting from the throttle body **(see illustration)**.

12.32 Disconnect the air ducting from the throttle body

**12.36a Pry out the snap-ring . . .**

**12.36b . . . to release the filter grille from the air mass meter**

**12.38 Disconnecting the electrical connector from the throttle position sensor (air ducting removed for clarity)**

33  Unscrew the securing screws, or release the clips (as applicable) securing the air mass meter to the air cleaner casing. Pull the air mass meter from the air cleaner casing.
34  Either release the two breather hoses from the air ducting and remove the air ducting complete with the air mass meter, or loosen the hose clamp and disconnect the air ducting from the air mass meter, leaving the ducting in position.
35  Withdraw the air mass meter assembly.

### Installation

36  Installation is a reversal of removal, bearing in mind the following points:
a) Check the condition of the filter grilles at the inlet and outlet of the air mass meter (the grilles are secured by large circlips). Replace the grilles if they are badly contaminated or damaged **(see illustrations)**.
b) Check the condition of the O-ring installed on the air cleaner casing outlet, and replace if necessary.

## Throttle position sensor

*Refer to illustration 12.38*

### Removal

37  Disconnect the battery negative cable.
**Caution:** *If the stereo in your vehicle is equipped with an anti-theft system, make sure you have the correct activation code before disconnecting the battery.*
38  Disconnect the electrical connector from the sensor **(see illustration)**.
39  Where applicable, to improve access to the switch, disconnect the electrical connector from the air temperature sensor mounted in the intake manifold.
40  Unscrew the two securing screws, and withdraw the sensor from the throttle body.

### Installation

41  Installation is a reversal of removal.

## Coolant temperature sensor

42  The sensor is located in the left-hand side of the cylinder head. Refer to Chapter 3 for removal and installation details.

## Inlet air temperature sensor

*Refer to illustration 12.46*
**Note:** *A new O-ring may be required on installation.*

### Removal

43  The sensor may be located in the side of the intake manifold, next to the throttle position sensor (mounted on the throttle body), or in the bottom of the intake manifold towards the rear of the engine.
44  Disconnect the battery negative cable.
**Caution:** *If the stereo in your vehicle is equipped with an anti-theft system, make sure you have the correct activation code before disconnecting the battery.*
45  On models where the sensor is mounted in the side of the manifold, proceed as follows.
a) Remove the air mass meter as described previously in this Section.
b) Disconnect the electrical connector from the throttle position sensor, then disconnect the electrical connector from the air temperature sensor.
c) Unscrew the air temperature sensor from the intake manifold.
46  On models where the sensor is mounted in the bottom of the manifold, proceed as follows **(see illustration)**.

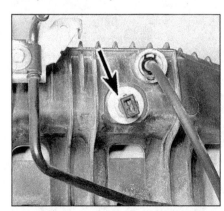

**12.46 Air temperature sensor (arrow) mounted in the bottom of the intake manifold - manifold removed for clarity**

a) Reach down under the rear of the manifold to locate the sensor.
b) Depress the securing clip, and disconnect the electrical connector from the sensor.
c) Using a box-end wrench or a long-reach socket, unscrew the sensor from the manifold. Note that it may be necessary to move certain wires and/or hoses to one side to improve access (note the locations of any wires and hoses to ensure correct routing on installation).

### Installation

47  Installation is a reversal of removal but, where applicable, use a new sealing ring when installing the sensor, and ensure that all wires and hoses are routed as noted before removal.

## Crankshaft position sensor - engines without VANOS

*Refer to illustrations 12.49a, 12.49b, 12.50 and 12.51*
**Note:** *A new sealing ring may be required on installation.*

### Removal

48  Locate the sensor, mounted beneath

**12.49a Pry off the securing clips . . .**

# Chapter 4 Part B  Fuel and exhaust systems - six-cylinder engines

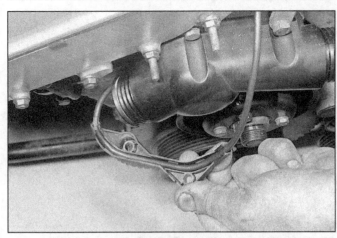

12.49b ... and withdraw the crankshaft position sensor wiring cover

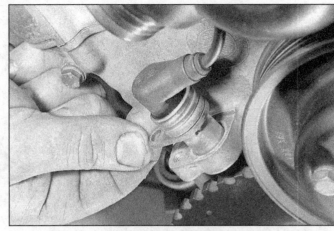

12.50 Withdrawing the crankshaft position sensor from its mounting bracket

the thermostat housing.
49  Pry off the securing clips and withdraw the sensor wiring cover from the upper timing chain cover studs. Release the wiring from the cover (see illustrations).
50  Unscrew the securing bolt and withdraw the sensor from its mounting bracket (see illustration).
51  Trace the sensor wiring back to the connector mounted on the bracket under the intake manifold. Note that there are two connectors mounted on the bracket, and the crankshaft position sensor connector is the lower connector. Disconnect the wiring connector (see illustration).
52  Note the routing of the wiring, then withdraw the sensor from the engine. Where applicable, recover the sealing ring.

### Installation

53  Installation is a reversal of removal, but ensure that the wiring is correctly routed as noted before removal. Where applicable, check the condition of the sealing ring and replace if necessary.

## Crankshaft position sensor - VANOS engines

Note: *New VANOS unit oil feed pipe sealing rings may be required on installation.*

### Removal

54  Locate the sensor, mounted beneath the thermostat housing.
55  Unscrew the union bolt, and disconnect the oil feed pipe from the VANOS adjustment unit. Be prepared for oil spillage, and plug or cover the open ends of the pipe and the VANOS adjustment unit. Recover the sealing rings.
56  Unscrew the securing nut and bolt and remove the engine lifting bracket from the front of the cylinder head. Note the location of the ground lead on the lower bracket mounting bolt.
57  Pry off the securing clips and withdraw the sensor wiring cover from the studs. Release the wiring from the cover.

58  Proceed as described in Steps 50 through 52.

### Installation

59  Installation is a reversal of removal, but ensure that the wiring is correctly routed a noted before removal. Before reconnecting the oil feed pipe to the VANOS adjustment unit, check the condition of the sealing rings and replace if necessary.

## Camshaft position sensor - engines without VANOS

Refer to illustration 12.61

### Removal

60  The sensor is located at the front left-hand corner of the cylinder head.
61  Unscrew the securing bolt, and withdraw the sensor from its housing (see illustration).
62  Trace the sensor wiring back to the connector mounted on the bracket under the intake manifold. Note that there are two connectors mounted on the bracket, and the camshaft position sensor connector is the upper connector (see illustration 12.51).
63  Note the routing of the wiring, then withdraw the sensor from the engine.

### Installation

64  Installation is a reversal of removal, but ensure that the wiring is routed as noted before removal.

## Camshaft position sensor - VANOS engines

Note: *A new VANOS solenoid valve sealing ring will be required on installation.*

### Removal

65  The sensor is located at the front left-hand corner of the cylinder head.
66  For access to the sensor, remove the VANOS solenoid valve as described in Chapter 2 Part B.
67  Proceed as described in Steps 61 to 63.

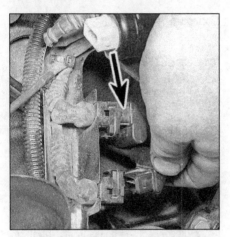

12.51 Disconnecting the crankshaft position sensor wiring connector - arrow points to the camshaft position sensor wiring connector

### Installation

68  Installation is a reversal of removal, but ensure that the wiring is routed as noted before removal, and install the VANOS solenoid valve using a new sealing ring.

12.61 Withdrawing the camshaft position sensor - viewed with intake manifold removed

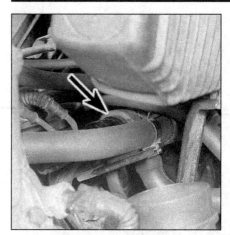

12.70a Idle speed control valve location (arrow)

12.70b Idle speed control valve location viewed with intake manifold removed. Note arrow on body must point up

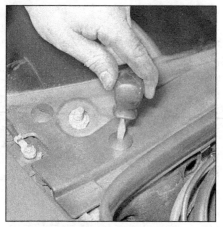

13.3a Remove the plastic securing screws . . .

### Oxygen sensor

69 Refer to Chapter 6.

### Idle speed control valve

*Refer to illustrations 12.70a and 12.70b*

#### Removal

70 The valve is mounted on a bracket under the intake manifold **(see illustrations)**.
71 Disconnect the battery negative cable. **Caution:** *If the stereo in your vehicle is equipped with an anti-theft system, make sure you have the correct activation code before disconnecting the battery.*
72 Reach under the intake manifold, and disconnect the hoses from the valve. Alternatively, disconnect the hoses from the bottom of the intake manifold (depress the clip to release the connector), the breather connector on the valve cover, and the air ducting, and leave the hoses attached to the valve.
73 Disconnect the electrical connector from the valve.
74 Push the valve rearwards (towards the engine compartment firewall) to release it from its mounting clamp.
75 Note the routing of the hoses, then manipulate the valve out from under the manifold.

#### Installation

76 Installation is a reversal of removal, but ensure that the arrow on the valve is pointing vertically upwards, and ensure that the hoses are securely reconnected and correctly routed.

### 13 Manifolds - removal and installation

#### Intake manifold

*Refer to illustrations 13.3a, 13.3b, 13.3c, 13.3d, 13.3e, 13.3f, 13.5, 13.7, 13.8a, 13.8b, 13.9, 13.10, 13.13, 13.17 and 13.18*

**Warning 1:** *Refer to the Warnings in Section 1 before proceeding.*
**Warning 2:** *Wait until the engine is completely cool before beginning this procedure.*
**Note:** *New sealing rings may be required on installation.*

#### Removal

1 Depressurize the fuel system as described in Section 7, then disconnect the battery negative cable. **Caution:** *If the stereo in your vehicle is equipped with an anti-theft system, make sure you have the correct activation code before disconnecting the battery.*
2 Remove the air cleaner/air mass meter assembly as described in Section 2.
3 To allow sufficient clearance for the manifold to be removed, raise the hood to its fully open position as described in Chapter 11, then remove the heater/ventilation inlet air ducting from the rear of the engine compartment as follows **(see illustrations)**.

 a) Lift the grille from the top of the ducting (on certain Coupe models, it will be necessary to remove the windshield wiper arms, then remove the plastic securing screws, peel back the weatherstrip, and lift off the complete cowl grille assembly).
 b) Working through the top of the ducting, remove the screws securing the cable ducting to the air ducting and move the cable ducting clear.
 c) Unscrew the nuts and/or screw(s) securing the air ducting to the firewall (where applicable, bend back the heat shielding for access).
 d) Remove the air ducting by pulling upwards.
 e) Move the previously removed cable ducting clear of the manifold.

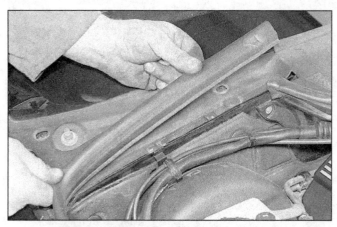

13.3b . . . then peel back the weatherstrip . . .

13.3c . . . and lift off the complete cowl grille assembly

**13.3d** Remove the screws securing the cable ducting

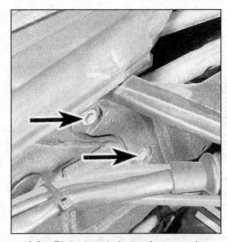

**13.3e** Right-hand air ducting securing screws (arrows)

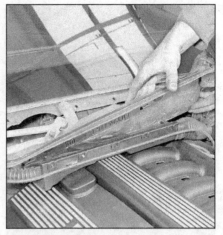

**13.3f** Removing the air ducting

**13.5** Disconnecting the brake booster vacuum hose

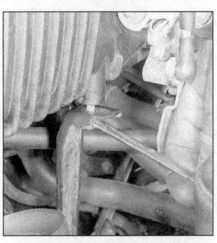

**13.7** Unscrew the bolts securing the manifold to the support brackets

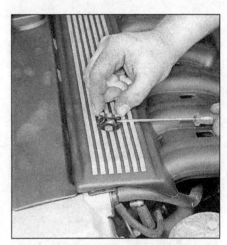

**13.8a** Pry out the cover plates to reveal the securing bolts . . .

4  Disconnect the throttle position sensor electrical connector.
5  Disconnect the brake booster vacuum hose from the manifold **(see illustration)**.
6  Disconnect the throttle cable from the manifold, with reference to Section 4.

7  Working under the manifold, unscrew the bolts securing the manifold to the two support brackets **(see illustration)**.
8  Pry out the cover plates, then unscrew the bolts and remove the plastic cover from the top of the fuel injectors **(see illustrations)**.

9  Remove the two nuts securing the engine wiring ducting to the fuel rail, then pull the ducting up to release the electrical connectors from the fuel injectors **(see illustration)**.
10  To improve access, loosen the securing clip and remove the alternator cooling duct

**13.8b** . . . and remove the cover from the fuel injectors

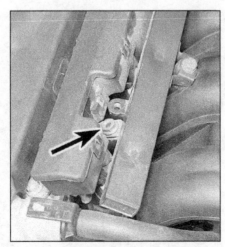

**13.9** Engine wiring duct securing nut (arrow)

**13.10** Loosening the alternator cooling duct securing clip

**13.13 Disconnect the fuel hoses (arrows)**

**13.17 Where applicable, disconnect the air temperature electrical connector as the manifold is removed**

(see illustration).
11  Disconnect the breather hose from the bottom of the manifold.
12  If the cooling system is not going to be drained at a later stage, clamp the two throttle body coolant hoses to prevent spillage, then loosen the clamps and disconnect the hoses.
13  Disconnect the two fuel hoses from the pipes at the rear of the manifold (see illustration). Be prepared for fuel spillage, and clamp or plug the hoses to prevent further fuel loss and dirt ingress.
14  Disconnect the smaller breather hose from the connector on the valve cover.
15  Reach under the manifold, and carefully pull the air hose from the top of the idle speed control valve.
16  Disconnect the electrical connector from the air temperature sensor. The sensor may be located in the side of the manifold, or underneath the manifold (in which case it may prove easier to disconnect the plug as the manifold is removed), depending on model.
17  Lift the manifold off the cylinder head studs and withdraw it from the engine compartment (see illustration). Recover the sealing rings if they are loose.

### Installation

18  Installation is a reversal of removal, bearing in mind the following points.
 a) Before installation, examine the condition of the sealing rings, and replace if necessary (see illustration).
 b) Reconnect and if necessary adjust the throttle cable as described in Section 4.
 c) On completion, prime the fuel system as described in Section 7.

## Exhaust manifold

*Refer to illustration 13.26*

**Note:** *New manifold-to-cylinder head and manifold-to-exhaust front section gaskets and new manifold nuts will be required on installation.*

### Removal

19  The engine has twin exhaust manifolds, each manifold serving three cylinders.
20  Remove the washer fluid reservoir, as described in Chapter 12.
21  To improve access, jack up the front of the vehicle and support securely on axle stands.
22  Working underneath the vehicle, unscrew the nuts securing the exhaust downpipes to the manifolds.
23  Working at the transmission exhaust bracket, unscrew the two bolts securing the two exhaust mounting clamp halves together.
24  Loosen the bolt securing the clamp halves to the bracket on the transmission, then lower the exhaust downpipes down from the manifold studs. Recover the gaskets.
25  Working in the engine compartment, unscrew the manifold securing nuts.
26  Withdraw the manifolds from the studs and recover the gaskets (see illustration).
27  It is possible that some of the manifold studs may be unscrewed from the cylinder head when the manifold securing nuts are unscrewed. In this event, the studs should be screwed back into the cylinder head once the manifolds have been removed, using two manifold nuts locked together.

### Installation

28  Installation is a reversal of removal, but use new gaskets and new manifold securing nuts.

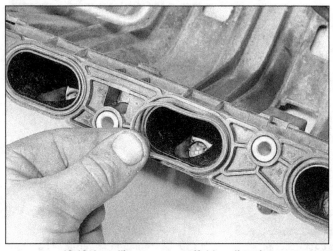

**13.18 Installing a new manifold sealing ring**

**13.26 Withdrawing an exhaust manifold**

# Chapter 4 Part B  Fuel and exhaust systems - six-cylinder engines      4B-15

14.5 Disconnecting the exhaust front sections from the manifolds

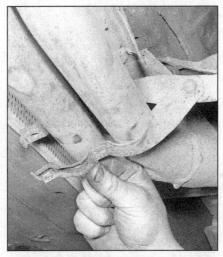

14.7 Pivot the clamp halves away from the exhaust system

7   Unscrew the clamp bolts securing the two halves of the transmission exhaust mounting bracket together, then unscrew the clamp pivot bolt, and pivot the clamp halves away from the exhaust system (see illustration).
8   Slide the rear exhaust mounting rubbers from the brackets on the exhaust system.
9   Working at the rear of the exhaust system, unscrew the nuts securing the rear muffler box mounting clamps.
10   Withdraw the complete exhaust system from under the vehicle.

### Installation

11   Installation is a reversal of removal, bearing in mind the following points (see illustrations).

   a) *Use new gaskets when reconnecting the exhaust front section to the manifolds. Also use new nuts, and coat the threads of the new nuts with copper grease.*
   b) *When installing the rear muffler box clamps, slide the clamps onto the muffler box and allow the system to hang so that the mountings are free from tension, then slide the clamps towards the rear of the muffler by 5/8-inch to give a preload, forcing the system towards the front of the vehicle.*
   c) *Check the position of the tailpipes in relation to the cut-out in the rear valance, and if necessary adjust the exhaust mountings to give sufficient clearance between the system and the valance.*
   d) *Once the mountings have been reconnected and tightened, loosen the two nuts and bolts securing the exhaust mounting bracket to the transmission bracket, and if necessary slide the bracket within the elongated holes to release any sideways tension on the sys-*

## 14  Exhaust system - general information, removal and installation

*Refer to illustrations 14.5, 14.7, 14.11a, 14.11b and 14.14*

### General information

1   The original equipment exhaust system consists of two sections. The front section incorporates the catalytic converter and the oxygen sensor and the center expansion box. The rear section incorporates the rear muffler.
2   The system is suspended throughout its length by rubber mountings and a metal bracket.

### Complete system

**Note:** *New exhaust front section-to-manifold gaskets and securing nuts will be required on installation.*

#### Removal

3   Jack up the vehicle and support securely on axle stands.
4   On models with a catalytic converter, turn the securing ring, and disconnect the oxygen sensor wiring connector at the bracket on the transmission crossmember.
5   Unscrew the securing nuts, and disconnect the exhaust front sections from the manifolds (see illustration). Recover the gaskets.
6   On models with automatic transmission, unscrew the securing bolts, and remove the front transmission crossmember from under the vehicle.

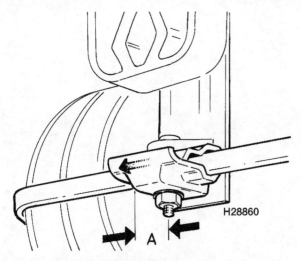

14.11a Preload the exhaust system by sliding the clamps towards the rear of the muffler box

A = 5/8-inch

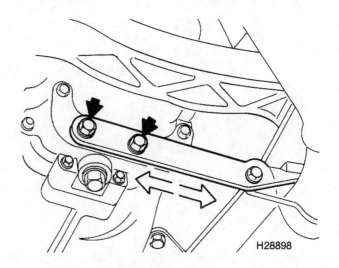

14.11b Loosen the bolts (arrows) and slide the bracket to release any tension on the exhaust system

tem. Once the system is correctly positioned, tighten the nuts and bolts.

### Front section
#### Removal
12  If desired, the exhaust front section can be removed leaving the rear section in place.
13  Follow the procedure described previously in Steps 3 through 8.
14  Counterhold the bolts, and unscrew the clamp nuts securing the exhaust front section to the rear section, then slide the front section from the rear section and remove the front section from under the vehicle **(see illustration)**.

#### Installation
15  Refer to Step 11.

### Rear section
#### Removal
16  Counterhold the bolts, and unscrew the clamp nuts securing the exhaust front section to the rear section.
17  Working at the rear of the exhaust system, unscrew the nuts securing the rear muffler box mounting clamps, then slide the rear section from the front section and withdraw the rear section from under the vehicle.

#### Installation
18  Installation is a reversal of removal, but coat the threads of the clamp mounting nuts and bolts with a little copper grease before fitting, and preload the rear mounting as described above in Step 11.

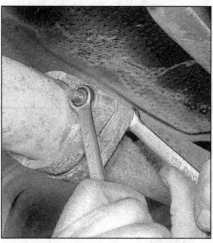

14.14  Counterhold the bolts and unscrew the exhaust clamp nuts

# Chapter 5 Part A
# Starting and charging systems

## Contents

| | Section | | Section |
|---|---|---|---|
| Alternator drivebelt - removal, installation and tensioning | 6 | Electrical fault finding - general information | 2 |
| Alternator - removal and installation | 7 | General information and precautions | 1 |
| Alternator - testing and overhaul | 8 | Ignition switch - removal and installation | 12 |
| Battery - removal and installation | 4 | Starter motor - removal and installation | 10 |
| Battery - testing and charging | 3 | Starter motor - testing and overhaul | 11 |
| Charging system - testing | 5 | Starting system - testing | 9 |

## Specifications

**System type** .................................................. 12-volt negative ground

### Alternator
Regulated voltage (at 1500 rpm engine speed with no electrical
equipment switched on) .................................................. 13.5 to 14.2 volts

### Starter motor
Current draw
   Four-cylinder models .................................................. 140 amps maximum
   Six-cylinder models .................................................. 170 amps maximum

### Torque specifications
**Ft-lbs** (unless otherwise indicated)
Starter motor-to-transmission nuts and bolts .................................................. 37
Starter motor support bracket-to-starter motor nuts .................................................. 48 in-lbs
Starter motor support bracket-to-engine bolts .................................................. 35

## 1 General information and precautions

### General information

The engine electrical system consists mainly of the charging and starting systems. Because of their engine-related functions, these components are covered separately from the body electrical devices such as the lights, instruments, etc. (which are covered in Chapter 12). Refer to Part B for information on the ignition system.

The electrical system is of the 12-volt negative ground type.

The battery is of the low maintenance or "maintenance-free" (sealed for life) type and is charged by the alternator, which is belt-driven from the crankshaft pulley.

The starter motor is of the pre-engaged type incorporating an integral solenoid. On starting, the solenoid moves the drive pinion into engagement with the flywheel ring gear before the starter motor is energized. Once the engine has started, a one-way clutch prevents the motor armature being driven by the engine until the pinion disengages from the flywheel.

### Precautions

Further details of the various systems are given in the relevant Sections of this Chapter. While some repair procedures are given, the usual course of action is to replace the component concerned. The owner whose interest extends beyond mere component replacement should obtain a copy of the *"Automotive Electrical Manual"*, available from the publishers of this manual.

# Chapter 5 Part A  Starting and charging systems

It is necessary to take extra care when working on the electrical system to avoid damage to semi-conductor devices (diodes and transistors), and to avoid the risk of personal injury. In addition to the precautions given in *"Safety first!"* at the beginning of this manual, observe the following when working on the system:

*Always remove rings, watches, etc. before working on the electrical system.* Even with the battery disconnected, capacitive discharge could occur if a component's live terminal is grounded through a metal object. This could cause a shock or nasty burn.

*Do not reverse the battery connections.* Components such as the alternator, electronic control units, or any other components having semi-conductor circuitry could be irreparably damaged.

If the engine is being started using jump leads and a remote battery, connect the batteries *positive-to-positive* and *negative-to-negative* (see *"Jump starting"*, at the beginning of this manual). This also applies when connecting a battery charger.

Never disconnect the battery terminals, the alternator, any electrical wiring or any test instruments when the engine is running.

Do not allow the engine to turn the alternator when the alternator is not connected.

Never "test" for alternator output by "shorting" the output lead to ground.

Never use an ohmmeter of the type incorporating a hand-cranked generator for circuit or continuity testing.

Always ensure that the battery negative lead is disconnected when working on the electrical system.

Before using electric-arc welding equipment on the car, disconnect the battery, alternator and components such as the fuel injection/ignition electronic control unit to protect them from the risk of damage.

If a radio/cassette unit with a built-in security code is fitted, note the following precautions. If the power source to the unit is cut, the anti-theft system will activate. Even if the power source is immediately reconnected, the radio/cassette unit will not function until the correct security code has been entered. Therefore, if you do not know the correct security code for the radio/cassette unit **do not** disconnect the battery or remove the radio/cassette unit from the vehicle.

## 2  Electrical fault finding - general information

Refer to Chapter 12.

## 3  Battery - testing and charging

**Note:** *The following is intended as a guide only. Always refer to the manufacturer's recommendations (often printed on a label attached to the battery) before charging a battery.*

1  All models are fitted with a maintenance-free battery in production, which should require no maintenance under normal operating conditions.

2  If the condition of the battery is suspect, remove the battery (see Section 4), and check that the electrolyte level in each cell is up to the "MAX" mark on the outside of the battery case (approximately 3/16-inch above the tops of the plates in the cells). If necessary, the electrolyte level can be topped up by removing the cell plugs from the top of the battery and adding distilled water (not acid).

3  An approximate check on battery condition can be made by checking the specific gravity of the electrolyte, using the following as a guide.

### Ambient temperature above 77°F
Fully-charged . . . . .  1.210 to 1.230
70% charged . . . . .  1.170 to 1.190
Fully-discharged . .  1.050 to 1.070

### Ambient temperature below 77°F
Fully-charged . . . . .  1.270 to 1.290
70% charged . . . . .  1.230 to 1.250
Fully-discharged . .  1.110 to 1.130

4  Use a hydrometer to make the check and compare the results with the following table. Note that the specific gravity readings assume an electrolyte temperature of 60°F; for every 48°F below 60°F subtract 0.007. For every 48°F above 60°F add 0.007.

5  If the battery condition is suspect, first check the specific gravity of electrolyte in each cell. A variation of 0.040 or more between any cells indicates loss of electrolyte or deterioration of the internal plates.

6  If the specific gravity variation is 0.040 or more, the battery should be renewed. If the cell variation is satisfactory but the battery is discharged, it should be charged in accordance with the manufacturer's instructions.

7  If testing the battery using a voltmeter, connect the voltmeter across the battery. A fully-charged battery should give a reading of 12.5 volts or higher. The test is only accurate if the battery has not been subjected to any kind of charge for the previous six hours. If this is not the case, switch on the headlights for 30 seconds, then wait four to five minutes before testing the battery after switching off the headlights. All other electrical circuits must be switched off, so check that the doors and tailgate are fully shut when making the test.

8  Generally speaking, if the voltage reading is less than 12.2 volts, then the battery is discharged, while a reading of 12.2 to 12.4 volts indicates a partially discharged condition.

9  If the battery is to be charged, remove it from the vehicle (Section 4) and charge it in accordance with the manufacturer's instructions.

## 4  Battery - removal and installation

**Caution:** *If the radio in your vehicle is equipped with an anti-theft system, make sure you have the correct activation code before disconnecting the battery.*

### Four-cylinder 3-Series models and 1996 Z3 models

#### Removal
*Refer to illustrations 4.2 and 4.6*

1  On four-cylinder models, the battery is located at the rear right-hand corner of the engine compartment.

2  Where applicable, release the securing clips and open the battery cover **(see illustration)**.

3  Loosen the clamp bolt and disconnect the clamp from the battery negative (ground) terminal.

4  Remove the insulation cover (where fitted) and disconnect the positive terminal lead in the same way.

5  Where applicable, side off the battery cover.

6  Unscrew the bolt and remove the battery retaining clamp **(see illustration)**.

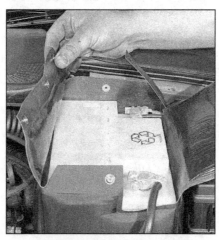

4.2  Opening the battery cover (four-cylinder 3-Series model shown)

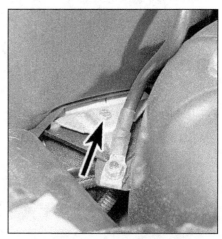

4.6  Battery retaining clamp bolt (arrow) (four-cylinder 3-Series model shown)

# Chapter 5 Part A  Starting and charging systems

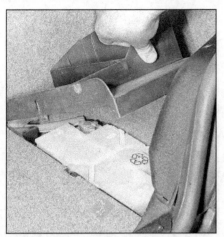

4.10  Unclip the battery cover/first aid kit tray to expose the battery (six-cylinder 3-Series model shown)

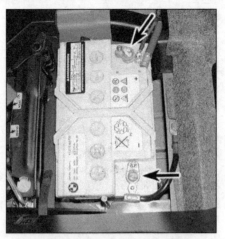

4.11  Disconnect the negative battery cable (lower arrow) first, and then disconnect the positive cable (upper arrow) (Z3 model shown)

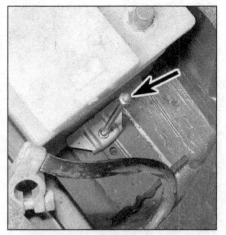

4.13  Battery clamp bolt (arrow) (six-cylinder 3-Series model shown)

7   Lift out the battery. Take care as the battery is heavy!

### Installation

8   Installation is a reversal of removal, but smear petroleum jelly on the terminals when reconnecting the leads, and always reconnect the positive lead first and the negative lead last.

## Six-cylinder 3-Series models and all 1997 and 1998 Z3 models

### Removal

*Refer to illustrations 4.10, 4.11 and 4.13*

9   On six-cylinder models, the battery is located beneath a cover on the right-hand side of the luggage compartment.
10  Open the deck lid and unclip the battery cover/first aid kit tray from the right-hand side of the luggage compartment **(see illustration)**.
11  Loosen the clamp bolt **(see illustration)** and disconnect the clamp from the battery negative (ground) terminal.
12  Remove the insulation cover (where fitted) and disconnect the positive terminal lead in the same way.
13  Unscrew the hold-down bolt **(see illustration)** and remove the battery hold-down clamp.
14  Lift the battery from its housing. Take care as the battery is heavy!

### Installation

15  Installation is a reversal of removal, but smear petroleum jelly on the terminals when reconnecting the leads, and always reconnect the positive lead first and the negative lead last.

### 5  Charging system - testing

**Note:** *Refer to the warnings given in "Safety first!" and in Section 1 of this Chapter before*

*starting work.*
1   If the ignition warning light fails to illuminate when the ignition is switched on, first check the alternator wiring connections for security. If satisfactory, check that the warning light bulb has not blown, and that the bulb holder is secure in its location in the instrument panel. If the light still fails to illuminate, check the continuity of the warning light feed wire from the alternator to the bulb holder. If all is satisfactory, the alternator is at fault and should be renewed or taken to an auto-electrician for testing and repair.
2   If the ignition warning light illuminates when the engine is running, stop the engine and check that the drivebelt is correctly tensioned (see Chapter 1) and that the alternator connections are secure. If all is so far satisfactory, have the alternator checked by an auto-electrician for testing and repair.
3   If the alternator output is suspect even though the warning light functions correctly, the regulated voltage may be checked as follows.

7.4  Pry the cover (arrow) from the rear of the alternator (four-cylinder 3-Series model shown)

4   Connect a voltmeter across the battery terminals and start the engine.
5   Increase the engine speed until the voltmeter reading remains steady; the reading should be approximately 12 to 13 volts, and no more than 14.2 volts.
6   Switch on as many electrical accessories (the headlights, heated rear window and heater blower) as possible, and check that the alternator maintains the regulated voltage at around 13 to 14 volts.
7   If the regulated voltage is not as stated, the fault may be due to worn alternator brushes, weak brush springs, a faulty voltage regulator, a faulty diode, a severed phase winding or worn or damaged slip rings. The alternator should be renewed or taken to an auto-electrician for testing and repair.

### 6  Alternator drivebelt - removal, installation and tensioning

Refer to the procedure given for the drivebelt(s) in Chapter 1.

### 7  Alternator - removal and installation

## Four-cylinder models

### Removal

*Refer to illustrations 7.4, 7.6a and 7.6b*
1   Disconnect the battery negative lead.
**Caution:** *If the radio in your vehicle is equipped with an anti-theft system, make sure you have the correct activation code before disconnecting the battery.*
2   Remove the air cleaner assembly and the air mass meter (see Chapter 4A).
3   Remove the drivebelt (see Chapter 1).
4   Pry the cover(s) from the rear of the alternator, then unscrew the nuts and bolt, and disconnect the wiring **(see illustration)**.
5   Where applicable, pry the cover from

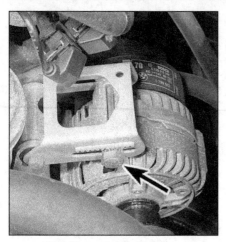  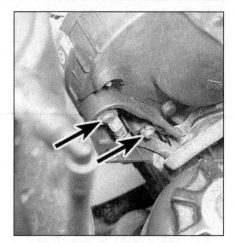

**7.6a Unscrew the upper ...**

**7.6b ... and lower alternator securing bolts (arrow)- four-cylinder model**

**7.12 Disconnect the wiring (arrows) from the rear of the alternator - six-cylinder model**

the center of the drivebelt guide pulley, then unbolt the pulley for access to the upper alternator securing bolt. Note the location of the pulley to ensure correct installation.

6 Counterhold the nuts and unscrew the upper and lower alternator securing through-bolts **(see illustrations)**.

7 Withdraw the alternator from the engine.

### Installation

8 Installation is a reversal of removal, but ensure that the drivebelt guide pulley is correctly fitted as noted before removal, and install the drivebelt (see Chapter 1).

## Six-cylinder models

### Removal

*Refer to illustrations 7.12, 7.13 and 7.15*

9 Disconnect the battery negative lead. **Caution:** *If the radio in your vehicle is equipped with an anti-theft system, make sure you have the correct activation code before disconnecting the battery.*

10 Remove the air cleaner assembly and the air mass meter (see Chapter 4B).

11 Remove the drivebelt (see Chapter 1).

12 Pull off the cover(s), then unscrew the nuts and disconnect the wiring from the rear of the alternator **(see illustration)**.

13 Pry the cover from the center of the drivebelt tensioner idler pulley, then unscrew the upper through-bolt securing the idler pulley and the alternator **(see illustration)**.

14 Counterhold the nut and unscrew the lower alternator securing through-bolt.

15 Withdraw the alternator from the engine **(see illustration)**.

### Installation

16 Installation is a reversal of removal, bearing in mind the following points.

a) *When installing the tensioner idler pulley, ensure that the lug on the rear of the pulley assembly engages with the corresponding cut-out in the mounting bracket.*

b) *Install the drivebelt (see Chapter 1).*

## 8  Alternator - testing and overhaul

If the alternator is thought to be suspect, it should be removed from the vehicle and taken to an auto-electrician for testing. Most auto-electricians will be able to supply and fit brushes at a reasonable cost. However, check on the cost of repairs before proceeding as it may prove more economical to obtain a new or exchange alternator.

## 9  Starting system - testing

**Note:** *Refer to the precautions given in "Safety first!" and in Section 1 of this Chapter before starting work.*

1 If the starter motor fails to operate when the ignition key is turned to the appropriate position, the following possible causes may be to blame.

a) *The battery is faulty.*
b) *The electrical connections between the switch, solenoid, battery and starter motor are somewhere failing to pass the necessary current from the battery through the starter to ground.*
c) *The solenoid is faulty.*
d) *The starter motor is mechanically or electrically defective.*

2 To check the battery, switch on the

**7.13 Unscrew the upper through-bolt securing the idler pulley - six-cylinder model**

**7.15 Withdraw the alternator from the engine**

# Chapter 5 Part A  Starting and charging systems

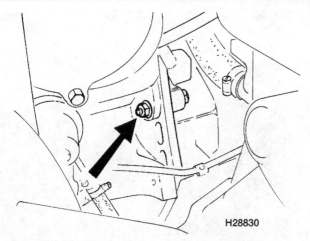

10.5  Working under the vehicle, unscrew the lower starter motor securing nut and bolt (arrow) - four-cylinder model

10.10  Disconnect the starter motor wiring (arrows) - six-cylinder model

headlights. If they dim after a few seconds, this indicates that the battery is discharged - recharge (see Section 3) or replace the battery. If the headlights glow brightly, operate the ignition switch and observe the lights. If they dim, then this indicates that current is reaching the starter motor, therefore the fault must lie in the starter motor. If the lights continue to glow brightly (and no clicking sound can be heard from the starter motor solenoid), this indicates that there is a fault in the circuit or solenoid - see following paragraphs. If the starter motor turns slowly when operated, but the battery is in good condition, then this indicates that either the starter motor is faulty, or there is considerable resistance somewhere in the circuit.

3  If a fault in the circuit is suspected, disconnect the battery leads (including the ground connection to the body), the starter/solenoid wiring and the engine/transmission ground strap. Thoroughly clean the connections, and reconnect the leads and wiring, then use a voltmeter or test lamp to check that full battery voltage is available at the battery positive lead connection to the solenoid, and that the ground is sound. Smear petroleum jelly around the battery terminals to prevent corrosion - corroded connections are among the most frequent causes of electrical system faults.

4  If the battery and all connections are in good condition, check the circuit by disconnecting the wire from the solenoid blade terminal. Connect a voltmeter or test lamp between the wire end and a good ground (such as the battery negative terminal), and check that the wire is live when the ignition switch is turned to the "start" position. If it is, then the circuit is sound - if not the circuit wiring can be checked (see Chapter 12).

5  The solenoid contacts can be checked by connecting a voltmeter or test lamp between the battery positive feed connection on the starter side of the solenoid, and ground. When the ignition switch is turned to the "start" position, there should be a reading or lighted bulb, as applicable. If there is no reading or lighted bulb, the solenoid is faulty and should be renewed.

6  If the circuit and solenoid are proved sound, the fault must lie in the starter motor. In this event, it may be possible to have the starter motor overhauled by a specialist, but check on the cost of a replacement unit before proceeding, as it may prove more economical to obtain a new or exchange motor.

## 10  Starter motor - removal and installation

### Four-cylinder models
#### Removal
*Refer to illustration 10.5*

1  Disconnect the battery negative lead.
**Caution:** *If the radio in your vehicle is equipped with an anti-theft system, make sure you have the correct activation code before disconnecting the battery.*
2  Unscrew the nuts and disconnect the wiring from the rear of the starter motor.
3  Working from the engine compartment, reach down behind the engine and unscrew the upper starter motor securing nut and bolt. For improved access to the bolt, unscrew the bolt securing the dipstick tube bracket to the engine, then pull the dipstick tube from the cylinder block.
4  Apply the parking brake, then jack up the front of the vehicle and support securely on axle stands.
5  Working underneath the vehicle, unscrew the lower starter motor securing nut and bolt **(see illustration)**.
6  Still working underneath the vehicle, pull the starter motor back to release it from the transmission bellhousing, then turn the motor so that the solenoid is at the top.
7  Pivot the rear of the motor down and manipulate it out from under the vehicle.

#### Installation
8  Installation is a reversal of removal, but check the condition of the O-ring at the bottom of the dipstick tube and replace if necessary.

### Six-cylinder models
#### Removal
*Refer to illustrations 10.10, 10.11, 10.12 and 10.13*

9  Remove the intake manifold (see Chapter 4B).
10  Unscrew the nuts and disconnect the wiring from the starter motor terminals **(see illustration)**.
11  Unscrew the bolt securing the starter motor front mounting bracket to the cylinder block, then unscrew the two nuts securing the bracket to the motor, and remove the bracket **(see illustration)**
12  Unscrew the two rear motor securing nuts and bolts, noting that an open-end

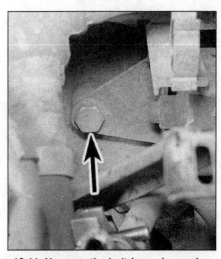

10.11  Unscrew the bolt (arrow) securing the starter motor bracket to the cylinder block - six-cylinder model

10.12 Lower starter motor securing nut (arrow) - six-cylinder model

10.13 Removing the starter motor - six-cylinder model

wrench may be required to counterhold the bolts, as there is insufficient clearance to use a socket **(see illustration)**.

13 Manipulate the starter motor out from the engine compartment **(see illustration)**.

## Installation

14 Installation is a reversal of removal, but install the intake manifold (see Chapter 4B).

## 11 Starter motor - testing and overhaul

If the starter motor is thought to be suspect, it should be removed from the vehicle and taken to an automotive electrical specialist for testing. Most auto-electricians will be able to supply and fit brushes at a reasonable cost. However, check on the cost of repairs before proceeding as it may prove more economical to obtain a new or exchange motor.

## 12 Ignition switch - removal and installation

The ignition switch is integral with the steering column lock, and can be removed (see Chapter 10).

# Chapter 5 Part B
# Ignition systems

## Contents

| | Section | | Section |
|---|---|---|---|
| Engine management system check | See Chapter 1 | Ignition system - testing | 2 |
| General information and precautions | 1 | Knock sensor - removal and installation | 4 |
| Ignition coil - removal and installation | 3 | Spark plug replacement | See Chapter 1 |

## Specifications

### Firing order
| | |
|---|---|
| Four-cylinder models | 1-3-4-2 |
| Six-cylinder models | 1-5-3-6-2-4 |

### Ignition timing
Electronically-controlled by DME - no adjustment possible

### Torque specifications
| | Ft-lbs |
|---|---|
| Spark plugs | 22 |
| Knock sensor securing bolt | 15 |

## 1 General information and precautions

### General information

The ignition system is controlled by the engine management system (see Chapter 4), known as DME (Digital Motor Electronics). The DME system controls all ignition and fuel injection functions using a central ECM (Engine Control Module).

The ignition timing is based on inputs provided to the ECM by various sensors supplying information on engine load, engine speed, coolant temperature sensor and intake air temperature (see Chapter 4).

Some engines are fitted with knock sensors to detect "knocking" (also known as "pinking" or pre-ignition). The knock sensors are sensitive to vibration and detect the knocking which occurs when a cylinder starts to pre-ignite. The knock sensor provides a signal to the ECM which in turn retards the ignition advance setting until the knocking ceases.

On all models, a distributorless ignition system is used, with a separate ignition coil for each cylinder. No distributor is used, and the coils provide the high voltage signal direct to each spark plug.

The ECM uses the inputs from the various sensors to calculate the required ignition advance and the coil charging time.

### Precautions

Refer to the precautions given in Chapter 5A.

Testing of ignition system components should be entrusted to a BMW dealer or other qualified repair shop. Improvised testing techniques are time-consuming and run the risk of damaging the engine management ECM.

## 2 Ignition systems - testing

1 If a fault appears in the engine management (fuel/injection) system, first ensure that the fault is not due to a poor electrical connection, or to poor maintenance, i.e. check that the air filter element is clean, that the spark plugs are in good condition and correctly gapped, and that the engine breather hoses are clear and undamaged.

2 Check the condition of the spark plug wires (if equipped) as follows:

a) Make sure that the leads are numbered to ensure correct installation, then pull the end of one of the leads from the spark plug.
b) Check inside the end fitting for signs of corrosion, which will look like a white crusty powder.
c) Push the end fitting back onto the spark plug, ensuring that it is a tight fit on the plug. If not, remove the lead again and use pliers to carefully crimp the metal connector inside the end fitting until it fits securely on the end of the spark plug.
d) Using a clean rag, wipe the entire length of the lead to remove any built up dirt and grease. Once the lead is clean, check for burns, cracks and other damage. Do not bend the lead excessively or pull the lead lengthwise - the conductor inside might break.
e) Disconnect the other end of the lead from the distributor cap (release the clips and pull off the cover for access to the leads), or coil, as applicable, and check the end fitting in the same manner as the spark plug end. Again ensure that the lead is identified to ensure correct installation.
f) Install the lead securely on completion.
g) Check the remaining leads one at a time in the same manner.

3 Check that the throttle cable is correctly adjusted (see Chapter 4).

4 If the engine is running very roughly, check the compression pressures (see Chapter 2).

5 If these checks fail to reveal the cause of the problem, then the vehicle should be taken

**3.4 Unclip the cover from the top of the coil assembly - four-cylinder model**

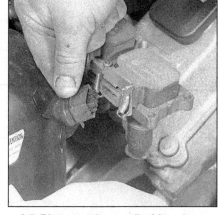

**3.5 Disconnecting a coil wiring plug - four-cylinder model**

**3.6 Unscrewing a coil securing nut - four-cylinder model**

to a BMW dealer or other qualified repair shop for testing using the appropriate specialist diagnostic equipment. The ECM incorporates a self-diagnostic function which stores fault codes in the system memory (note that stored fault codes are erased if the battery is disconnected). These fault codes can be read using the appropriate BMW diagnosis equipment. Improvised testing techniques are time-consuming and run the risk of damaging the engine management ECM.

### 3 Ignition coil - removal and installation

*Refer to illustrations 3.4, 3.5 and 3.6*

## Four-cylinder models

### Removal

1 Each spark plug is fed by its own coil, and coils are mounted together on the right-hand suspension tower in the engine compartment.

2 Disconnect the battery negative lead. **Caution:** *If the radio in your vehicle is equipped with an anti-theft system, make sure you have the correct activation code before disconnecting the battery.*

3 On early models, disconnect the spark plug wires from the coils, noting their locations to ensure correct installation (mark the leads if necessary), then unclip the cover from the top of the coils.

4 On later models, unclip the cover from the top of the coil assembly and disconnect the spark plug wire from the coil **(see illustration)**. If all the coils are to be removed, ensure that the spark plug wires are marked to avoid confusion on installation.

5 Lift the securing clip and disconnect the wiring plug from the relevant coil **(see illustration)**. Again if all the coils are to be removed, ensure that the wiring is marked.

6 Unscrew the coil securing nuts and/or bolts, noting the location of any ground wires or brackets secured by the bolts, then withdraw the coil(s) **(see illustration)**.

### Installation

7 Installation is a reversal of removal, ensuring that the spark plug wires and coil wiring plugs are correctly reconnected. Where applicable, ensure that any ground leads and/or brackets are in place on the coil securing bolts.

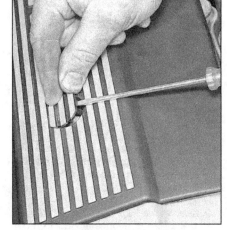

**3.11a Pry out the covers for access to the securing nuts . . .**

**3.11b . . . then remove the plastic cover from the cylinder head - six-cylinder model**

**3.12 Disconnecting a coil wiring plug - six-cylinder model**

# Chapter 5 Part B  Ignition systems

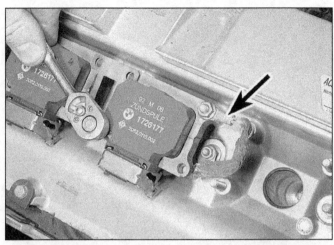

3.13a Unscrewing a coil securing nut. Note the location of the ground lead (arrow) - six-cylinder model

3.13b Where applicable remove the bracket . . .

## Six-cylinder models

*Refer to illustrations 3.11a, 3.11b, 3.12, 3.13a, 3.13b and 3.14*

### Removal

8  Each spark plug is fed by its own coil, and the coils are mounted directly on top of the spark plugs, in the valve cover.
9  Disconnect the battery negative lead.
**Caution:** *If the radio in your vehicle is equipped with an anti-theft system, make sure you have the correct activation code before disconnecting the battery.*
10  Remove the engine oil filler cap.
11  Remove the plastic cover from the top of the valve cover. To remove the cover, pry out the cover plates and unscrew the two securing nuts, then lift and pull the cover forwards. Manipulate the cover over the oil filler neck **(see illustrations)**.
12  Lift the securing clip, and disconnect the wiring plug from the relevant coil **(see illustration)**. If all the coils are to be removed, disconnect all the wiring connectors, then unscrew the nut securing the coil wiring ground lead to the stud on the front of the timing chain cover - the wiring harness can then be unclipped from the camshaft cover and moved to one side.
13  Unscrew the two coil securing nuts, noting the locations of any ground leads and/or brackets secured by the nuts (note that where one of the nuts also secures a metal wiring bracket, it may be necessary to unscrew the bracket securing nut from the adjacent coil, allowing the bracket to be removed to enable coil removal) **(see illustrations)**. Note that the coil connectors are spring-loaded, so the top of the coil will lift as the nuts are unscrewed.
14  Pull the coil from the camshaft cover and spark plug, and withdraw it from the engine **(see illustration)**.

### Installation

15  Installation is a reversal of removal, but ensure that any ground leads and brackets are in position as noted before removal. The ground leads connect to the studs for No. 3 and 6 cylinder coils.

## 4  Knock sensor - removal and installation

*Refer to illustrations 4.5 and 4.13*

### Four-cylinder models
### Removal

1  Two knock sensors are fitted, screwed into the left-hand side of the cylinder block. One sensor detects knocking in No. 1 and 2 cylinders, and the other sensor detects knocking in No. 3 and 4 cylinders.
2  Disconnect the battery negative lead.
**Caution:** *If the radio in your vehicle is equipped with an anti-theft system, make sure you have the correct activation code before disconnecting the battery.*
3  Release the securing clips, and pull the wiring ducting from the fuel injectors.
4  Remove the upper section of the intake manifold (see Chapter 4A).
5  Unscrew the bolts securing the wiring ducting to the lower section of the intake manifold **(see illustration)**.

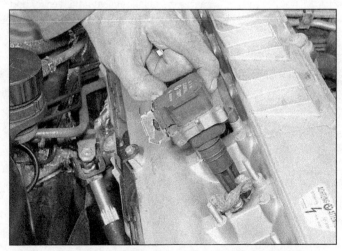

3.14 . . . then withdraw the coil - six-cylinder model

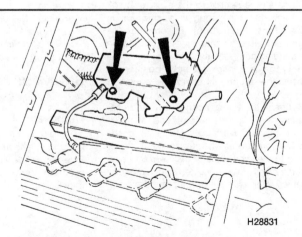

4.5 Wiring ducting-to-manifold securing bolts (arrows) - four-cylinder model

# Chapter 5 Part B Ignition systems

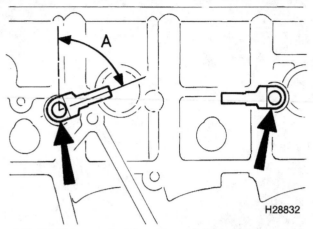

**4.13 Knock sensor locations (arrows) - four-cylinder model**

A  Ensure that the wiring is positioned at an angle of 70° as shown

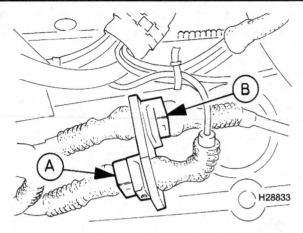

**4.20 Knock sensor wiring plug locations - six-cylinder model**

A  Plug for No. 1 to 3 cylinders knock sensor
B  Plug for No. 4 to 6 cylinders knock sensor

6  Release the securing clips and pull the wiring ducting from the knock sensor wiring connectors. Move the wiring ducting to one side.
7  Remove the idle speed control valve (see Chapter 4A).
8  Disconnect the wiring from the starter motor and the alternator.
9  Trace the wiring back from the relevant sensor and release the wiring plug from the clip. **Warning:** *If both sensors are being removed, take note of the plug locations, as if the plugs are mixed up and incorrectly reconnected, engine damage could result.*
10  Disconnect the coolant hose, and move it to one side for access to the sensors. Be prepared for coolant spillage, and plug the openings to prevent further coolant loss.
11  Unscrew the securing bolt and withdraw the sensor from the engine. Note the angle of the wiring in relation to the cylinder block.

## Installation

12  Commence installation by thoroughly cleaning the mating faces of the sensor and the cylinder block.
13  Install the sensor to the cylinder block ensuring that the wiring is positioned as shown **(see illustration)**. Ensure that the wiring is routed so that it will not chafe on surrounding components.
14  With the sensor correctly positioned, tighten the securing bolt to the specified torque.
15  Install the wiring connector to the clip, ensuring that it is correctly positioned as noted before removal. **Warning:** *If the plugs for the two sensors are incorrectly connected (i.e., mixed up), engine damage may result. Ensure that the plugs are reconnected as noted before removal. Normally the connectors are posi-*

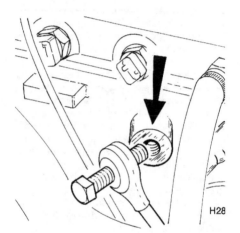

**4.22 Removing the knock sensor (arrow) for cylinders 1 to 3 - six-cylinder model**

*tioned so that they cannot be mixed up.*
16  Further installation is a reversal of removal, bearing in mind the following points.
  a) *Install the idle speed control valve with reference to Chapter 4A.*
  b) *Install the upper section of the intake manifold with reference to Chapter 4A.*
  c) *On completion, check and if necessary top-up the coolant level (see Chapter 1).*

## Six-cylinder models

*Refer to illustrations 4.20 and 4.22*

### Removal

17  Two knock sensors are fitted, screwed into the left-hand side of the cylinder block. One sensor detects knocking in No. 1 to 3 cylinders, and the other sensor detects knocking in No. 4 to 6 cylinders.
18  Disconnect the battery negative lead. **Caution:** *If the radio in your vehicle is equipped with an anti-theft system, make sure you have the correct activation code before disconnecting the battery.*
19  Remove the intake manifold (see Chapter 4B).
20  Locate the sensor connector bracket which is located beneath the idle speed control valve **(see illustration)**. **Warning:** *If both knock sensors are to be removed, mark the wiring connectors to ensure correct installation. Incorrect reconnection may result in engine damage.*
21  Disconnect the sensor wiring connector(s).
22  Unscrew the securing bolt and remove the knock sensor, noting the routing of the wiring. The sensor for cylinders 1 to 3 is located beneath the temperature sensors in the cylinder head **(see illustration)**. The sensor for cylinders 4 to 6 is located to the rear of the sensor wiring connector bracket.

### Installation

23  Commence installation by thoroughly cleaning the mating faces of the sensor and the cylinder block.
24  Install the sensor to the cylinder block, tightening the securing bolt to the specified torque.
25  Route the wiring as noted before removal, then reconnect the connector(s) to the bracket, ensuring that the connectors are positioned as noted before removal. **Warning:** *Ensure that the wiring connectors are correctly connected as noted before removal. If the connectors are incorrectly connected (i.e., mixed up), engine damage may result.*
26  Install the intake manifold (see Chapter 4B).

# Chapter 6
# Emission control systems

## Contents

| | Section | | Section |
|---|---|---|---|
| Catalytic converter - general information and precautions | 3 | General information | 1 |
| Emission control systems - component replacement | 2 | On-board diagnostics - general information | 4 |

## Specifications

**Torque specification**     Ft-lbs

Oxygen sensor-to-exhaust system ....................... 37

## 1 General information

1   All models have various built-in fuel system features which help to minimize emissions, and all models have a crankcase emissions-control system (described in the following paragraphs). A catalytic converter, EGR and evaporative emission control systems are also present.

### Crankcase emission control

2   To reduce the emission of unburned hydrocarbons from the crankcase into the atmosphere, the engine is sealed, and the blow-by gases and oil vapor are drawn from the crankcase and the valve cover, through an oil separator, into the intake tract, to be burned by the engine during normal combustion.

3   Under conditions of high manifold vacuum (idling, deceleration) the gases will be sucked positively out of the crankcase. Under conditions of low manifold vacuum (acceleration, full-throttle running) the gases are forced out of the crankcase by the (relatively) higher crankcase pressure; if the engine is worn, the raised crankcase pressure (due to increased blow-by) will cause some of the flow to return under all manifold conditions.

### Exhaust emission control

4   To minimize the amount of pollutants which escape into the atmosphere, these models are equipped with a catalytic converter in the exhaust system. The fuel-injection/engine control system is of the "closed-loop" type; an oxygen sensor in the exhaust system provides the ECM with constant feedback, enabling the ECM to adjust the mixture to provide the best possible conditions for the converter to operate.

5   The oxygen sensor has a built-in heating element, controlled by the ECM, to quickly bring the sensor's tip to an efficient operating temperature. The sensor's tip is sensitive to oxygen, and sends the ECM a varying voltage depending on the amount of oxygen in the exhaust gases. If the intake air/fuel mixture is too high, the exhaust gases are low in oxygen, so the sensor sends a low-voltage signal. The voltage rises as the mixture weakens and the amount of oxygen in the exhaust gases rises. Peak conversion efficiency of all major pollutants occurs if the intake air/fuel mixture is maintained at the chemically-correct ratio for the complete combustion of gasoline - 14.7 parts (by weight) of air to 1 part of fuel (the "stoichiometric" ratio). The sensor output voltage alters in a large step at this point, the ECM using the signal change as a reference point, and correcting the intake air/fuel mixture accordingly by altering the fuel injector pulse width (the length of time that the injector is open).

### Evaporative emission control

6   An evaporative emissions control system minimizes the escape of unburned hydrocarbons into the atmosphere. The fuel tank filler cap is sealed, and a charcoal canister, mounted in the engine compartment, collects the gasoline vapors generated in the tank when the car is parked. The canister stores them until they can be cleared from the canister (under the control of the ECM) via the purge solenoid valve. When the valve is opened, the fuel vapors pass into the intake tract, to be burned by the engine during normal combustion.

7   To ensure that the engine runs correctly when it is cold and/or idling, the ECM does not open the purge control valve until the engine has warmed up and is under load; the valve solenoid is then modulated on and off, to allow the stored vapor to pass into the intake tract.

### Secondary air injection

8   All OBD-II six-cylinder engines, and January 1997 and later four-cylinder engines, are equipped with a secondary air injection system. The system consists of an electric air pump, an electric solenoid valve, a one-way check valve, a secondary air check-valve and the hoses, pipes and tubing connecting these components. When the engine is started, the ECM opens the electric solenoid valve and turns on the electric pump. The open solenoid valve allows air from the intake manifold to be drawn through the one-way check valve and the solenoid valve, and then to the pump, from which it is pumped through the secondary air check valve into the exhaust manifolds.

9   While the engine is warming up, the air/fuel mixture is richer than normal. Not all the fuel is burned in the combustion chamber. In a fully warmed up engine, any unburned hydrocarbons are catalyzed into less harmful substances in the catalytic converter. But during engine warm-up, the catalytic converter is not yet heated up to its normal operating temperature. The extra air introduced into the exhaust by the air pump increases the amount of oxygen in the exhaust stream, which helps unburned hydrocarbons to oxidize before they pass through the catalytic converter. Some of the oxygen also bonds with carbon monoxide, forming carbon dioxide and water.

10   The air injection pump and the solenoid valve are turned off by the ECM after a timed interval.

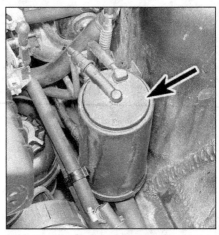

2.2 Charcoal canister location (arrow) - six-cylinder model

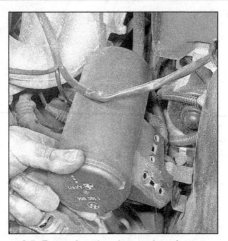

2.5 Removing the charcoal canister - four-cylinder model

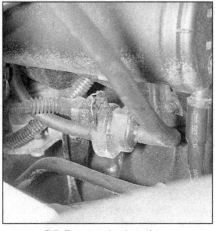

2.7 Purge valve location - four-cylinder engine

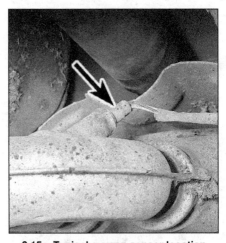

2.15a Typical oxygen sensor location (arrow) in the front section of the exhaust system (six-cylinder model shown)

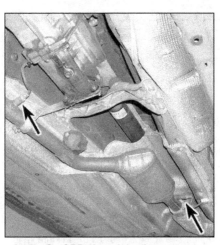

2.15b On OBD-II models, there are two oxygen sensors (arrows), one ahead of and one behind the converter

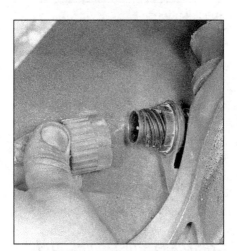

2.18 Disconnecting the oxygen sensor electrical connector - six-cylinder model

## 2 Emission control systems - component replacement

### Crankcase emission control

1  The components of this system require no routine attention, other than to check that the hoses are clear and undamaged at regular intervals.

### Evaporative emission control

#### Charcoal canister - replacement

*Refer to illustrations 2.2 and 2.5*

2  The canister may be located in the engine compartment, on a bracket attached to the left-hand suspension tower, or under the air intake ducting **(see illustration)**.
3  Move any surrounding pipes and hoses to one side to improve access to the canister.
4  Disconnect the hoses from the canister. If the hose is secured by a plastic locking clamp, squeeze the ends of the clamp to release it from the connection on the canister. Note the hose locations to ensure correct installation.

5  Unscrew the securing screws, and withdraw the canister/bracket assembly from the engine compartment **(see illustration)**.
6  Installation is a reversal of removal, but ensure that the hoses are correctly reconnected as noted before removal, and make sure that the hose securing clamps are correctly engaged.

#### Purge valve (solenoid valve) - replacement

*Refer to illustration 2.7*

7  The valve is located on a bracket, next to the charcoal canister, or under the air intake ducting **(see illustration)**.
8  Disconnect the battery negative cable. **Caution:** *If the radio in your vehicle is equipped with an anti-theft system, make sure you have the correct activation code before disconnecting the battery.*
9  If necessary, to improve access, loosen the clamps and remove the air ducting. Note the locations of any breather hoses connected to the ducting.
10  Disconnect the hoses from the valve, noting their locations to ensure correct installation.
11  Disconnect the electrical connector from the valve.
12  Pull the valve from its rubber mounting.
13  Installation is a reversal of removal, but make sure that all hoses are correctly reconnected as noted before removal.

### Exhaust emission control

#### Catalytic converter - replacement

14  The catalytic converter is integral with the front section of the exhaust system. Refer to Part A or B of Chapter 4 (as applicable) for details of removal and installation.

#### Oxygen sensor - replacement

*Refer to illustrations 2.15a, 2.15b and 2.18*
**Note:** *Ensure that the exhaust system is cold before attempting to remove the oxygen sensor.*

15  On pre-OBD-II models, the oxygen sensor is screwed into the front section of the exhaust system under the vehicle **(see illustration)**. On OBD-II-equipped models, there is a second oxygen sensor, behind the con-

# Chapter 6 Emission control systems

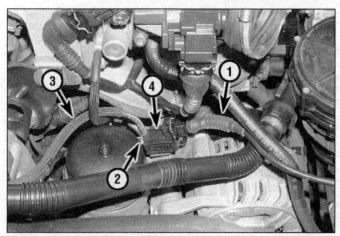

**2.23 Typical electric solenoid valve in a secondary air injection system (Z3 four-cylinder model shown, other systems similar)**

1. Electrical lead
2. Intake manifold-to-solenoid line
3. Solenoid-to-check valve line
4. Electric solenoid valve

**2.27 Typical secondary air pump (Z3 four-cylinder model shown, other systems similar)**

1. Pump-to-check valve hose
2. Secondary air pump

verter **(see illustration)**.
16 Disconnect the battery negative cable.
**Caution:** *If the radio in your vehicle is equipped with an anti-theft system, make sure you have the correct activation code before disconnecting the battery.*
17 Apply the parking brake, then jack up the front of the vehicle and support securely on axle stands.
18 Trace the wiring back from the sensor to the electrical connector under the vehicle, and disconnect the connector **(see illustration)**.
19 Unscrew the sensor and remove it from the exhaust pipe.
20 Installation is a reverse of the removal procedure. Tighten the sensor to the specified torque. Check that the wiring is correctly routed, and in no danger of contacting the exhaust system.

## Secondary air injection

### One-way check valve

21 The one-way check valve is located in the air hose connecting the intake manifold to the electric solenoid. Using a flashlight and a mirror to locate the valve, trace the line from the underside of the intake manifold to the solenoid. Disconnect the lines from both ends of the check valve and remove the valve.
22 Installation is the reverse of removal. Make sure that the tube connections are snug.

### Solenoid valve

*Refer to illustration 2.23*
23 Unplug the solenoid valve electrical connector **(see illustration)**.
24 Disconnect the inlet line (from the one-way check valve) and outlet line (to the secondary air check valve) from the solenoid valve and remove the valve.
25 Installation is the reverse of removal. Make sure that the tube connections are snug.

### Secondary air pump

*Refer to illustration 2.27*
26 Unplug the air pump electrical connector.
27 Disconnect the air pump-to-secondary air check valve hose **(see illustration)** and remove the pump.
28 Installation is the reverse of removal. Make sure that the hose connections are snug.

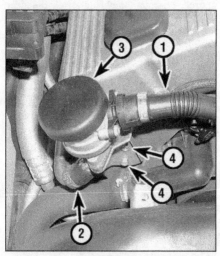

**2.29 Typical secondary air check-valve (Z3 four-cylinder model shown, other systems similar)**

1. Pump-to-check valve hose
2. Check valve-to-exhaust manifold hose and line
3. Secondary air check valve
4. Check valve mounting bracket bolts (it's not necessary to remove the bracket unless you're reconditioning or replacing the cylinder head; the check valve can be detached from the bracket by removing two nuts - not visible in this photo)

### Secondary air check valve

*Refer to illustration 2.29*
29 Disconnect the hoses from the check valve **(see illustration)**.
30 Remove the check valve mounting nuts, remove the pump from its mounting bracket and discard the old gasket.
31 Installation is the reverse of removal. Be sure to use a new gasket and make sure that the hose connections are snug.

## 3 Catalytic converter - general information and precautions

The catalytic converter is a reliable and simple device, which needs no maintenance in itself, but there are some facts of which an owner should be aware, if the converter is to function properly for its full service life.

a) DO NOT use leaded gasoline in a car equipped with a catalytic converter - the lead will coat the precious metals, reducing their converting efficiency, and will eventually destroy the converter.
b) Always keep the ignition and fuel systems well-maintained in accordance with the manufacturer's schedule.
c) If the engine develops a misfire, do not drive the car at all (or at least as little as possible) until the fault is cured.
d) DO NOT push- or tow-start the car - this will soak the catalytic converter in unburned fuel, causing it to overheat when the engine does start.
e) DO NOT switch off the ignition at high engine speeds.
f) DO NOT use fuel or engine oil additives - these may contain substances harmful to the catalytic converter.
g) DO NOT continue to use the car if the engine burns oil to the extent of leaving a visible trail of blue smoke.

h) Remember that the catalytic converter operates at very high temperatures. DO NOT, therefore, park the car in dry undergrowth, or over long grass or piles of dead leaves after a long run.

l) Remember that the catalytic converter is FRAGILE - do not strike it with tools during servicing work.

j) In some cases, a sulfurous smell (like that of rotten eggs) may be noticed from the exhaust. This is common to many catalytic converter-equipped cars, and once the car has covered a few thousand miles the problem should disappear.

k) The catalytic converter, used on a well-maintained and well-driven car, should last for between 50,000 and 100,000 miles - if the converter is no longer effective, it must be replaced.

## 4 On-Board Diagnostics - general information

1   All models covered by this manual are equipped with an engine management system. The engine management system consists of an array of information sensors, an Engine Control Module (ECM) and various output actuators. The information sensors collect data from various parts of the engine (exhaust gas oxygen content, intake air temperature, throttle position, engine speed, etc.) and transmit this data to the ECM in the form of analog voltage signals. The ECM converts this analog information into digital data and "processes" it, i.e. compares it to a program built into the computer. If the incoming data doesn't agree with the program, the ECM produces a digital output command (which is converted back to an analog voltage signal) to the appropriate output actuator, which alters the air-fuel ratio, the ignition advance, etc. The ECM is capable of processing the data from a dozen information sensors simultaneously and issuing just as many simultaneous commands to the actuators many times a second. The end result is a smooth-running engine which produces good power, gets good fuel economy and produces very low emissions.

2   But as a vehicle ages, its emission control, fuel injection and ignition systems can occasionally malfunction, causing driveability problems. Because these systems are all highly interrelated, and because the problems are often intermittent, it isn't always easy to track down the real cause of a problem. That's why, in addition to its control function, the ECM also has the ability to monitor itself (on-board diagnostics). Each information sensor circuit operates at a certain voltage and each sensor is designed to produce a specified number of voltage signals per second. When the engine is running, the ECM constantly monitors the operating characteristics of each sensor circuit and compares its monitoring data with the program. If a circuit operates outside its operating range (too slow, too weak, intermittent, nonexistent, etc.), the ECM notes this anomaly. The program has certain thresholds for each sensor circuit: a sensor circuit must not produce an intermittent or poor signal more than a specified number of times, or it must not produce an intermittent or poor signal even once that lasts more than a specified interval. There are different thresholds for different sensor circuits, but when a circuit exceeds its threshold, the ECM stores a diagnostic trouble code (DTC). A DTC doesn't necessarily tell you exactly what the problem is. But it does indicate the sensor circuit where a problem is occurring, if it is occurring in a monitored circuit. (Not all circuits are monitored; problems occurring in unmonitored circuits will go unnoticed by the ECM, or will cause it to store a code for a related circuit that is affected by the problems) This self-monitoring capability is known as On-Board Diagnostics (OBD).

3   The vehicles covered in this manual are equipped with one of the two versions of OBD, depending on the model year: 1992 through 1995 models are equipped with OBD I; 1996 and later models are equipped with OBD II. The two systems are similar in concept, but the OBD II system is capable of storing many more DTCs than an OBD I system.

4   The DTCs can only be accessed by a special BMW diagnostic device known as a scanner. There are also a few aftermarket scanners capable of accessing BMW DTCs. If the Malfunction Indicator Light (MIL) on the dash comes on, take the vehicle to a dealer service department or other qualified repair shop and have it diagnosed with the special equipment needed to do the job.

# Chapter 7 Part A
# Manual transmission

## Contents

| | Section | | Section |
|---|---|---|---|
| Back-up light switch - testing, removal and installation | 5 | Manual transmission oil level check | 2 |
| Gear selector components - removal and installation | 3 | Manual transmission oil replacement | See Chapter 1 |
| General information | 1 | Manual transmission - removal and installation | 6 |
| Manual transmission overhaul - general information | 7 | Oil seals - replacement | 4 |

## Specifications

### Lubrication
| | |
|---|---|
| Recommended oil | See Chapter 1 |
| Capacity (from dry) | See Chapter 1 |

### Torque specifications — Ft-lbs
| | |
|---|---|
| Transmission-to-engine bolts | |
|   Hexagon head bolts | |
|     M8 bolts | 18 |
|     M10 bolts | 36 |
|     M12 bolts | 55 |
|   Torx head bolts | |
|     M8 bolts | 16 |
|     M10 bolts | 32 |
|     M12 bolts | 53 |
| Oil drain plug | 37 |
| Oil filler/level plug | 37 |
| Output flange-to-output shaft nut* | |
|   Step 1 | 125 |
|   Step 2 | Fully loosen nut |
|   Step 3 | 89 |
| Transmission crossmember-to-body bolts | |
|   M8 bolts | 15 |
|   M10 bolts | 31 |
| Transmission mount-to-transmission nuts | |
|   M8 nuts | 15 |
|   M10 nuts | 31 |

*Coat the threads of the nut with thread-locking compound.*

## 1 General information

The manual transmission is a 5-speed unit, and is contained in a cast-alloy casing bolted to the rear of the engine.

Drive is transmitted from the crankshaft via the clutch to the input shaft, which has a splined extension to accept the clutch friction plate. The output shaft transmits the drive via the driveshaft to the rear differential.

The input shaft runs in line with the output shaft. The input shaft and output shaft gears are in constant mesh with the layshaft gear cluster. Gears are selected by sliding synchromesh hubs, which lock the appropriate output shaft gears to the output shaft.

Gear selection is via a floor-mounted lever and selector mechanism. The selector mechanism causes the appropriate selector fork to move its respective synchro-sleeve along the shaft, to lock the gear pinion to the synchro-hub. Since the synchro-hubs are splined to the output shaft, this locks the pinion to the shaft, so that drive can be transmitted. To ensure that gear-changing can be made quickly and quietly, a synchro-mesh system is fitted to all forward gears, consisting of balk rings and spring-loaded fingers, as well as the gear pinions and synchro-hubs. The synchro-mesh cones are formed on the mating faces of the balk rings and gear pinions.

# Chapter 7 Part A  Manual transmission

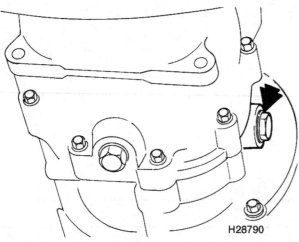

**2.2  Manual transmission oil level/filler plug location (arrow)**

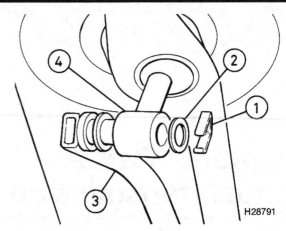

**3.4  Pry off the securing clip and disconnect the gear selector rod pin from the shift lever**

1  Securing clip
2  Washer
3  Selector rod
4  Shift lever eye

## 2  Manual transmission oil level check

*Refer to illustration 2.2*

1  To improve access, jack up the vehicle and support on axle stands. Ensure that the vehicle is level.
2  Unscrew the transmission oil level/filler plug from the right-hand side of the transmission casing **(see illustration)**.
3  The oil level should be up to the bottom of the level/filler plug hole.
4  If necessary, top-up the oil level, using the specified type of fluid (see Chapter1) until the oil overflows from the filler/level plug hole.
5  Wipe away any spilled oil, then install the filler/level plug and tighten to the specified torque.
6  Lower the vehicle to the ground.

## 3  Gear selector components - removal and installation

*Refer to illustrations 3.4, 3.5 and 3.9*

### Shift lever

**Note:** *A new shift lever bearing will be required on installation.*

#### Removal

1  Jack up the vehicle and support it securely on axle stands.
2  Remove the knob from the shift lever by pulling it sharply upwards.
3  Unclip the shift lever boot from the center console and withdraw the boot over the shift lever. Where applicable, remove the foam insulation.
4  Working under the vehicle, pry the securing clip from the end of the gear selector rod pin. Withdraw the selector rod pin from eye on the end of the shift lever, and recover the washers **(see illustration)**.

5  It is now necessary to release the shift lever lower bearing retaining ring from the gear selector arm. A special tool is available for this purpose, but two screwdrivers, with the tips engaged in opposite slots in the bearing ring can be used instead. To unlock the bearing ring, turn it a quarter-turn counterclockwise **(see illustration)**.
6  The bearing can now be pushed up through the housing, and the shift lever can be withdrawn from inside the vehicle.
7  If desired, the bearing can be removed from the shift lever ball by pressing it downwards. To withdraw the bearing over the lever eye, rotate the bearing until the eye passes through the slots provided in the bearing.
8  Fit a new bearing using a reversal of the removal process. Ensure that the bearing is pressed securely into position on the shift lever ball.

#### Installation

9  Install the lever using a reversal of the removal process, bearing in mind the following points.

a) Grease the contact faces of the bearing before installation.
b) Lower the shift lever into position, ensuring that the arrow on the shift lever grommet points towards the front of the vehicle.
c) Make sure that the shift lever grommet is correctly engaged with the gear selector arm and with the opening in the vehicle floor **(see illustration)**.
d) When engaging the bearing with the selector arm, make sure that the arrows or tabs (as applicable) on the top of the bearing point towards the rear of the vehicle.
e) To lock the bearing in position in the selector arm, press down on the top of the bearing retaining tab locations until the tabs are heard to click into position.
f) Grease the selector rod pin before engaging it with the shift lever eye.

### Gear selector shaft eye

*Refer to illustration 3.13*

**Note:** *A new selector shaft eye securing roll-pin will be required on installation.*

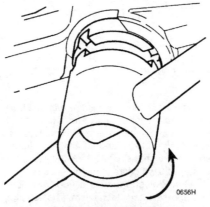

**3.5  Turn the bearing ring counterclockwise - special tool shown**

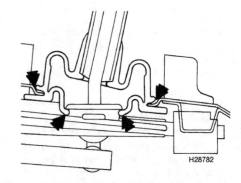

**3.9  Shift lever grommet correctly engaged with selector arm and vehicle floor**

## Chapter 7 Part A  Manual transmission

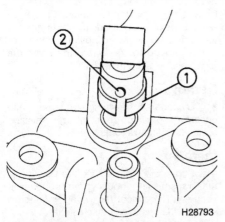

3.13 Slide back the locking sleeve (1), then drive out the roll-pin (2) securing the gear selector shaft eye

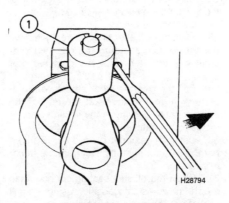

3.19 Levering the gear selector arm rear mounting sleeve (1) from the body bracket

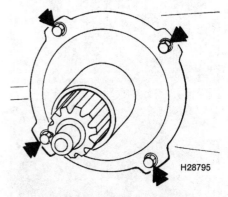

4.3 Clutch release bearing guide sleeve securing bolts (arrows)

### Removal

10  Jack up the vehicle and support securely on axle stands.
11  Disconnect the driveshaft from the transmission flange, and support it clear of the transmission using wire or string. Refer to Chapter 8 for details.
12  Pry the retaining clip from the end of the gear selector rod pin. Withdraw the selector rod pin from the selector shaft eye, and recover the washers.
13  Slide back the locking sleeve, then drive out the roll-pin securing the gear selector shaft eye to the end of the gear selector shaft **(see illustration)**.
14  Pull the gear selector shaft eye off the end of the selector shaft.

### Installation

15  Installation is a reversal of removal, bearing in mind the following points.
  a) Before installation, check the condition of the rubber washer in the end of the selector shaft eye and replace if necessary.
  b) Use a new roll-pin to secure the eye to the selector shaft.
  c) Grease the selector rod pin.
  d) Reconnect the driveshaft to the transmission flange (see Chapter 8).

## Gear selector arm rear mounting

*Refer to illustration 3.19*

### Removal

16  Jack up the vehicle and support securely on axle stands.
17  Disconnect the driveshaft from the transmission flange, and support it clear of the transmission using wire or string. Refer to Chapter 8 for details.
18  Remove the shift lever as described previously in this Section.
19  Using a screwdriver or a small pin-punch, lever the mounting sleeve from the bracket on the body **(see illustration)**.
20  Pull the mounting from the selector arm.

### Installation

21  Grease the mounting, then push the mounting onto the selector arm, with the cut-out facing the rear of the vehicle and the arrow pointing upwards.
22  Clip the mounting into position in the bracket, making sure that the mounting is securely located.
23  Reconnect the driveshaft to the transmission flange (see Chapter 8), then lower the vehicle to the ground.

## 4  Oil seals - replacement

*Refer to illustration 4.3*

### Input shaft oil seal

1  With the transmission removed (see Section 6), proceed as follows.
2  Remove the clutch release bearing and lever (see Chapter 8).
3  Unscrew the securing bolts and withdraw the clutch release bearing guide sleeve from the transmission bellhousing **(see illustration)**.
4  Note the fitted depth of the now-exposed input shaft oil seal.
5  Drill two small holes in the oil seal (two small pilot holes should be provided at opposite points on the seal - a sharp instrument can be used to extend these holes all the way though the seal).
6  Using a small drift, tap one side of the seal (adjacent to one of the holes) into the bellhousing as far as the stop.
7  Screw a small self-tapping screw into the opposite side of the seal, and use pliers to pull out the seal.
8  Clean the oil seal seating surface.
9  Wind a length of tape over the splines on the input shaft to prevent damage to the new seal as it is slid over the shaft.
10  Lubricate the lips of the new oil seal with a little clean transmission oil, then carefully slide the seal over the input shaft into position in the bellhousing.
11  Remove the tape from the input shaft then, using a tube of the correct diameter, tap the oil seal into the bellhousing to the previously noted depth.
12  Install the clutch release lever and bearing (see Chapter 8).
13  Install the transmission (see Section 6), then check the transmission oil level (see Section 2).

### Output flange oil seal

**Note:** *Thread-locking compound will be required for the transmission flange nut on installation.*

14  Jack up the vehicle and support securely on axle stands.
15  Disconnect the driveshaft from the transmission flange, and support it clear of the transmission using wire or string. Refer to Chapter 8 for details.
16  Where applicable, pry the transmission flange nut cover plate from the flange using a screwdriver. Discard the cover plate - it is not required on installation.
17  Counterhold the transmission flange by bolting a forked or two-legged tool to two of the flange bolt holes, then unscrew the flange securing nut using a socket and extension bar. **Warning:** *The nut is very tight.*
18  Using a puller, draw the flange from the end of the transmission output shaft. Be prepared for oil spillage.
19  Note the fitted depth of the oil seal then, again using a puller (take care to avoid damage to the transmission output shaft), pull the oil seal from the transmission casing.
20  Clean the oil seal seating surface.
21  Lubricate the lips of the new oil seal with a little clean transmission oil, then tap the seal into the transmission casing to the to the previously noted depth.
22  Install the flange to the output shaft. **Caution:** *When working on S5D 260Z or S5D 310Z-type transmissions, the flange must be heated to a temperature of 175-degrees F before fitting. The transmission type can be identified from the shape of the output flange - on S5D 260Z and S5D 310Z units the flange*

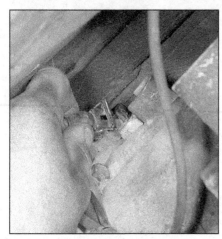

**5.3 Disconnecting the wiring plug from the back-up light switch**

has three arms with a bolt hole in each - on other transmissions the flange has a triangular plate with a bolt hole at each corner of the triangle. Consult a BMW dealer if there is any doubt about the type of transmission fitted. **Warning:** *If the transmission flange is heated, take precautions against burns - the metal will stay hot for some time.*

23  Coat the threads of the flange nut with thread-locking compound, then tighten the nut to the specified torque in the three stages given in the Specifications. Counterhold the flange as during removal.
24  If a flange nut cover plate was originally fitted, discard it. There is no need to fit a cover plate on installation.
25  Reconnect the driveshaft to the transmission flange (see Chapter 8), then check the transmission oil level (see Section 2), and lower the vehicle to the ground.

## Gear selector shaft oil seal

**Note:** *A new selector shaft eye securing roll-pin will be required on installation.*
26  Jack up the vehicle and support securely on axle stands.
27  Disconnect the driveshaft from the transmission flange, and support it clear of the transmission using wire or string. Refer to Chapter 8 for details.
28  Slide back the locking sleeve, then drive out the roll-pin securing the gear selector shaft eye to the end of the gear selector shaft.
29  Pull the gear selector shaft eye (complete with gear linkage) off the end of the selector shaft, and move the linkage clear of the selector shaft.
30  Using a small flat-bladed screwdriver, pry the selector shaft oil seal from the transmission casing.
31  Clean the oil seal seating surface, then tap the new seal into position using a small socket or tube of the correct diameter.
32  Check the condition of the rubber washer in the end of the selector shaft eye and replace if necessary.
33  Push the selector shaft eye back onto the end of the selector shaft, then align the holes in the eye and shaft and secure the eye to the shaft using a new roll-pin.
34  Slide the locking sleeve into position over the roll-pin.
35  Reconnect the driveshaft to the transmission flange (see Chapter 8).
36  Check the transmission oil level (see Section 2), then lower the vehicle to the ground.

## 5  Back-up light switch - testing, removal and installation

*Refer to illustration 5.3*

### Testing

1  The back-up light circuit is controlled by a plunger-type switch screwed into the left-hand side of the transmission casing. If a fault develops in the circuit, first ensure that the circuit fuse has not blown.
2  To test the switch, disconnect the wiring connector, and use a multimeter (set to the resistance function) or a battery-and-bulb test circuit to check that there is continuity between the switch terminals only when reverse gear is selected. If this is not the case, and there are no obvious breaks or other damage to the wires, the switch is faulty, and must be replaced.

### Removal

3  Disconnect the wiring connector, then unscrew the switch from the transmission casing **(see illustration)**.

### Installation

4  Screw the switch back into position in the transmission housing and tighten it securely. Reconnect the wiring connector, and test the operation of the circuit.

## 6  Manual transmission - removal and installation

*Refer to illustrations 6.6a, 6.6b, 6.9, 6.10, 6.18, 6.20, 6.21 and 6.22*

### Removal

**Note:** *This is an involved operation. Read through the procedure thoroughly before starting work, and ensure that adequate lifting tackle and/or jacking/support equipment is available.*
1  Open the hood, then raise the hood to its fully open position (see Chapter 11).
2  Disconnect the battery negative lead.
**Caution:** *If the radio in your vehicle is equipped with an anti-theft system, make sure you have the correct activation code before disconnecting the battery.*
3  Jack up the vehicle and support securely on axle stands. Note that the vehicle must be raised sufficiently to allow clearance for the transmission to be removed from under the vehicle.
4  Remove the starter motor (see Chapter 5 Part A).
5  Remove the driveshaft (see Chapter 8).
6  Working under the vehicle, pry the retaining clip from the end of the gear selec-

**6.6a Remove the retaining clips from the gear selector rod pins . . .**

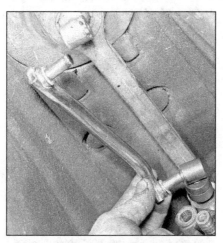

**6.6b . . . then withdraw the selector rod**

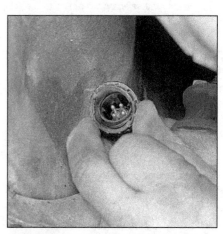

**6.9 Releasing the oxygen sensor wiring connector from the bracket on the transmission**

# Chapter 7 Part A Manual transmission

6.10 Unbolt the exhaust mounting bracket from the rear of the transmission

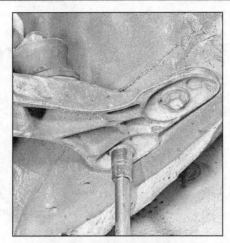

6.18 Unscrewing a transmission crossmember securing bolt

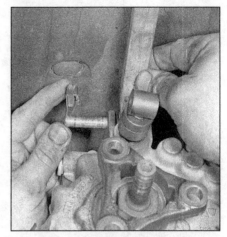

6.20 Releasing the gear selector arm from the transmission

tor rod pin. Withdraw the selector rod pin from the eye on the end of the transmission selector shaft, and recover the washers. Similarly, disconnect the selector rod pin from the end of the shift lever, and withdraw the selector rod **(see illustrations)**.

7 Working at the transmission bellhousing, unscrew the securing nuts, and withdraw the clutch release cylinder from the studs on the bellhousing. Support the release cylinder clear of the working area, taking care not to strain the hose.

8 Disconnect the wiring from the back-up light switch, located in the left-hand side of the transmission casing, and release the switch wiring from the clips on the transmission.

9 Separate the two halves of the oxygen sensor wiring connector, then withdraw the clamping ring to release the connector from the bracket, and unbolt the connector bracket from the transmission **(see illustration)**.

10 Where applicable, unbolt the exhaust mounting bracket from the rear of the transmission and move the bracket clear **(see illustration)**. If necessary, loosen the clamp on the exhaust to allow the bracket to be pivoted clear of the working area.

11 If necessary, to provide additional clearance, unbolt the bracing tube from the floor of the vehicle.

12 On six-cylinder models, if not already done, remove the heater/ventilation inlet air ducting from the rear of the engine compartment as follows.

   a) *Lift the grille from the top of the ducting (on certain Coupe models, it will be necessary to remove the securing screws and lift off the complete cowl grille assembly).*
   b) *Working through the top of the ducting, remove the screws securing the cable ducting to the air ducting and move the cable ducting clear.*
   c) *Unscrew the nuts and/or screw(s) securing the air ducting to the firewall (where applicable, bend back the heat shielding for access).*
   d) *Remove the air ducting by pulling upwards.*
   e) *Move the previously removed cable ducting clear of the valve cover.*

13 On four-cylinder models, unbolt the wiring ducting from the rear of the engine, then connect an engine hoist and lifting tackle to the engine lifting eye (incorporated in the rear flange of the cylinder block casting) at the rear left-hand corner of the cylinder block. Raise the lifting tackle to just take the weight of the engine.

14 As an alternative, support the engine using a floor jack under the oil pan, with a block of wood between the jack and oil pan to spread the load. Raise the jack to just touch the oil pan.

15 On six-cylinder models, connect the lifting tackle to the engine lifting eye at the rear left-hand corner of the cylinder block (incorporated in the rear flange of the cylinder block casting).

16 Place a floor jack under the transmission casing, just forward of the bellhousing. Use a block of wood to spread the load, then raise the jack to just take the weight of the transmission.

17 Check to ensure that the engine and transmission are adequately supported then, working under the vehicle, unscrew the nuts securing the transmission rubber mountings to the lugs on the transmission casing.

18 Remove the bolts securing the transmission crossmember to the body, then withdraw the crossmember from under the vehicle **(see illustration)**. If necessary, bend back or unbolt the exhaust heat shield for access to the crossmember bolts.

19 Using the jack(s) and engine hoist (where applicable), lower the engine and transmission until the rear of the engine cylinder head/manifold assembly is almost touching the engine compartment firewall. Check that the assembly is not resting against the heater hose connections on the firewall.

20 Working at the top of the transmission, pry up the clip securing the gear selector arm pivot pin to the transmission casing, then pull out the pivot pin to release the selector arm from the transmission **(see illustration)**.

21 Where applicable, unscrew the bolt securing the engine/transmission adapter plate to the right-hand side of the transmission bellhousing **(see illustration)**.

22 Unscrew the engine-to-transmission bolts, and recover the washers, then slide the transmission rearwards to disengage the

6.21 Engine/transmission adapter plate (arrow)

input shaft from the clutch **(see illustration)**. Take care during this operation to ensure that the weight of the transmission is not allowed to hang on the input shaft. As the transmission is released from the engine, check to make sure that the engine is not forced against the heater hose connections or the firewall.

23  Lower the transmission and carefully withdraw it from under the vehicle. If the transmission is to be removed for some time, ensure that the engine is adequately supported in the engine compartment.

## Installation

24  Commence installation by checking that the clutch friction disc is centralized (see Chapter 8, Section 2).

25  Before installing the transmission, it is advisable to inspect and grease the clutch release bearing and lever (see Chapter 8).

26  The remainder of the installation procedure is a reversal of removal, bearing in mind the following points.

 a) Check that the transmission positioning dowels are securely in place at the rear of the engine.
 b) Make sure that the washers are in place on the engine-to-transmission bolts.
 c) Tighten all fasteners to the specified torque.
 d) Lightly grease the gear selector arm pivot pin and the gear selector rod pin before installation.
 e) Reconnect the driveshaft to the transmission flange (see Chapter 8).
 f) Install the starter motor (see Chapter 5 Part A).
 g Check the lubricant level (see Chapter 1).

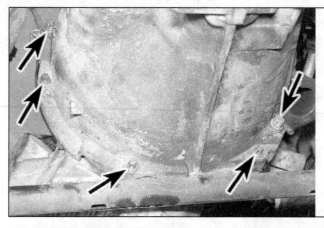

6.22 Lower engine-to-transmission bolts (arrows)

## 7  Manual transmission overhaul - general information

Overhauling a manual transmission is a difficult and involved job for the DIY home mechanic. In addition to dismantling and reassembling many small parts, clearances must be precisely measured and, if necessary, changed by selecting shims and spacers. Internal transmission components are also often difficult to obtain, and in many instances, extremely expensive. Because of this, if the transmission develops a fault or becomes noisy, the best course of action is to have the unit overhauled by a specialist repairer, or to obtain an exchange reconditioned unit. Be aware that some transmission repairs can be carried out with the transmission in the car.

Nevertheless, it is possible for the more experienced mechanic to overhaul the transmission, provided the special tools are available, and the job is done in a deliberate step-by-step manner so that nothing is overlooked.

The tools necessary for an overhaul include internal and external snap-ring pliers, bearing pullers, a slide hammer, a set of pin punches, a dial test indicator and possibly a hydraulic press. In addition, a large, sturdy workbench and a vise will be required.

While dismantling the transmission, make careful notes of how each component is fitted, to make reassembly easier and more accurate.

Before dismantling the transmission, it will help if you have some idea what area is malfunctioning. Certain problems can be closely related to specific areas in the transmission, which can make component examination and replacement easier. Refer to the *"Troubleshooting"* Section at the beginning of this manual for more information.

# Chapter 7 Part B
# Automatic transmission

## Contents

| | Section | | Section |
|---|---|---|---|
| Automatic transmission fluid level check | See Chapter 1 | Fluid seals - replacement | 4 |
| Automatic transmission fluid replacement | See Chapter 1 | Gear selector cable - removal, installation and adjustment | 3 |
| Automatic transmission overhaul - general information | 6 | Gear selector lever - removal and installation | 2 |
| Automatic transmission - removal and installation | 5 | General information | 1 |

## Specifications

### Torque specifications

| | Ft-lbs |
|---|---|
| Engine-to-transmission bolts | |
| Hexagon bolts | |
| M8 bolts | 18 |
| M10 bolts | 33 |
| M12 bolts | 61 |
| Torx bolts | |
| M8 bolts | 15 |
| M10 bolts | 31 |
| M12 bolts | 46 |
| Engine/transmission adapter plate bolt | 17 |
| Transmission crossmember-to-body bolts | |
| M8 bolts | 15 |
| M10 bolts | 31 |
| Transmission mount-to-transmission nuts | |
| M8 nuts | 15 |
| M10 nuts | 31 |
| Driveplate-to-torque-converter bolts | |
| M8 bolts | 19 |
| M10 bolts | 36 |

## 1 General information

Models covered in this manual are equipped with a four-speed fully-automatic transmission, consisting of a torque converter, a planetary geartrain and hydraulically operated clutches and brakes.

The torque converter provides a fluid coupling between engine and transmission, which acts as a clutch, and also provides a degree of torque multiplication when accelerating.

The planetary geartrain provides either of the four forward or one reverse gear ratio, according to which of its component parts are held stationary or allowed to turn. The components of the geartrain are held or released by brakes and clutches which are activated by a hydraulic control unit. A fluid pump within the transmission provides the necessary hydraulic pressure to operate the brakes and clutches.

Driver control of the transmission is by a seven-position selector lever, and a four-position switch. The transmission has a "drive" position, and a "hold" facility on the first three gear ratios. The "drive" position ("D") provides automatic changing throughout the range of all forward gear ratios, and is the position selected for normal driving. An automatic kickdown facility shifts the transmission down a gear if the accelerator pedal is fully depressed. The "hold" facility is very similar, but limits the number of gear ratios available - i.e., when the selector lever is in the "3" position, only the first three ratios can be selected; in the "2" position, only the first two can be selected, and so on. The lower ratio "hold" is useful when traveling down steep gradients, or for preventing unwanted selection of high gear on curved roads. Three driving programs are provided for selection by the switch; "economy," "sport," and "manual."

Due to the complexity of the automatic transmission, any repair or overhaul work must be left to a BMW dealer with the necessary special equipment for fault diagnosis and repair. The contents of the following Sections are therefore confined to supplying general information, and any service information and instructions that can be used by the owner.

# 7B-2   Chapter 7 Part B   Automatic transmission

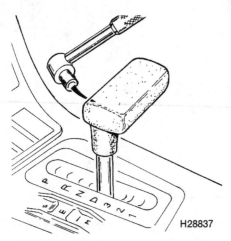

2.2 Unscrewing the selector lever handle securing screw

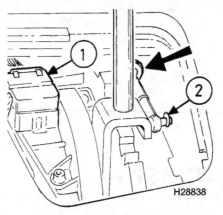

2.6a Disconnect the wiring plug (1), pry off the cable clip (2) and unscrew the cable sleeve securing nut (arrow) . . .

2.6b . . . then pull the cable forwards

## 2   Gear selector lever - removal and installation

*Refer to illustrations 2.2, 2.6a, 2.6b, 2.7 and 2.8*

### Selector lever - models without "interlock" system

#### Removal

1   Disconnect the battery negative lead.
**Caution:** *If the radio in your vehicle is equipped with an anti-theft system, make sure you have the correct activation code before disconnecting the battery.*
2   Unscrew the securing screw from the front of the selector lever handle, then pull off the handle **(see illustration)**.
3   Remove the cigarette lighter and housing, (see Chapter 12).
4   Unclip the selector lever cover and disconnect the wiring from the switch and the light mounted in the cover.
5   Disconnect the wiring plug from the lever assembly.
6   Pry off the clip securing the selector cable to the lever, then unscrew the cable sleeve securing nut, and pull the cable forwards. Unhook the cable end from the lever **(see illustrations)**.
7   Unscrew the three mounting bolts, then tilt the lever assembly, and withdraw upwards **(see illustration)**. Recover the seal between the lever assembly and the floor panel.

#### Installation

8   Installation is a reversal of removal, bearing in mind the following points.
  a) Ensure that the seal is correctly fitted between the lever assembly and the floor panel.
  b) When reconnecting the selector cable, ensure that the lug on the lever assembly engages with the hole in the cable sleeve plate.
  c) When installing the lever handle, ensure that the pin on the button under the handle engages with the hole in the top of the lever. Fit the handle with the button released **(see illustration)**.
  d) On completion, adjust the selector cable (see Section 3).

### Selector lever - models with "interlock" system

*Refer to illustrations 2.10 and 2.24*

#### Removal

9   Follow the procedure outlined in Steps 1 through 6.
10   Unscrew the interlock cable clamp bolt, and remove the clamp **(see illustration)**.
11   Disconnect the end of the interlock cable from the pin on the lever.
12   Disconnect the interlock cable from the steering lock (see Chapter 10).
13   Working at the linkage, separate the two halves of the interlock wiring connector.
14   Unscrew the three mounting bolts, then tilt the lever assembly, and withdraw upwards. Recover the seal between the lever assembly and the floor.

#### Installation

15   Reconnect the cable to the steering lock (see Chapter 10).
16   Manipulate the lever assembly into position, then install and tighten the securing bolts. Ensure that the seal is correctly fitted between the lever assembly and the floor panel.

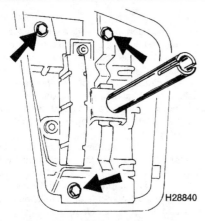

2.7 Selector lever assembly mounting bolts (arrows)

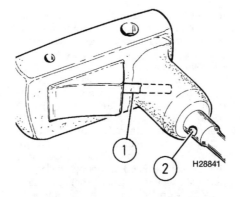

2.8 Ensure that the pin (1) engages with the hole (2) in the lever

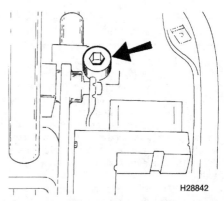

2.10 Unscrew the interlock cable clamp bolt (arrow)

## Chapter 7 Part B  Automatic transmission

17  Reconnect the selector cable to the lever, ensuring that the lug on the lever assembly engages with the hole in the cable sleeve plate. Tighten the securing nut.
18  Reconnect the wiring plug to the lever assembly.
19  Move the selector lever into position "P" (the forward position).
20  Reconnect the end of the interlock cable to the pin on the lever.
21  Lay the cable sleeve in position, then install the clamp and tighten the clamp bolt by hand only. It should be possible for the cable to slide easily through the clamp.
22  Turn the ignition key to the "lock" position, and remove the key.
23  Check that the interlock mechanism operates correctly as follows.
  a) Turn the ignition key to the center (accessory) position.
  b) It should be possible to move the selector lever in and out of all positions.
  c) It should only be possible to turn the ignition key back to the "lock" position and remove the key with the selector in position "P".
  d) It must not be possible to move the selector lever our of position "P" with the ignition key in the "lock" position or removed.
24  Press the locking lever down and tighten the cable clamp bolt (see illustration).
25  Again, check the operation of the interlock mechanism (see Step 23).
26  Further installation is a reversal of removal, but when installing the lever handle, ensure that the pin on the button under the handle engages with the hole in the top of the lever. Fit the handle with the button released.
27  On completion, adjust the selector cable (see Section 3).

### 3  Gear selector cable - removal, installation and adjustment

Refer to illustrations 3.3 and 3.11

#### Removal
1  Disconnect the cable from the selector lever assembly (see Section 2).
2  To improve access, apply the parking brake, then jack up the front of the vehicle and support securely on axle stands.
3  Working at the end of the cable, counterhold the clamp bolt, and loosen the securing nut. Take care not to bend the end of the cable (see illustration).
4  Loosen the securing nut and release the cable from the bracket on the transmission.
5  Slide the end of the cable from the end fitting.
6  Withdraw the cable down from under the vehicle, noting its routing to ensure correct installation.

#### Installation
7  Installation is a reversal of removal,

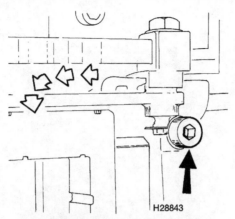

2.24  Press the locking lever down and tighten the clamp bolt (arrows)

bearing in mind the following points.
  a) Do not tighten the cable end securing nut and bolts until the cable has been adjusted.
  b) Reconnect the cable to the selector lever assembly with reference to Section 2.
  c) On completion, adjust the cable outlined in the following steps.

#### Adjustment
8  Move the selector lever to position "P".
9  If not already done, counterhold clamp bolt and loosen the clamp nut securing the cable to the end fitting (the vehicle should be raised for access).
10  Push the operating lever on the transmission away from the cable bracket on the

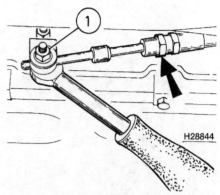

3.3  Counterhold the clamp bolt, and loosen the selector cable securing nut (1)

Cable-to-bracket securing nut arrow

transmission (towards the "Park" position).
11  Press the end of the cable in the opposite direction (towards the cable bracket), then release the cable and tighten the clamp nut (see illustration).
12  Check that the cable is correctly adjusted by starting the engine, applying the brakes firmly, and moving the selector lever through all the selector positions.

### 4  Fluid seals - replacement

#### Torque converter seal
1  Remove the transmission and the torque converter (see Section 5).
2  Using a hooked tool, pry the old oil seal

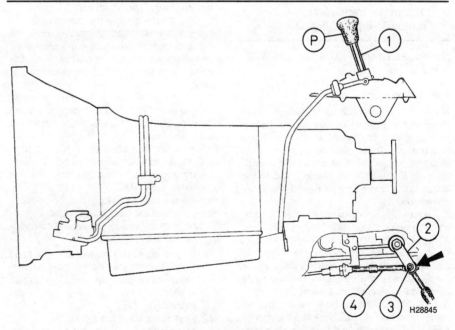

3.11  Adjusting the selector cable

| P | Park position | 2 | Operating lever | 4 | Selector cable |
| 1 | Selector lever | 3 | Clamp nut | | |

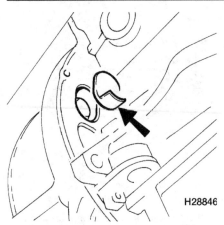

**5.13a Pry the plug (arrow) from the engine/transmission adapter plate ...**

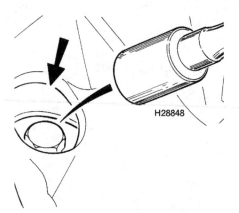

**5.13b ... or from the aperture in the crankcase**

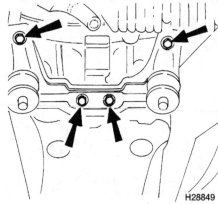

**5.21 Where applicable, unscrew the securing bolts (arrows) and remove the transmission front mounting assembly**

from the transmission bellhousing. Alternatively, drill a small hole, then screw a self-tapping screw into the seal and use pliers to pull out the seal.
3 Lubricate the lip of the new seal with clean fluid, then carefully drive it into place using a large socket or section of pipe.
4 Install the torque converter and transmission (see Section 5).

### Output flange oil seal

5 Replacement of the oil seal involves partial dismantling of the transmission, which is a complex operation (see Section 6). Oil seal replacement should be entrusted to a BMW dealer or other qualified repair shop.

### 5 Automatic transmission - removal and installation

*Refer to illustrations 5.13a, 5.13b, 5.21, 5.30, 5.36 and 5.40*

### Removal

**Note:** *This is an involved operation. Read through the procedure thoroughly before starting work, and ensure that adequate lifting tackle and/or jacking/support equipment is available. A suitable tool will be required to align the torque converter when installing the transmission, and new fluid pipe O-rings may be required.*
1 Open the hood, then raise the hood to its fully open position (see Chapter 11).
2 Disconnect the battery negative cable.
**Caution:** *If the radio in your vehicle is equipped with an anti-theft system, make sure you have the correct activation code before disconnecting the battery.*
3 Jack up the vehicle and support securely on axle stands. Note that the vehicle must be raised sufficiently to allow clearance for the transmission to be removed from under the vehicle.
4 Remove the starter motor (see Chapter 5 Part A).

5 Where applicable, unbolt the exhaust mounting crossmember from under the vehicle.
6 Remove the driveshaft (see Chapter 8).
7 Drain the automatic transmission fluid (see Chapter 1).
8 Where applicable, unscrew the union nut and remove the fluid filler pipe from the transmission fluid pan.
9 Disconnect the selector cable from the transmission with reference to Section 3.
10 Disconnect the transmission wiring harness plug(s). Release the wiring harness from the brackets and clips on the transmission.
11 Where applicable, release the oxygen sensor from the bracket on the transmission.
12 Unbolt the fluid cooler pipe brackets and clamps, then pull the fluid cooler pipes from the transmission fluid pan.
13 Pry the plug from the aperture in the engine/transmission adapter plate, above the oil pan, or from the aperture in the crankcase, depending on model, for access to the torque converter securing bolts **(see illustrations)**.
14 Unscrew the three torque converter bolts, turning the crankshaft using a wrench or socket on the pulley hub bolt, for access to each bolt in turn.
15 Support the transmission using a floor jack and interposed block of wood. **Caution:** *The transmission is heavy, so ensure that it is adequately supported.*
16 On six-cylinder models, if not already done, remove the heater/ventilation inlet air ducting from the rear of the engine compartment as follows.

 a) *Lift the grille from the top of the ducting (on certain Coupe models, it will be necessary to remove the securing screws and lift off the complete cowl grille assembly).*
 b) *Working through the top of the ducting, remove the screws securing the cable ducting to the air ducting and move the cable ducting clear.*
 c) *Unscrew the nuts and/or screw(s) securing the air ducting to the firewall (where*

 *applicable, bend back the heat shielding for access).*
 d) *Remove the air ducting by pulling upwards.*
 e) *Move the previously removed cable ducting clear of the valve cover.*

17 On four-cylinder models, unbolt the wiring ducting from the rear of the engine, then connect an engine hoist and lifting tackle to the engine lifting eye (incorporated in the rear flange of the cylinder block casting) at the rear left-hand corner of the cylinder block. Raise the lifting tackle to just take the weight of the engine.
18 As an alternative, support the engine using a floor jack under the oil pan, with a block of wood between the jack and oil pan to spread the load. Raise the jack to just touch the oil pan.
19 On six-cylinder models, connect the lifting tackle to the engine lifting eye at the rear left-hand corner of the cylinder block (incorporated in the rear flange of the cylinder block casting).
20 Unbolt the crosstube from the vehicle floor, under the transmission bellhousing.
21 Where applicable, unbolt the transmission front mount assembly **(see illustration)**.
22 Check to ensure that the engine and transmission are adequately supported then, working under the vehicle, unscrew the nuts securing the transmission rubber mountings to the lugs on the transmission casing.
23 Remove the bolts securing the transmission crossmember to the body, then withdraw the crossmember from under the vehicle. If necessary, bend back or unbolt the exhaust heat shield for access to the crossmember bolts.
24 Using the jack(s) and engine hoist (where applicable), lower the engine and transmission until the rear of the engine cylinder head/manifold assembly is almost touching the engine compartment firewall. Check that the assembly is not resting against the heater hose connections on the firewall.
25 Where applicable, unscrew the bolt securing the engine/transmission adapter

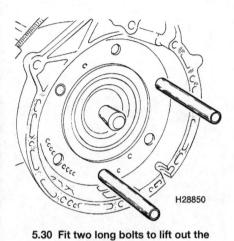

**5.30 Fit two long bolts to lift out the torque converter**

plate to the right-hand side of the transmission bellhousing.

26  Unscrew the engine-to-transmission bolts, and recover the washers, then slide the transmission rearwards.

27  Insert a suitable metal or wooden lever through the slot in the bottom of the bellhousing to retain the torque converter. As the transmission is released from the engine, check to make sure that the engine is not forced against the heater hose connections or the firewall.

28  Lower the transmission and carefully withdraw it from under the vehicle, making sure that the torque converter is held in position. If the transmission is to be removed for some time, ensure that the engine is adequately supported in the engine compartment.

29  To remove the torque converter, first remove the retaining lever.

30  Fit two long bolts to two of the torque converter securing bolt holes, and use the bolts to pull the torque converter from the transmission **(see illustration)**. Pull evenly on both bolts. Be prepared for fluid spillage.

## Installation

31  Where applicable, install the torque converter, using the two bolts to manipulate the converter into position.

32  Ensure that the transmission locating dowels are in position on the engine.

33  Before mating the transmission with the engine, it is essential that the torque converter is perfectly aligned with the driveplate. Once the engine and transmission have been mated, it is no longer possible to turn the torque converter to allow re-alignment.

34  To align the driveplate with the torque converter, BMW recommends a special tapered tool which screws into the converter. It may be possible to improvise a suitable tool using an old driveplate-to-torque converter bolt with the head cut off, or a length of threaded bar - note that the end of the bolt or bar must either have a slot cut in the end, or flats machined on it to allow it to be unscrewed once the engine and transmission have been mated.

35  Turn the driveplate to align one of the driveplate-to-torque converter bolt holes with the aperture in the bottom of the oil pan/bellhousing (for access to the oil pan securing bolt). This is essential to enable the alignment stud to be removed after the engine and transmission have been mated.

36  Screw the alignment tool into the relevant hole in the converter **(see illustration)**.

37  Where applicable, remove the retaining lever from the torque converter.

38  Ensure that the transmission is adequately supported, and maneuver it into position under the vehicle.

39  Turn the torque converter to align one of the driveplate bolt holes with the alignment tool fitted to the converter, then move the transmission into position.

40  Ensure that the alignment tool passes through the hole in the driveplate, then install and tighten the engine-to-transmission bolts, ensuring that the washers are in place **(see illustration)**.

41  Unscrew the alignment tool from the converter, then install the driveplate-to-torque converter bolt. Tighten the bolt to the specified torque.

42  Turn the crankshaft as during removal for access to the remaining two driveplate-to-torque converter bolt locations. Install and tighten the bolts.

43  Further installation is a reversal of removal, bearing in mind the following points.

  a) Tighten all fasteners to the specified torque values, where applicable.
  b) Check the condition of the transmission fluid pipe O-rings and replace if necessary.
  c) Install the driveshaft (see Chapter 8).
  d) Install the starter motor (see Chapter 5 Part A).
  e) Reconnect and adjust the selector cable (see Section 3).
  f) On completion, refill the transmission with fluid (see Chapter 1).

## 6  Automatic transmission overhaul - general information

In the event of a fault occurring with the transmission, it is first necessary to determine whether it is of an electrical, mechanical or hydraulic nature, and to do this special test equipment is required. It is therefore essential to have the work carried out by a BMW dealer if a transmission fault is suspected.

Do not remove the transmission from the vehicle for possible repair before professional fault diagnosis has been carried out, since most tests require the transmission to be in the vehicle.

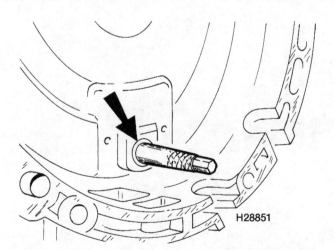

**5.36 Alignment tool (arrow) screwed into the torque converter, aligned with aperture in bottom of oil pan/bellhousing**

**5.40 Ensure that the alignment tool (1) passes through the hole (2) in the driveplate**

**Notes**

# Chapter 8
# Clutch and driveline

## Contents

| | Section | | Section |
|---|---|---|---|
| Clutch assembly - removal, inspection and installation | 2 | Differential oil replacement | See Chapter 1 |
| Clutch check | See Chapter 1 | Differential oil seals - replacement | 10 |
| Clutch fluid level check | See Chapter 1 | Differential - removal and installation | 9 |
| Clutch hydraulic system - bleeding | 6 | Driveaxle boots - replacement | 12 |
| Clutch master cylinder - removal, inspection and installation | 5 | Driveaxle - removal and installation | 11 |
| Clutch pedal - removal and installation | 7 | Driveshaft - removal and installation | 13 |
| Clutch release bearing and lever - removal, inspection and installation | 3 | Driveshaft rubber coupling - check and replacement | 14 |
| | | Driveshaft support bearing - check and replacement | 15 |
| Clutch release cylinder - removal, inspection and installation | 4 | Driveshaft universal joints - check and replacement | 16 |
| Clutch start switch - check and replacement | 8 | General information | 1 |

## Specifications

### Clutch disc
Minimum thickness .................................................. 19/64-inch

### Torque specifications
**Ft-lbs** (unless otherwise indicated)

| | |
|---|---|
| Clutch cover-to-flywheel bolts | 18 |
| Clutch release cylinder nuts | 16 |
| Clutch master cylinder bolts | 16 |
| Hydraulic fittings | 144 in-lbs |
| Differential mounting bolts | |
|   Front bolt | 70 |
|   Rear bolts | 59 |
| Driveshaft flange retaining nut (approximate - see text for tightening procedure) | |
|   M20 nut | 131 |
|   M22 nut | 138 |
| Driveshaft | |
|   Front universal joint-to-transmission | 48 |
|   Rubber coupling-to-transmission/driveshaft | |
|     M10 (Grade 8.8) bolts | 35 |
|     M10 (Grade 10.9) bolts | 48 |
|     M12 (Grade 8.8) bolts | 60 |
|     M12 (Grade 10.9) bolts | 74 |
|   Support bearing bracket nuts | 15 |
|   Rear universal joint-to-differential | |
|     Nyloc nut ("squeeze nut") | 48 |
|     Ribbed nut | 67 |
|     Threaded sleeve | 84 in-lbs |
| Driveaxle | |
|   Shaft to differential flange bolts: | |
|     Allen bolts: | |
|       M10 bolts: | |
|         Bolts with locking teeth | 71 |
|         Bolts without locking teeth | 61 |
|       M12 bolts | 81 |
|     Torx bolts: | |
|       M10 bolts: | |
|         Bolts with locking teeth | 74 |
|         Bolts without locking teeth | 61 |
|       M8 bolts | 48 |
|   Driveaxle retaining nut | 186 |
| Wheels | |
|   Wheel bolts | See Chapter 1 |

## 1 General information

### Clutch

All models with a manual transmission are equipped with a single dry plate clutch, which consists of five main components; friction disc, pressure plate, diaphragm spring, cover and release bearing.

The friction disc is free to slide along the splines of the transmission input shaft and is held in position between the flywheel and the pressure plate by the pressure exerted on the pressure plate by the diaphragm spring. Friction lining material is riveted to both sides of the friction disc, and cushioning springs between the friction surfaces help to absorb transmission shocks and ensure a smooth take-up of power as the clutch is engaged. Two types of friction disc are installed on models covered by this manual, depending on the type of flywheel. On engines with a conventional flywheel, the friction disc has integral damping springs (visible in the hub, between the center of the hub and the friction material). On engines with a "dual-mass" flywheel, the damping springs are integrated into the flywheel itself. The damping springs serve a similar purpose to the cushioning springs described previously.

The diaphragm spring is mounted on pins, and is held in place in the cover by annular fulcrum rings.

The release bearing is located on a guide sleeve at the front of the transmission, and the bearing is free to slide on the sleeve, under the action of the release arm which pivots inside the clutch bellhousing.

The release mechanism is operated by the clutch pedal, using hydraulic pressure. The pedal acts on the clutch master cylinder pushrod, and a release cylinder, mounted on the transmission bellhousing, operates the clutch release lever via a pushrod.

When the clutch pedal is depressed, the release arm pushes the release bearing forwards, to bear against the center of the diaphragm spring, thus pushing the center of the diaphragm spring inwards. The diaphragm spring acts against the fulcrum rings in the cover, and so as the center of the spring is pushed in, the outside of the spring is pushed out, so allowing the pressure plate to move backwards away from the friction disc.

When the clutch pedal is released, the diaphragm spring forces the pressure plate into contact with the friction linings on the friction disc, and simultaneously pushes the friction disc forwards on its splines, forcing it against the flywheel. The friction disc is now firmly sandwiched between the pressure plate and the flywheel, and drive is taken up.

The clutch is self-adjusting. As wear takes place on the friction disc over a period of time, the pressure plate automatically moves closer to the friction plate to compensate.

### Driveline

Power is transmitted from the transmission to the rear axle by a two-piece driveshaft, joined behind the center bearing by a "slip joint," a sliding, splined coupling. The slip joint allows slight fore-and-aft movement of the driveshaft. The forward end of the driveshaft is attached to the output flange of the transmission either by a flexible rubber coupling or a universal flange joint. On some models, a vibration damper is mounted between the front of the driveshaft and coupling. The middle of the driveshaft is supported by the center bearing which is bolted to the vehicle body. Universal joints are located at the center bearing and at the rear end of the driveshaft, to compensate for movement of the transmission and differential on their mountings and for any flexing of the chassis.

The differential assembly includes the drive pinion, the ring gear, the differential and the output flanges. The drive pinion, which drives the ring gear, is also known as the differential input shaft and is connected to the driveshaft via an input flange. The differential is bolted to the ring gear and drives the rear wheels through a pair of output flanges bolted to driveaxles with constant velocity (CV) joints at either end. The differential allows the wheels to turn at different speeds when cornering.

The driveaxles deliver power from the differential output flanges to the rear wheels. The driveaxles are equipped with Constant Velocity (CV) joints at each end. The inner CV joints are bolted to the differential flanges. The outer CV joints engage the splines of the wheel hubs and are secured by a large nut.

Major repair work on the differential assembly components (drive pinion, ring-and-pinion, and differential) requires many special tools and a high degree of expertise, and therefore should not be attempted by the home mechanic. If major repairs become necessary, we recommend that they be performed by a BMW service department or other qualified repair shop.

## 2 Clutch assembly - removal, inspection and installation

**Warning:** *Dust created by clutch wear and deposited on the clutch components may contain asbestos, which is a health hazard. DO NOT blow it out with compressed air, or inhale any of it. DO NOT use gasoline (or petroleum-based solvents) to clean off the dust. Brake system cleaner should be used to flush the dust into a suitable receptacle. After the clutch components are wiped clean with rags, dispose of the contaminated rags and cleaner in a sealed, marked container.*

### Removal

*Refer to illustration 2.4*

1  Remove the transmission (see Chapter 7A).

2.4 Removing the clutch assembly

2  If the original clutch is to be reinstalled, make alignment marks between the clutch cover and the flywheel, so that the clutch can be reinstalled in its original position.

3  Progressively unscrew the bolts securing the clutch cover to the flywheel, and where applicable recover the washers.

4  Withdraw the clutch cover from the flywheel **(see illustration)**. Be prepared to catch the clutch friction disc, which may drop out of the cover as it is withdrawn, and note which way the friction disc is installed - the two sides of the disc are normally marked "Engine side" and "Transmission side". The greater projecting side of the hub faces away from the flywheel.

### Inspection

5  With the clutch assembly removed, clean off all traces of dust using brake system cleaner. Although most friction discs now have asbestos-free linings, some do not; *asbestos dust is harmful, and must not be inhaled.*

6  Examine the linings of the friction disc for wear and loose rivets, and the disc for distortion, cracks, broken damping springs (where applicable) and worn splines. The surface of the friction linings may be highly glazed, but, as long as the friction material pattern can be clearly seen, this is satisfactory. If there is any sign of oil contamination, indicated by a continuous, or patchy, shiny black discoloration, the disc must be replaced. The source of the contamination must be traced and rectified before installing new clutch components; typically, a leaking crankshaft rear oil seal or transmission input shaft oil seal - or both - will be to blame (replacement procedures are given in the relevant Part of Chapter 2, and Chapter 7 Part A respectively). The disc must also be replaced if the lining thickness has worn down to, or just above, the level of the rivet heads. The manufacturer specifies a minimum friction material thickness (see Specifications) - on models with a conventional flywheel (and a friction disc with damping springs, see Section 1), the measurement should be made

# Chapter 8  Clutch and driveline

2.12  Install the clutch disc with the damper springs (or the greater projection) facing away from the flywheel

2.13  Ensure that the clutch cover locates over the dowels (arrow) on the flywheel

2.15  Using a clutch alignment tool to center the friction disc

with the edge of the friction disc at the measurement point compressed in a vise.

7  Check the machined faces of the flywheel and pressure plate. If either is grooved, or heavily scored, replacement is necessary. The pressure plate must also be replaced if any cracks are apparent, or if the diaphragm spring is damaged or its pressure suspect.

8  With the clutch removed, it is advisable to check the condition of the release bearing (see Section 3).

9  Check the pilot bearing in the end of the crankshaft. Make sure that it turns smoothly and quietly. If the transmission input shaft contact face on the bearing is worn or damaged, install a new bearing (see Chapter 2C).

## Installation

*Refer to illustrations 2.12, 2.13 and 2.15*

10  If new clutch components are to be installed, where applicable, ensure that all anti-corrosion preservative is cleaned from the friction material on the disc, and the contact surfaces of the pressure plate.

11  It is important to ensure that no oil or grease gets onto the friction disc linings, or the pressure plate and flywheel faces. It is advisable to install the clutch assembly with clean hands, and to wipe down the pressure plate and flywheel faces with a clean rag before assembly begins.

12  Apply a smear of moly-based grease to the splines of the friction disc hub, then position the disc on the flywheel, with the greater projecting side of the hub facing away from the flywheel (most friction discs will have an "Engine side" marking which should face the flywheel) **(see illustration)**. Hold the friction disc against the flywheel while the cover/pressure plate assembly is guided into position.

13  Install the clutch cover assembly, where applicable aligning the marks on the flywheel and clutch cover. Ensure that the clutch cover locates over the dowels on the flywheel **(see illustration)**. Insert the securing bolts and washers, and tighten them finger-tight, so that the friction disc is gripped, but can still be moved.

14  The friction disc must now be centralized, so that when the engine and transmission are mated, the transmission input shaft splines will pass through the splines in the friction disc hub.

15  Center the clutch disc by inserting a clutch alignment tool through the splined hub and into the pilot bearing in the crankshaft **(see illustration)**. Wiggle the tool up, down or side-to-side as needed to bottom the tool in the pilot bearing. Tighten the pressure plate-to-flywheel bolts a little at a time, working in a criss-cross pattern to prevent distorting the cover. After all of the bolts are snug, tighten them to the torque listed in this Chapter's Specifications. Remove the alignment tool.

16  Install the transmission (see Chapter 7A).

## 3  Clutch release bearing and lever - removal, inspection and installation

**Warning:** *Dust created by clutch wear and deposited on the clutch components may contain asbestos, which is a health hazard. DO NOT blow it out with compressed air or inhale any of it. DO NOT use gasoline (or petroleum-based solvents) to clean off the dust. Brake system cleaner should be used to flush the dust into a suitable receptacle. After the clutch components are wiped clean with rags, dispose of the contaminated rags and cleaner in a sealed, marked container.*

### Release bearing

#### Removal

*Refer to illustration 3.2*

1  Remove the transmission (see Chapter 7A).

2  Pull the bearing forwards, and slide it from the guide sleeve in the transmission bellhousing **(see illustration)**.

#### Inspection

3  Spin the release bearing, and check it for excessive roughness. Hold the outer race and attempt to move it laterally against the inner race. If any excessive movement or roughness is evident, replace the bearing. If a new clutch has been installed, it is wise to replace the release bearing as a matter of course.

#### Installation

4  Clean and then lightly grease the release bearing contact surfaces on the release lever. **Do not** apply grease to the guide sleeve.

5  Slide the bearing into position on the guide sleeve, ensuring that the bearing engages correctly with the release lever.

6  Install the transmission (see Chapter 7A).

### Release lever

#### Removal

*Refer to illustrations 3.8a and 3.8b*

7  Remove the release bearing, as described previously in this Section.

3.2  Removing the clutch release bearing

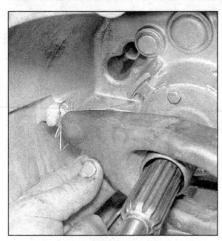

3.8a Slide the clutch release lever sideways to release the spring clip . . .

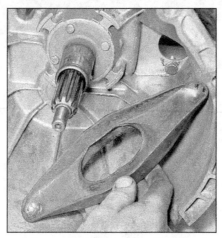

3.8b . . . then slide the lever from the guide sleeve

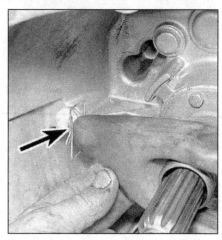

3.12 Ensure that the retaining spring clip (arrow) engages correctly

8  Slide the release lever sideways to release it from the retaining spring clip and pivot, then pull the lever forwards from the guide sleeve (see illustrations).

### Inspection

9  Inspect the release bearing, pivot and release cylinder pushrod contact faces on the release lever for wear. Replace the lever if excessive wear is evident.

10  Check the condition of the release lever retaining spring clip, and replace if necessary. It is advisable to replace the clip as a matter of course.

### Installation

*Refer to illustration 3.12*

11  Clean and then lightly grease the release bearing contact surfaces on the release lever. **Do not** apply grease to the guide sleeve.

12  Slide the release lever into position over the guide sleeve, then push the end of the lever over the pivot, ensuring that the retaining spring clip engages correctly over the end of the release lever (see illustration).

13  Install the release bearing as described previously in this Section.

### 4  Clutch release cylinder - removal, inspection and installation

**Warning:** *Brake fluid is poisonous; wash it off immediately and thoroughly in the case of skin contact, and seek immediate medical advice if any fluid is swallowed or gets into the eyes. Certain types of brake fluid are flammable, and may ignite when allowed into contact with hot components; when servicing any hydraulic system, it is safest to assume that the fluid is flammable, and to take precautions against the risk of fire as though it is gasoline that is being handled. Brake fluid is also an effective paint stripper, and will attack plastics; if any is spilled, it should be washed off immediately using large quantities of fresh water. Finally, it is hygroscopic (it absorbs moisture from the air) - old fluid may be contaminated and unfit for further use. When topping-up or replacing the fluid, always use the recommended type, and ensure that it comes from a freshly opened container.*

### Removal

*Refer to illustrations 4.5a and 4.5b*

1  Remove the brake fluid reservoir cap, and siphon out sufficient brake fluid so that the fluid level is below the level of the reservoir fluid hose connection to the clutch master cylinder (the brake fluid reservoir feeds both the brake and clutch hydraulic systems). **Do not** empty the reservoir, as this will draw air into the brake hydraulic circuits.

2  To improve access, jack up the vehicle, and support it securely on axle stands.

3  Where applicable, remove the underbody shield for access to the transmission bellhousing.

4  Place a container beneath the hydraulic pipe connection on the clutch release cylinder to catch escaping brake fluid. Unscrew the fitting nut and disconnect the fluid pipe.

5  Unscrew the two securing nuts and withdraw the release cylinder from the mounting studs on the bellhousing (see illustrations).

### Inspection

6  Inspect the release cylinder for fluid leaks and damage, and replace if necessary. No spare parts are available for the release cylinder, and if faulty, the complete unit must be replaced.

4.5a Unscrew the securing nuts . . .

4.5b . . . and remove the clutch release cylinder

## Installation

7  Installation is a reversal of removal, bearing in mind the following points.
 a) Before installation, clean and then lightly grease the end of the release cylinder pushrod.
 b) Tighten the mounting nuts to the specified torque.
 c) On completion, top-up the brake fluid level and bleed the clutch hydraulic circuit (see Section 6).

## 5  Clutch master cylinder - removal, inspection and installation

**Warning:** *Brake fluid is poisonous; wash it off immediately and thoroughly in the case of skin contact, and seek immediate medical advice if any fluid is swallowed or gets into the eyes. Certain types of brake fluid are flammable, and may ignite when allowed into contact with hot components; when servicing any hydraulic system, it is safest to assume that the fluid is flammable, and to take precautions against the risk of fire as though it is gasoline that is being handled. Brake fluid is also an effective paint stripper, and will attack plastics; if any is spilled, it should be washed off immediately, using large quantities of fresh water. Finally, it is hygroscopic (it absorbs moisture from the air) - old fluid may be contaminated and unfit for further use. When topping-up or replacing the fluid, always use the recommended type, and ensure that it comes from a freshly opened container.*

### Removal

*Refer to illustration 5.6*

1  Remove the brake fluid reservoir cap, and siphon out sufficient brake fluid so that the fluid level is below the level of the reservoir fluid hose connection to the clutch master cylinder (the brake fluid reservoir feeds both the brake and clutch hydraulic systems). **Do not** empty the reservoir, as this will draw air into the brake hydraulic circuits.
2  Disconnect the clutch master cylinder hose from the brake fluid reservoir. Be prepared for fluid spillage, and plug the open end of the hose to prevent dirt entry.
3  Working inside the vehicle, remove the securing screws, and remove the driver's side lower dash trim panel.
4  Where applicable, disconnect the clutch pedal return spring, using a suitable pair of pliers, to enable the clutch master cylinder pushrod-to-pedal clevis pin to be removed.
5  Pry off the clip securing the master cylinder pushrod-to-pedal clevis pin, then withdraw the clevis pin.
6  Unscrew the two bolts and nut securing the master cylinder to the pedal bracket in the footwell, noting that the bolts also secure the brake light mounting bracket **(see illustration)**. Release the master cylinder from the bracket, and ease the fluid hose through the firewall grommet, taking care not to strain the pipe.

## Inspection

7  Inspect the master cylinder for fluid leaks and damage, and replace if necessary. No spare parts are available for the master cylinder, and if faulty, the complete unit must be replaced.

## Installation

**Note:** *A new master cylinder self-locking securing nut should be used on installation.*

8  Installation is a reversal of removal, bearing in mind the following points.
 a) Take care not to strain the master cylinder fluid pipe during installation.
 b) Use a new master cylinder self-locking securing nut.
 c) On completion, top-up the level in the brake fluid reservoir, then bleed the clutch hydraulic system (see Section 6).

## 6  Clutch hydraulic system - bleeding

**Warning:** *Brake fluid is poisonous; wash it off immediately and thoroughly in the case of skin contact, and seek immediate medical advice if any fluid is swallowed or gets into the eyes. Certain types of brake fluid are flammable, and may ignite when allowed into contact with hot components; when servicing any hydraulic system, it is safest to assume that the fluid is flammable, and to take precautions against the risk of fire as though it is gasoline that is being handled. Brake fluid is also an effective paint stripper, and will attack plastics; if any is spilled, it should be washed off immediately, using large quantities of fresh water. Finally, it is hygroscopic (it absorbs moisture from the air) - old fluid may be contaminated and unfit for further use. When topping-up or replacing*

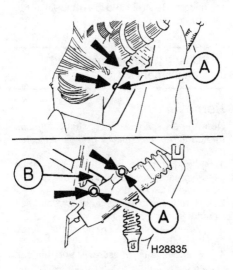

5.6  Clutch master cylinder securing bolts (A) and fluid hose (B)

*the fluid, always use the recommended type, and ensure that it comes from a freshly opened container.*

**Note:** *The manufacturer recommends that pressure-bleeding equipment is used to bleed the clutch hydraulic system.*

### General

1  The correct operation of any hydraulic system is only possible after removing all air from the components and circuit; this is achieved by bleeding the system.
2  During the bleeding procedure, add only clean, unused brake fluid of the recommended type; never re-use fluid that has already been bled from the system. Ensure that sufficient fluid is available before starting work.
3  If there is any possibility of incorrect fluid being already in the system, the brake and clutch components and circuit must be flushed completely with uncontaminated, correct fluid, and new seals should be installed on the various components.
4  If brake fluid has been lost from the system, or air has entered because of a leak, ensure that the fault is cured before proceeding further.
5  To improve access, apply the parking brake, then jack up the front of the vehicle and support it securely on axle stands.
6  Where applicable, remove the underbody shield for access to the transmission bellhousing.
7  Check that the clutch hydraulic pipe(s) and hose(s) are secure, that the fittings are tight, and that the bleed screw on the rear of the clutch release cylinder (mounted under the vehicle on the lower left-hand side of the transmission bellhousing) is closed. Clean any dirt from around the bleed screw.
8  Unscrew the brake fluid reservoir cap, and top the fluid up to the "MAX" level line; install the cap loosely, and remember to maintain the fluid level at least above the "MIN" level line throughout the procedure, or there is a risk of further air entering the system. Note that the brake fluid reservoir feeds both the brake and clutch hydraulic systems.
9  It is recommended that pressure-bleeding equipment is used to bleed the system. Alternatively, there are a number of one-man, do-it-yourself brake bleeding kits currently available from auto parts stores. These kits greatly simplify the bleeding operation, and also reduce the risk of expelled air and fluid being drawn back into the system. If such a kit is not available, the basic (two-man) method must be used, which is described in detail below.
10  If pressure-bleeding equipment or a one-man kit is to be used, prepare the vehicle as described previously, and follow the equipment/kit manufacturer's instructions, as the procedure may vary slightly according to the type being used; generally, they are as outlined below in the relevant sub-section.
11  Whichever method is used, the same basic process must be followed to ensure removal of all air from the system.

## Bleeding - basic (two-man) method

12  Obtain a clean glass or plastic jar, a length of plastic or rubber tubing which is a tight fit over the bleed screw, and a box-end wrench to fit the screw. The help of an assistant will also be required.
13  Where applicable, remove the dust cap from the bleed screw. Place the wrench and tube on the screw, place the other end of the tube in the jar, and pour in sufficient fluid to cover the end of the tube.
14  Ensure that the reservoir fluid level is maintained at least above the "MIN" level line throughout the procedure.
15  Open the bleed screw, then have the assistant fully depress the clutch pedal and hold it to the floor. Fluid and air will flow into the jar.
16  When the flow stops, tighten the bleed screw and have the assistant release the pedal slowly, then recheck the reservoir fluid level.
17  Repeat the steps given in Steps 15 and 16 until the fluid emerging from the bleed screw is free from air bubbles.
18  When no more air bubbles appear, tighten the bleed screw securely. Do not overtighten the bleed screw.
19  Temporarily disconnect the bleed tube from the bleed screw, and move the container of fluid to one side.
20  Unscrew the two securing nuts, and withdraw the release cylinder from the bellhousing, taking care not to strain the fluid hose.
21  Reconnect the bleed tube to the bleed screw, and submerge the end of the tube in the container of fluid.
22  With the bleed screw pointing vertically upwards, unscrew the bleed screw (approximately one turn), and slowly push the release cylinder pushrod into the cylinder until no more air bubbles appear in the fluid.
23  Hold the pushrod in position, then tighten the bleed screw.
24  Slowly allow the pushrod to return to its rest position. Do not allow the pushrod to return quickly, as this could cause air to enter the release cylinder.
25  Remove the tube and wrench, and install the dust cap on the bleed screw.
26  Install the release cylinder and tighten the nuts to the specified torque.

## Bleeding - using a one-way valve kit

27  As their name implies, these kits consist of a length of tubing with a one-way valve to prevent expelled air and fluid being drawn back into the system; some kits include a translucent container, which can be positioned so that the air bubbles can be more easily seen flowing from the end of the tube.
28  The kit is connected to the bleed screw, which is then opened. The user returns to the driver's seat, depresses the clutch pedal with a smooth, steady stroke, and slowly releases it; this is repeated until the expelled fluid is clear of air bubbles.
29  Note that these kits simplify work so much that it is easy to forget the reservoir fluid level; ensure that this is maintained at least above the "MIN" level line at all times.

## Bleeding - using a pressure-bleeding kit

30  These kits are usually operated by the reservoir of pressurized air contained in the spare tire. However, note that it will probably be necessary to reduce the pressure to a lower level than normal; refer to the instructions supplied with the kit.
31  By connecting a pressurized, fluid-filled container to the fluid reservoir, bleeding can be carried out simply by opening the bleed screw, and allowing the fluid to flow out until no more air bubbles can be seen in the expelled fluid.
32  This method has the advantage that the large reservoir of fluid provides an additional safeguard against air being drawn into the system during bleeding.

## All methods

33  When bleeding is complete, and firm pedal feel is restored, wash off any spilled fluid, check that the bleed screw is tightened securely, and install the dust cap.
34  Check the brake fluid level in the reservoir, and top-up if necessary (Chapter 1).
35  Discard any brake fluid that has been bled from the system; it will not be fit for re-use.
36  Check the feel of the clutch pedal. If it feels at all spongy or soft, air must still be present in the system, and further bleeding is required. Failure to bleed satisfactorily after a reasonable repetition of the bleeding procedure may be due to worn master or release cylinder seals.
37  On completion, where applicable install the underbody shield and lower the vehicle to the ground.

---

## 7  Clutch pedal - removal and installation

### Removal

**Note:** *A new self-locking nut should be used to secure the clutch master cylinder on installation.*

1  Working inside the vehicle, remove the securing screws, and withdraw the driver's side lower dash panel.
2  Remove the brake light switch (see Chapter 9).
3  Where applicable, remove the clutch pedal switch.
4  Remove the two screws and the nut securing the clutch master cylinder, noting that they also secure the brake light switch/clutch switch mounting bracket, and remove the switch bracket.
5  On models equipped with a conven-

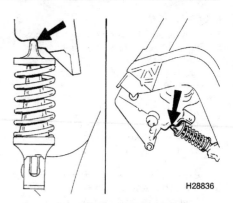

**7.10  Ensure that the helper spring is correctly located when installing the clutch pedal**

tional clutch pedal return spring, carefully disconnect the return spring from the pedal using a pair of pliers. Pry off the clevis pin securing clip, and remove the clevis pin securing the clutch master cylinder pushrod to the pedal.
6  On models where the clutch pedal is equipped with an over-center helper spring, pry off the securing clip from the clutch master cylinder-to-pedal clevis pin. Hold the clutch pedal tightly, then push out the clevis pin. Pull the clutch pedal slowly upwards, then release the top of the helper spring from the pedal mounting bracket.
7  Move the clutch master cylinder to one side, taking care not to strain the fluid pipe.
8  Pry off the clip securing the pedal to the pivot shaft, then slide off the clutch pedal. Recover the pivot bushings if they are loose.

### Installation

*Refer to illustration 7.10*

9  Before installation the pedal to the pivot shaft, check the condition of the pivot bushings, and replace if necessary.
10  Installation is a reversal of removal, bearing in mind the following points.
  a) *Before installation the pedal, lightly grease the pedal pivot pin.*
  b) *Where applicable, ensure that the end of the over-center helper spring is correctly located in the pedal bracket* **(see illustration).**
  c) *Use a new self-locking nut to secure the clutch master cylinder.*
  d) *Take care not to strain the clutch master cylinder fluid pipe during installation.*

---

## 8  Clutch start switch - check and replacement

*Refer to illustration 8.1*

1  The clutch start switch **(see illustration)**, located at the top of the clutch pedal, prevents the engine from being started unless the clutch pedal is depressed. The switch, which is part of the starting circuit, is normally open. When the clutch pedal is

# Chapter 8 Clutch and driveline 8-7

8.1 A typical clutch start switch (Z3 model shown); to remove it, unplug the electrical connector on the bottom of the switch, pull up the lock (A), squeeze the locking tangs (B) and pull the switch out of the mounting bracket

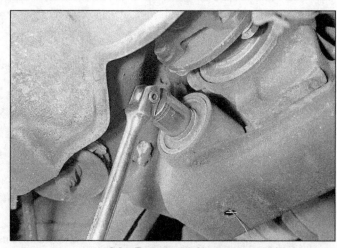

9.8a Remove the front . . .

depressed, it depresses a small pushrod in the switch, closing the circuit and allowing the starter to operate.

2  If the vehicle starts in gear, the switch is malfunctioning. To check its operation, remove the pedal cover (see illustration 27.16 in Chapter 11), unplug the electrical connector and hook up an ohmmeter across the terminals of the connector. Verify that there's no continuity when the clutch pedal is released and continuity when the pedal is depressed. If the switch doesn't operate as described, replace it.

3  To replace the clutch start switch, remove the pedal cover (see illustration 27.16 in Chapter 11), unplug the electrical connector, pull up the red lock around the plunger, depress the locking tangs and disengage the switch from its mounting bracket.

4  Installation is the reverse of removal. After the new switch snaps into place in its bracket, be sure to depress the red lock.

## 9  Differential - removal and installation

### Removal

Refer to illustrations 9.8a and 9.8b

1  Chock the front wheels. Jack up the rear of the vehicle and support it on axle stands. Remove both rear wheels. If necessary, drain the differential (see Chapter 1).

2  Using paint or a suitable marker pen, make alignment marks between the driveshaft and differential flange. Unscrew the nuts securing the driveshaft to the differential and discard them; new ones must be used on installation.

3  Loosen and remove the retaining bolts and plates securing the right-hand driveaxle to the differential flange and support the driveaxle by tying it to the vehicle underbody using a piece of wire. Note: *Do not allow the driveaxle to hang under its own weight as the CV joint may be damaged.* Discard the bolts, new ones should be used on installation.

4  Disconnect the left-hand driveaxle from the differential (see Step 3).

5  Disconnect the wiring connector from the speedometer drive on the rear of the differential.

6  Loosen and remove the left- and right-hand stabilizer bar mountings (see Chapter 10).

7  Move a jack and block of wood into position and raise it so that it is supporting the weight of the differential.

8  Making sure the differential is safely supported, loosen and remove the two bolts securing the rear of the unit in position and the single bolt securing the front of the unit in position (see illustrations).

9  Carefully lower the differential out of position and remove it from underneath the vehicle. Examine the differential mounting rubbers for signs of wear or damage and replace if necessary.

### Installation

Note: *New driveshaft rear coupling nuts and driveaxle retaining bolts will be required on installation.*

10  Installation is a reversal of removal noting the following.

  a) Raise the differential into position and

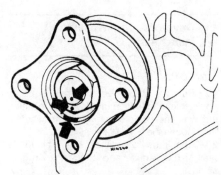

10.3 Scribe or paint alignment marks (arrows) on the flange, the pinion shaft and nut to ensure proper reassembly

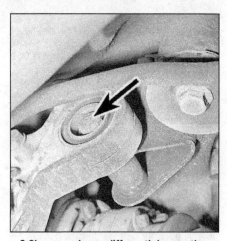

9.8b . . . and rear differential mounting bolts (arrow)

   engage it with the driveshaft rear joint, making sure the marks made prior to removal are correctly aligned.

  b) Tighten the differential mounting bolts to the specified torque setting.

  c) Install the new driveshaft joint nuts and tighten them to the specified torque.

  d) Install the stabilizer bar mountings (see Chapter 10).

  e) Install the new driveaxle joint retaining bolts and plates and tighten them to the specified torque.

  f) On completion, refill/top-up the differential with oil (see Chapter 1).

## 10  Differential oil seals - replacement

### Driveshaft flange oil seal

Refer to illustration 10.3

Note: *A new flange nut retaining plate will be required.*

1  Drain the differential (see Chapter 1).

2  Remove the differential (see Section 2) and secure the unit in a vise.

3  Remove the retaining plate and make

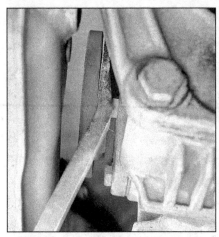

**10.13 Using a prybar to remove the driveaxle flange from the differential**

**10.15 Remove the flange retaining snap-ring from the center of the differential gear**

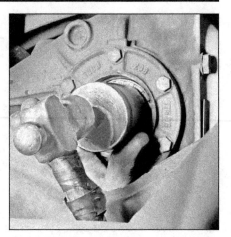

**10.17 Tap the new seal into position with a hammer and socket which bears only on the outer edge of the seal**

alignment marks between the propeller flange nut, the drive flange and pinion **(see illustration)**. Discard the retaining plate a new one must be used on installation.

4   Hold the drive flange stationary by bolting a length of metal bar to it, then unscrew the nut noting the exact number of turns necessary to remove it.

5   Using a suitable puller, draw the drive flange from the pinion and remove the dust cover. If the dust cover shows signs of wear, replace it.

6   Lever the oil seal from the differential casing with a screwdriver. Wipe clean the oil seal seating.

7   Smear a little oil on the sealing lip of the new oil seal, then press it squarely into the casing until flush with the outer face. If necessary the seal can be tapped into position using a metal tube which bears only on its hard outer edge.

8   Install the dust cover and locate the drive flange on the pinion aligning the marks made on removal. Install the flange nut, screwing it on by the exact number of turns counted on removal, so that the alignment marks align. **Warning:** *Do not overtighten the flange nut. If the nut is overtightened, the collapsible spacer behind the flange will be deformed necessitating its replacement. This is a complex operation requiring the differential to be disassembled* (see Section 1).

9   Secure the nut in position with the new retaining plate, tapping it squarely into position.

10   Install the differential (see Section 2) and refill it with oil (see Chapter 1).

## Driveaxle flange oil seal

*Refer to illustrations 10.13, 10.15 and 10.17*

**Note:** *New driveaxle joint retaining bolts and a driveaxle flange snap-ring will be required.*

11   Drain the differential oil (see Chapter 1).

12   Loosen and remove the bolts securing the driveaxle constant velocity joint to the differential and recover the retaining plates. Position the driveaxle clear of the flange and tie it to the vehicle underbody using a piece of wire. **Note:** *Do not allow the driveaxle to hang under its own weight as the CV joint may be damaged.*

13   Using a suitable lever, carefully pry the driveaxle flange out from the differential taking care not to damage the dust seal or casing **(see illustration)**. Remove the flange and recover dust seal. If the dust seal shows signs of damage, replace it.

14   Carefully lever the oil seal out from the differential. Wipe clean the oil seal seating.

15   With the seal removed, pry out the snap-ring from the center of the differential gear **(see illustration)**.

16   Install a new snap-ring, making sure its is correctly located in the differential groove.

17   Smear a little oil on the sealing lip of the new oil seal, then press it squarely into the casing until it reaches its stop. If necessary the seal can be tapped into position using a socket or metal tube which bears only on its hard outer edge **(see illustration)**.

18   Install the dust cover and insert the drive flange. Push the drive flange fully into position and check that it is securely retained by the snap-ring.

19   Align the driveaxle with the flange and install the new retaining bolts and plates, tightening them to the specified torque.

20   Refill the differential with oil (see Chapter 1).

## 11  Driveaxle - removal and installation

### Removal

1   Remove the wheel trim/hub cap (as applicable) and loosen the driveaxle retaining nut with the vehicle resting on its wheels. Also loosen the wheel bolts.

2   Chock the front wheels, then jack up the rear of the vehicle and support it on axle stands.

3   Remove the relevant rear wheel.

4   If the left-hand driveaxle is to be removed, remove the exhaust system tailpipe to improve access (see the relevant Part of Chapter 4).

5   Loosen and remove the left- and right-hand stabilizer bar mountings and pivot the bar downwards (see Chapter 10).

6   Unscrew and remove the driveaxle retaining nut.

7   Loosen and remove the bolts securing the driveaxle constant velocity joint to the differential and recover the retaining plates. Position the driveaxle clear of the flange and tie it to the vehicle underbody using a piece of wire. **Note:** *Do not allow the driveaxle to hang under its own weight as the CV joint may be damaged.*

8   Withdraw the driveaxle outer constant velocity joint from the hub assembly. The outer joint will be very tight, tap the joint out of the hub using a soft-faced mallet. If this fails to free it from the hub, the joint will have to be pressed out using a suitable tool which is bolted to the hub.

9   Remove the driveaxle from underneath the vehicle.

### Installation

**Note:** *A new driveaxle retaining nut and bolts will be required on installation.*

10   Installation is the reverse of removal noting the following points.

  a) *Lubricate the threads of the new driveaxle nut with clean engine oil prior to fitting and tighten it to the specified torque. If necessary, wait until the vehicle is lower to the ground and then tighten the nut to the specified torque.*

  b) *Install the inner joint new retaining bolts and plates and tighten them to the specified torque.*

## 12  Driveaxle boots - replacement

*Refer to illustrations 12.3, 12.4, 12.5, 12.6, 12.7, 12.20a and 12.20b*

1   Remove the driveaxle (see Section 11).

2   Clean the driveaxle and mount it in a vise.

# Chapter 8 Clutch and driveline

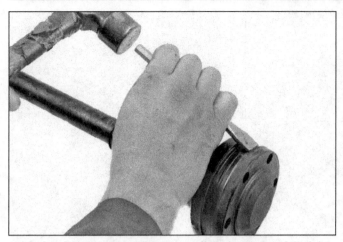

12.3 Carefully tap the sealing cover off the inner end of the joint

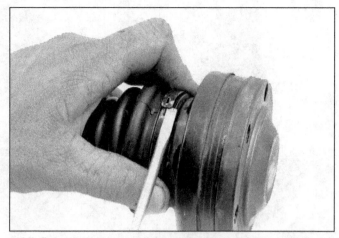

12.4 Release the boot retaining clamps and slide the boot down the shaft

3   Tap off the sealing cover from the end of the inner constant velocity (CV) joint **(see illustration)**.

4   Release the two inner joint boot retaining clips and free the boot and dust cover from the joint **(see illustration)**.

5   Scoop out excess grease and remove the inner joint snap-ring from the end of the driveaxle **(see illustration)**.

6   Securely support the joint inner member and tap the driveaxle out of position using a hammer and suitable drift **(see illustration)**. If the joint is a tight fit, a suitable puller will be required to draw off the joint. Do not disassemble the inner joint.

7   With the joint removed, slide the inner boot and dust cover off from the end of the driveaxle **(see illustration)**.

8   Release the outer joint boot retaining clips then slide the boot along the shaft and remove it.

9   Thoroughly clean the constant velocity joints using solvent, then dry them thoroughly. Carry out a visual inspection as follows.

10   Move the inner splined driving member from side to side to expose each ball in turn at the top of its track. Examine the balls for cracks, flat spots or signs of surface pitting.

11   Inspect the ball tracks on the inner and outer members. If the tracks have widened, the balls will no longer be a tight fit. At the same time check the ball cage windows for wear or cracking between the windows.

12   If on inspection any of the constant velocity joint components are found to be worn or damaged, it must be replaced. The inner joint is available separately but if the outer joint is worn it will be necessary to replace the complete joint and driveaxle assembly. If the joints are in satisfactory condition, obtain new boot repair kits which contain boots, retaining clamps, an inner constant velocity joint snap-ring and the correct type and quantity of grease required.

13   Tape over the splines on the end of the driveaxle.

14   Slide the new outer boot onto the end of the driveaxle.

15   Pack the outer joint with the specified type of grease. Work the grease well into the bearing tracks while twisting the joint, and fill the rubber boot with any excess.

16   Ease the boot over the joint and ensure

12.5 Removing the inner joint snap-ring from the driveaxle

that the boot lips are correctly located on both the driveaxle and constant velocity joint. Lift the outer sealing lip of the boot to equalize air pressure within the boot.

17   Install the large metal retaining clamp on the boot. Pull the retaining clamp tight then

12.6 Tap the axleshaft out of the CV joint ...

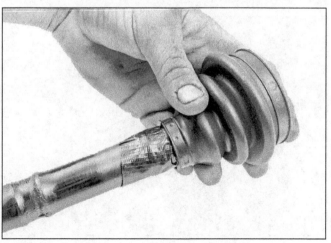

12.7 ... and slide off the boot

## Chapter 8 Clutch and driveline

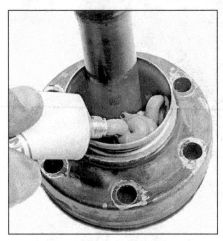

12.20a Fill the inner joint with the grease supplied . . .

12.20b . . . and work it into the bearing tracks

13.4 On models with a rubber coupling, remove the bolts securing the coupling to the transmission flange

bend it back to secure it in position and cut off any excess clamp. Secure the small retaining clamp using the same procedure.

18  Engage the new inner boot with its dust cover and slide the assembly onto the driveaxle.

19  Remove the tape from the driveaxle splines and install the inner constant velocity joint. Press the joint fully onto the shaft and secure it in position with a new snap-ring.

20  Work the grease supplied fully into the inner joint and fill the boot with any excess **(see illustrations)**.

21  Slide the inner boot into position and press the dust cover onto the joint, making sure the retaining bolt holes are correctly aligned. Lift the outer sealing lip of the boot, to equalize air pressure within the boot, and secure it in position with the retaining clips (see paragraph 17).

22  Apply a smear of suitable sealant (the manufacturer recommends BMW sealing gel) and press the new sealing cover fully onto the end of the inner joint.

23  Check that both constant velocity joints are free to move easily then install the driveaxle (see Section 11).

### 13  Driveshaft - removal and installation

#### Removal

*Refer to illustrations 13.4, 13.5 and 13.7*

1  Chock the front wheels. Jack up the rear of the vehicle and support it on axle stands.

2  Remove the exhaust system and heatshield (see Chapter 4). Where necessary, unbolt the exhaust system mounting bracket(s) in order to gain the necessary clearance required to remove the driveshaft.

3  On models where the front of the driveshaft is bolted straight onto the transmission output flange, make alignment marks between the shaft and transmission then loosen and remove the retaining nuts. Discard the nuts, new ones should be used on installation.

4  On models where a rubber coupling is installed between the front end of the driveshaft and transmission output flange, make alignment marks between the shaft, transmission and (where necessary) vibration damper. Loosen and remove the nuts and bolts securing the coupling to the transmission **(see illustration)**. Discard the nuts, new ones should be used on installation.

5  Using a large open-ended wrench, loosen the threaded sleeve nut, which is situated near the support bearing, through a couple of turns **(see illustration)**.

6  Using paint or a suitable marker pen, make alignment marks between the driveshaft and differential flange. Unscrew the nuts securing the driveshaft to the differential and discard them; new ones must be used on installation.

7  With the aid of an assistant, support the driveshaft then unscrew the center support bearing bracket retaining nuts **(see illustration)**. Slide the two halves of the shaft towards each other then lower the center of the shaft and disengage it from the transmission and differential. Remove the shaft from underneath the vehicle. **Note:** *Do not separate the two halves of the shaft without first making alignment marks. If the shafts are incorrectly joined, the driveshaft assembly may become unbalanced, leading to noise and vibration during operation.*

8  Inspect the rubber coupling (if equipped), the support bearing and shaft universal joints (see Sections 7, 8 and 9). Inspect

13.5 Unscrew the large threaded sleeve nut by a couple of turns

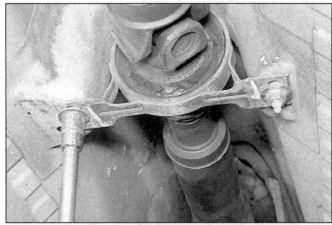

13.7 Unscrew the center bearing bracket retaining nuts and remove the driveshaft from underneath the vehicle

## Chapter 8 Clutch and driveline

14.5 Coupling-to-driveshaft retaining bolts (arrows)

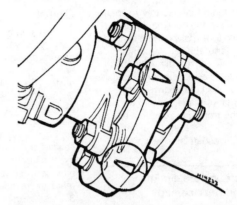

14.7 If the coupling has directional arrows, make sure the arrows are pointing towards the driveshaft/transmission flanges and not the bolt heads

the transmission flange locating pin and driveshaft bushing for signs of wear or damage and replace as necessary.

### Installation

**Note:** *New driveshaft front and rear coupling nuts will be required on installation.*

9  Apply a smear of molybdenum disulfide grease (the manufacturer recommends Molykote Longterm 2) to the transmission pin and shaft bushing and maneuver the shaft into position.
10  Align the marks made prior to removal and engage the shaft with the transmission and differential flanges. With the marks correctly aligned, install the support bracket retaining nuts, tightening them lightly only at this stage.
11  Install new retaining nuts to the rear coupling of the driveshaft and tighten them to the specified torque.
12  On models where the driveshaft is bolted straight onto the transmission flange, install the new retaining nuts and tighten them to the specified torque.
13  On models with a rubber coupling, insert the bolts and install the new retaining nuts. Tighten them to the specified torque, noting that the nut/bolt should only be rotated on the flange side to avoid stressing the rubber coupling.
14  Tighten the driveshaft threaded sleeve to the specified torque.
15  Loosen the center bearing bracket nuts. Slide the bracket forwards to remove all freeplay, then preload the bearing by moving the bracket forwards a further 5/32 to 1/4-inch. Hold the bracket in this position and tighten its retaining nuts to the specified torque.
16  Install the exhaust system and associated components (see Chapter 4).

### 14 Driveshaft rubber coupling - check and replacement

**Note:** *A rubber coupling is not present on all models. On some models, a universal joint is installed on the front of the driveshaft instead (see Section 9).*

### Check

1  Firmly apply the parking brake, then jack up the front of the vehicle and support it on axle stands.
2  Closely examine the rubber coupling, links the driveshaft to the transmission, looking for signs of damage such as cracking or splitting or for signs of general deterioration. If necessary, replace the coupling as follows.

### Replacement

*Refer to illustrations 14.5 and 14.7*

**Note:** *New driveshaft coupling nuts will be required.*

3  Follow the procedures outlined in steps 1, 2, 4 and 5 of Section 6.
4  Slide the front half of the driveshaft to the rear then disengage it from the transmission locating pin and pivot it downwards.
5  Loosen and remove the nuts securing the coupling to the shaft and remove it **(see illustration)**. If necessary, also remove the vibration damper; the damper should also be replaced if it shows signs of wear or damage.
6  Aligning the marks made on removal, install the vibration damper (if equipped) to the driveshaft.
7  Install the new rubber coupling noting that the arrows on the side of the coupling must point towards the driveshaft/transmission flanges **(see illustration)**. Install the new retaining nuts and tighten them to the specified torque.
8  Apply a smear of molybdenum disulfide grease (the manufacturer recommends Molykote Longterm 2) to the transmission pin and shaft bushing and maneuver the shaft into position.
9  Align the marks made prior to removal and engage the shaft with the transmission flange. With the marks correctly aligned, insert the bolts and install the new retaining nuts. Tighten them to the specified torque, noting that the nut/bolt should only be rotated on the flange side to avoid stressing the rubber coupling.
10  Tighten the driveshaft threaded sleeve to the specified torque.
11  Install the exhaust system and associated components (see Chapter 4).

### 15 Driveshaft support bearing - check and replacement

### Check

1  Wear in the support bearing will lead to noise and vibration when the vehicle is driven. The bearing is best checked with the driveshaft removed (see Section 13). To gain access to the bearing with the shaft in position, remove the exhaust system and heatshields (see Chapter 4).
2  Rotate the bearing and check that it turns smoothly with no sign of freeplay; if it's difficult to turn, or if it has a gritty feeling, replace it. Also inspect the rubber portion. If it's cracked or deteriorated, replace it.

### Replacement

3  Remove the driveshaft (see Section 13).
4  Make alignment marks between the front and rear sections of the driveshaft then unscrew the threaded sleeve nut and separate the two halves. Recover the sleeve nut, washer and bushing noting their locations.
5  Remove the snap-ring and slide off the support bearing rear dust cover.
6  Draw the support bearing off from the driveshaft using a suitable puller then remove the front dust cover in the same way.
7  Firmly support the support bearing bracket and press out the bearing with a suitable tubular spacer.
8  Install the new bearing to the bracket and press it into position using a tubular spacer which bears only on the bearing outer race.
9  Thoroughly clean the shaft splines and carefully press the new front dust seal onto the driveshaft, making sure it is installed the correct way.
10  Press the support bearing fully onto the driveshaft using a tubular spacer which bears only on the bearing inner race.

11 Check that the bearing is free to rotate smoothly, then install the new rear dust seal.
12 Apply a smear of molybdenum disulfide grease (the manufacturer recommends Molykote Longterm 2) to the splines and install the threaded sleeve, washer and bushing to the front section of the driveshaft.
13 Align the marks made prior to separation and joint the front and rear sections of the driveshaft.
14 Install the driveshaft (see Section 13).

## 16 Driveshaft universal joints - check and replacement

**Note:** *Some models have a rubber coupling between the driveshaft and transmission instead of a universal joint* (see Section 14).

### Check

1 Wear in the universal joints is characterized by vibration in the transmission, noise during acceleration, and metallic squeaking and grating sounds as the bearings disintegrate. The joints can be checked with the driveshaft in place, but it will be necessary to remove the exhaust system and heat shields (see Chapter 4) to gain access.
2 If the driveshaft is in position on the vehicle, try to turn the driveshaft while holding the transmission/differential flange. Free play between the driveshaft and the front or rear flanges indicates excessive wear.
3 If the driveshaft is already removed, you can check the universal joints by holding the shaft in one hand and turning the yoke or flange with the other. If the axial movement is excessive, replace the driveshaft.

### Replacement

4 At the time of writing, no spare parts were available to enable replacement of the universal joints to be carried out. Therefore, if any joint shows signs of damage or wear the complete driveshaft assembly must be replaced. Consult your BMW dealer for the latest information on parts availability.
5 If replacement of the driveshaft is necessary, it may be worthwhile seeking the advice of an automotive machine shop. They may be able to repair the original shaft assembly or supply a reconditioned shaft on an exchange basis.

9-1

# Chapter 9  Brakes

## Contents

| | Section |
|---|---|
| Anti-lock Braking System (ABS) components - removal and installation | 20 |
| Anti-lock Braking System (ABS) - general information | 19 |
| Brake booster check valve - removal, testing and installation | 13 |
| Brake booster - general information | 12 |
| Brake fluid level check | See Chapter 1 |
| Brake fluid replacement | See Chapter 1 |
| Brake hydraulic system - bleeding | 2 |
| Brake light switch - removal and installation | 18 |
| Brake lines and hoses - replacement | 3 |
| Brake pedal - removal and installation | 11 |
| Front brake caliper - removal, overhaul and installation | 5 |
| Front brake disc - inspection, removal and installation | 6 |

| | Section |
|---|---|
| Front brake pad wear check | See Chapter 1 |
| Front brake pads - replacement | 4 |
| General information | 1 |
| Master cylinder - general information | 10 |
| Parking brake - adjustment | 14 |
| Parking brake cables - removal and installation | 16 |
| Parking brake lever - removal and installation | 15 |
| Parking brake shoes - removal and installation | 17 |
| Rear brake caliper - removal, overhaul and installation | 8 |
| Rear brake disc - inspection, removal and installation | 9 |
| Rear brake pads - replacement | 7 |
| Rear brake pad wear check | See Chapter 1 |

## Specifications

### Front brakes

| | |
|---|---|
| Disc minimum thickness | Refer to the dimension marked on the disc |
| Maximum disc runout | 0.007 inch |
| Parallelism | 0.0005 inch |
| Brake pad friction material minimum thickness | See Chapter 1 |

### Rear brakes

| | |
|---|---|
| Disc minimum thickness | Refer to the dimension marked on the disc |
| Maximum disc runout | 0.007 inch |
| Brake pad friction material minimum thickness | See Chapter 1 |
| Parking brake drum diameter (maximum) | 6.304 inches (160 mm) |

### Torque specifications

| | Ft-lbs (unless otherwise indicated) |
|---|---|
| ABS wheel sensor retaining bolts | 84 in-lbs |
| Brake booster mounting nuts | 16 |
| Brake disc retaining screw | 144 in-lbs |
| Brake drum retaining screw | 144 in-lbs |
| Brake hose unions | |
|   M10 thread | 132 in-lbs |
|   M12 thread | 144 in-lbs |
| Drum brake backing plate bolts | 48 |
| Front brake caliper | |
|   Guide pin bolts | 26 |
|   Mounting bracket bolts | 81 |
| Master cylinder mounting nuts | 19 |
| Rear brake caliper | |
|   Guide pin bolts | 26 |
|   Mounting bracket bolts | 49 |
| Wheel cylinder bolts | 84 in-lbs |
| Wheel bolts | 74 |

## 1 General information

The braking system is of the power-assisted, dual-circuit hydraulic type. Under normal circumstances, both circuits operate in unison. However, if there is hydraulic failure in one circuit, full braking force will still be available at two wheels.

All models have disc brakes on all four wheels. ABS is available as standard equipment (refer to Section 19 for further information on ABS operation). **Note:** *On models equipped with All Season Traction (AST), the ABS system also operates the traction control side of the system.*

The brakes are actuated by single-piston sliding type calipers, which ensure that equal pressure is applied to each disc pad.

The parking brake consists of a small drum brake fitted in the center of each rear brake disc.

**Note:** *When servicing any part of the system, work carefully and methodically; also observe scrupulous cleanliness when overhauling any part of the hydraulic system. Always replace components (in axle sets, where applicable) if in doubt about their condition. Note the warnings given in "Safety first" and at relevant points in this Chapter concerning the dangers of asbestos dust and brake fluid.*

## 2 Brake hydraulic system - bleeding

**Warning 1:** *Brake fluid is poisonous; wash it off immediately and thoroughly in the case of skin contact, and seek immediate medical advice if any fluid is swallowed or gets into the eyes. Certain types of brake fluid are flammable, and may ignite when allowed into contact with hot components; when servicing any hydraulic system, it is safest to assume that the fluid IS flammable, and to take precautions against the risk of fire as though it is gasoline that is being handled. Brake fluid is also an effective paint stripper, and will attack plastics; if any is spilled, it should be washed off immediately using large amounts of fresh water. Finally, it is hygroscopic (it absorbs moisture from the air) - old fluid may be contaminated and unfit for further use. When topping-up or replacing the fluid, always use the recommended type, and ensure that it comes from a freshly-opened container.*

**Warning 2:** *BMW specifies pressure bleeding as the only acceptable method for bleeding the brake system. Do NOT rely on any other bleeding method.*

**Warning 3:** *On models with All Season Traction (AST), bleeding of the brakes should be entrusted to a BMW dealer or other qualified repair shop.*

### General

1 The correct operation of any hydraulic system is only possible after removing all air from the components and circuit; this is achieved by bleeding the system.

2 During the bleeding procedure, add only clean, unused brake fluid of the recommended type; never re-use fluid that has already been bled from the system. Ensure that sufficient fluid is available before starting work.

3 If there is any possibility of incorrect fluid being already in the system, the brake components and circuit must be flushed completely with uncontaminated, correct fluid, and new seals should be installed on the various components.

4 If brake fluid has been lost from the system, or air has entered because of a leak, ensure that the fault is cured before continuing further.

5 Park the vehicle on level ground, switch off the engine and select first or reverse gear, then chock the wheels and release the parking brake.

6 Check that all lines and hoses are secure, unions tight and bleed screws closed. Clean any dirt from around the bleed screws.

7 Unscrew the master cylinder reservoir cap, and top the master cylinder reservoir up to the "MAX" level line; install the cap loosely, and remember to maintain the fluid level at least above the "MIN" level line throughout the procedure, or there is a risk of further air entering the system.

### Bleeding sequence

*Refer to illustration 2.13*

8 If the system has been only partially disconnected, and suitable precautions were taken to minimize fluid loss, it should be necessary only to bleed that part of the system.

9 If the complete system is to be bled, then it should be done working in the following sequence:

a) Right-hand rear brake.
b) Left-hand rear brake.
c) Right-hand front brake.
d) Left-hand front brake.

10 Conventional brake bleeding relies on brake pedal pressure. Pressure bleeding uses air pressure on the brake fluid in the master cylinder to push the fluid from the reservoir into the master cylinder, through the brake hoses and lines, and out to the calipers. There are two general types of pressure bleeders: those that use a hand-operated pump, and more expensive models that use compressed air. Pressure bleeders might also be available at some tool rental yards.

11 The following procedure is only a general guide to pressure bleeding. Be sure to read and follow the manufacturer's instructions for your specific pressure bleeder.

12 Make sure that the master cylinder reservoir is filled to the MAX mark on the side of the reservoir, then hook up the pressure bleeder in accordance with the manufacturer's instructions.

13 Place a box wrench over the bleeder screw nut on the right rear caliper, hook up a short section of clear plastic tubing to the bleeder screw and insert the other end of the tube into a suitable container **(see illustra-**

**2.13 When bleeding the brakes, a hose is connected to the bleeder valve at the caliper and then submerged in brake fluid - air will be seen as bubbles in the container and in the tube (all air must be expelled before continuing to the next wheel)**

**tion)**. Do not open the bleeder screw yet.

14 Pressurize the system to about 14.5-psi. **Caution:** *Do not pressurize the system beyond 29 psi. Higher pressure will damage the reservoir.*

15 With an assistant holding down the brake pedal, open the bleeder screw with the wrench. Have your assistant slowly pump the brake pedal about a dozen times with the bleeder screw open. Tell your assistant to hold down the pedal on the last stroke. When the fluid being expelled from the caliper is free of bubbles, close the bleeder screw. Your assistant can now release the brake pedal. **Caution:** *While the bleeder screw is open, make sure that the plastic tube remains immersed in the fluid in the container. Failure to do so might allow air to enter the system.*

16 Detach the plastic tube from the bleeder screw, remove the box wrench and move them to the left rear caliper. Refill the brake fluid reservoir and repressurize the system to 14.5 psi and repeat this procedure for the right rear caliper.

17 After the left rear caliper has been bled, repeat this procedure for the right front caliper, and then for the left front caliper.

18 When bleeding is complete, and firm pedal feel is restored, wash off any spilled fluid, tighten the bleed screws securely, and install their dust caps.

19 Check the brake fluid level in the master cylinder reservoir, and top-up if necessary (Chapter 1).

20 Discard any brake fluid that has been bled from the system; it will not be suitable for re-use.

21 Check the feel of the brake pedal. If it feels at all spongy, air must still be present in the system, and further bleeding is required. Failure to bleed satisfactorily after a reasonable repetition of the bleeding procedure may be due to worn master cylinder seals.

# Chapter 9  Brakes

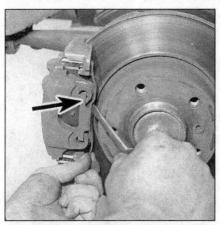

4.2 Using a large screwdriver, carefully unclip the anti-rattle spring from the caliper

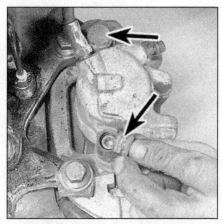

4.4 Remove the plastic plugs (arrows) to gain access to the guide pin bolts

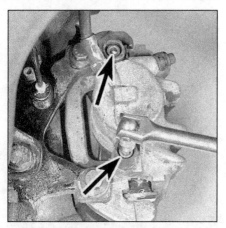

4.5a Unscrew the guide pin bolts (arrows) . . .

## 3  Brake lines and hoses - replacement

**Warning:** *On models with ABS and AST, under no circumstances should the hydraulic lines/hoses linking the master cylinder, hydraulic unit and (where attached) the accumulator be disturbed. If these unions are disturbed and air enters the high-pressure hydraulic system, bleeding of the system can only be safely carried out by a BMW dealer or other qualified repair shop using the special service tester.*

**Note:** *Before starting work, refer to the warnings at the beginning of Section 2.*

1  If any line or hose is to be replaced, minimize fluid loss by first removing the master cylinder reservoir cap, then tightening it down onto a piece of cellophane to obtain an airtight seal. Brake line unions should be plugged (if care is taken not to allow dirt into the system) or capped immediately after they are disconnected. Place a bundle of rags under any union that is to be disconnected, to catch any spilled fluid.

2  If a flexible hose is to be disconnected, unscrew the brake line fitting nut before removing the spring clip which secures the hose to its mounting bracket.

3  To unscrew the fitting nuts, it is preferable to obtain a flare-nut wrench of the correct size; these are available from most large auto parts stores. Failing this, a close-fitting open-ended wrench will be required, though if the nuts are tight or corroded, their flats may be rounded-off if the wrench slips. In such a case, using self-locking pliers is often the only way to unscrew a stubborn union, but it follows that the line and the damaged nuts must be replaced on reassembly. Always clean a union and surrounding area before disconnecting it. If disconnecting a component with more than one union, make a careful note of the connections before disturbing any of them.

4  If a brake line is to be replaced, it can be obtained from BMW dealers and some auto parts stores. All that is then necessary is to bend it to shape, following the line of the original, before fitting it to the car. Alternatively, most auto parts stores can make up brake lines from kits, but this requires very careful measurement of the original, to ensure that the replacement is of the correct length. The safest answer is usually to take the original to the shop as a pattern.

5  On installation, do not overtighten the fitting nuts. It is not necessary to exercise brute force to obtain a sound joint.

6  Ensure that the lines and hoses are correctly routed, with no kinks, and that they are secured in the clips or brackets provided. After fitting, remove the cellophane from the reservoir, and bleed the hydraulic system (see Section 2). Wash off any spilled fluid, and check carefully for fluid leaks.

## 4  Front brake pads - replacement

*Refer to illustrations 4.2, 4.4, 4.5a, 4.5b, 4.6a and 4.6b*

**Warning:** *Replace both sets of front brake pads at the same time - NEVER replace the pads on only one wheel, as uneven braking may result. Note that the dust created by wear of the pads may contain asbestos, which is a health hazard. Never blow it out with compressed air, and do not inhale any of it. An approved filtering mask should be worn when working on the brakes. DO NOT use gasoline or petroleum-based solvents to clean brake parts; use brake cleaner or clean brake fluid only.*

1  Apply the parking brake, then jack up the front of the vehicle and support it on axle stands. Remove the front wheels.

2  Unclip the anti-rattle spring from the side of the brake caliper, noting its installed position **(see illustration)**.

3  Unclip the brake pad wear sensor (if equipped) and remove it from the caliper aperture.

4  Remove the plastic plugs from the caliper guide bushings to gain access to the guide pin bolts **(see illustration)**.

5  Loosen and remove the guide pin bolts, noting that an Allen socket may be needed. Lift the caliper away from the caliper mounting bracket, and tie it to the suspension strut using a suitable piece of wire **(see illustrations)**. Do not allow the caliper to hang unsupported on the flexible brake hose.

6  Unclip the inner brake pad from the

4.5b . . . and lift the caliper away from the disc

4.6a Unclip the inner pad from the caliper piston . . .

4.6b ... and remove the outer pad from the caliper mounting bracket

5.7 With the caliper padded to catch the piston, use compressed air to ease the piston out of its bore - make sure your fingers are not between the piston and caliper frame

5.8 To remove the seal from the caliper bore, use a plastic or wooden tool, such as a pencil

caliper piston, and withdraw the outer pad from the caliper mounting bracket **(see illustrations)**.

7  First measure the thickness of each brake pad's friction material. If either pad is worn at any point to the specified minimum thickness or less, all four pads must be replaced. Also, the pads should be replaced if any are fouled with oil or grease; there is no satisfactory way of degreasing friction material, once contaminated. If any of the brake pads are worn unevenly, or are fouled with oil or grease, trace and rectify the cause before reassembly.

8  If the brake pads are still serviceable, clean them with brake system cleaner. Clean out the grooves in the friction material (where applicable), and pick out any large embedded particles of dirt or debris. Carefully clean the pad locations in the caliper body/mounting bracket.

9  Prior to installing the pads, check that the guide pin bolts are a light, sliding fit in the caliper body bushings, with little sign of freeplay. Inspect the dust seal around the piston for damage, and the piston for evidence of fluid leaks, corrosion or damage. If attention to any of these components is necessary, refer to Section 10.

10  If new brake pads are to be installed, the caliper piston must be pushed back into the cylinder to make room for them. Either use a C-clamp or similar tool, or use suitable pieces of wood as levers. Provided that the master cylinder reservoir has not been overfilled with brake fluid, there should be no spillage, but keep a careful watch on the fluid level while retracting the piston. If the fluid level rises above the "MAX" level line at any time, the surplus should be siphoned off or ejected through a plastic tube connected to the bleed screw (see Section 2). **Note:** *Do not siphon the fluid by mouth, as it is poisonous; use a syringe or an old poultry baster.* **Warning:** *If a baster is used, never again use it for preparing food.*

11  Apply a smear of brake grease (the manufacturer recommends Plastilube lubricant) to the backing plate of each pad; do not apply excess grease, nor allow the grease to contact the friction material.

12  Install the outer pad to the caliper mounting bracket, ensuring that its friction material is against the brake disc.

13  Clip the inner pad into the caliper piston, and maneuver the caliper assembly into position.

14  Install the caliper guide pin bolts, and tighten them to the specified torque setting. Install the plugs to the ends of the caliper guide bushings.

15  Clip the pad wear sensor back into position in the outer pad, making sure its wiring is correctly routed.

16  Clip the anti-rattle spring into position in the caliper. Depress the brake pedal repeatedly, until the pads are pressed into firm contact with the brake disc, and normal (non-assisted) pedal pressure is restored.

17  Repeat the above procedure on the remaining front brake caliper.

18  Install the wheels, then lower the vehicle to the ground and tighten the wheel bolts to the specified torque setting. **Note:** *New pads will not give full braking efficiency until they have bedded in. Be prepared for this, and avoid hard braking as far as possible for the first hundred miles or so after pad replacement.*

## 5  Front brake caliper - removal, overhaul and installation

**Note:** *Before starting work, refer to the note at the beginning of Section 2 concerning the dangers of brake fluid, and to the warning at the beginning of Section 4 concerning the dangers of asbestos dust.*

### Removal

1  Apply the parking brake, then jack up the front of the vehicle and support it on axle stands. Remove the appropriate wheel.

2  Minimize fluid loss by first removing the master cylinder reservoir cap, and then tightening it down onto a piece of cellophane, to obtain an airtight seal.

3  Clean the area around the union, then loosen the brake hose fitting nut.

4  Remove the brake pads (see Section 4).

5  Unscrew the caliper from the end of the brake hose and remove it from the vehicle. Plug the hose to prevent fluid leakage.

### Overhaul

*Refer to illustrations 5.7 and 5.8*

6  With the caliper on the bench, wipe away all traces of dust and dirt, but *avoid inhaling the dust, as it is a health hazard*.

7  Withdraw the partially ejected piston from the caliper body, and remove the dust seal. **Note:** *If the piston cannot be withdrawn by hand, it can be pushed out by applying compressed air to the brake hose union hole* **(see illustration)**. *Place a block of wood between the piston and the caliper frame to act as a cushion. Only low pressure should be required. As the piston is expelled, take great care not to trap your fingers between the piston and caliper.*

8  Using a wood or plastic tool, extract the piston hydraulic seal (metal tools may cause bore damage) **(see illustration)**.

9  Thoroughly clean all components, using brake system cleaner. Never use petroleum-based solvents.

10  Check all components, and replace any that are worn or damaged. Check particularly the cylinder bore and piston; these should be replaced (note that this means the replacement of the complete body assembly) if they are scratched, worn or corroded in any way. Similarly check the condition of the guide pins and their bushings; both pins should be undamaged and (when cleaned) a reasonably tight sliding fit in the bushings. If there is any doubt about the condition of any component, replace it.

# Chapter 9  Brakes

6.3  Measuring the brake disc thickness with a micrometer

6.4  To check disc runout, mount a dial indicator as shown and rotate the disc

6.7a  Unscrew the retaining screw . . .

6.7b  . . . and remove the brake disc from the hub

11  If the assembly is fit for further use, obtain the appropriate repair kit. All rubber seals should be replaced as a matter of course; these should never be re-used.
12  On reassembly, ensure that all components are clean and dry.
13  Soak the piston and the new piston (fluid) seal in clean brake fluid. Smear clean fluid on the cylinder bore surface.
14  Install the new piston (fluid) seal, using only your fingers (no tools) to manipulate it into the cylinder bore groove.
15  Install the new dust seal to the piston. Locate the rear of the seal in the recess in the caliper body, and install the piston to the cylinder bore using a twisting motion. Ensure that the piston enters squarely into the bore, and press it fully into the bore.

## Installation

16  Screw the caliper fully onto the flexible hose union.
17  Install the brake pads (see Section 4).
18  Securely tighten the brake line fitting nut.
19  Remove the brake hose clamp or cellophane, as applicable, and bleed the hydraulic system (see Section 2). Note that, providing the precautions described were taken to minimize brake fluid loss, it should only be necessary to bleed the relevant front brake.
20  Install the wheel, then lower the vehicle to the ground and tighten the wheel bolts to the specified torque. On completion, check the brake fluid level (see Chapter 1).

## 6  Front brake disc - inspection, removal and installation

**Note:** *Before starting work, refer to the note at the beginning of Section 4 concerning the dangers of asbestos dust.*

## Inspection

Refer to illustrations 6.3 and 6.4
**Note:** *If either disc requires replacement, BOTH should be replaced at the same time, to ensure even and consistent braking. New brake pads should also be installed.*
1  Apply the parking brake, then jack up the front of the car and support it on axle stands. Remove the appropriate front wheel.
2  Slowly rotate the brake disc so that the full area of both sides can be checked; remove the brake pads if better access is required to the inboard surface. Light scoring is normal in the area swept by the brake pads, but if heavy scoring or cracks are found, the disc must be replaced.
3  It is normal to find a lip of rust and brake dust around the disc's perimeter; this can be scraped off if required. If, however, a lip has formed due to excessive wear of the brake pad swept area, then the disc's thickness must be measured using a micrometer **(see illustration)**. Take measurements at several places around the disc, at the inside and outside of the pad swept area; if the disc has worn at any point to the specified minimum thickness or less, the disc must be replaced. Otherwise, it can be refinished by an automotive machine shop.
4  If the disc is thought to be warped, it can be checked for run-out. Use a dial indicator mounted on any convenient fixed point and slowly rotate the disc **(see illustration)**. If the measurements obtained are at the specified maximum or beyond, the disc is excessively warped and must be machined or replaced; however, it is worth checking first that the hub bearing is in good condition (Chapters 1 and/or 10).
5  Check the disc for cracks, especially around the wheel bolt holes, and any other wear or damage, and replace if necessary.

## Removal

Refer to illustrations 6.7a and 6.7b
6  Unscrew the two bolts securing the brake caliper mounting bracket to the steering knuckle, then slide the caliper assembly off the disc. Using a piece of wire or string, tie the caliper to the front suspension coil spring, to avoid placing any strain on the hydraulic brake hose.
7  Use chalk or paint to mark the relationship of the disc to the hub, then remove the screw securing the brake disc to the hub, and remove the disc **(see illustrations)**. If it is tight, lightly tap its rear face with a rubber or plastic mallet.

## Installation

8  Installation is the reverse of the removal procedure, noting the following points:
  a) Ensure that the mating surfaces of the disc and hub are clean and flat.
  b) Align (if applicable) the marks made on removal, and tighten the disc retaining screw to the specified torque.
  c) If a new disc has been fitted, use a suitable solvent to wipe any preservative coating from the disc, before installing the caliper.
  d) Slide the caliper into position over the disc, making sure the pads pass either side of the disc. Lightly oil the threads of the caliper bracket mounting bolts prior to installation, and tighten them to the specified torque setting.

e) Install the wheel, then lower the vehicle to the ground and tighten the wheel bolts to the specified torque. On completion, repeatedly depress the brake pedal until normal (non-assisted) pedal pressure returns.

## 7 Rear brake pads - replacement

**Warning:** *Replace both sets of rear brake pads at the same time - NEVER replace the pads on only one wheel, as uneven braking may result. Note that the dust created by wear of the pads may contain asbestos, which is a health hazard. Never blow it out with compressed air, and do not inhale any of it. An approved filtering mask should be worn when working on the brakes. DO NOT use gasoline or petroleum-based solvents to clean brake parts; use brake system cleaner only.*

The rear brake calipers are very similar to those at the front. Refer to Section 4 for pad inspection and replacement details.

## 8 Rear brake caliper - removal, overhaul and installation

**Note:** *Before starting work, refer to the note at the beginning of Section 2 concerning the dangers of brake fluid, and to the warning at the beginning of Section 7 concerning the dangers of asbestos dust.*

### Removal

1 Chock the front wheels, then jack up the rear of the vehicle and support on axle stands. Remove the relevant rear wheel.
2 Minimize fluid loss by first removing the master cylinder reservoir cap, and then tightening it down onto a piece of cellophane, to obtain an airtight seal.
3 Clean the area around the union, then loosen the brake hose fitting nut.
4 Remove the brake pads (see Section 7).
5 Unscrew the caliper from the end of the flexible hose, and remove it from the vehicle. Plug the hose to prevent fluid leakage,

### Overhaul

6 Refer to Section 5, noting that the dust seal is secured in position with a snap-ring.

### Installation

7 Screw the caliper fully onto the flexible hose union.
8 Install the brake pads (see Section 7).
9 Securely tighten the brake line fitting nut.
10 Remove the brake hose clamp or cellophane, as applicable, and bleed the hydraulic system (see Section 2). Note that, providing the precautions described were taken to minimize brake fluid loss, it should only be necessary to bleed the relevant rear brake.

9.3 Slide the rear caliper assembly off the disc

11 Install the wheel, then lower the vehicle to the ground and tighten the wheel bolts to the specified torque. On completion, check the brake fluid level (see Chapter 1).

## 9 Rear brake disc - inspection, removal and installation

**Note:** *Before starting work, refer to the note at the beginning of Section 5 concerning the dangers of asbestos dust.*

### Inspection

**Note:** *If either disc requires replacement, BOTH should be replaced at the same time, to ensure even and consistent braking. New brake pads should also be installed.*
1 Firmly chock the front wheels, then jack up the rear of the car and support it on axle stands. Remove the appropriate rear wheel.
2 Inspect the disc (see Section 6).

### Removal

*Refer to illustrations 9.3 and 9.4*
3 Unscrew the two bolts securing the brake caliper mounting bracket in position, then slide the caliper assembly off the disc. Using a piece of wire or string, tie the caliper to the rear suspension coil spring, to avoid placing any strain on the hydraulic brake hose **(see illustration)**.
4 Loosen and remove the brake disc retaining screw and release the parking brake **(see illustration)**.
5 It should now be possible to withdraw the brake disc from the stub axle by hand. If it is tight, lightly tap its rear face with a hide or plastic mallet. If the parking brake shoes are binding, first check that the parking brake is fully released, then continue as follows.
6 Fully loosen the parking brake adjustment, to obtain maximum free play in the cable (see Section 14).

9.4 Unscrew the retaining screw and remove the rear disc

7 Insert a screwdriver through one of the wheel bolt holes in the brake disc, and rotate the adjuster knurled wheel on the upper pivot to retract the shoes **(see illustrations 14.3a and 14.3b)**. The brake disc can then be withdrawn.

### Installation

8 If a new disc is being installed, use a suitable solvent to wipe any preservative coating from the disc.
9 Align (if applicable) the marks made on removal, then install the disc and tighten the retaining screw to the specified torque.
10 Slide the caliper into position over the disc, making sure the pads pass either side of the disc. Tighten the caliper bracket mounting bolts to the specified torque setting.
11 Adjust the parking brake shoes and cable (see Section 14).
12 Install the wheel, then lower the vehicle to the ground, and tighten the wheel bolts to the specified torque. On completion, repeatedly depress the brake pedal until normal (non-assisted) pedal pressure returns. Recheck the parking brake adjustment.;

## 10 Master cylinder - general information

1 It is not possible for the home mechanic to remove the master cylinder. If the hydraulic lines are disconnected from the master cylinder, air will enter the high-pressure hydraulic system linking the master cylinder and hydraulic unit. Bleeding of the high-pressure system can only be safely carried out by a BMW dealer who has access to the service tester (see Section 2). Master cylinder removal and installation should therefore be entrusted to a BMW dealer or other qualified repair shop.

# Chapter 9 Brakes

## 11 Brake pedal - removal and installation

### Removal

1  Disconnect the battery negative terminal.
2  Remove the brake light switch (see Section 18).
3  Using a pair of pliers, carefully unhook the return spring from the brake pedal.
4  Slide off the retaining clip and remove the clevis pin securing the brake pedal to the brake booster pushrod.
5  Slide off the pedal pivot pin retaining clip and remove the pedal from the pivot.
6  Carefully clean and inspect all components, replacing any that are worn or damaged.

### Installation

7  Installation is the reverse of removal. Apply a smear of multi-purpose grease to the pedal pivot and clevis pin.

## 12 Brake booster - general information

1  It is not possible for the home mechanic to remove the brake booster. If the hydraulic unions are disconnected from the master cylinder, air will enter the high-pressure hydraulic system linking the master cylinder, hydraulic unit. Bleeding of the high-pressure system can only be safely carried out by a BMW dealer who has access to the service tester (see Section 2). The brake booster can be tested (see steps 3 through 5 in Section 13), but removal and installation should be entrusted to a BMW dealer or other qualified repair shop.

## 13 Brake booster check valve - removal, testing and installation

### Removal

1  Disconnect the vacuum hose from the booster check valve.
2  Carefully ease the check valve out of the booster, taking care not to displace the grommet.

### Testing

3  Examine the check valve for signs of damage, and replace if necessary.
4  The valve may be tested by blowing through it in both directions; air should flow through the valve in one direction only - when blown through from the booster end of the valve. Replace the valve if this is not the case.
5  Examine the booster rubber sealing grommet for signs of damage or deterioration, and replace as necessary.

### Installation

6  Ensure that the sealing grommet is cor-

14.3a  Position one of the wheel bolt holes as shown, then insert a screwdriver through the hole . . .

rectly installed on the booster.
7  Ease the valve into position in the servo, taking great care not to displace or damage the grommet.
8  Reconnect the vacuum hose securely to the valve.
9  On completion, start the engine and ensure there are no air leaks at the check valve-to-booster connection.

## 14 Parking brake - adjustment

*Refer to illustrations 14.3a, 14.3b and 14.6*

1  Applying normal moderate pressure, pull the parking brake lever to the fully applied position, counting the number of clicks emitted from the parking brake ratchet mechanism. If adjustment is correct, there should be approximately 7 or 8 clicks before the parking brake is fully applied. If there are more than 10 clicks, adjust as follows.
2  Loosen and remove one wheel bolt from each rear wheel then chock the front wheels, jack up the rear of the vehicle and support it on axle stands.
3  Starting on the left-hand rear wheel, fully release the parking brake and position the wheel/disc so the exposed bolt hole is positioned 65-degrees clockwise from the vertical position. Make sure the parking brake lever is fully released, then insert a screwdriver in through the bolt hole and fully expand the parking brake shoes by rotating the adjuster. When the wheel/disc can no longer be turned, back the ring off by 18 teeth (clicks) so that the wheel is free to rotate easily **(see illustrations)**.
4  Repeat Step 9 on the right-hand wheel.
5  Adjust the parking brake cables as follows.
6  Access to the parking brake cable adjusting nuts can be gained by removing the parking brake lever boot from the center console **(see illustration)**. If greater access is required, the rear section of the center

14.3b  . . . and rotate the parking brake shoe adjuster (shown with disc removed)

14.6  Unclip the parking brake boot from the center console to gain access to the parking brake cable adjustment nuts (arrows)

console will have to be removed (see Chapter 11).
7  With the parking brake set on the sixth notch of the ratchet mechanism, loosen the locknuts and rotate the adjusting nuts equally until it is difficult to turn both rear wheels/drums. Once this is so, fully release the parking brake lever, and check that the wheels/hubs rotate freely. Slowly apply the parking brake, and check that the brake shoes start to contact the drums when the parking brake is set to the second notch of the ratchet mechanism. Check the adjustment by applying the parking brake fully, counting the clicks emitted from the parking brake ratchet and, if necessary, re-adjust.
8  Once adjustment is correct, hold the adjusting nuts and securely tighten the locknuts. Check the operation of the parking brake warning light switch, then install the center console section/parking brake lever boot (as applicable). Install the wheels, then lower the vehicle to the ground and tighten the wheel bolts to the specified torque.

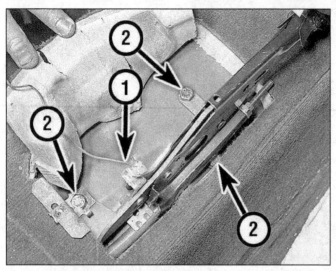

**15.3 Parking brake lever switch wire (1) and lever retaining bolts (2)**

**16.7a On rear disc brake models, unfold the expander then withdraw the pin (arrow) . . .**

## 15 Parking brake lever - removal and installation

### Removal

*Refer to illustration 15.3*

1  Remove the rear section of the center console (see Chapter 11) to gain access to the parking brake lever.
2  Loosen and remove both the parking brake cable locknuts and adjusting nuts, and detach the cables from the compensator plate.
3  Disconnect the wiring connector from the warning light switch then unscrew the retaining nuts/bolts, and remove the lever from the vehicle **(see illustration)**.

### Installation

4  Installation is a reversal of the removal. Prior to installing the center console, adjust the parking brake (see Section 14).

## 16 Parking brake cables - removal and installation

### Removal

*Refer to illustrations 16.7a and 16.7b*

1  Remove the rear section of the center console (see Chapter 11) to gain access to the parking brake lever. The parking brake cable consists of two sections, a right- and a left-hand section, which are linked to the lever by a compensator plate. Each section can be removed individually.
2  Loosen and remove the relevant parking brake cable locknut and adjusting nut, and disengage the inner cable from the parking brake compensator plate.
3  Firmly chock the front wheels, then jack up the rear of the vehicle and support it on axle stands.
4  Referring to the relevant Part of Chapter 4, remove the exhaust system heat shield to gain access to the parking brake cables. Note that on some models, it may also be necessary to remove part of the exhaust system.
5  Free the front end of the outer cable from the body, and withdraw the cable from its support guide.
6  Working back along the length of the cable, noting its correct routing, and free it from all the relevant retaining clips.
7  Remove the parking brake shoes (see Section 17). Unfold the expander, then withdraw the cable pivot pin and detach the parking brake shoe expander from the cable end. Release the cable from the back plate, and remove it from the vehicle **(see illustrations)**.

### Installation

8  Installation is a reversal of the removal procedure; on disc brake models, apply a smear of grease to the cable pivot pin prior to installation the parking brake shoes. Prior to installing the center console, adjust the parking brake (see Section 14).

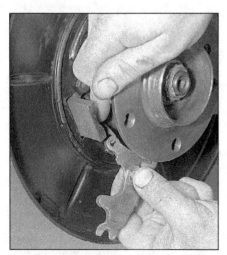

**16.7b . . . and detach the expander from the end of the parking brake cable**

**17.2a Using pliers, unhook and remove the parking brake shoe front . . .**

**17.2b . . . and rear return springs**

# Chapter 9 Brakes

17.3a Using a suitable Allen key, rotate the retainer pins through 90°...

17.3b ... then remove the pins and springs ...

17.4 ... and parking brake shoes from the back plate

## 17 Parking brake shoes - removal and installation

### Removal

*Refer to illustrations 17.2a, 17.2b, 17.3a, 17.3b, 17.4 and 17.6*

1  Remove the rear brake disc (see Section 9), and make a note of the correct fitted position of all components.
2  Using a pair of pliers, carefully unhook and remove the parking brake shoe return springs **(see illustrations)**.
3  Release the shoe retaining pins by depressing them and rotating them through 90-degrees, then remove the pins and springs **(see illustrations)**.
4  Remove both parking brake shoes, and recover the shoe adjuster mechanism, noting which way it is installed **(see illustration)**.
5  Inspect the parking brake shoes for signs of wear or contamination, and replace if necessary. It is recommended that the return springs are replaced as a matter of course.

The manufacturer does not state any wear limit for the shoe friction material thickness, but anything less than 1/16-inch is not ideal.
6  While the shoes are removed, clean and inspect the condition of the shoe adjuster and expander mechanisms, replace them if they show signs of wear or damage. If all is well, apply a fresh coat of brake grease (the manufacturer recommends Molykote Paste G) to the threads of the adjuster and sliding surfaces of the expander mechanism **(see illustration)**. Do not allow the grease to contact the shoe friction material.

### Installation

*Refer to illustration 17.9*

7  Prior to installation, clean the back plate, and apply a thin smear of high-temperature brake grease or anti-seize compound to all those surfaces of the back plate which bear on the shoes. Do not allow the lubricant to foul the friction material.
8  Offer up the parking brake shoes, and secure them in position with the retaining pins and springs.
9  Make sure the lower ends of the shoes are correctly engaged with the expander, then slide the adjuster mechanism into position between the upper ends of the shoes **(see illustration)**.
10  Check all components are correctly fitted, and install the upper and lower return springs using a pair of pliers.
11  Centralize the parking brake shoes, and install the brake disc (see Section 9).
12  Prior to installation the wheel, adjust the parking brake as (see Section 14).

## 18 Brake light switch - removal and installation

### Removal

*Refer to illustrations 18.1, 18.4a and 18.4b*

1  The brake light switch **(see illustration)** is located on the pedal bracket up under the dash.

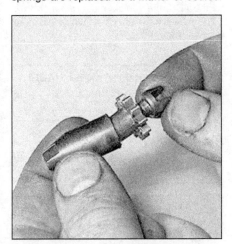

17.6 Clean the adjuster assembly and lubricate its moving parts with fresh brake grease

17.9 Install the adjuster assembly, making sure it is correctly engaged with both parking brake shoes

18.1 The brake light switch (arrow) is mounted on a bracket at the upper end of the brake pedal (Z3 model shown, other models similar)

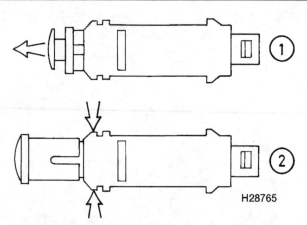

**18.4a Depress the brake pedal and withdraw the plunger sleeve from the switch body (1), then depress the retaining clips (2) ...**

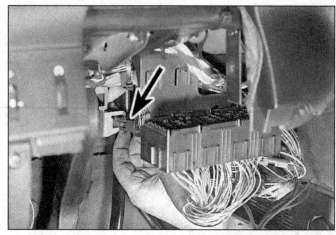

**18.4b ... and remove the switch from its mounting bracket**

2  Loosen and remove the retaining screws securing the driver's side lower dash panel. Unclip the panel and remove it from the vehicle.

3  Reach up behind the dash and disconnect the wiring connector from the switch.

4  Depress the brake pedal fully, then withdraw the plunger sleeve from the front of the switch to gain access to the switch retaining clips. Depress the clips and withdraw the switch from the pedal bracket (see illustrations).

## Installation

5  Ensure that the brake light switch plunger and sleeve are fully withdrawn from the switch body.

6  Fully depress the brake pedal and hold it down, then maneuver the switch into position. Hold the switch fully in position, then slowly release the brake pedal and allow it to return to its stop. This will automatically adjust the brake light switch. **Note:** *If the pedal is released too quickly, the switch will be incorrectly adjusted.*

7  Reconnect the wiring connector, and check the operation of the brake lights. The brake lights should illuminate after the brake pedal has traveled approximately 13/64-inch. If the switch is not functioning correctly, it is faulty and must be replaced; no other adjustment is possible.

8  On completion, install the driver's side lower dash panel.

## 19  Anti-lock braking system (ABS) - general information

**Note:** *On models equipped with traction control, the ABS unit is a dual function unit, and works both the anti-lock braking system (ABS) and traction control function of the Automatic Stability Control plus Traction (ASC+T) system.*

1  ABS is installed on most models as standard, and was available as an option on all others. The system comprises a hydraulic block which contains the hydraulic solenoid valves and the electrically-driven return pump, the four wheel sensors (one installed on each wheel), the electronic control unit (ECU) and the brake pedal position sensor. The purpose of the system is to prevent the wheel(s) from locking during heavy braking. This is achieved by automatic release of the brake on the relevant wheel, followed by re-application of the brake.

2  The solenoids are controlled by the ECU, which itself receives signals from the four wheel sensors (one fitted on each hub), which monitor the speed of rotation of each wheel. By comparing these signals, the ECU can determine the speed at which the vehicle is traveling. It can then use this speed to determine when a wheel is decelerating at an abnormal rate, compared to the speed of the vehicle, and therefore predicts when a wheel is about to lock. During normal operation, the system functions in the same way as a non-ABS braking system. In addition to this, the brake pedal position sensor (which is installed on the vacuum brake booster) also informs the ECU of how hard the brake pedal is being depressed.

3  If the ECU senses that a wheel is about to lock, it operates the relevant solenoid valve in the hydraulic unit, which then isolates the brake caliper on the wheel which is about to lock from the master cylinder, effectively sealing-in the hydraulic pressure.

4  If the speed of rotation of the wheel continues to decrease at an abnormal rate, the ECU switches on the electrically-driven return pump operates, and pumps the brake fluid back into the master cylinder, releasing pressure on the brake caliper so that the brake is released. Once the speed of rotation of the wheel returns to an acceptable rate, the pump stops; the solenoid valve opens, allowing the hydraulic master cylinder pressure to return to the caliper, which then re-applies the brake. This cycle can be carried out at up to 10 times a second.

5  The action of the solenoid valves and return pump creates pulses in the hydraulic circuit. When the ABS system is functioning, these pulses can be felt through the brake pedal.

6  The operation of the ABS system is entirely dependent on electrical signals. To prevent the system responding to any inaccurate signals, a built-in safety circuit monitors all signals received by the ECU. If an inaccurate signal or low battery voltage is detected, the ABS system is automatically shut down, and the warning light on the instrument panel is illuminated, to inform the driver that the ABS system is not operational. Normal braking should still be available, however.

7  If a fault does develop in the ABS system, the vehicle must be taken to a BMW dealer for fault diagnosis and repair.

8  On models equipped with AST, an accumulator is also incorporated into the hydraulic system. As well as performing the ABS function as described above, the hydraulic unit also works the traction control side of the AST system. If the ECU senses that the wheels are about to lose traction under acceleration, the hydraulic unit momentarily applies the rear brakes to prevent the wheel(s) spinning. In the same way as the ABS, the vehicle must be taken to a BMW dealer for testing if a fault develops in the AST system.

## 20  Anti-lock braking system (ABS) components - removal and installation

### Hydraulic unit

1  It is not possible for the home mechanic to remove the hydraulic unit. If the hydraulic unions are disconnected from the unit, air will enter the high-pressure hydraulic system linking the master cylinder and hydraulic unit. Bleeding of the high-pressure system can only be safely carried out by a BMW dealer who has access to the service tester (see Section 2). Hydraulic unit removal and installation should therefore be entrusted to a BMW dealer or other qualified repair shop.

# Chapter 9 Brakes

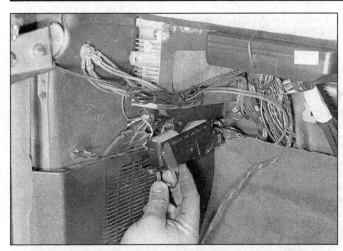

20.4 Release the retaining clip and disconnect the wiring connector from the ABS ECU

20.5 Unscrew the retaining bolts and remove the ABS ECU and mounting bracket from the underneath the dash

## Accumulator - models with All Season Traction (AST)

2  It is not possible for the home mechanic to remove the accumulator unit. If the hydraulic unions are disconnected from the unit, air will enter the high-pressure hydraulic system linking the accumulator and hydraulic unit. Bleeding of the high-pressure system can only be safely carried out by a technician who has access to the service tester (see Section 2). Accumulator removal and installation should therefore be entrusted to a BMW dealer or other qualified repair shop.

## Electronic control unit (ECU)

**Note:** *On vehicles equipped with All Season Traction (AST), if the ECU is disconnected, on reconnection it must be initialized using BMW diagnostic equipment. Note that the traction control system will be disabled until the ECU has been initialized.*

### Removal

*Refer to illustrations 20.4 and 20.5*

3  Disconnect the battery negative terminal and remove the glove box (see Section 27 in Chapter 11). Unclip the dash undercover and remove it. **Caution:** *If the stereo in your vehicle is equipped with an anti-theft system, make sure you have the correct activation code before disconnecting the battery.*

4  Lift the ECU wiring connector locking clip, and carefully disconnect the wiring connector **(see illustration)**.

5  Loosen and remove the nuts/bolts securing the ECU mounting bracket in position, and remove it from the vehicle **(see illustration)**.

### Installation

6  Installation is a reversal of the removal procedure, ensuring that the ECU wiring connector is correctly and securely reconnected.

## Front wheel sensor

### Removal

*Refer to illustrations 20.7 and 20.8*

7  Chock the rear wheels, then firmly apply the parking brake, jack up the front of the vehicle and support on axle stands. Remove the appropriate front wheel. Trace the wiring back from the sensor to the connector which is situated in a protective plastic box. Unclip the lid, then free the wiring connector and disconnect it from the main harness **(see illustration)**.

8  Loosen and remove the bolt securing the sensor to the steering knuckle, and remove the sensor and lead assembly from the vehicle **(see illustration)**.

### Installation

9  Prior to installation, apply a thin coat of multi-purpose grease to the sensor tip.

10  Ensure that the sensor and steering knuckle sealing faces are clean, then install the sensor to the hub. Install the retaining bolt and tighten it to the specified torque.

11  Ensure that the sensor wiring is correctly routed and retained by all the necessary clips, and reconnect it to its wiring connector. Install the sensor into the box and securely clip the lid in position.

12  Install the wheel, then lower the vehicle to the ground and tighten the wheel bolts to the specified torque.

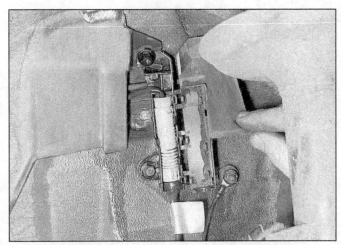

20.7 Unclip the cover then free the wheel sensor from its protective box and disconnect it

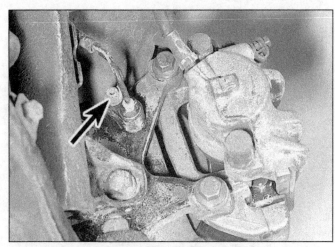

20.8 Unscrew the retaining bolt (arrow) and withdraw the front wheel sensor from the steering knuckle

## Rear wheel sensor

### Removal
13  Chock the front wheels, then jack up the rear of the vehicle and support it on axle stands. Remove the appropriate wheel.
14  Remove the sensor (see Steps 7 and 8).

### Installation
15  Install the sensor (see Steps 9 through 12).

## Front reluctor rings
16  The front reluctor rings are fixed onto the rear of wheel hubs. Examine the rings for damage such as chipped or missing teeth. If replacement is necessary, the complete hub assembly must be disassembled and the bearings replaced (see Chapter 10).

## Rear reluctor rings
17  The rear reluctor rings are pressed onto the driveaxle outer joints. Examine the rings for signs of damage such as chipped or missing teeth, and replace as necessary. If replacement is necessary, the driveaxle assembly must be replaced (see Chapter 8).

## Brake pedal position sensor
18  Release the vacuum from inside the brake booster by depressing the brake pedal several times.
19  Disconnect the battery negative terminal, then disconnect the wiring connector from the pedal position sensor. **Caution:** *If the stereo in your vehicle is equipped with an anti-theft system, make sure you have the correct activation code before disconnecting the battery.*
20  Using a small screwdriver, carefully lever off the sensor retaining clip, then withdraw the sensor from the front of the booster unit. Recover the sealing ring and snap-ring.

### Installation
21  If a new sensor is being installed, note the color of the spacer installed on the original sensor, and install the relevant color spacer to the new sensor. This is vital to ensure that the correct operation of the anti-lock braking system.
22  Install the new snap-ring in the groove on the front of the brake booster, positioning its end gap over the booster sensor locating slot.
23  Install the new sealing ring to the sensor, lubricating it with a smear of oil to aid installation.
24  Install the sensor to the booster, aligning its locating notch with the booster upper groove. Push the sensor until it clicks into position, and check that it is securely retained by the snap-ring.
25  Reconnect the sensor wiring, and connect the battery negative terminal.

# Chapter 10
# Suspension and steering

## Contents

| | Section |
|---|---|
| Front hub assembly - removal and installation | 2 |
| Front stabilizer bar - removal and installation | 7 |
| Front stabilizer bar connecting link - removal and installation | 8 |
| Front suspension and steering check | See Chapter 1 |
| Front suspension control arm - removal, overhaul and installation | 5 |
| Front suspension control arm balljoint - removal and installation | 6 |
| Front suspension strut - removal, overhaul and installation | 4 |
| General information | 1 |
| Ignition switch/steering column lock - removal and installation | 19 |
| Power steering fluid level check | See Chapter 1 |
| Power steering pump - removal and installation | 22 |
| Power steering pump drivebelt check, adjustment and replacement | See Chapter 1 |
| Power steering system - bleeding | 23 |
| Rear hub assembly - removal and installation | 9 |
| Rear hub bearings - replacement | 10 |
| Rear stabilizer bar - removal and installation | 16 |

| | Section |
|---|---|
| Rear suspension coil spring - removal and installation | 12 |
| Rear suspension lower control arm - removal, overhaul and installation | 15 |
| Rear suspension shock absorber - removal, overhaul and installation | 11 |
| Rear suspension trailing arm - removal, overhaul and installation | 13 |
| Rear suspension upper control arm - removal, overhaul and installation | 14 |
| Steering column (3-Series models) - removal, inspection and installation | 18 |
| Steering column intermediate shaft - removal and installation | 20 |
| Steering gear assembly - removal, overhaul and installation | 21 |
| Steering gear boots - replacement | 24 |
| Steering knuckle - removal and installation | 3 |
| Steering wheel - removal and installation | 17 |
| Tie-rod - replacement | 26 |
| Tie-rod end - removal and installation | 25 |
| Wheel alignment - general information | 27 |
| Wheel and tire maintenance and tire pressure checks | See Chapter 1 |

## Specifications

### Front suspension
Type .................. Independent, with MacPherson struts incorporating coil springs, telescopic shock absorbers and a stabilizer bar

### Rear suspension
Type
All except 318ti and Z3 models .................. Independent, trailing arms located by upper and lower control arms with coil springs, shock absorbers and a stabilizer bar
318ti and Z3 models .................. Independent, with semi-trailing arms, coil springs, shock absorbers and a stabilizer bar

### Steering
Type .................. Power-assisted rack and pinion

### Torque specifications                                     Ft-lbs
### Front suspension
Stabilizer bar connecting link nuts
  Models with connecting link mounted on the strut .................. 43
  Models with connecting link mounted on control arm .................. 34
Stabilizer bar clamp nuts .................. 16
Hub nut .................. 215
Control arm balljoint nut .................. 45
Control arm rear mounting bracket bolts .................. 34
Control arm-to-crossmember nut .................. 63
Strut mounting plate nut:
  M12 thread
    Piston with an external hexagon end (retain with socket) .................. 47
    Piston with an internal hexagon end (retain with Allen key) .................. 32
  M14 thread .................. 47
Strut-to-steering knuckle bolt nut .................. 79
Strut upper mounting nuts .................. 16

## Torque specifications (continued)      Ft-lbs

### Rear suspension
Control arm-to-subframe pivot bolt .................................................. 57
Shock absorber lower mounting bolt ................................................. 74
Shock absorber upper mounting nuts ................................................ 16
Trailing arm-to-control arm pivot bolts .............................................. 81
Trailing arm mounting bracket-to-body bolts ..................................... 57
Trailing arm-to-mounting bracket pivot bolt ...................................... 81

### Steering
Intermediate shaft clamp bolts ........................................................... 13
Power steering pipe fitting bolts
   M14 bolt ............................................................................................ 26
   M16 bolt ............................................................................................ 30
Power steering pump bolts .................................................................. 16
Steering gear mounting nuts ............................................................... 31
Steering wheel bolt ............................................................................... 46
Tie-rod .................................................................................................. 53
Tie-rod end
   Retaining nut .................................................................................... 33
   Locknut ............................................................................................. 11

### Wheels
Wheel bolts ........................................................................................... See Chapter 1

## 1  General information

*Refer to illustrations 1.1a, 1.1b and 1.3*

The independent front suspension **(see illustrations)** is of the MacPherson strut type, incorporating coil springs and integral telescopic shock absorbers. The MacPherson struts are located by transverse control arms. The steering knuckles, which carry the brake calipers and the hub/disc assemblies, are bolted to the MacPherson struts, and connected to the control arms through balljoints. A stabilizer bar reduces body roll. The stabilizer bar is rubber-mounted and is connected to both suspension struts/control arms (as applicable) by connecting links.

The fully-independent rear suspension on all 3-Series models (except 318ti models) consists of trailing arms, which are linked to the rear axle carrier by upper and lower control arms. Coil springs are installed between the upper control arms and vehicle body, and shock absorbers are connected to the vehicle body and trailing arms. A stabilizer bar minimizes body roll. The stabilizer bar is rubber-mounted and is connected to the upper control arms by connecting links.

The fully-independent rear suspension **(see illustration)** on 318ti and Z3 models is of the semi-trailing arm type. Coil springs and telescopic shock absorbers are positioned between the trailing arms and the body. A

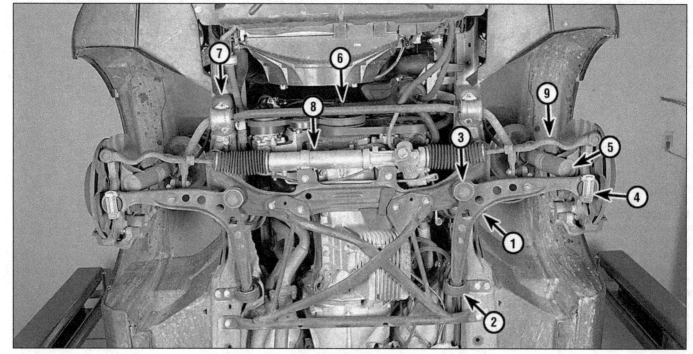

**1.1a  Typical front suspension (Z3 shown, other models similar)**

1  Control arm
2  Control arm bushing
3  Center balljoint
4  Outer balljoint
5  Strut assembly
6  Stabilizer bar
7  Stabilizer bar bushing
8  Steering gear
9  Outer tie-rod

## Chapter 10  Suspension and steering

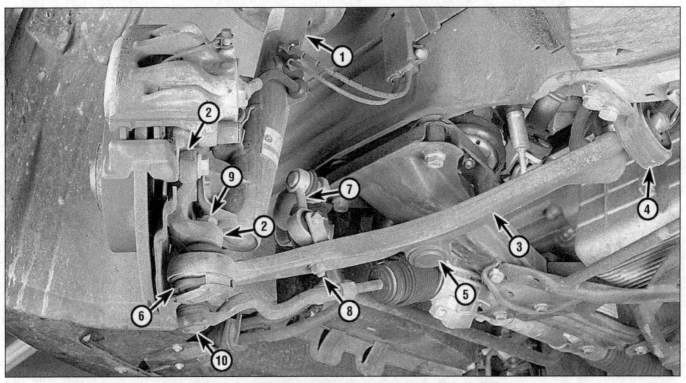

**1.1b  Front suspension details (Z3 shown, other models similar)**

| | | |
|---|---|---|
| 1  Strut assembly | 4  Control arm bushing | 7  Stabilizer bar link |
| 2  Steering knuckle | 5  Center balljoint | 8  Stabilizer link bracket nut |
| 3  Control arm | 6  Outer balljoint | 9  Balljoint nut |
| | | 10  Outer tie-rod |

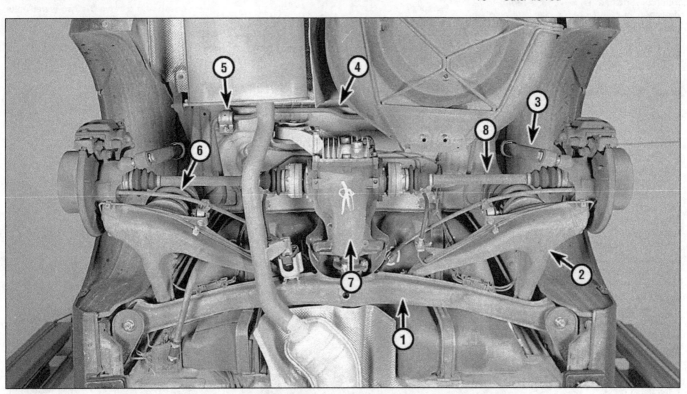

**1.3  Typical rear 318 and Z3 rear suspension (Z3 shown, 318ti similar)**

| | | |
|---|---|---|
| 1  Rear axle carrier | 4  Stabilizer bar | 7  Differential |
| 2  Trailing arm | 5  Stabilizer bar bushing | 8  Driveaxle assembly |
| 3  Shock absorber | 6  Coil spring | |

stabilizer bar is attached to the trailing arms via links and to the body with clamps.

The steering column is connected to the steering gear by an intermediate shaft, which incorporates a universal joint.

The steering gear is mounted onto the front subframe, and is connected by two tie-rods, with balljoints at their outer ends, to the steering arms projecting forwards from the steering knuckles. The tie-rod ends are threaded, to facilitate adjustment.

Power-assisted steering is standard. The hydraulic steering system is powered by a belt-driven pump, which is driven off the crankshaft pulley.

**Note:** *The information contained in this Chapter is applicable to the standard suspension set-up. On M3 models, slight differences will be found. Refer to your BMW dealer for details.*

## 2 Front hub assembly - removal and installation

### Removal

*Refer to illustrations 2.2, 2.3 and 2.4*

**Note:** *The hub assembly should not be removed unless it is to be replaced. The hub bearing inner race is a press-fit on the steering knuckle, and removal of the hub will almost certainly damaged the bearings; the manufacturer states that the hub assembly must be replaced whenever it is removed. A new hub nut and grease cap will also be required on installation.*

1  Remove the front brake disc (see Chapter 9). On models with ABS, also remove the front wheel sensor.
2  Tap the grease cap out from the center of the hub **(see illustration)**.
3  Using a hammer and chisel, tap up the staking securing the hub retaining nut in position, then loosen and remove the nut **(see illustration)**.
4  Attach a suitable puller to the hub assembly and draw the assembly off the steering knuckle. If the bearing inner race remains on the steering knuckle, a knife-edge type puller will be required to remove it **(see illustration)**. Note that the hub bearings are not available separately - the hub must be replaced as a complete assembly.
5  Inspect the steering knuckle shaft for signs of damage, and replace it if necessary (see Section 3). If required, remove the retaining screws and remove the disc guard from the steering knuckle.

### Installation

*Refer to illustration 2.7*

6  Where removed, install the disc guard to the steering knuckle, and securely tighten its retaining screws.
7  Ensure the dust cover is correctly installed on the rear of the hub assembly, and locate the hub on the steering knuckle. Tap or press the hub assembly fully onto the steering knuckle shaft using a tubular spacer which bears only on the bearing inner race **(see illustration)**.
8  Screw the new hub nut onto the steering knuckle. Tighten the nut to the specified torque, and secure it in position by staking it firmly into the knuckle groove using a hammer and punch.
9  Check that the hub rotates freely, and press the new grease cap into the hub center.
10  Install the brake disc (see Chapter 9). Where necessary, also install the ABS wheel sensor.

## 3 Steering knuckle - removal and installation

### Removal

1  Firmly apply the parking brake, then jack up the front of the vehicle and support it on axle stands. Remove the relevant front wheel.
2  If the steering knuckle is to be replaced, remove the hub assembly (see Section 2).

2.2 Tap out the grease cap from the center of the hub

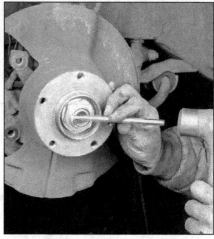

2.3 Using a hammer and pointed-nosed chisel, tap out the hub nut staking

3  If the steering knuckle assembly is to be reinstalled, loosen and remove the two bolts securing the brake caliper mounting bracket to the knuckle then slide the caliper assembly off the disc. Using a piece of wire or string, tie the caliper to the front suspension coil spring, to avoid placing any strain on the hydraulic brake hose. On models with ABS, also remove the wheel sensor (see Chapter 9).
4  Loosen and remove the nut securing the steering gear tie-rod end to the steering knuckle, and release the balljoint tapered shank using a universal balljoint separator.
5  Loosen and remove the two lower bolts securing the suspension strut to the steering knuckle, and the upper nut and bolt.
6  Unscrew the control arm balljoint and remove the steering knuckle from the vehicle. If necessary, release the steering knuckle from the control arm using the balljoint separator.
7  Examine the knuckle for signs of wear or damage, and replace if necessary.

### Installation

**Note:** *New suspension strut-to-steering*

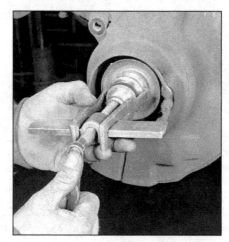

2.4 If the hub bearing inner race remains on the steering knuckle, use a puller to get it off

2.7 Tap the hub assembly into position using a socket which bears only on the inner race of the new bearing

# Chapter 10 Suspension and steering

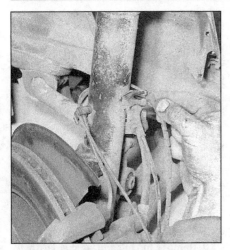

**4.3 Unclip the brake hose and wiring harness from its retaining clips on the base of the strut**

**4.6 Removing a front suspension strut**

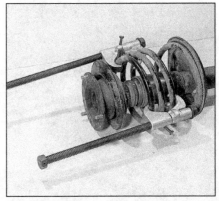

**4.8 Compress the suspension strut coil spring with a suitable spring compressor**

knuckle lower bolts, and tie-rod end, control arm balljoint, and strut-to-knuckle upper bolt nuts, will be required on installation.

8   Prior to installation, clean the threads of the strut-to-steering knuckle lower bolt holes by running a tap of the correct thread size and pitch down them.

9   Engage the knuckle with the control arm bushing stud, and install the new retaining nut.

10   Locate the knuckle correctly with the suspension strut, and insert the upper retaining bolt and new nut. Install the two new lower bolts securing the strut to the knuckle, and tighten both the lower and upper bolt to the specified torque.

11   Tighten the control arm balljoint nut to the specified torque.

12   Engage the tie-rod end in the steering knuckle, then install a new retaining nut and tighten it to the specified torque.

13   Install the new hub assembly (see Section 2).

14   On models where the hub was not disturbed, slide the caliper into position over the disc, making sure the pads pass either side of the disc. Lightly oil the threads of the caliper bracket mounting bolts prior to installation, and tighten them to the specified torque (see Chapter 9).

15   Install the wheel, then lower the vehicle to the ground and tighten the wheel bolts to the specified torque.

## 4   Front suspension strut - removal, overhaul and installation

**Note:** *Always replace the struts and/or coil springs in pairs.*

### Removal

*Refer to illustrations 4.3 and 4.6*

1   Chock the rear wheels, apply the parking brake, then jack up the front of the vehicle and support on axle stands. Remove the appropriate wheel.

2   To prevent the control arm assembly from hanging down while the strut is removed, screw a wheel bolt into the hub, then wrap a piece of wire around the bolt and tie it to the vehicle body. This will support the weight of the hub assembly. Alternatively, support the control arm with a jack.

3   Unclip the brake hose and wiring harness from its clips on the base of the strut **(see illustration)**.

4   On models where the stabilizer bar connecting link is mounted onto the suspension strut, loosen and remove the retaining nut and washer, and position the connecting link clear of the strut.

5   On all models, remove the two lower bolts securing the suspension strut to the steering knuckle, and also the upper nut and bolt.

6   From within the engine compartment, unscrew the strut upper mounting nuts, then carefully lower the strut assembly out from underneath the fender **(see illustration)**.

### Overhaul

*Refer to illustrations 4.8, 4.9a and 4.9b*

**Warning:** *Before attempting to disassemble the front suspension strut, some kind of spring compressor must be obtained. Adjustable coil spring compressors are readily available and are recommended for this operation. Any attempt to disassemble the strut without such a tool is likely to result in damage or personal injury.*

**Note:** *A new mounting plate nut will be required.*

7   With the strut removed from the car, clean away all external dirt, then mount it upright in a vise.

8   Install the spring compressor, and compress the coil spring until all tension is relieved from the upper spring seat **(see illustration)**.

9   Remove the cap from the top of the strut to gain access to the strut upper mount retaining nut. Loosen the nut while retaining the strut piston with an Allen wrench **(see illustrations)**.

10   Remove the mounting nut and washer, and lift off the rubber mounting plate.

11   Remove the gasket and dished washer followed by the upper spring plate and upper spring seat.

12   On all models, lift off the coil spring and remove the lower spring seat.

13   Slide the rubber seat (if equipped), rubber damper stop and piston dust cover off the strut.

**4.9a Remove the cap from the top of the strut mounting . . .**

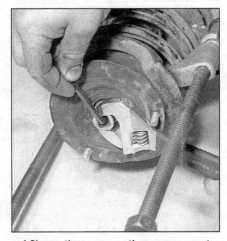

**4.9b . . . then remove the upper mount retaining nut**

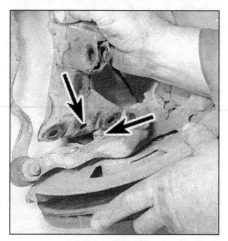

4.26a Ensure the strut is correctly engaged with the steering knuckle peg (arrows) . . .

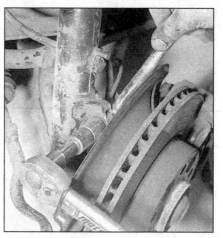

4.26b . . . then install the strut-to-knuckle bolts and tighten them to the specified torque

5.3 Remove the nut and washer (arrow) securing the connecting link to the lower arm

14  With the strut assembly now completely disassembled, examine all the components for wear, damage or deformation, and check the upper mounting bearing for smoothness of operation. Replace any of the components as necessary.

15  Examine the strut for signs of fluid leakage. Check the strut piston for signs of pitting along its entire length, and check the strut body for signs of damage. While holding it in an upright position, test the operation of the strut by moving the piston through a full stroke, and then through short strokes of two to four inches. In both cases, the resistance felt should be smooth and continuous. If the resistance is jerky, or uneven, or if there is any visible sign of wear or damage to the strut, replacement is necessary.

16  If any doubt exists about the condition of the coil spring, carefully remove the spring compressors, and check the spring for distortion and signs of cracking. Replace the spring if it is damaged or distorted, or if there is any doubt as to its condition.

17  Inspect all other components for signs of damage or deterioration, replacing any that are suspect.

18  Slide the rubber damper and piston boot onto the strut piston and (where necessary) install the rubber seat.

19  Install the spring seat and coil spring onto the strut, making sure the seat ridge and spring end are correctly located against the strut stop.

20  Install the upper spring seat so that the spring end is against the seat stop.

21  Install the upper spring plate, aligning its stop with that of the seat, and install the dished washer and gasket followed by the upper mounting plate.

22  Locate the washer on the strut piston, then install the new mounting plate nut and tighten it to the specified torque.

23  Ensure the spring ends and seats are correctly located, then carefully release the compressor and remove it from the strut. Install the cap to the top of the strut.

## Installation

*Refer to illustrations 4.26a and 4.26b*

**Note:** *New suspension strut upper mounting nuts, a strut-to-knuckle upper bolt nut and lower retaining bolts will be required on installation. On models where the stabilizer bar is mounted to the strut, a connecting link retaining nut will also be required.*

24  Prior to installation, clean the threads of the strut-to-steering knuckle lower bolt holes by running a tap of the correct thread size and pitch down them. **Note:** *If a suitable tap is not available, clean out the holes using one of the old bolts with slots cut in its threads.*

25  Maneuver the strut assembly into position, and install the new upper mounting nuts.

26  Locate the knuckle correctly with the suspension strut, and insert the upper retaining bolt and new nut. Install the two new lower bolts securing the strut to the knuckle, and tighten both the lower and upper bolts to the specified torque **(see illustrations)**.

27  Tighten the strut upper mounting nuts to the specified torque.

28  Where necessary, engage the stabilizer bar connecting link with the strut. Make sure the flat on the balljoint shank is correctly

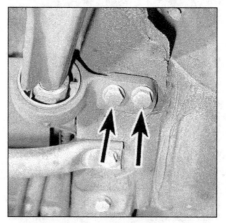

5.5 Lower arm rear mounting bracket retaining bolts (arrows)

located against the lug on the strut, then install the washer and new retaining nut and tighten to the specified torque.

29  Clip the hose/wiring back onto the strut, then install the wheel. Lower the vehicle to the ground and tighten the wheel bolts to the specified torque.

## 5  Front suspension control arm - removal, overhaul and installation

### Removal

*Refer to illustrations 5.3, 5.5 and 5.6*

1  Chock the rear wheels, firmly apply the parking brake, then jack up the front of the vehicle and support on axle stands.

2  Remove the appropriate front wheel.

3  On models where the stabilizer bar is connected to the control arm, remove the nut and washer securing the connecting link to the arm **(see illustration)**.

4  Unscrew the control arm balljoint retaining nut, and release the arm from the steering knuckle. **Note:** *You might not be able to unscrew the nut completely until the balljoint stud is separated from the steering knuckle. If necessary release the steering knuckle from the arm using the balljoint separator.*

5  Loosen and remove the two bolts securing the control arm rear mounting to the vehicle body **(see illustration)**.

6  Unscrew the nut from the control arm front mounting stud, and remove the control arm assembly from underneath the vehicle **(see illustration)**. Note that the stud may be a tight fit in the crossmember and may need to be tapped out of position.

### Overhaul

7  Thoroughly clean the control arm and the area around the arm mountings, removing all traces of dirt and underseal if necessary, then check carefully for cracks, distor-

# Chapter 10 Suspension and steering

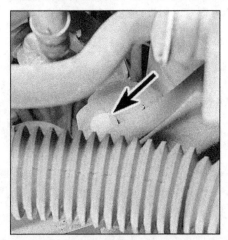

5.6 Control arm front mounting stud nut (arrow)

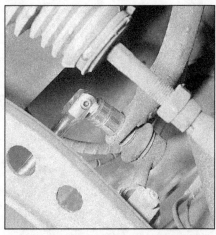

7.2 Unscrew the retaining nuts and free the connecting links from the stabilizer bar

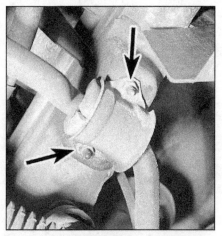

7.3 Stabilizer bar mounting clamp retaining nuts (arrows)

tion or any other signs of wear or damage, paying particular attention to the mounting bushings and balljoints. If either the bushing or the outer balljoint requires replacement, the arm should be taken to a BMW dealer or other qualified repair shop. A hydraulic press and suitable spacers are required to press the bushings out of position and install the new ones. **Note:** *The center balljoint is integral with the control arm and can't be replaced separately.*

## Installation

**Note:** *New control arm rear mounting and balljoint nuts will be required on installation. On models where the stabilizer bar is connected to the lower arm, a connecting link retaining nut will also be required.*

8   Ensure the mounting studs are clean and dry, then install the control arm.
9   Locate the front mounting stud in the crossmember, and engage the balljoint stud with the steering knuckle. Where necessary, also align the stabilizer bar connecting link with the arm hole. If necessary, press the front mounting bushing stud into place using a jack positioned beneath the arm.
10   Install a new nut to the front mounting stud and tighten it to the specified torque.
11   Install a new nut to the balljoint shank, and tighten it to the specified torque.
12   Install the control arm rear mounting bracket bolts, and tighten them to the specified torque.
13   Where necessary, install the washer and new retaining nut to the stabilizer bar connecting link, and tighten it to the specified torque.
14   Install the wheel, then lower the vehicle to the ground and tighten the wheel bolts to the specified torque.

## 6   Front suspension control arm balljoint - replacement

**Note:** *This applies to the outer balljoint only;* *the center balljoint is integral with the control arm and can't be replaced separately.*

Front suspension control arm balljoint replacement requires the use of a hydraulic press and several suitable spacers if it is to be carried out safely and successfully. If replacement is necessary, then the arm should be removed (Section 5) and taken to a BMW dealer or other qualified repair shop. **Note:** *Technicians with the necessary tools can replace the balljoint without removing the arm from the vehicle.*

## 7   Front stabilizer bar - removal and installation

### Removal

*Refer to illustrations 7.2 and 7.3*
1   Chock the rear wheels, firmly apply the parking brake, then jack up the front of the vehicle and support on axle stands. Remove both front wheels.
2   Unscrew the retaining nuts and free the connecting link from each end of the stabilizer bar **(see illustration)**.
3   Make alignment marks between the mounting bushings and stabilizer bar, then loosen the two stabilizer bar mounting clamp retaining nuts **(see illustration)**.
4   Remove both clamps from the subframe, then maneuver the stabilizer bar out from underneath the vehicle. Remove the bushings from the bar.
5   Carefully examine the stabilizer bar components for signs of wear, damage or deterioration, paying particular attention to the bushings. Replace worn components as necessary.

### Installation

**Note:** *New mounting clamp retaining nuts and connecting link nuts will be required on installation.*
6   Install the rubber bushings to the stabilizer bar, aligning them with the marks made prior to removal. Rotate each bushing so that its flat surface is facing up.
7   Maneuver the stabilizer bar into position. Install the mounting clamps, ensuring that their ends are correctly located in the hooks on the subframe, and install the new retaining nuts. Ensure that the bushing markings are still aligned with the marks on the bars, then tighten the mounting clamp retaining nuts to the specified torque.
8   Engage the stabilizer bar connecting links with the bar. Make sure the flats on the balljoint shank are correctly located against the lugs on the bar then install the new retaining nuts and tighten to the specified torque.
9   Install the wheels then lower the vehicle to the ground and tighten the wheel bolts to the specified torque.

## 8   Front stabilizer bar connecting link - removal and installation

### Removal

#### Models where the stabilizer bar is connected to the lower arms

1   Firmly apply the parking brake, then jack up the front of the car and support it on axle stands.
2   Unscrew the retaining nut, and free the connecting link from the stabilizer bar.
3   Loosen and remove the nut and bolt securing the link to its control arm mounting bracket, and remove the link.
4   Inspect the connecting link balljoint and bushing for signs of wear or damage. Check that the balljoint is free to move easily, and that its rubber boot is undamaged. Replace the link/mounting bracket if they are damaged, noting that all self-locking nuts should be replaced as a matter of course.

#### Models where the stabilizer bar is connected to the suspension struts

5   Carry out the operations described in Steps 1 and 2.

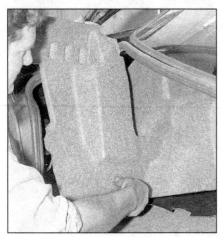

**11.2a To gain access to the rear shock absorber upper mounting, remove the luggage compartment side trim panel ...**

**11.2b ... and lift out the insulation panel (3-Series models)**

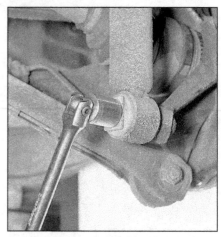

**11.4a Remove the shock absorber lower mounting bolt (3-Series models)**

6   Unscrew the retaining nut and washer securing the link to the strut, and remove the link from the vehicle.
7   Check the connecting link balljoints for signs of wear. Check that each balljoint is free to move easily, and that the rubber boots are undamaged. If necessary replace the connecting link.

### Installation

**Note:** *New connecting link nuts will be required on installation.*

#### Models where the stabilizer bar is connected to the lower arms

8   Installation is a reverse of the removal sequence, using new nuts and tightening them to the specified torque.

#### Models where the stabilizer bar is connected to the suspension struts

9   Installation is the reverse of removal, ensuring the balljoint shank flats are correctly engaged with the strut/stabilizer bar lugs. Install the new retaining nuts and tighten them to the specified torque.

## 9   Rear hub assembly - removal and installation

### Removal

**Note:** *The hub assembly should not be removed unless it, or the hub bearing, is to be replaced. The hub is a press fit in the bearing inner race, and removal of the hub will almost certainly damage the bearings. If the hub is to be removed, be prepared to replace the hub bearing at the same time.*

1   Remove the relevant driveaxle (see Chapter 8).
2   Remove the brake disc (see Chapter 9).
3   Bolt a slide hammer to the hub flange, and use the hammer to draw the hub out from the bearing. If the bearing inner race stays attached to the hub, a puller will be required to draw it off.
4   With the hub removed, closely examine the hub bearing for signs of damage. Check that the bearing rotates freely and easily, without any sign of roughness. If the inner race remains attached to the hub, or there is any doubt about its condition, replace the bearing (see Section 10).

### Installation

**Note:** *A long bolt/length of threaded bar and suitable washers will be required on installation.*

5   Apply a smear of oil to the hub surface, and locate it in the bearing inner race.
6   Draw the hub into position using a long bolt or threaded length of bar and two nuts. Install a large washer to either end of the bolt/bar, so the inner one bears against the bearing inner race, and the outer one against the hub. Slowly tighten the nut(s) until the hub is pulled fully into position. **Note:** *Do not be tempted to knock the hub into position with a hammer and drift, as this will almost certainly damage the bearing.*
7   Remove the bolt/threaded bar and washers (as applicable), and check that the hub bearing rotates smoothly and easily.
8   Install the brake disc (see Chapter 9).
9   Install the driveaxle (see Chapter 8).

## 10   Rear hub bearings - replacement

1   Remove the rear hub (see Section 9).
2   Remove the hub bearing retaining snap-ring from the trailing arm.
3   Tap the hub bearing out from the trailing arm using a hammer and suitable punch.
4   Thoroughly clean the trailing arm bore, removing all traces of dirt and grease, and polish away any burrs or raised edges which might hinder reassembly. Replace the snap-ring if there is any doubt about its condition.
5   On reassembly, apply a light film of clean engine oil to the bearing outer race to aid installation.
6   Locate the bearing in the trailing arm and tap it fully into position, ensuring that it enters the arm squarely, using a suitable tubular spacer which contacts the bearing only on the outer race.
7   Secure the bearing in position with the snap-ring, making sure it is correctly located in the trailing arm groove.
8   Install the rear hub (see Section 9).

## 11   Rear suspension shock absorber - removal, overhaul and installation

**Note:** *Always replace the shock absorbers in pairs.*

### Removal

*Refer to illustrations 11.2a, 11.2b, 11.4a, 11.4b, 11.5a and 11.5b*

1   Check the front wheels, then jack up the rear of the vehicle and support it on axle stands. To improve access, remove the rear wheel.
2   If you're working on a 3-Series model, remove the rear speaker (see Chapter 12). Release the rear light access cover fastener by rotating it 90-degrees and removing the cover. Pry out the retaining clips securing the luggage compartment side trim cover in position, and remove the trim and insulation panel to gain access to the shock absorber upper mount **(see illustrations)**.
3   If you're working on a Z3 model, lower the convertible top (but don't lock it down), then pull out the clips retaining the carpet behind the front seats. Peel the carpet back for access to the shock absorber upper mount.
4   Position a jack underneath the rear of the trailing arm and raise the jack so that it is supporting the weight of the arm and the force of the coil spring. **Warning:** *The jack must remain in this position throughout the entire procedure.* Remove the bolt securing

# Chapter 10 Suspension and steering

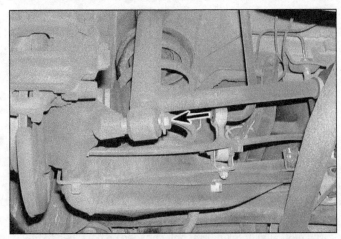

11.4b  Shock absorber mounting bolt (arrow) (Z3 models)

11.5a  Upper mounting nuts (arrows) (3-Series models)

the shock absorber to the trailing arm **(see illustrations)**.

5  Unscrew the upper mounting nuts **(see illustrations)**. Lower the shock absorber out from underneath the vehicle, and recover the gasket which is installed between the upper mount and the body.

## Overhaul

**Note:** *A new piston nut will be required.*

6  Remove the trim cap from the top of the shock absorber, then remove all traces of dirt. Loosen and remove the piston nut and dished washer, noting which way it is installed.

7  Lift off the upper mounting plate and remove the dust cover.

8  Slide the spacer and rubber stop off the shock absorber piston.

9  Examine the shock absorber for signs of fluid leakage. Check the piston for signs of pitting along its entire length, and check the body for signs of damage. While holding it in an upright position, test the operation of the shock absorber by moving the piston through a full stroke, and then through short strokes of two to four inches. In both cases, the resistance felt should be smooth and continuous. If the resistance is jerky, or uneven, or if there is any visible sign of wear or damage, replacement is necessary.

10  Inspect all other components for signs of damage or deterioration, replacing any that are suspect.

11  Slide the rubber stop and spacer onto the strut piston, then install the dust cover.

12  Install the upper mounting plate and dished washer, and screw on the new piston nut and tighten it securely. Install the trim cap.

## Installation

**Note:** *New shock absorber upper mounting nuts and a new mounting gasket will be required on installation.*

13  Ensure the upper mounting plate and body contact surfaces are clean and dry, and install a new gasket to the upper mounting plate.

14  Maneuver the shock absorber into position, and install the new upper mounting nuts.

15  Ensure the lower end of the shock absorber is positioned with the mounting bushing spacer thrustwasher facing towards the bolt. Screw in the lower mounting bolt, tightening it by hand only at this stage.

16  Tighten the upper mounting nuts to the specified torque then install all components removed previously for access.

17  Install the wheel and lower the vehicle to the ground. With the vehicle resting on its wheels, tighten the shock absorber lower mounting bolt and wheel bolts to the specified torque.

## 12  Rear suspension coil spring - removal and installation

**Note:** *Always replace the coil springs in pairs.*

### Removal

1  Chock the front wheels, then jack up the rear of the vehicle and support it on axle stands. Remove the relevant wheel.

2  Referring to Chapter 8, loosen and remove the retaining bolts and plates securing the right-hand driveaxle to the differential flange. Free the driveaxle and support it by tying it to the vehicle underbody using a piece of wire. **Note:** *Do not allow the driveaxle to hang under its own weight, as the CV joint may be damaged.*

3  Detach the stabilizer bar links from the trailing arms (see Section 16).

4  Position a jack underneath the rear of the trailing arm, and support the weight of the arm.

5  Loosen and remove the shock absorber lower mounting bolt.

6  Slowly lower the trailing arm, keeping watch on the brake pipe/hose to ensure no excess strain is placed on them, until it is possible to withdraw the coil spring.

7  Recover the spring seats from the vehicle body and control arm or trailing arm. If the vehicle is to be left for some time, raise the trailing arm back up and install the shock absorber lower mounting bolt.

11.5b  Upper mounting nuts (arrows) (Z3 models)

8  Inspect the spring closely for signs of damage, such as cracking, and check the spring seats for signs of wear. Replace worn components as necessary.

### Installation

9  Install the upper and lower spring seats, making sure they are correctly located on the pegs.

10  Apply a smear of grease to the spring ends and engage the spring with its upper seat.

11  Hold the spring in position and carefully raise the trailing arm while aligning the coil spring with its lower seat.

12  Raise the arm fully and install the shock absorber lower mounting bolt, tightening it by hand only at this stage.

13  Reconnect the stabilizer bar links to the trailing arms (see Section 16).

14  Referring to Chapter 8, connect the driveaxle to the differential, then install the retaining plates and bolts and tighten them to the specified torque.

15  Install the wheel then lower the vehicle to the ground. Tighten the wheel bolts and shock absorber lower mounting bolt to the specified torque.

## Chapter 10 Suspension and steering

**13.8 Mark the position of the trailing arm bracket on the body before loosening the mounting bolts (arrows)**

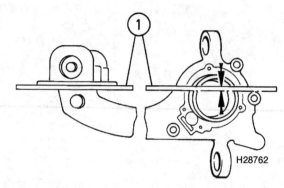

**13.14 To position the mounting bracket correctly in relation to the trailing arm, place an 8 mm (5/16-inch) rod (1) against the mounting bracket and rest it on the trailing arm as shown. The mounting bracket pivot bolt can then be tightened to the specified torque**

## 13 Rear suspension trailing arm - removal, overhaul and installation

### 3-Series models (except 318ti)

#### Removal

*Refer to illustration 13.8*

1 Chock the front wheels, then jack up the rear of the vehicle and support it on axle stands. Remove the relevant wheel.
2 Remove the relevant driveaxle (see Chapter 8) and continue as described under the relevant sub-heading.
3 Unscrew the two bolts securing the brake caliper mounting bracket in position, then slide the caliper assembly off the disc. Using a piece of wire or string, tie the caliper to the upper control arm, to avoid placing any strain on the hydraulic brake hose.
4 Disconnect the parking brake cable from the rear of the back plate. On models with ABS, also remove the rear wheel sensor (see Chapter 9).
5 Unscrew the retaining bolts and release the brake line bracket from the trailing arm.
6 Remove the coil spring (see Section 12).
7 Using paint or a suitable marker pen, make alignment marks between the lower control arm and pivot bolt eccentric washer.
8 Also make alignment marks between the trailing arm front mounting bracket and the vehicle underbody **(see illustration)**. This is necessary to ensure that the rear wheel alignment and camber are correct on installation.
9 Loosen and remove the nut and washer from the lower control arm pivot bolt. Withdraw the pivot bolt, then slowly lower the arm and remove the jack.
10 Loosen and remove the nut and pivot bolt securing the upper control arm to the trailing arm.
11 Unscrew the three bolts securing the trailing arm mounting bracket to the vehicle body and remove the trailing arm. **Note:** *Do not loosen the trailing arm pivot bushing bolt unless replacement of the bushing/mounting bracket is necessary.*

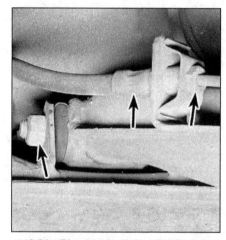

**13.24a Disconnect the rear brake line fitting (right arrow) from the hose (center arrow) at this bracket on the trailing arm, then plug the line and hose immediately to prevent fluid leaks; the left arrow points to the nut for the inner pivot bolt (318ti models)**

#### Overhaul

*Refer to illustration 13.14*

12 Loosen and remove the nut and pivot bolt and separate the front mounting bracket and trailing arm.
13 Thoroughly clean the trailing arm and the area around the arm mountings, removing all traces of dirt and underseal if necessary. Check carefully for cracks, distortion or any other signs of wear or damage, paying particular attention to the mounting bushings. If either bushing requires replacement, the lower arm should be taken to a BMW dealer or suitably-equipped garage. A hydraulic press and suitable spacers are required to press the bushings out of position and install the new ones. Inspect the pivot bolts for signs of wear or damage and replace as necessary.
14 Install the mounting bracket to trailing arm, and install the pivot bolt and nut. Position the bracket as shown, using an 8 mm (5/16-inch) rod, and tighten the pivot bolt to the specified torque **(see illustration)**.

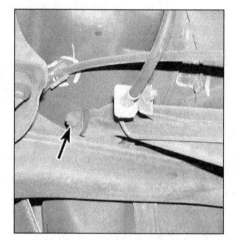

**13.24b Inner pivot nut/bolt (arrow) on a Z3 trailing arm**

#### Installation

15 Place the trailing arm assembly into position and install the mounting bracket retaining bolts. Align the marks made prior to removal, then tighten the mounting bracket bolts to the specified torque.
16 Engage the control arms with the trailing arm and install the pivot bolts and nuts. Tighten the bolts by hand only at this stage. Make sure the marks on the lower control arm and eccentric washer are aligned.
17 Install the coil spring (see Section 12), then install the shock absorber lower mounting bolt. Tighten the bolt by hand only.
18 Install the brake line retaining bracket to the trailing arm, and securely tighten the retaining bolts.
19 Referring to Chapter 9, reconnect the parking brake cable to the expander lever and (where necessary) install the ABS wheel sensor. Slide the caliper into position over the disc, making sure the pads pass either side of the disc, and tighten the caliper bracket mounting bolts to the specified torque.
20 Install the driveaxle (see Chapter 8) and lower the vehicle to the ground.
21 With the vehicle on its wheels, rock the

# Chapter 10 Suspension and steering

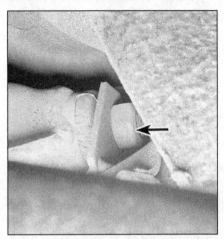

13.27a Nut (arrow) for the outer pivot bolt (318ti models)

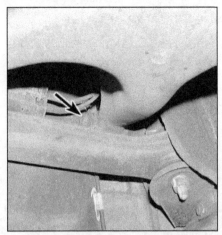

13.27b Outer pivot nut/bolt (arrow) on a Z3 trailing arm

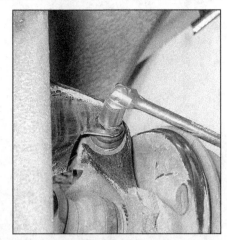

14.3 Remove the upper control arm-to-trailing arm pivot bolt . . .

car to settle the disturbed components in position, then tighten the shock absorber lower mounting bolt and the upper control arm pivot bolts to their specified torques. Check that the lower arm eccentric washer is still correctly aligned with the mark, then tighten it to the specified torque. **Note:** *On completion, it is advisable to have the camber angle and wheel alignment checked and, if necessary, adjusted.*

## 318ti and Z3 models

*Refer to illustrations 13.24a, 13.24b, 13.27a and 13.27b*

22  Loosen the wheel lug bolts, raise the rear of the vehicle and support it securely on axle stands. Remove the wheel.
23  Remove the driveaxle (see Chapter 8).
24  Disconnect the rear brake hose from the metal brake line at the bracket on the trailing arm **(see illustrations)**. **Note:** *For information on disconnecting brake hose-to-metal line connections, see Chapter 9.* Plug the line and hose to prevent brake fluid from leaking out.
25  Disconnect the parking brake cable (see Chapter 9).
26  Remove the coil spring (see Section 12).
27  Remove the trailing arm pivot bolts **(see illustrations)** and remove the trailing arm.
28  Inspect the pivot bolt bushings. If they're cracked, hardened or otherwise deteriorated, take the trailing arm to a BMW dealer or other qualified repair shop and have them replaced. Each bushing has a larger diameter shoulder on one end. Make sure this larger diameter shoulder on each bushing faces away from the trailing arm (the inner bushing shoulder faces the center of the vehicle and the outer bushing shoulder faces away from the vehicle).
29  Installation is the reverse of removal. Support the trailing arm with a floor jack and raise it to simulate normal ride height, then tighten the fasteners to the torque listed in this Chapter's Specifications. Be sure to bleed the brakes (see Chapter 9).

## 14  Rear suspension upper control arm - removal, overhaul and installation

**Note:** *This procedure does not apply to 318ti and Z3 models.*

### Removal

*Refer to illustrations 14.3 and 14.5*

1  Remove the coil spring (see Section 12).
2  Release the wiring from its retaining clips on the side of the upper control arm.
3  Loosen and remove the pivot bolt securing the control arm to the trailing arm **(see illustration)**.
4  Referring to Chapter 8, support the weight of the unit with a jack, and remove the differential mounting bolts.
5  Loosen and remove the nut from the control arm to the rear subframe pivot bolt. Withdraw the bolt, moving the differential slightly to the rear, and remove the control arm from underneath the vehicle **(see illustration)**. Note that on some models it may be necessary to detach the driveshaft from the differential in order to gain the clearance required to remove the pivot bolt. **Note:** *If the vehicle is to be left for some time, install the differential mounting bolts and tighten them securely.*

### Overhaul

6  Thoroughly clean the control arm and the area around the arm mounts, removing all traces of dirt and underseal if necessary. Check carefully for cracks, distortion or any other signs of wear or damage, paying particular attention to the mounting bushing. If the bushing requires replacement, the arm should be taken to a BMW dealer or other qualified repair shop. A hydraulic press and suitable spacers are required to press the bushings out of position and install the new ones.
7  Inspect the pivot bolts for signs of wear or damage and replace as necessary. The control arm-to-subframe bolt and nut should be replaced as a matter of course.

### Installation

**Note:** *A new control arm-to-rear subframe pivot bolt and nut will be required on installation.*

8  Maneuver the control arm into position, and install the new arm to subframe pivot bolt and nut. Only tighten the nut lightly at this stage.
9  Referring to Chapter 8, maneuver the differential into position, and tighten its mounting bolts to the specified torque. Where necessary, reconnect the driveshaft to the differential.
10  Install the pivot bolt and nut securing the control arm to the trailing arm, tightening it lightly only at this stage.
11  Clip the wiring back into position on the upper control arm.
12  Install the coil spring (see Section 12).
13  On completion, lower the vehicle to the ground and rock the vehicle to settle all disturbed components. With the vehicle resting on its wheels tighten the wheel bolts, shock absorber lower mounting bolt and the control

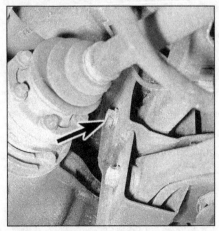

14.5 . . . and the control arm-to-subframe pivot bolt (arrow)

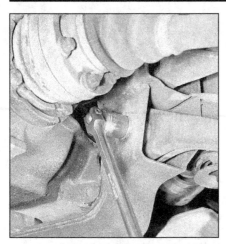

15.5a Remove the lower control arm pivot bolt . . .

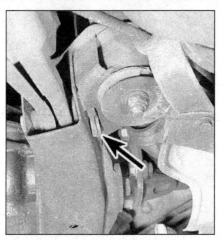

15.5b . . . and recover the special nut (arrow) from the subframe

16.2 Rear stabilizer bar connecting link-to-control arm nut (arrow)

arm pivot bolts to their specified torques. **Note:** *On completion, it is advisable to have the camber angle and wheel alignment checked and, if necessary, adjusted.*

## 15 Rear suspension lower control arm - removal, overhaul and installation

**Note:** *This procedure does not apply to 318ti and Z3 models (the trailing arm for these models, sometimes called the control arm, can be found in Section 13).*

### Removal

*Refer to illustrations 15.5a and 15.5b*

1 Chock the front wheels, then jack up the rear of the vehicle and support it on axle stands. To improve access, remove the rear wheel.
2 Using paint or a suitable marker pen, make alignment marks between the lower control arm pivot bolt eccentric washer and the trailing arm. This is necessary to ensure that the rear wheel alignment and camber are correct on installation.
3 Support the trailing arm with a jack, then loosen and remove the nut and washer from the lower control arm pivot bolt. Withdraw the pivot bolt.
4 Referring to Chapter 8, support the weight of the differential with a jack, and remove the unit mounting bolts.
5 Loosen and remove the pivot bolt securing the control arm to the rear subframe. Withdraw the bolt, moving the differential slightly to the rear, and remove the control arm from underneath the vehicle. Recover the special nut from the subframe **(see illustrations)**. **Note:** *On some models it may be necessary to detach the driveshaft from the differential in order to gain the clearance required to remove the pivot bolt. If the vehicle is to be left for some time, install the differential mounting bolts and tighten securely.*

### Overhaul

6 Refer to Steps 6 and 7 of Section 14.

### Installation

**Note:** *A new control arm-to-rear subframe pivot bolt and nut will be required on installation.*

7 Locate the special nut in the subframe cut-out, and maneuver the control arm into position. Install the new pivot bolt, tightening it lightly only at this stage.
8 Referring to Chapter 8, maneuver the differential into position and tighten its mounting bolts to the specified torque. Where necessary, reconnect the driveshaft to the differential.
9 Install the lower arm-to-trailing arm pivot bolt, eccentric washer and nut. Align the washer with the mark made prior to removal and lightly tighten it.
10 Install the wheel and lower the vehicle to the ground.
11 With the vehicle on its wheels, rock the car to settle the disturbed components in position. Check that the lower arm eccentric washer is still correctly aligned with the mark, then tighten both the control arm pivot bolts to the specified torque. Also tighten the wheel bolts to the specified torque. **Note:** *On completion, it is advisable to have the wheel alignment checked and, if necessary, adjusted.*

## 16 Rear stabilizer bar - removal and installation

### Removal

*Refer to illustrations 16.2 and 16.3*

1 Chock the front wheels, then jack up the rear of the vehicle and support it on axle stands. To improve access, remove the rear wheels.
2 Loosen and remove the nut securing each connecting link to the upper control arms **(see illustration)**.
3 Make alignment marks between the mounting bushings and stabilizer bar, then loosen the two stabilizer bar mounting clamp retaining nuts and bolts **(see illustration)**.
4 Remove both clamps from the subframe, and maneuver the stabilizer bar and connecting link assembly out from underneath the vehicle. Remove the mounting bushings and connecting links from the bar.
5 Carefully examine the stabilizer bar components for signs of wear, damage or deterioration, paying particular attention to the mounting bushings. Replace worn components as necessary.

### Installation

**Note:** *New mounting clamp nuts and connecting link nuts will be required on installation.*

6 Install the rubber mounting bushings to the stabilizer bar, aligning them with the marks made prior to removal. Rotate each bushing so that its flat surface is facing forwards.
7 Maneuver the stabilizer bar into position. Locate the connecting links in the upper control arms, and install the new retaining nuts and tighten securely.

16.3 Loosen and remove the rear stabilizer bar clamp retaining nuts and bolts

# Chapter 10  Suspension and steering

17.3  After removing the airbag module, remove the steering wheel retaining bolt (arrow)

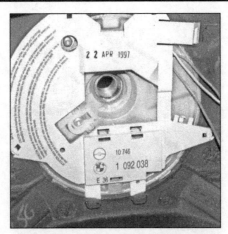

17.4a  After removing the steering wheel retaining bolt (but before removing the wheel), mark the relationship of the steering wheel to the steering shaft . . .

17.4b  . . . and then lift off the steering wheel. Note the locating pin (arrow) on the column; when the wheel is installed again, make sure that this pin engages the hole in the contact ring on the backside of the steering wheel

8  Install the mounting clamps, ensuring that their ends are correctly located in the hooks on the subframe, and install the bolts and new retaining nuts. Ensure that the bushing markings are still aligned with the marks on the bars, then securely tighten the mounting clamp retaining nuts.
9  Install the wheels then lower the vehicle to the ground and tighten the wheel bolts to the specified torque.

## 17  Steering wheel - removal and installation

### Removal

Refer to illustrations 17.3, 17.4a and 17.4b
**Warning:** *All models covered by this manual are equipped with Supplemental Restraint Systems (SRS), more commonly known as airbags. To avoid the possibility of accidental deployment of the airbag, which could cause serious injury, always disable the airbag system (see Section 24 in Chapter 12) before working in the vicinity of the impact sensors, steering column or instrument panel. Do not use electrical test equipment on any of the airbag system wiring or tamper with the airbag components in any way. Do NOT rotate the steering shaft while the steering wheel is removed; doing so can disable the driver's side airbag. An electrical contact ring on the backside of the steering wheel is held in the centered position by a plastic lock. Do NOT remove this lock! Doing so will allow the contact ring to rotate.*
1  Set the front wheels in the straight-ahead position and release the steering lock by inserting the ignition key. Disable the airbag system (see Section 24 in Chapter 12), remove the lower steering column cover (see Section 27 in Chapter 11) and unplug the airbag and contact ring connectors (see Section 25 in Chapter 12).
2  Remove the airbag unit from the center of the steering wheel (see Section 25 in Chapter 12). **Warning:** *When carrying the airbag module, keep the trim side of it away from your body, and when you place it on the bench, have the trim side facing up.*
3  Loosen and remove the steering wheel retaining bolt **(see illustration)**.
4  Mark the steering wheel and steering column shaft in relation to each other, then lift the steering wheel off the column splines **(see illustrations)**. If it is tight, tap it up near the center, using the palm of your hand, or twist it from side to side, while pulling upwards to release it from the shaft splines. Unplug the electrical connector.
5  The airbag contact unit will automatically be locked in position as the wheel is removed; do not attempt to rotate it while the wheel is removed, and don't allow the steering shaft to turn either. **Warning:** *Once the steering wheel has been removed, make SURE that the steering shaft is not turned. If it is, be sure to point the wheels straight ahead again before installing the steering wheel. Failure to do so could cause the airbag system to fail to deploy in an accident.*

### Installation

6  Installation is the reverse of removal, noting the following points.
 a) The front wheels must be pointing straight ahead. If the steering shaft has been accidentally turned, make sure that it's rotated so that the wheels are pointing straight ahead before proceeding.
 b) The contact ring on the on the backside of the steering wheel should remain locked in its centered position when the steering wheel is removed. If the contact ring was accidentally rotated while the wheel was removed, re-center it by rotating it counterclockwise until it stops. Then rotate it clockwise until it stops, counting the number of turns. Divide this number by two and rotate the contact ring counterclockwise by that amount. The ring should now be centered again.
 c) Prior to installation, ensure the indicator switch stalk is in the central (OFF) position. Failure to do so could lead to the steering wheel lug breaking the switch tab.
 d) Coat the steering wheel horn contact ring with a smear of petroleum jelly and install the wheel, making sure the contact unit wiring is correctly routed.
 e) Make sure that the locating pin on the column **(see illustration 17.4b)** engages correctly with the hole in the contact ring as the steering wheel is reinstalled.
 f) Engage the wheel with the column splines, aligning the marks made on removal, and tighten the steering wheel retaining bolt to the specified torque.
 g) Install the airbag unit, plug in the connector for the contact ring and plug in the airbag system connector (see Chapter 25 in Chapter 12). Install the lower steering column cover (see Section 27 in Chapter 11).

## 18  Steering column (3-Series models) - removal, inspection and installation

### Removal

Refer to illustrations 18.6 and 18.10
**Warning:** *If the vehicle is equipped with an airbag, disable the airbag system before starting this procedure (see Chapter 12).*
1  Disconnect the battery negative terminal. **Caution:** *If the radio in your vehicle is equipped with an anti-theft system, make sure you have the correct activation code before disconnecting the battery.*
2  Remove the steering wheel (see Section 17).
3  Remove the steering column combination switches (see Chapter 12).
4  Working in the engine compartment, using paint or a marking pen, make alignment

# Chapter 10 Suspension and steering

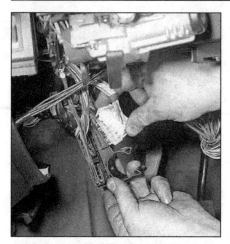

**18.6 Unclip the wiring connectors from the column and disconnect the ignition switch connector**

**18.10 Remove the steering column and recover the rubber mountings and collars**

marks between the lower end of the steering column and the intermediate shaft upper joint.

5   Loosen and remove the nut and clamp bolt, and disengage the shaft from the column.

6   Disconnect the wiring connectors from the ignition switch and free the harness from its retaining clips on the column **(see illustration)**.

7   Release the steering column lower fixing ring by rotating it counterclockwise.

8   Where necessary, loosen and remove the nut and bolt securing the column support bracket to the firewall and withdraw the spacer from the column. Also unscrew the interlock cable (if equipped) from the steering lock.

9   The steering column is secured in position with shear-bolts. The shear-bolts can be extracted using a hammer and chisel to tap the bolt heads around until they can be unscrewed by hand. Alternatively, drill a hole in the center of each bolt head and extract them using a bolt/stud extractor (sometimes called an "easy-out").

10   Pull the column upwards and away from the firewall, and slide off the rubber mounting, mounting seat, washer and fixing ring off from the column lower end. Remove the collars and rubber mountings from the column mountings **(see illustration)**.

## Inspection

11   The steering column incorporates a telescopic safety feature. In the event of a front-end crash, the shaft collapses and prevents the steering wheel injuring the driver. Before installing the steering column, examine the column and mountings for signs of damage and deformation, and replace as necessary.

12   Check the steering shaft for signs of free play in the column bushings. If any damage or wear is found on the steering column bushings, the column should be overhauled. Overhaul of the column is a complex task requiring several special tools, and should be entrusted to a BMW dealer or other qualified repair shop.

## Installation

*Refer to illustrations 18.14, 18.15 and 18.16*
**Note:** *New steering column shear-bolts, and an intermediate shaft clamp bolt nut, will be required on installation.*

13   Ensure the mounting rubbers are in position, and install the collars to the rear of the mounting rubbers.

14   Slide the fixing ring, washer, mounting seat and rubber mounting onto the base of the steering column **(see illustration)**.

15   Maneuver the column into position and engage it with the intermediate shaft splines, aligning the marks made prior to removal **(see illustration)**.

16   Locate the lower end of the column in its seat and screw in the new shear-bolts, but only tighten them lightly at this stage **(see illustration)**.

17   Where necessary, install the spacer, washer and bolt securing the column support bracket to the firewall and install its retaining nut.

18   Secure the lower end of the column in position by rotating the fixing ring clockwise until it clicks into position.

19   Tighten the column shear-bolts evenly until both their heads break off. Where necessary, also securely tighten the support bracket bolt.

20   Reconnect the wiring connectors to the ignition switch, and secure the wiring to the column, ensuring it is correctly routed.

21   Ensure the intermediate shaft and column marks are correctly aligned, and insert the clamp bolt. Install the new clamp bolt nut and tighten it to the specified torque.

22   Where necessary, reconnect the interlock cable to the switch and secure it in position.

23   Install the combination switches (see Chapter 12).

24   Install the steering wheel (see Section 17).

## 19  Ignition switch/steering column lock - removal and installation

### Lock assembly

1   Replacement of the lock assembly requires the steering column to be disassembled. This task requires the use of several special tools, and for this reason should be entrusted to a BMW dealer.

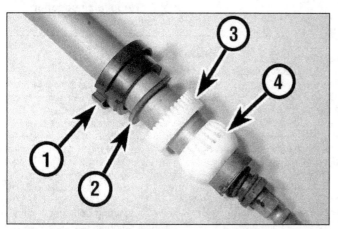

**18.14 Slide the fixing ring (1), washer (2), mounting seat (3) and rubber mounting (4) onto the base of the steering column . . .**

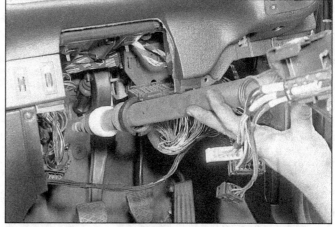

**18.15 . . . and guide the column into place**

# Chapter 10 Suspension and steering

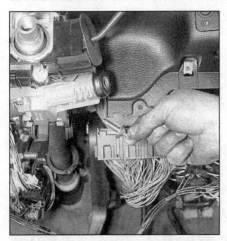

18.16 Seat the column in its lower mounting, and install the new shear-bolts

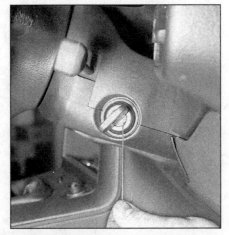

19.3a Position the lock cylinder as shown, and insert the rod into the hole . . .

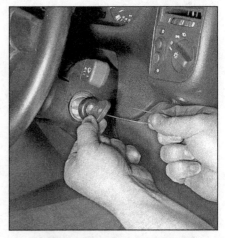

19.3b . . . depress the detent and withdraw the lock cylinder

## Lock cylinder

### Removal

*Refer to illustrations 19.3a and 19.3b*

2 Disconnect the battery negative terminal. **Caution:** *If the radio in your vehicle is equipped with an anti-theft system, make sure you have the correct activation code before disconnecting the battery.* Insert the key into the lock and release the steering lock.

3 Position the lock cylinder as shown, then insert a 1.2 mm (3/64-inch) diameter rod into the hole in the cylinder. Depress the lock cylinder detent, and slide the lock cylinder out of position **(see illustrations)**.

### Installation

4 Position the lock cylinder as shown in Step 3 and insert the cylinder into the housing until it clicks in to position.

## Ignition switch block

### Removal

*Refer to illustration 19.8*

5 Disconnect the battery negative terminal. **Caution:** *If the radio in your vehicle is equipped with an anti-theft system, make sure you have the correct activation code before disconnecting the battery.*

6 Loosen and remove the retaining screws securing the driver's side lower dash panel. Unclip the panel and remove it from the vehicle.

7 Unscrew the steering column shroud lower fastener screw and pull out the fastener. Unclip the lower half of the shroud and remove it from the column.

8 Disconnect the wiring connector from the switch, then unscrew the two set screws and remove the switch block from the lock assembly **(see illustration)**.

### Installation

9 Installation is the reverse of removal, noting the following points:

a) Apply varnish or a non-hardening thread locking compound to the switch set screws prior to installation, to lock them in position.
b) Reconnect the battery and check the operation of the switch prior to installation the steering column shroud.

## 20 Steering column intermediate shaft - removal and installation

### Removal

1 Chock the rear wheels, firmly apply the parking brake, then jack up the front of the vehicle and support on axle stands. Set the front wheels in the straight-ahead position.

### Models with a one-piece intermediate shaft

*Refer to illustration 20.3*

2 Using paint or a suitable marker pen, make alignment marks between the intermediate shaft universal joint and the steering column, the shaft and flexible coupling and the flexible coupling and the steering gear pinion.

3 Loosen and remove the nuts and clamp bolts, then slide the two halves of the shaft together and remove the shaft assembly from the vehicle **(see illustration)**.

4 Inspect the intermediate shaft universal joint for signs of roughness in its bearings and ease of movement. Also examine the shaft rubber coupling for signs of damage or deterioration, and check that the rubber is securely bonded to the flanges. If the universal joint or rubber coupling are suspect, the complete intermediate shaft should be replaced.

### Models with a two-piece intermediate shaft

5 Using paint or a suitable marker pen, make alignment marks between the intermediate shaft universal joint and the steering column, and the flexible coupling and steering gear pinion.

6 Loosen and remove all three intermediate shaft clamp bolts and nuts.

7 Slide the shaft downwards and disen-

19.8 Removing the ignition switch wiring block

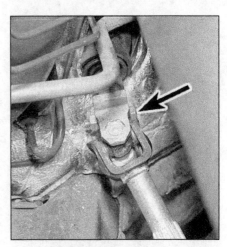

20.3 Steering column-to-intermediate shaft clamp bolt (arrow)

gage its upper end from the steering column. Release its lower end from the rubber coupling.
8  Remove the rubber coupling from the steering gear.
9  Inspect the intermediate shaft universal joint for signs of roughness in its bearings and ease of movement. If the universal joint is worn, replace the intermediate shaft. Examine the rubber coupling for signs of damage or deterioration, and check that the rubber is securely bonded to the flanges. If the rubber coupling is suspect, it should be replaced.

## Installation

**Note:** *New intermediate shaft clamp bolt nuts will be required on installation.*

### Models with a one-piece intermediate shaft

10  Check that the front wheels are still in the straight-ahead position, and that the steering wheel is correctly positioned.
11  Align the marks made on removal, and engage the intermediate shaft joint with the steering column and the coupling with the steering gear.
12  Insert the clamp bolts, then install the new nuts and tighten them to the specified torque. Lower the vehicle to the ground.

### Models with a two-piece intermediate shaft

13  Check that the front wheels are still in the straight-ahead position and the steering wheel is correctly positioned.
14  Align the marks made prior to removal and engage the rubber coupling with the steering gear pinion.
15  Insert the intermediate shaft into the coupling, then engage the shaft joint with the steering gear, aligning the marks made on removal.
16  Ensure all the alignment marks are correctly positioned then insert the clamp bolts.
17  Install a new nut to each clamp bolt, tightening them to the specified torque, and lower the vehicle to the ground.

## 21  Steering gear assembly - removal, overhaul and installation

### Removal

*Refer to illustration 21.5*

1  Chock the rear wheels, firmly apply the parking brake, then jack up the front of the vehicle and support on axle stands. Remove both front wheels.
2  Loosen and remove the nuts securing the steering gear tie-rod ends to the steering knuckles, and release the balljoint tapered shanks using a universal balljoint separator.
3  Using paint or a suitable marker pen, make alignment marks between the intermediate shaft flexible coupling and the steering gear pinion.

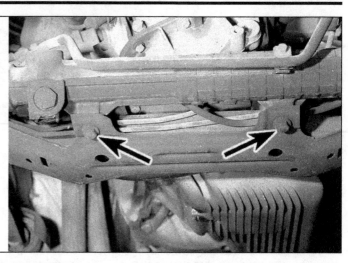

21.5 Steering gear mounting bolts (arrows)

4  On models with power-assisted steering, using brake hose clamps, clamp both the supply and return hoses near the power steering fluid reservoir. This will minimize fluid loss. Mark the fittings to ensure they are correctly positioned on reassembly, then loosen and remove the feed and return pipe fitting bolts and recover the sealing washers. Be prepared for fluid spillage, and position a suitable container beneath the pipes while unscrewing the bolts. Plug the pipe ends and steering gear orifices, to prevent fluid leakage and to keep dirt out of the hydraulic system.
5  Loosen and remove the steering gear mounting bolts and nuts, and remove the steering gear from underneath the vehicle **(see illustration).**

### Overhaul

6  Examine the steering gear assembly for signs of wear or damage, and check that the rack moves freely throughout the full length of its travel, with no signs of roughness or excessive free play between the steering gear pinion and rack. It is not possible to overhaul the steering gear assembly components; if it is faulty, the assembly must be replaced. The only components which can be replaced individually are the steering gear boots, the tie-rod ends and the tie-rods. These procedures are covered later in this Chapter.

### Installation

**Note:** *New tie-rod end nuts, steering gear mounting nuts, and an intermediate shaft clamp bolt nut, will be required on installation. New sealing washers for the fluid pipe fitting bolt will also be needed.*

7  Install up the steering gear, and insert the mounting bolts. Install new nuts to the bolts, and tighten them to the specified torque.
8  On power-assisted steering models, position a new sealing washer on each side of the pipe hose fittings and install the fitting bolts. Tighten the fitting bolts to the specified torque.
9  Align the marks made on removal, and connect the intermediate shaft coupling to the steering gear. Insert the clamp bolt then install the new nut and tighten it to the specified torque.
10  Locate the tie-rod ends in the steering knuckles, then install the new nuts and tighten them to the specified torque.
11  Install the wheels, then lower the vehicle to the ground and tighten the wheel bolts to the specified torque.
12  On models with power steering, bleed the hydraulic system (see Section 23).

## 22  Power steering pump - removal and installation

### Removal

1  Chock the front wheels, then jack up the rear of the vehicle and support it on axle stands.
2  Release the drivebelt tension and unhook the drivebelt from the pump pulley, noting that the pulley retaining bolts should be loosened prior to releasing the tension (see Chapter 1).
3  Unscrew the retaining bolts and remove the pulley from the power steering pump, noting which way it is installed.
4  Using brake hose clamps, clamp both the supply and return hoses near the power steering fluid reservoir. This will minimize fluid loss during subsequent operations.
5  Mark the fittings to ensure they are correctly positioned on reassembly, then loosen and remove the feed and return pipe fitting bolts and recover the sealing washers. Be prepared for fluid spillage, and position a suitable container beneath the pipes while unscrewing the bolts. Plug the pipe ends and steering pump orifices, to prevent fluid leakage and to keep dirt out of the hydraulic system.
6  On 4-cylinder engines, loosen and remove the bolt securing the pump to the adjuster strap and the pump pivot bolt, then remove the pump from the engine unit.
7  On 6-cylinder engines, loosen and remove the bolts securing the power steering pump front and rear mounting brackets in position, and remove the pump from the engine.

## Chapter 10 Suspension and steering

8 If the power steering pump is faulty, seek the advice of your BMW dealer as to the availability of spare parts. If spares are available, it may be possible to have the pump overhauled by a suitable specialist, or alternatively obtain an exchange unit. If not, the pump must be replaced.

### Installation

**Note:** *New feed pipe fitting bolt sealing washers will be required on installation.*

9 Where necessary, transfer the rear mounting bracket to the new pump, and securely tighten its mounting bolts.
10 Prior to installation, ensure that the pump is primed by injecting the specified type of fluid in through the supply hose fitting and rotating the pump shaft.
11 On 4-cylinder engines, maneuver the pump into position and install the pivot bolt and adjuster strap bolt, tightening them lightly only at this stage.
12 On 6-cylinder engines, maneuver the pump into position and install its mounting bolts, tightening them to the specified torque.
13 Position a new sealing washer on each side of the pipe hose fittings and install the fitting bolts. Tighten the fitting bolts to the specified torque.
14 Remove the hose clamps and install the pump pulley. Ensure the pulley is the right way around, and securely tighten its retaining bolts.
15 Install the drivebelt and adjust the drivebelt tension (see Chapter 1).
16 On completion, lower the car to the ground and bleed the hydraulic system (see Section 23).

### 23 Power steering system - bleeding

1 With the engine stopped, fill the fluid reservoir right up to the top with the specified type of fluid.
2 With the engine stopped, slowly move the steering from lock-to-lock several times to purge out the trapped air, then top-up the level in the fluid reservoir. Repeat this procedure until the fluid level in the reservoir does not drop any further.
3 Have an assistant start the engine, while you keep watch on the fluid level. Be prepared to add more fluid as the engine starts as the fluid level is likely to drop quickly. The fluid level must be kept above the "MIN" mark at all times.
4 With the engine running at idle speed, turn the steering wheel slowly two or three times approximately 45-degrees to the left and right of the center, then turn the wheel twice from lock to lock. Do not hold the wheel on either lock, as this imposes strain on the hydraulic system. Repeat this procedure until bubbles cease to appear in fluid reservoir.
5 If, when turning the steering, an abnormal noise is heard from the fluid lines, it indicates that there is still air in the system. Check this by turning the wheels to the straight-ahead position and switching off the engine. If the fluid level in the reservoir rises, then air is present in the system and further bleeding is necessary.
6 Once all traces of air have been removed from the power steering hydraulic system, turn the engine off and allow the system to cool. Once cool, check that fluid level is up to the maximum mark on the power steering fluid reservoir, topping-up if necessary (see Chapter 1).

### 24 Steering gear boots - replacement

1 Remove the tie-rod end (see Section 25).
2 Note the correct installed position of the boot on the tie-rod, then release the retaining clip(s) and slide the boot off the steering gear housing and tie-rod end.
3 Thoroughly clean the tie-rod and the steering gear housing, using fine abrasive paper to polish off any corrosion, burrs or sharp edges, which might damage the new boot's sealing lips on installation. Scrape off all the grease from the old boot, and apply it to the tie-rod inner balljoint. (This assumes that grease has not been lost or contaminated as a result of damage to the old boot. Use fresh grease if in doubt.)
4 Carefully slide the new boot onto the tie-rod end, and locate it on the steering gear housing. Position the outer edge of the boot on the tie-rod, as was noted prior to removal.
5 Make sure the boot is not twisted, then lift the outer sealing lip of the boot to equalize air pressure within the boot. Secure the boot in position with the new retaining clip(s).
6 Install the tie-rod end (see Section 25).

### 25 Tie-rod end - removal and installation

#### Removal

1 Apply the parking brake, then jack up the front of the vehicle and support it on axle stands. Remove the appropriate front wheel.
2 Make a mark on the tie-rod and measure the distance from the mark to the center of the balljoint. Note this measurement down, as it will be needed to ensure the wheel alignment remains correctly set when the balljoint is installed.
3 Hold the tie-rod and unscrew the balljoint locknut.
4 Loosen and remove the nut securing the tie-rod end to the steering knuckle, and release the balljoint tapered shank using a universal balljoint separator.
5 Counting the exact number of turns necessary to do so, unscrew the balljoint from the tie-rod end.
6 Carefully clean the balljoint and the threads. Replace the balljoint if its movement is sloppy or too stiff, if excessively worn, or if damaged in any way; carefully check the stud taper and threads. If the balljoint boot is damaged, the complete balljoint assembly must be replaced; it is not possible to obtain the boot separately.

#### Installation

**Note:** *A new balljoint retaining nut will be required on installation.*

7 If necessary, transfer the locknut and collar to the new tie-rod end.
8 Screw the balljoint onto the tie-rod by the number of turns noted on removal. This should position the balljoint at the relevant distance from the tie-rod mark that was noted prior to removal. Tighten the locknut securely.
9 Install the balljoint shank to the steering knuckle, then install a new retaining nut and tighten it to the specified torque.
10 Install the wheel, then lower the vehicle to the ground and tighten the wheel bolts to the specified torque.
11 Have the front end alignment checked and, if necessary, adjusted.

### 26 Tie-rod - replacement

**Note:** *A new tie-rod locking plate and steering boot retaining clip(s) will be required.*

1 Remove the steering gear boot (see Section 24).
2 Using a pair of pliers, bend back the tie-rod locking plate. **Note:** *Do not use a hammer and chisel, as this could damage the steering gear.*
3 Unscrew the tie-rod from the end of the steering rack.
4 Install the new locking plate to the steering rack, making sure its locating tabs are correctly seated in the steering rack grooves.
5 Screw in the tie-rod and tighten it to the specified torque. Secure the tie-rod in position by bending down the locking plate with a pair of pliers.
6 Install the steering gear boot (see Section 24).

### 27 Wheel alignment - general information

1 A car's steering and suspension geometry is defined in four basic settings - all angles are expressed in degrees (toe settings are also expressed as a measurement); the steering axis is defined as an imaginary line drawn through the axis of the suspension strut, extended where necessary to contact the ground.
2 Camber is the angle between each wheel and a vertical line drawn through its center and tire contact patch, when viewed from the front or rear of the car. Positive camber is when the wheels are tilted outwards from the vertical at the top; negative camber is when they are tilted inwards.

3 The front camber angle is not adjustable, and is given for reference only (see Step 5). The rear camber angle is adjustable and can be adjusted using a camber angle gauge.

4 Caster is the angle between the steering axis and a vertical line drawn through each wheel's center and tire contact patch, when viewed from the side of the car. Positive caster is when the steering axis is tilted so that it contacts the ground ahead of the vertical; negative caster is when it contacts the ground behind the vertical.

5 Caster is not adjustable, and is given for reference only; while it can be checked using a caster checking gauge, if the figure obtained is significantly different from that specified, the vehicle must be taken for careful checking by a professional, as the fault can only be caused by wear or damage to the body or suspension components.

6 Toe is the difference, viewed from above, between lines drawn through the wheel centers and the car's center-line. "Toe-in" is when the wheels point inwards, towards each other at the front, while "toe-out" is when they splay outwards from each other at the front.

7 The front wheel toe setting is adjusted by screwing the tie-rod in or out of its balljoint, to alter the effective length of the tie-rod assembly.

8 Rear wheel toe setting is also adjustable. The toe setting is adjusted by loosening the trailing arm mounting bracket bolts and repositioning the bracket.

# Chapter 11  Body

## Contents

| Section | | Section |
|---|---|---|
| Body exterior fittings - removal and installation | 23 | Front seat belt tensioning mechanism - general information ... 25 |
| Bumpers - removal and installation | 6 | General information ... 1 |
| Center console - removal and installation | 28 | Grille and front panel - removal and installation ... 19 |
| Dash panel assembly - removal and installation | 29 | Hood lock(s) - removal and installation ... 10 |
| Dash trim panels - removal and installation | 27 | Hood release cable - removal and installation ... 9 |
| Door handle and lock components - removal and installation | 13 | Hood - removal, installation and adjustment ... 8 |
| Door - removal, installation and adjustment | 11 | Maintenance - bodywork and underframe ... 2 |
| Door trim panel - removal and installation | 12 | Maintenance - upholstery and carpets ... 3 |
| Door window glass and regulator and rear vent window (3-Series coupe models) - removal and installation | 15 | Major body damage - repair ... 5 |
| | | Minor body damage - repair ... 4 |
| Door window glass and regulator (3-Series sedan models) - removal and installation | 14 | Seat belt components - removal and installation ... 26 |
| | | Seats - removal and installation ... 24 |
| Door window glass and regulator (Z3 models) - removal, installation and adjustment | 16 | Sunroof (3-Series models) - general information ... 22 |
| | | Trunk lid and support struts - removal and installation ... 17 |
| Exterior mirrors and associated components - removal and installation | 20 | Trunk lid lock components - removal and installation ... 18 |
| | | Windshield and back glass - general information ... 21 |
| Fenders - removal and installation | 7 | |

## Specifications

### Torque specifications

| | |
|---|---|
| Door exterior handle nut/bolt | 84 in-lbs |
| Door lock retaining bolts | 72 in-lbs |
| Door window glass and regulator mounting bolts (sedan models) | |
|     Regulator stop bolt | 72 in-lbs |
|     Regulator mounting bolts | 48 in-lbs |
|     Window guide bolts | 84 in-lbs |
| Exterior mirror bolts | 48 in-lbs |
| Front seat belt height adjustment mechanism bolts | 18 ft-lbs |
| Front seat mounting nuts/bolts | 41 ft-lbs |
| Rear vent window hinge bolts/screw - coupe models | 48 in-lbs |
| Seat belt mounting bolts | 35 ft-lbs |

## 1  General information

The bodyshell is made of pressed-steel sections. Most components are welded together, but some use is made of structural adhesives.

The hood, door and some other vulnerable panels are made of zinc-coated metal, and are further protected by being coated with an anti-chip primer before being sprayed.

Extensive use is made of plastic materials, mainly in the interior, but also in exterior components. The front and rear bumpers and front grille are injection-molded from a synthetic material that is very strong and yet light. Plastic components such as wheel arch liners are fitted to the underside of the vehicle to improve the body's resistance to corrosion.

## 2  Maintenance - bodywork and underframe

1  The general condition of a vehicle's bodywork is the one thing that significantly affects its value. Maintenance is easy, but needs to be regular. Neglect, particularly after minor damage, can lead quickly to further deterioration and costly repair bills. It is important also to keep watch on those parts of the vehicle not immediately visible, for instance the underside, inside all the wheel arches, and the lower part of the engine compartment.

2  The basic maintenance routine for the bodywork is washing - preferably with a lot of water, from a hose. This will remove all the loose solids which may have stuck to the vehicle. It is important to flush these off in such a way as to prevent grit from scratching the finish. The wheel arches and underframe need washing in the same way, to remove any accumulated mud which will retain moisture and tend to encourage rust. Oddly enough, the best time to clean the underframe and wheel arches is in wet weather, when the mud is thoroughly wet and soft. In very wet weather, the underframe is usually cleaned of large accumulations automatically, and this is a good time for inspection.

3  Periodically, except on vehicles with a wax-based underbody protective coating, it is a good idea to have the whole of the underframe of the vehicle steam-cleaned, engine compartment included, so that a thorough inspection can be carried out to see what minor repairs and renovations are necessary. Steam cleaning is available at many garages, and is necessary for the removal of the accumulation of oily grime, which sometimes is allowed to become thick in certain

areas. If steam-cleaning facilities are not available, there are some excellent grease solvents available which can be brush-applied; the dirt can then be simply hosed off. Note that these methods should not be used on vehicles with wax-based underbody protective coating, or the coating will be removed. Such vehicles should be inspected annually, preferably just before winter, when the underbody should be washed down, and repair any damage to the wax coating. Ideally, a completely fresh coat should be applied. It would also be worth considering the use of such wax-based protection for injection into door panels, sills, box sections, etc., as an additional safeguard against rust damage, where such protection is not provided by the vehicle manufacturer.

4   After washing paintwork, wipe off with a chamois leather to give an unspotted clear finish. A coat of clear protective wax polish will give added protection against chemical pollutants in the air. If the paintwork sheen has dulled or oxidized, use a cleaner/polisher combination to restore the brilliance of the shine. This requires a little effort, but such dulling is usually caused because regular washing has been neglected. Care needs to be taken with metallic paintwork, as special non-abrasive cleaner/polisher is required to avoid damage to the finish. Always check that the door and ventilator opening drain holes and pipes are completely clear, so that water can be drained out. Chrome should be treated in the same way as paintwork. The windshield and windows can be kept clear of the smeary film which often appears, by proprietary glass cleaner. Never use any form of wax or other body or chrome polish on glass.

### 3   Maintenance - upholstery and carpets

Mats and carpets should be brushed or vacuum-cleaned regularly, to keep them free of grit. If they are badly stained, remove them from the vehicle for scrubbing or sponging, and make quite sure they are dry before installation. Seats and interior trim panels can be kept clean by wiping with a damp cloth and a proprietary brand of cleaner. If they do become stained (which can be more apparent on light-colored upholstery), use a little liquid detergent and a soft nail brush to scour the grime out of the grain of the material. Do not forget to keep the headlining clean in the same way as the upholstery. When using liquid cleaners inside the vehicle, do not overwet the surfaces being cleaned. Excessive moisture could get into the seams and padded interior, causing stains, offensive odors or even rot. If the inside of the vehicle gets wet accidentally, it is worthwhile taking some trouble to dry it out properly, particularly where carpets are involved. *Do not leave oil or electric heaters inside the vehicle for this purpose.*

### 4   Minor body damage - repair

## *Repairs of minor scratches in bodywork*

1   If the scratch is very superficial, and does not penetrate to the metal of the bodywork, repair is very simple. Lightly rub the area of the scratch with a paintwork renovator or a very fine cutting paste to remove loose paint from the scratch, and to clear the surrounding bodywork of wax polish. Rinse the area with clean water.

2   Apply touch-up paint to the scratch using a fine paint brush; continue to apply fine layers of paint until the surface of the paint in the scratch is level with the surrounding paintwork. Allow the new paint at least two weeks to harden, then blend it into the surrounding paintwork by rubbing the scratch area with a paintwork renovator or a very fine cutting paste. Finally, apply wax polish.

3   Where the scratch has penetrated right through to the metal of the bodywork, causing the metal to rust, a different repair technique is required. Remove any loose rust from the bottom of the scratch with a penknife, then apply rust-inhibiting paint to prevent the formation of rust in the future. Using a rubber or nylon applicator, fill the scratch with glaze-type filler. If required, this paste can be mixed with cellulose thinners to provide a very thin paste which is ideal for filling narrow scratches. Before the stopper-paste in the scratch hardens, wrap a piece of smooth cotton rag around the top of a finger. Dip the finger in cellulose thinners, and quickly sweep it across the surface of the stopper-paste in the scratch; this will ensure that the surface of the stopper-paste is slightly hollowed. The scratch can now be painted over as described earlier in this Section.

## *Repairs of dents in bodywork*

4   When deep denting of the vehicle's bodywork has taken place, the first task is to pull the dent out, until the affected bodywork almost attains its original shape. There is little point in trying to restore the original shape completely, as the metal in the damaged area will have stretched on impact, and cannot be reshaped fully to its original contour. It is better to bring the level of the dent up to a point which is about 3 mm below the level of the surrounding bodywork. In cases where the dent is very shallow anyway, it is not worth trying to pull it out at all. If the underside of the dent is accessible, it can be hammered out gently from behind, using a mallet with a wooden or plastic head. While doing this, hold a suitable block of wood firmly against the outside of the panel, to absorb the impact from the hammer blows and thus prevent a large area of the bodywork from being "belled-out".

5   Should the dent be in a section of the bodywork which has a double skin, or some other factor making it inaccessible from behind, a different technique is called for. Drill several small holes through the metal inside the area - particularly in the deeper section. Then screw long self-tapping screws into the holes, just sufficiently for them to gain a good purchase in the metal. Now the dent can be pulled out by pulling on the protruding heads of the screws with a pair of pliers.

6   The next stage of the repair is the removal of the paint from the damaged area, and from an inch or so of the surrounding "sound" bodywork. This is accomplished most easily by using a wire brush or abrasive pad on a power drill, although it can be done just as effectively by hand, using sheets of abrasive paper. To complete the preparation for filling, score the surface of the bare metal with a screwdriver or the tang of a file, or alternatively, drill small holes in the affected area. This will provide a good "key" for the filler paste.

7   To complete the repair, see the Section on filling and respraying.

## *Repairs of rust holes or gashes in bodywork*

8   Remove all paint from the affected area, and from an inch or so of the surrounding "sound" bodywork, using an abrasive pad or a wire brush on a power drill. If these are not available, a few sheets of abrasive paper will do the job most effectively. With the paint removed, you will be able to judge the severity of the corrosion, and therefore decide whether to replace the whole panel (if this is possible) or to repair the affected area. New body panels are not as expensive as most people think, and it is often quicker and more satisfactory to fit a new panel than to attempt to repair large areas of corrosion.

9   Remove all fittings from the affected area, except those which will act as a guide to the original shape of the damaged bodywork (e.g. headlamp shells etc.). Then, using tin snips or a hacksaw blade, remove all loose metal and any other metal badly affected by corrosion. Hammer the edges of the hole inwards, to create a slight depression for the filler paste.

10   Wire-brush the affected area to remove the powdery rust from the surface of the remaining metal. Paint the affected area with rust-inhibiting paint; if the back of the rusted area is accessible, treat this also.

11   Before filling can take place, it will be necessary to block the hole in some way. This can be achieved with aluminum or plastic mesh, or aluminum tape.

12   Aluminum or plastic mesh, or fiberglass matting, is probably the best material to use for a large hole. Cut a piece to the approximate size and shape of the hole to be filled, then position it in the hole so that its edges are below the level of the surrounding bodywork. It can be retained in position by several blobs of filler paste around its periphery.

13   Aluminum tape should be used for small or very narrow holes. Pull a piece off the roll, trim it to the approximate size and shape required, then pull off the backing paper (if

# Chapter 11  Body

used) and stick the tape over the hole; it can be overlapped if the thickness of one piece is insufficient. Burnish down the edges of the tape with the handle of a screwdriver or similar, to ensure that the tape is securely attached to the metal underneath.

## Bodywork repairs - filling and respraying

14  Before using this Section, see the Sections on dent, deep scratch, rust holes and gash repairs.

15  Many types of bodyfiller are available, but generally speaking, those proprietary kits which contain a tin of filler paste and a tube of resin hardener are best for this type of repair which can be used directly from the tube. A wide, flexible plastic or nylon applicator will be found invaluable for imparting a smooth and well-contoured finish to the surface of the filler.

16  Mix up a little filler on a clean piece of card or board - measure the hardener carefully (follow the maker's instructions on the pack), otherwise the filler will set too rapidly or too slowly. Using the applicator, apply the filler paste to the prepared area; draw the applicator across the surface of the filler to achieve the correct contour and to level the surface. When a contour that approximates to the correct one is achieved, stop working the paste - if you carry on too long, the paste will become sticky and begin to "pick-up" on the applicator. Continue to add thin layers of filler paste at 20-minute intervals, until the level of the filler is just proud of the surrounding bodywork.

17  Once the filler has hardened, the excess can be removed using a metal plane or file. From then on, progressively-finer grades of abrasive paper should be used, starting with a 40-grade production paper, and finishing with a 400-grade wet-and-dry paper. Always wrap the abrasive paper around a flat rubber, cork, or wooden block - otherwise the surface of the filler will not be completely flat. During the smoothing of the filler surface, the wet-and-dry paper should be periodically rinsed in water. This will ensure that a very smooth finish is imparted to the filler at the final stage.

18  At this stage, the "dent" should be surrounded by a ring of bare metal, which in turn should be encircled by the finely "feathered" edge of the good paintwork. Rinse the repair area with clean water, until all the dust produced by the rubbing-down operation has gone.

19  Spray the whole area with a light coat of primer - this will show up any imperfections in the surface of the filler. Repair these imperfections with fresh filler paste or bodystopper, and again smooth the surface with abrasive paper. If bodystopper is used, it can be mixed with cellulose thinners, to form a thin paste which is ideal for filling small holes. Repeat this spray-and-repair procedure until you are satisfied that the surface of the filler, and the feathered edge of the paintwork, are perfect. Clean the repair area with clean water, and allow to dry fully.

20  The repair area is now ready for final spraying. Paint spraying must be carried out in a warm, dry, windless and dust-free atmosphere. This condition can be created artificially if you have access to a large indoor working area, but if you are forced to work in the open, you will have to pick your day very carefully. If you are working indoors, dousing the floor in the work area with water will help to settle the dust which would otherwise be in the atmosphere. If the repair area is confined to one body panel, mask off the surrounding panels; this will help to minimize the effects of a slight mis-match in paint colors. Bodywork fittings (e.g. chrome strips, door handles etc.) will also need to be masked off. Use genuine masking tape, and several thickness of newspaper, for the masking operations.

21  Before starting to spray, agitate the aerosol can thoroughly, then spray a test area (an old tin, or similar) until the technique is mastered. Cover the repair area with a thick coat of primer; the thickness should be built up using several thin layers of paint, rather than one thick one. Using 400 grade wet-and-dry paper, rub down the surface of the primer until it is smooth. While doing this, the work area should be thoroughly doused with water, and the wet-and-dry paper periodically rinsed in water. Allow to dry before spraying on more paint.1

22  Spray on the top coat, again building up the thickness by using several thin layers of paint. Start spraying in the center of the repair area, and then, using a circular motion, work outwards until the whole repair area and about 2 inches of the surrounding original paintwork is covered. Remove all masking material 10 to 15 minutes after spraying on the final coat of paint.

23  Allow the new paint at least two weeks to harden, then, using a paintwork renovator or a very fine cutting paste, blend the edges of the paint into the existing paintwork. Finally, apply wax polish.

## Plastic components

24  With the use of more and more plastic body components by the vehicle manufacturers (e.g. bumpers. spoilers, and in some cases major body panels), rectification of more serious damage to such items has become a matter of either entrusting repair work to a specialist in this field, or renewing complete components. Repair of such damage by the home mechanic is not feasible, owing to the cost of the equipment and materials required for effecting such repairs. The basic technique involves making a groove along the line of the crack in the plastic, using a rotary burr in a power drill. The damaged part is then welded back together, using a hot air gun to heat up and fuse a plastic filler rod into the groove. Any excess plastic is then removed, and the area rubbed down to a smooth finish. It is important that a filler rod of the correct plastic is used, as body components can be made of a variety of different types (e.g. polycarbonate, ABS, polypropylene).

25  Damage of a less serious nature (abrasions, minor cracks etc.) can be repaired by the home mechanic using a two-part epoxy filler repair material which can be used directly from the tube. Once mixed in equal proportions, this is used in similar fashion to the bodywork filler used on metal panels. The filler is usually cured in twenty to thirty minutes, ready for sanding and painting.

26  If the owner is renewing a complete component himself, or if he has repaired it with epoxy filler, he will be left with the problem of finding a suitable paint for finishing which is compatible with the type of plastic used. At one time, the use of a universal paint was not possible, owing to the complex range of plastics met with in body component applications. Standard paints, generally speaking, will not bond to plastic or rubber satisfactorily, but professional matched paints, to match any plastic or rubber finish, can be obtained from some dealers. However, it is now possible to obtain a plastic body parts finishing kit which consists of a pre-primer treatment, a primer and colored top coat. Full instructions are normally supplied with a kit, but basically the method of use is to first apply the pre-primer to the component concerned, and allow it to dry for up to 30 minutes. Then the primer is applied, and left to dry for about an hour before finally applying the special-colored top coat. The result is a correctly colored component, where the paint will flex with the plastic or rubber, a property that standard paint does not normally possess.

## 5  Major body damage - repair

Where serious damage has occurred, or large areas need replacement due to neglect, it means that complete new panels will need welding-in, and this is best left to professionals. If the damage is due to impact, it will also be necessary to check completely the alignment of the bodyshell, and this can only be carried out accurately by a body repair shop using special jigs. If the body is left misaligned, it is primarily dangerous, as the car will not handle properly, and secondly, uneven stresses will be imposed on the steering, suspension and possibly transmission, causing abnormal wear, or complete failure, particularly to such items as the tires.

## 6  Bumpers - removal and installation

### 3-Series models

#### Front bumper

*Refer to illustrations 6.2a, 6.2b, 6.5, 6.6a and 6.6b*

1  Apply the parking brake, then jack up the front of the vehicle and support it on axle stands.

These photos illustrate a method of repairing simple dents. They are intended to supplement *Body repair - minor damage* in this Chapter and should not be used as the sole instructions for body repair on these vehicles.

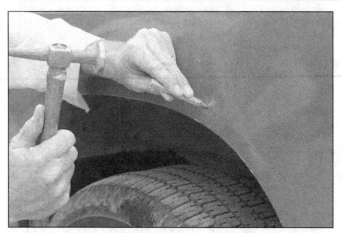

1 If you can't access the backside of the body panel to hammer out the dent, pull it out with a slide-hammer-type dent puller. In the deepest portion of the dent or along the crease line, drill or punch hole(s) at least one inch apart . . .

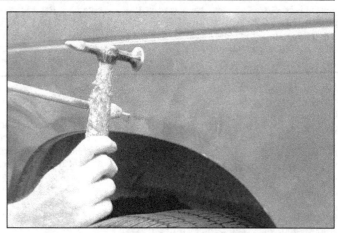

2 . . . then screw the slide-hammer into the hole and operate it. Tap with a hammer near the edge of the dent to help 'pop' the metal back to its original shape. When you're finished, the dent area should be close to its original contour and about 1/8-inch below the surface of the surrounding metal

3 Using coarse-grit sandpaper, remove the paint down to the bare metal. Hand sanding works fine, but the disc sander shown here makes the job faster. Use finer (about 320-grit) sandpaper to feather-edge the paint at least one inch around the dent area

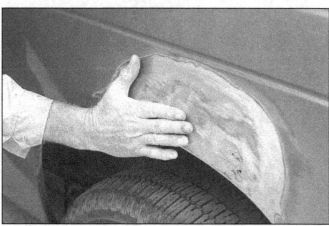

4 When the paint is removed, touch will probably be more helpful than sight for telling if the metal is straight. Hammer down the high spots or raise the low spots as necessary. Clean the repair area with wax/silicone remover

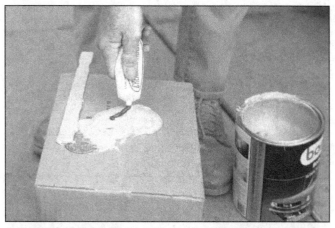

5 Following label instructions, mix up a batch of plastic filler and hardener. The ratio of filler to hardener is critical, and, if you mix it incorrectly, it will either not cure properly or cure too quickly (you won't have time to file and sand it into shape)

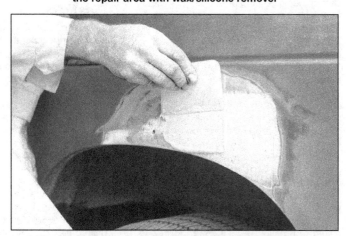

6 Working quickly so the filler doesn't harden, use a plastic applicator to press the body filler firmly into the metal, assuring it bonds completely. Work the filler until it matches the original contour and is slightly above the surrounding metal

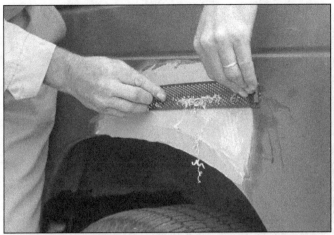

7 Let the filler harden until you can just dent it with your fingernail. Use a body file or Surform tool (shown here) to rough-shape the filler

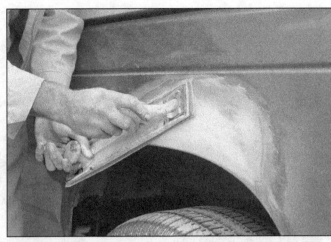

8 Use coarse-grit sandpaper and a sanding board or block to work the filler down until it's smooth and even. Work down to finer grits of sandpaper - always using a board or block - ending up with 360 or 400 grit

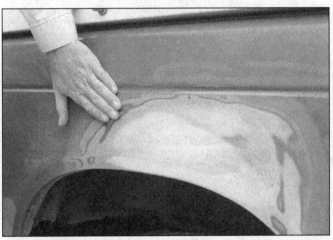

9 You shouldn't be able to feel any ridge at the transition from the filler to the bare metal or from the bare metal to the old paint. As soon as the repair is flat and uniform, remove the dust and mask off the adjacent panels or trim pieces

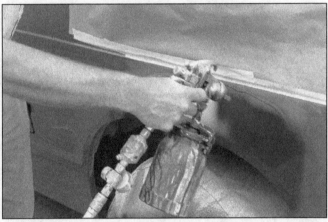

10 Apply several layers of primer to the area. Don't spray the primer on too heavy, so it sags or runs, and make sure each coat is dry before you spray on the next one. A professional-type spray gun is being used here, but aerosol spray primer is available inexpensively from auto parts stores

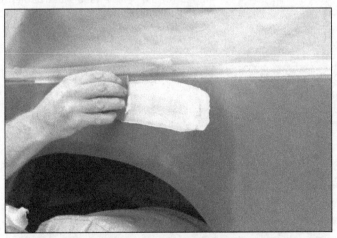

11 The primer will help reveal imperfections or scratches. Fill these with glazing compound. Follow the label instructions and sand it with 360 or 400-grit sandpaper until it's smooth. Repeat the glazing, sanding and respraying until the primer reveals a perfectly smooth surface

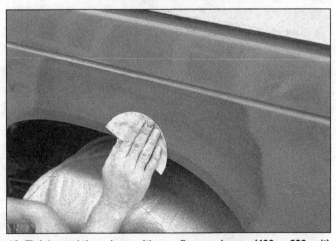

12 Finish sand the primer with very fine sandpaper (400 or 600-grit) to remove the primer overspray. Clean the area with water and allow it to dry. Use a tack rag to remove any dust, then apply the finish coat. Don't attempt to rub out or wax the repair area until the paint has dried completely (at least two weeks)

## Chapter 11 Body

**6.2a Unclip the trim strip from the left-hand end of the bumper . . .**

**6.2b . . . and remove the towing eye hole cover**

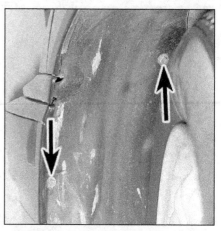

**6.5 Remove the screws securing the wheel arch liner in position (arrows)**

2  To gain access to the bumper mounting nuts, using a flat-bladed screwdriver, carefully unclip the rubber trim strip from the left-hand end of the bumper and the towing eye cover **(see illustrations)**. Take great care not to the damage the painted finish of the bumper.

3  On models with front foglights, remove the access covers from the base of the bumper and disconnect the wiring connectors from the foglights.

4  Where necessary, disconnect the wiring connector(s) from the temperature sensor/thermostatic switch which is/are clipped into the brake cooling duct(s).

5  Loosen and remove the bumper end retaining screws which are accessed from underneath the left and right-hand wheel arches **(see illustration)**.

6  Unscrew the bumper mounting nuts and remove the bumper forwards and away from the vehicle **(see illustrations)**.

7  Inspect the bumper mountings for signs of damage and replace if necessary.

8  Installation is a reverse of the removal procedure, ensuring that the bumper mounting nuts and screws are securely tightened.

### Rear bumper

*Refer to illustrations 6.10a, 6.10b, 6.12a and 6.12b*

9  To improve access, chock the front wheels, then jack up the rear of the vehicle and support it on axle stands.

10  Remove the fasteners and carefully unclip the panel from the base of the rear bumper. The fasteners are released by pulling out their center pins **(see illustrations)**.

11  Undo the screws securing the wheel arch liners to the bumper ends. Pry out the retaining clip center pins and remove the clips securing the liner to the bumper.

12  Loosen and remove the bumper lower mounting bolts and remove the bumper from the rear of the vehicle **(see illustrations)**. Where necessary, disconnect the wiring connectors from the bumper distance sensors.

13  Installation is a reverse of the removal procedure ensuring that the bumper ends are

**6.6a Loosen and remove the mounting nuts (arrows) . . .**

**6.6b . . . and remove the bumper from the vehicle**

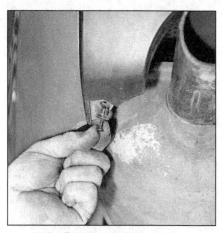

**6.10a Remove the fastener from each end . . .**

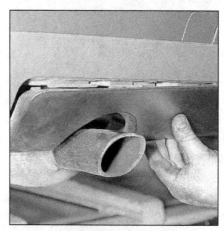

**6.10b . . . and unclip the trim panel from the base of the rear bumper**

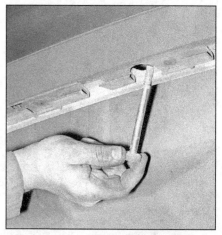

**6.12a Unscrew the mounting bolts . . .**

# Chapter 11 Body

6.12b ...and remove the bumper from the vehicle

6.15a To detach the engine splash shield, remove these three bolts (arrows) from each end (left end of splash shield shown)

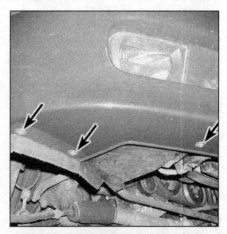

6.15b To remove the two filler panels from the bumper cover, remove these three bolts (arrows) from each panel (right filler panel shown)

correctly engaged with their slides. Apply locking compound to the bumper mounting bolts and tighten them securely.

## Z3 models

**Note:** *The front and rear bumper covers are each attached to the bumper by 16 rivets. If either cover is removed from the bumper, some of these rivets might be damaged. It's a good idea to have a supply of new rivets handy for reattaching the cover to the bumper. Do not reuse damaged rivets.*

### Front bumper

*Refer to illustrations 6.15a, 6.15b, 6.16, 6.17, 6.19a and 6.19b*

14  Raise the hood. Loosen the wheel bolts, raise the front of the vehicle and support it securely on jackstands. Remove the front wheels.
15  Remove the three mounting screws from each end of the engine splash shield and then remove the two filler panels **(see illustrations)**.
16  Unplug the electrical connectors from the fog lights **(see illustration)**.
17  Remove the left and right Torx 50 bumper mounting bolts from the bumper mounts **(see illustration)**.

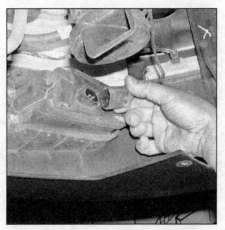

6.16 Unplug the electrical connectors from the fog lights (right fog light connector shown)

6.17 To detach the front bumper/bumper cover assembly from the vehicle, remove the two Torx 50 bumper mounting bolts (right bolt shown)

18  Pull the bumper cover/bumper assembly forward to separate it from the bumper mounts.
19  To detach the bumper cover from the bumper, remove the 16 rivets, eight on top, eight on the bottom **(see illustrations)**.
20  Installation is the reverse of removal. Be

sure to replace any damaged rivets. After the bumper has been installed, check the gap between the bumper cover and the body. The gap should be about 5 mm and it should be even. If it isn't, adjust the bumper (see below).

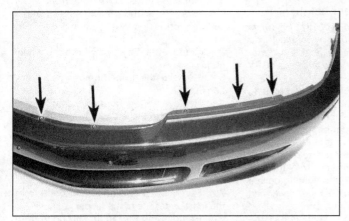

6.19a The front bumper cover is attached to the front bumper by 16 rivets, eight on the top (arrows)...

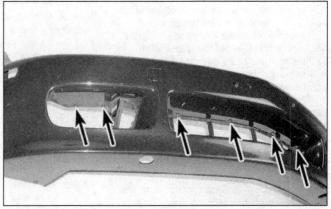

6.19b ...and the eight on the bottom (arrows) (not all rivets visible in these photos)

6.21 To detach the license plate support, remove these four retaining screws (arrows)

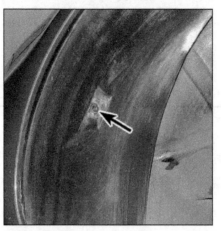

6.23 To detach the wheel liners from the rear bumper assembly, remove this rivet (arrow) from each liner

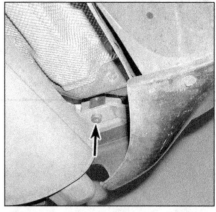

6.25 To detach the rear bumper assembly from the vehicle, remove the Torx 50 bumper mounting bolt (arrow) from the underside of each end of the bumper (left bolt shown)

7.12a To detach the inner liner from the front wheel well, remove these fasteners (arrows) . . .

7.12b . . . and these (arrows)

### Rear bumper

*Refer to illustrations 6.21, 6.23 and 6.25*

21  Remove the license plate support retaining screws **(see illustration)** and remove the license plate lights from the bumper.
22  Loosen the rear wheel bolts, raise the rear of the vehicle and place it securely on jackstands. Remove the rear wheels.
23  Remove the rivets that attach the wheel well liners to the bumper **(see illustration)**.
24  Remove the bolts that attach the wheel well liners to the rear bumper cover **(see illustration 7.20)**.
25  Remove the left and right Torx 50 bumper mounting bolts from the underside of the bumper **(see illustration)**.
26  Have an assistant push down on the bumper cover while you pull the bumper to the rear and slide it out from under the taillights.
27  To detach the bumper cover from the bumper, remove the 16 rivets (eight on top, eight on bottom).
28  Installation is the reverse of removal. Be sure to replace any damaged rivets. After the bumper has been installed, check the gap between the bumper cover and the body. The gap should be about 5 mm and it should be even. If it isn't, adjust the bumper (see below).

### Adjusting the bumper height

**Note:** *A special BMW tool (51 1 130) is needed to adjust the bumper height. If you don't have this tool or a suitable substitute, have the bumper height adjusted by a BMW dealer.*

29  Remove the Torx 50 bumper mounting bolts, attach the BMW special tool or a suitable substitute to a 12-inch extension, and insert the tool up through the recess for the bumper mount. Turn the tool clockwise or counterclockwise as necessary to change the height of the adjusting collar. Alternate between the left and right adjusting collars until the gap between the bumper cover and the body is about 5 mm and even.
30  Remove the special tool, install the bumper mounting bolts and tighten them securely.

### 7  Fenders - removal and installation

### 3-Series models

1  Raise the hood. Loosen the front wheel lug nuts. Raise the vehicle and place it securely on jackstands. Remove the front wheel.
2  Remove the inner liner from the wheel well.
3  Remove the front bumper (see Section 6).
4  Remove the four front grille panel retaining bolts (two at the left end, two at the right) from underneath the ends of the bumper assembly.
5  Remove the front turn signal assembly and the side marker light assembly (see Chapter 12).
6  Remove the two mounting bolts from the front underside of the fender.
7  Remove the four mounting bolts from the top edge of the fender. Loosen the fifth bolt, near the windshield.

# Chapter 11 Body

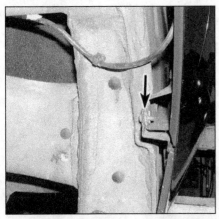

7.14 Remove the retaining bolt (arrow) from the inner front of the fender

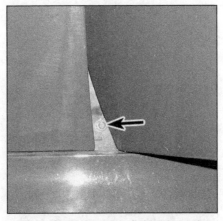

7.15 Open the door all the way and remove the retaining screw (arrow) from the bottom of the fender

7.16 Remove the two upper fender mounting bolts (arrows)

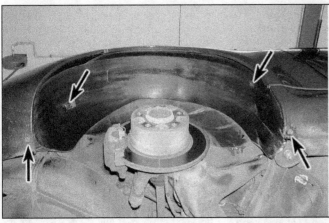

7.20 To detach the rear wheel well inner liner, remove these fasteners (arrows)

7.21 Scribe the location of the trunk lid hinge onto the fender and then remove the hinge

9   Open the front door. Remove the two mounting bolts from the doorjamb. Remove the fender.
10  Installation is the reverse of removal. Before installing the fender, be sure to remove all of the old protective coating and sealant from the mounting surfaces. Don't tighten any of the fender mounting bolts until all of them have been loosely installed and the fender has been aligned with the front hood and the door (the gap should be about 5.5 mm and even all the way around). Reseal the mounting surfaces and reapply the protective coating.

## Z3 models

### Front fender

*Refer to illustrations 7.12a, 7.12b, 7.14, 7.15 and 7.16*

11  Raise the hood. Loosen the front wheel lug nuts. Raise the front of the vehicle and place it securely on jackstands. Remove the front wheel.
12  Remove the inner liner from the wheel well **(see illustrations)**.
13  Remove the side turn signal light from the fender (see Chapter 12).
14  Remove the retaining bolt from the inside front of the fender **(see illustration)**.
15  Open the door all the way and remove the retaining screw from the bottom of the

fender **(see illustration)**.
16  Remove the two upper fender mounting bolts **(see illustration)**. Remove the fender.
17  Installation is the reverse of removal.

### Rear fender

*Refer to illustrations 7.20, 7.21, 7.22, 7.23, 7.24a, 7.24b, 7.24c, 7.25 and 7.26*

18  Remove the trunk lid (see Section 16).
19  Loosen the rear wheel lug nuts. Raise the rear of the vehicle and place it securely on jackstands. Remove the rear wheel.

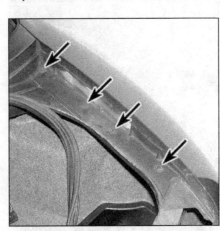

7.22 Remove the four retaining bolts (arrows) from the trunk sill

20  Remove the inner liner from the wheel well **(see illustration)**.
21  Scribe the location of the trunk lid hinge onto the fender **(see illustration)**. Remove the hinge.
22  Remove the four retaining bolts from the trunk sill **(see illustration)**.
23  Remove the retaining clip from the bumper overlay **(see illustration)**.

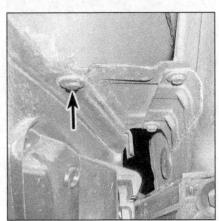

7.23 Remove the retaining clip (arrow) from the bumper overlay

11-10　Chapter 11　Body

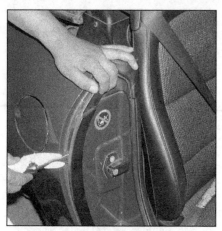

7.24a  Pry off the doorframe finish trim here . . .

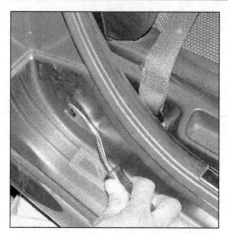

7.24b  . . . remove this retaining pin . . .

7.24c  . . . lift up the door sill trim and remove the doorframe finish trim

24  Open the door and pry off the doorframe finish trim **(see illustrations)**.
25  Remove the forward fender bolts **(see illustration)**.
26  Remove the retainer screw from the wheel arch side panel **(see illustration)**.
27  Remove the rear fender.
28  Installation is the reverse of removal. Replace any damaged retaining clips (briefly soaking the trim clips in hot water will facilitate their installation). Don't tighten any of the fender mounting bolts until all of them have been loosely installed and the fender has been aligned with the door, the trunk lid and the rear bumper cover (the gap should be about 5 mm and even all the way around).

8　Hood - removal, installation and adjustment

### 3-Series models

#### Removal

**Note:** *The hood can be raised to the near-vertical position to enable improved access when*

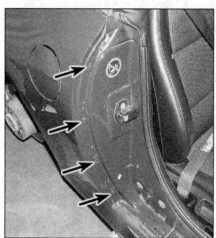

7.25  To detach the forward end of the rear fender from the vehicle, remove these fasteners (arrows)

*carrying out work in the engine compartment. To fully raise the hood, proceed as follows.* **Caution:** *Do not operate the windshield wipers with the hood in the fully open position, as the wiper blades will damage the hood.*

a) *On sedan models, either release the safety catch(es), or remove the safety bolt(s) (as applicable) from the hood strut upper mounting brackets. Lift the hood, and pivot the mounting bracket hinges down over the dead-center point to lock the hood in the fully open (near vertical) position. Check that the hood is locked securely in position.*

b) *On coupe models, have an assistant hold the hood in the open position, then release the securing catches and remove the hood struts. Unbolt the hood hinge ground strap(s), then remove the safety bolts from the hood hinges. Lift the hood and pull the hinge levers forwards to raise the hood to the fully open (near vertical) position. Support the hood in this position using two strong metal rods shaped to engage in the slots provided in the hood and body panels.*

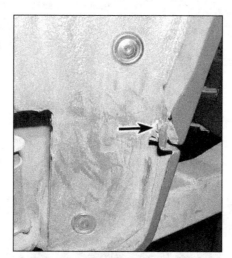

7.26  Remove the retainer screw (arrow) from the wheel arch side panel

1  Open the hood and have an assistant support it. Using a pencil or felt tip pen, mark the outline of each hood hinge relative to the hood, to use as a guide on installation.
2  Disconnect the hose from the washer jets. On models with heated jets also disconnect the wiring connectors and free the wiring from the hood.
3  With the aid of an assistant, support the hood in the open position then remove the retaining clips and detach the support struts from the hood.
4  Loosen and remove the left and right-hand hinge to hood rear bolts and loosen the front bolts. Slide the hood forwards to disengage it from the hinges and remove it from the vehicle. Recover any shims which are fitted between the hinge and hood.
5  Inspect the hood hinges for signs of wear and free play at the pivots, and if necessary replace. Each hinge is secured to the body by two bolts. Mark the position of the hinge on the body then undo the retaining bolts and remove it from the vehicle. On installation, align the new hinge with the marks and securely tighten the retaining bolts.

#### Installation and adjustment

6  Install the shims (if equipped) to the hinge and, with the aid of an assistant, engage the hood with the hinges. Install the rear bolts and tighten them by hand only. Align the hinges with the marks made on removal, then tighten the retaining bolts securely.
7  Close the hood, and check for alignment with the adjacent panels. If necessary, loosen the hinge bolts and re-align the hood to suit. Once the hood is correctly aligned, securely tighten the hinge bolts. Once the hood is correctly aligned, check that the hood fastens and releases satisfactorily.

### Z3 models

*Refer to illustrations 8.8a, 8.8b, 8.8c, 8.11 and 8.12*

8  Raise the hood. **Note:** *The hood can be*

# Chapter 11 Body

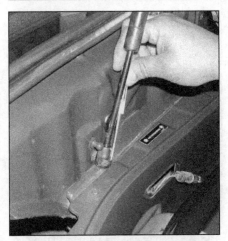

8.8a To put the hood in its service position, remove the clip from each support piston ball socket . . .

8.8b . . . have an assistant support the hood while you pry the left and right hood supports off their sockets . . .

8.8c . . . raise the hood all the way and press the supports onto the service position (upper) sockets

raised to a near-vertical position to facilitate access when servicing something in the engine compartment. To fully raise the hood, proceed as follows:

a) Remove the clip from each support piston ball socket **(see illustration)**.
b) Have an assistant support the hood while you use a wood stick to pry the left and right hood supports off their sockets **(see illustration)**.
c) Raise the hood all the way and press the supports onto the service position (upper) sockets **(see illustration)**.

9  Detach the washer nozzles from the washer.
10  Have an assistant help support the hood. **Caution:** *Do not try to remove the hood by yourself. It's too heavy.*
11  Disconnect the hood support struts from the hood **(see illustration)**.
12  Mark the hood hinges **(see illustration)**, then remove the hood hinge bolts. Remove the hood. **Caution:** *Do NOT place the hood on the roof. It's too heavy.*
13  Installation is the reverse of removal. Do not fully tighten the hinge bolts yet.
14  After installation, make sure that the striker is correctly aligned with the lock. If it isn't, loosen the striker retaining bolts, move the striker slightly and tighten the bolts.
15  Once the striker and lock are correctly aligned, close the hood and make sure that the gap between the hood and the bumper cover, fenders and A-pillar is even and about 5 mm in width.
16  Finally, make sure that the hood is resting on the detent buffers. One way to verify this is to apply grease to the tops of the buffers, close the hood, open it and note whether the buffers have smeared grease onto the underside of the hood. If they haven't, back both buffers out an equal amount and recheck. Continue doing this until the buffers contact the hood.

8.11 Pry off the retainer clips and disconnect the hood support struts from the hood

## 9  Hood release cable - removal and installation

### 3-Series models

#### Removal

*Refer to illustrations 9.3a, 9.3b, 9.3c, 9.3d and 9.3e*

1  The hood release cable is in two sections, the main cable linking the release lever to the first lock assembly and the joining cable which links both lock assemblies.
2  Remove the driver's side hood lock (see Section 10).
3  To remove the main cable, work back along the length of the cable, noting its correct routing, and free it from the retaining clips and ties. Tie a length of string to the end of the cable. From inside the vehicle, remove the two retaining screws then unclip and remove the driver's side lower dash panel. Remove the retaining screw and remove the hood release lever. Peel the door sealing strip away from the footwell side panel then

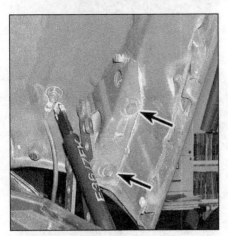

8.12 Mark the hood hinges, then remove the hood hinge bolts (arrows)

release the trim panel fastener by rotating it through 90-degrees. Unclip the panel. Remove the two retaining bolts and withdraw the cable assembly from the firewall **(see illustrations)**. Once the cable is free, untie the string and leave it in position in the vehicle; the string can then be used to draw the

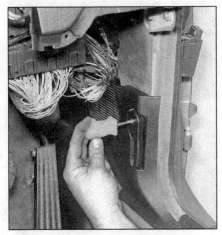

9.3a Remove the retaining screw . . .

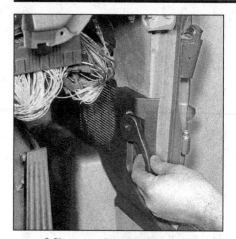

9.3b ... and remove the hood release lever

9.3c Remove the panel fastener ...

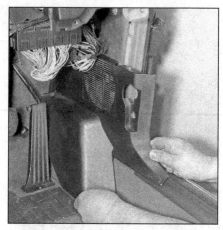

9.3d ... then unclip the panel and remove it from the driver's side footwell

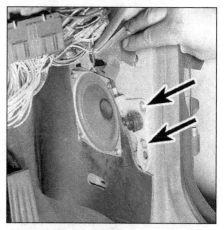

9.3e Unscrew the two retaining bolts (arrows) and withdraw the release cable from the firewall

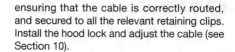

9.6 To detach the hood catches from the crossmember, remove these retaining screws (arrows)

new cable back into position.

4  To remove the joining cable, remove the second lock assembly (see Section 10), then release the cable from its retaining clips and remove it from the vehicle.

### Installation

5  Installation is the reverse of removal

ensuring that the cable is correctly routed, and secured to all the relevant retaining clips. Install the hood lock and adjust the cable (see Section 10).

### Z3 models

*Refer to illustrations 9.6, 9.7, 9.8, 9.9a, 9.9b, 9.9c and 9.9d*

6  Raise the hood and remove the retaining screws from two hood catches **(see illustration)**. Remove the catches.

7  Disconnect the center cable from the right hood catch **(see illustration)**.

8  Disconnect the main cable from the left catch **(see illustration)**.

9  Inside the vehicle, remove the hood release handle retaining screw, remove the left kick panel and remove the cable clamp screw **(see illustrations)**. Pry out the rubber grommet where the cable goes through the firewall.

10  Detach the cable guides inside the left

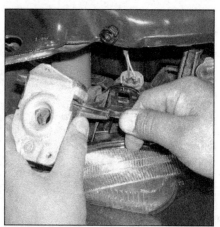

9.7 Disconnect the center cable from the right hood catch

9.8 Disconnect the main cable from the left hood catch

9.9a Inside the vehicle, remove the hood release handle retaining screw (arrow) ...

# Chapter 11 Body

9.9b ... remove the kick panel retaining screw ...

9.9c remove the left kick panel ...

9.9d and remove the cable clamp screws (arrows)

wheel well. Pull the cable through the firewall.
11  Installation is the reverse of removal. After you're done, make sure that the lock catches function correctly before shutting the hood.

## 10  Hood lock(s) - removal and installation

**Note:** *This procedure does not apply to Z3 models.*

### Removal

1  Open up the hood then remove the retaining screws and detach the plastic cover from the center of the hood lock crossmember.
2  To gain access to the rear of the lock(s), remove the radiator described in Chapter 3. Remove the retaining screws then release the clips and withdraw the radiator cooling duct.
3  Using a suitable marker pen, mark the outline of the relevant hood lock retaining screws on the crossmember.
4  Loosen and remove the lock retaining screws then free the release outer cable(s) from the lock lever then detach the inner cable(s) from the lock bracket and remove the lock from the vehicle.

### Installation

5  Locate the hood release inner cable(s) in the lock bracket and reconnect the outer cable(s) to the lever. Seat the lock on the crossmember.
6  Align the lock with the marks made prior to removal then install the bolts and tighten them securely.
7  Check that the locks operate smoothly when the release lever is moved, without any sign of undue resistance. Check that the hood fastens and releases satisfactorily. If necessary, adjust the cable using the threaded adjuster which is located in the center of the joining cable.
8  Once the locks are operating correctly,

install the cooling duct and panel and install the radiator (see Chapter 3).

## 11  Door - removal, installation and adjustment

**Warning:** *If the vehicle is equipped with side-impact airbags, be sure to disable the airbag system before beginning this procedure (see Chapter 12).*

### 3-Series models

#### Removal

*Refer to illustrations 11.2a, 11.2b, 11.3 and 11.4*

1  Disconnect the battery negative terminal. **Caution:** *If the radio in your vehicle is equipped with an anti-theft system, make sure you have the correct activation code before disconnecting the battery.*
2  Open up the door and loosen and remove the bolts securing the wiring connector block to the pillar. Withdraw the connector block, then release the securing clip by pulling it upwards and disconnect the wiring **(see illustrations)**.

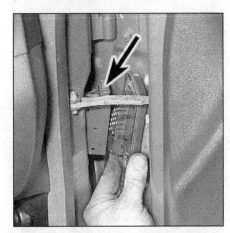

11.2b ... then lift up the securing clip (arrow) and disconnect the door wiring connector

3  Slide off the retaining clip and withdraw the pivot pin securing the check link to the pillar. Where the check link is secured in position with a roll pin, tap the pin out of position with a hammer and punch. Recover the rubber from the door pillar **(see illustration)**.

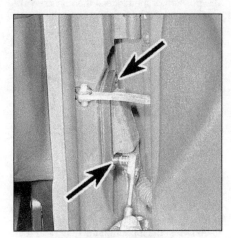

11.2a Remove the retaining screws (arrows) ...

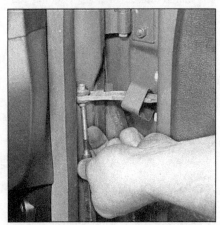

11.3 Where the check link is secured in position with a roll pin, tap it out using a hammer and punch

## Chapter 11 Body

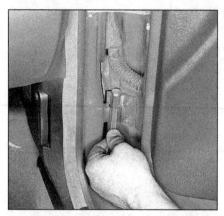

**11.4 Unscrew the hinge pivot bolts then lift the door upwards and away from the vehicle**

**11.9 Secure the check link in position and seat the rubber on the door pillar**

**12.3a On models with manual windows, pry off the trim cover . . .**

**12.3b . . . then remove the retaining screw . . .**

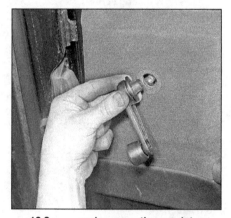

**12.3c . . . and remove the regulator handle and spacer from the door**

4   Unscrew the hinge pivot bolts from both the upper and lower door hinges **(see illustration)**.
5   With the aid of an assistant, lift the door upwards and away from the vehicle.
6   Examine the hinges for signs of wear or damage. If replacement is necessary, mark the position of the hinge(s) then unscrew the retaining bolts and remove them from the vehicle. Recover any shims which are fitted between the hinge and door/pillar. Install the new hinge(s), complete with shims (where fitted), align with the marks made before removal and lightly tighten the retaining bolts.

### Installation

*Refer to illustration 11.9*

7   Ensure the hinge pins are clean and apply a smear of fresh multi-purpose to them.
8   Maneuver the door into position and engage it with the hinges.
9   Slide the rubber onto the check link then align the link with its bracket and install the pivot pin and retaining clip/roll pin (as applicable). Seat the rubber on the door pillar **(see illustration)**.
10   Reconnect the door wiring connector and fasten the securing clip. Seat the connector block in the pillar and securely tighten its retaining bolts.
11   Check the door alignment and, if necessary, adjust then reconnect the battery negative terminal. If the paint work around the hinges has been damaged, paint the affected area with a suitable touch-up brush to prevent corrosion.

### Adjustment

12   Close the door and check the door alignment with surrounding body panels. If necessary, slight adjustment of the door position can be made by loosening the hinge retaining bolts and repositioning the hinge/door as necessary. Once the door is correctly positioned, securely tighten the hinge bolts. If the paint work around the hinges has been damaged, paint the affected area with a suitable touch-up brush to prevent corrosion.

### Z3 models

13   Open the door and put a floor jack under the door to support it. Put a rubber pad or some other suitably soft material between the jack head and the lower door edge to protect the finish. Raise the jack just enough to support the door.
14   Peel back the rubber door seal from the door check. Remove the door check retainer clip and remove the pin by tapping it out from the lower end **(see illustration 11.3)**.
15   Unscrew the upper and lower door hinge pins **(see illustration 11.4)** so that they're loose, but don't remove them yet.
16   Unplug the electrical connector for the door harness.
17   With an assistant supporting the door, remove the hinge pins and remove the door by lifting it off the hinges.
18   If you're replacing the door half of either hinge, scribe the position of the hinge on the door, then unbolt and remove the hinge. Using your scribe marks, position the new hinge(s), install the hinge bolts and tighten them securely. (The body side of each hinge is welded to the body and should only be replaced by a professional body shop.)
19   Installation is the reverse of removal. The door need not be aligned as long as the position of the hinges has not been moved in relation to the door.

## 12   Door trim panel - removal and installation

### 3-Series models

#### Removal

**Front door**

*Refer to illustrations 12.3a, 12.3b, 12.3c, 12.4, 12.5, 12.6a, 12.6b, 12.7a and 12.7b*

1   Disconnect the battery negative terminal then open the door. **Caution:** *If the radio in your vehicle is equipped with an anti-theft system, make sure you have the correct activation code before disconnecting the battery.*
2   On models with electric mirrors, remove the mirror switch (driver's door only) (see Chapter 12).
3   On models with manual windows, unclip the trim cover from the regulator handle then remove the screw and remove the handle and spacer **(see illustrations)**.
4   Unscrew and pull off the door lock inner operating knob from its rod **(see illustration)**.
5   Lift the door lock inner handle and unclip the trim cover from around the handle by sliding it forwards **(see illustration)**.
6   Unclip the trim caps from the door handle and loosen and remove the handle screws **(see illustrations)**.

# Chapter 11 Body

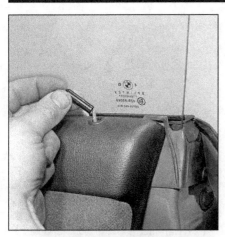

12.4 Removing the lock inner operating knob

12.5 Removing the inner handle trim cover

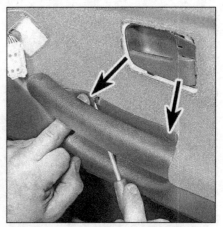

12.6a Pry out the trim caps (arrows) ...

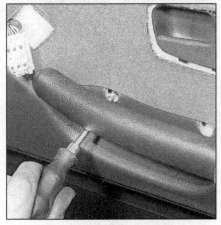

12.6b ... then remove the retaining screws

12.7a Carefully unclip the trim panel and remove it from the door ...

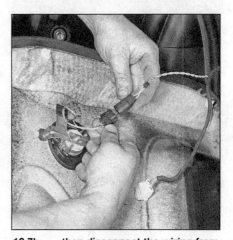

12.7b ... then disconnect the wiring from the speakers

7  Release the door trim panel clips, carefully levering between the panel and door with a flat-bladed screwdriver or a trim panel tool. Work around the outside of the panel, and when all the studs are released, ease the panel away from the door, disconnecting the wiring connector from the speaker as it becomes accessible **(see illustrations)**.

### Rear door

8  Disconnect the battery negative terminal then remove the trim panel (see steps 3 through 7), unclipping the panel from the top first. **Caution:** *If the radio in your vehicle is equipped with an anti-theft system, make sure you have the correct activation code before disconnecting the battery.*

### Installation

9  Installation of the trim panel is the reverse of removal. Before installation, check whether any of the trim panel retaining clips were broken on removal, and replace them as necessary.

## Z3 models

*Refer to illustrations 12.12, 12.13 and 12.14*

10  Open the door all the way and lower the window.

11  On models with power mirrors, carefully pry the power mirror switch out of the upper front end of the door handle with a small screwdriver, unplug the switch from the electrical connector and set the switch aside (see Section 4 in Chapter 12). On models without

power mirrors, pry out the trim plate.
12  Remove the screw from the forward end of the door handle **(see illustration)**.
13  Slide the door latch trim plate forward **(see illustration)** and remove the plate.
14  Disengage the door trim panel from the

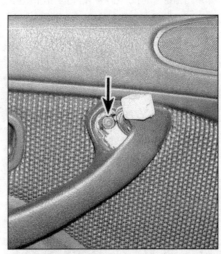

12.12 Remove the screw (arrow) from the forward end of the door handle

12.13 Slide the door latch trim plate forward and remove the plate

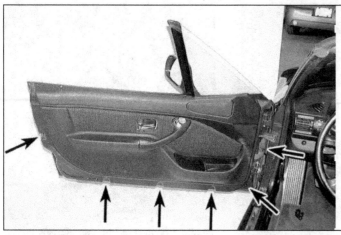

**12.14 To detach the door trim panel, disengage it from these six retaining clips (arrows)**

**13.1 Carefully peel the weathershield away from the door**

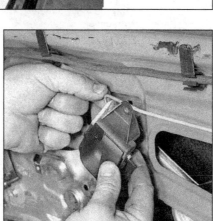

**13.2a Remove the retaining screw . . .**

**13.2b . . . then detach the interior handle from the link rod and remove it from the door**

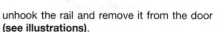

**13.4a On sedan models, remove the retaining bolt (arrow) . . .**

six retaining clips, one at the rear, three along the lower edge, two on the front edge **(see illustration)**.

15   Lift the door trim panel up and pull it off the door. Unplug the electrical connectors for the speaker and any other electrical accessories.

16   Installation is the reverse of removal. Be sure to replace any damaged or broken trim panel clips.

### 13  Door handle and lock components - removal and installation

## 3-Series models

### Removal
*Refer to illustration 13.1*

1   Remove the door inner trim panel (see Section 12). Carefully peel the plastic weathershield off from the door to gain access to the lock components and continue as described under the relevant sub-heading **(see illustration)**. If the weathershield is damaged on removal it must be replaced.

### Interior door handle
*Refer to illustrations 13.2a and 13.2b*

2   Remove the retaining screw then free the handle from the link rod and remove it from the door **(see illustrations)**.

### Front door lock assembly
*Refer to illustrations 13.4a, 13.4b, 13.5, 13.6, 13.7a, 13.7b and 13.7c*

3   Unscrew the retaining screw and remove the interior door handle from the door.

4   On sedan models, loosen and remove the window rear guide rail retaining bolt then

**13.4b . . . and remove the window rear guide rail from the body**

unhook the rail and remove it from the door **(see illustrations)**.

5   Release the lock assembly wiring retaining clips and disconnect the wiring connector(s) **(see illustration)**.

6   Unhook the lock cylinder link rod from the rear of the lock. Unclip the interior handle link rod guide from the door so the rod is free to be removed with the lock assembly **(see illustration)**.

7   Loosen and remove the lock assembly retaining screws then detach the lock assem-

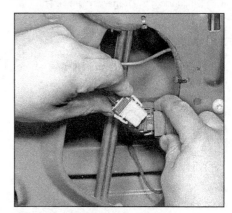

**13.5 Disconnect the wiring connectors from the door lock**

# Chapter 11  Body

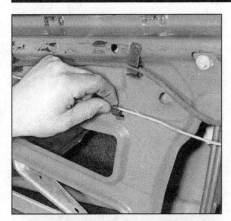

13.6  Release the link rod guide from the door . . .

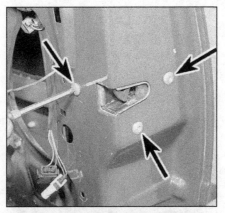

13.7a  . . . then remove the retaining screws (arrows) . . .

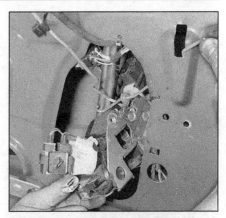

13.7b  . . . and maneuver the lock assembly out of position (3-Series coupe shown)

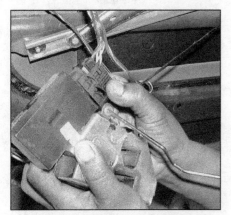

13.7c  If the vehicle is equipped with central locking, unplug the electrical connector from the locking actuator

13.10  Remove the rubber plug from the rear edge of the door . . .

13.11a  . . . then release the retaining clip with a piece of wire/screwdriver . . .

bly from the exterior handle linkage and maneuver it out from the door **(see illustrations)**. If the vehicle is equipped with central locking, unplug the electrical connector from the locking actuator **(see illustration)**. (To separate the locking actuator from the lock, refer to Section 28 in Chapter 12.)

### Front door exterior handle

*Refer to illustrations 13.10, 13.11a, 13.11b, 13.12, 13.13a, 13.13b, 13.14 and 13.15*

**Note:** *Since the exterior handle cannot be removed without first removing the window glass. On coupe models this task should be entrusted to a BMW dealer (see Section 15). With the glass removed the lock is removed and refitted as described below.*

8   Remove the window glass (see Section 14).
9   Remove the lock assembly (see steps 3 through 7).
10  Pry out the rubber plug from the rear edge of the door to gain access to the door handle cover retaining clip **(see illustration)**.

11  Insert a screwdriver or piece of wire in through the door access hole then release the handle trim cover retaining clip by pushing it into the door. Remove the handle trim cover and seal from the outside of the door **(see illustrations)**.
12  Where necessary, disconnect the wiring connector(s) from the handle assembly **(see illustration)**.
13  Slide off the retaining clip and withdraw the handle retaining pin **(see illustrations)**.

13.11b  . . . and remove the handle trim cover and seal

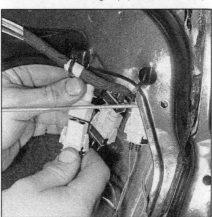

13.12  Disconnect the wiring connectors from the handle

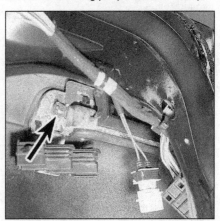

13.13a  Slide off the retaining clip (arrow) . . .

13.13b ... and withdraw the handle retaining pin

13.14 Unscrew the handle ring nut using a pair of snap-ring pliers

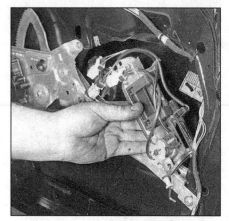

13.15 Removing the exterior handle from the door

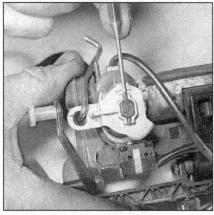

13.17a Tap out the roll pin ...

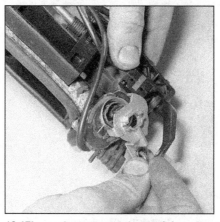

13.17b ... then remove the retaining plate from the rear of the handle ...

13.18 ... and withdraw the lock cylinder

14  Using a pair of snap-ring pliers, unscrew the ring nut from the outside of the lock cylinder and recover the rubber seal **(see illustration)**.
15  Maneuver the handle assembly out from inside the door **(see illustration)**.

#### Front door lock cylinder

*Refer to illustrations 13.17a, 13.17b and 13.18*

16  Remove the exterior handle (see steps 8 through 15).
17  With the key in the lock, tap out the roll pin and remove the retaining plate and link rod plate from the rear of the lock cylinder, noting their correct fitted locations **(see illustrations)**.
18  Withdraw the lock cylinder from the lock assembly **(see illustration)**. Recover the sealing ring from the cylinder and replace it if it is damaged.

#### Rear door lock

*Refer to illustrations 13.22a, 13.22b, 13.24a and 13.24b*

19  Remove the window glass (see Section 14).
20  Remove the sealing strip (if not already done so) then loosen and remove the window guide rail retaining screw and bolts and reposition the guide clear of the lock.

21  Loosen and remove the retaining screw and remove the interior door handle from the link rod.
22  Remove the screw securing the link rod pivot to the door and unhook it from the link rod. Release the link rod guide from the door so the rod is free to be removed with the lock **(see illustrations)**.
23  Where necessary, release the retaining clip and disconnect the wiring connector from the lock.

13.22a Remove the retaining screw ...

24  Loosen and remove the retaining screws then detach the lock assembly from the handle and remove it from the door **(see illustrations)**.

#### Rear door exterior handle

*Refer to illustrations 13.28a and 13.28b*

25  Remove the lock assembly as described above in paragraphs 19 to 24.
26  Pry out the rubber plug from the rear edge of the door to gain access to the door

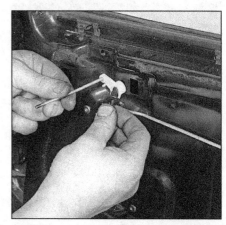

13.22b ... then detach the link rod pivot and remove it from the door

# Chapter 11 Body

13.24a Loosen and remove the three retaining screws . . .

13.24b . . . and maneuver the lock assembly out from the door

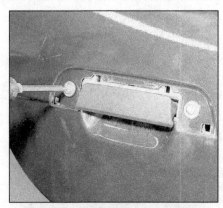

13.28a Loosen and remove the two retaining screws and seals . . .

### Front door lock cylinder

*Refer to illustration 13.43*

42  Lubricate the outside of the lock cylinder with a suitable grease.
43  Fit a new sealing ring to the lock cylinder and insert the cylinder into the handle (see illustration).
44  Install the link rod plate and retaining plate to the rear of the lock cylinder. Make sure all components are correctly fitted and secure them in position with the roll pin.
45  Install the exterior handle (see steps 37 to 41).

### Rear door lock

46  Prior to installation, remove all traces of old locking compound from the lock retaining screws.
47  Maneuver the lock assembly into position, making sure the link rods are correctly positioned, and hook it onto the exterior handle linkage.
48  Where necessary, reconnect the wiring connector(s) and secure the wiring in position with the relevant clips.
49  Apply fresh locking compound to the lock screws (BMW recommends Loctite 270) then install them and tighten to the specified torque.
50  Clip the link rod guide into the door then install the link rod pivot and securely tighten its retaining screw.
51  Connect the interior handle to the link rod and securely tighten the handle retaining screw.
52  Seat the window guide rail in position and securely tighten its retaining screw and bolts. Install the sealing strip.
53  Check the operation of the lock assembly then install the window glass (see Section 14).

### Rear door exterior handle

54  Install the handle assembly and tighten its retaining screws to the specified torque.
55  Fit the handle trim cover and seal to the door and secure it in position by pulling the retaining clip back using a hooked piece of wire. Make sure the trim cover is securely held and install the rubber plug to the door.
56  Install the lock assembly (see steps 46 through 53).

13.28b . . . and remove the handle assembly from the rear door - sedan models

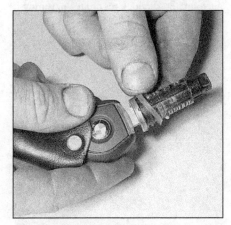

13.43 On installation do not forget to fit the seal to the lock cylinder

handle cover retaining clip.
27  Insert a screwdriver in through the door aperture and release the handle trim cover retaining clip by pushing it into the door. Remove the handle trim cover and seal from the outside of the door.
28  Loosen and remove the retaining screws and seals and remove the handle assembly from the door (see illustrations).

## Installation

### Interior door handle

29  Engage the handle with the link rod and securely tighten its retaining screw. Make sure the handle operates correctly then stick the weathershield to the door and install the trim panel (see Section 12).

### Front door lock assembly

30  Prior to installation, remove all traces of old locking compound from the lock retaining screws.
31  Maneuver the lock assembly into position making sure all the link rods are correctly positioned. Hook the lock onto the exterior handle linkage and reconnect the link rod(s).
32  Apply fresh locking compound to the lock screws (BMW recommends Loctite 270) then install them and tighten them to the specified torque.
33  Reconnect the wiring connector(s) and secure the wiring in position with the relevant clips.
34  On sedan models, hook the top of the guide rail into the door and secure it in position with the retaining bolt.
35  Clip the link rod guide into the door then install the interior handle and securely tighten the retaining screw.
36  Check the operation of the lock assembly then stick the plastic weathershield to the door. Install the trim panel (see Section 12).

### Front door exterior handle

37  Maneuver the handle assembly into position. Fit the rubber seal and ring nut to the lock cylinder and tighten it securely.
38  Fit the handle pin and secure it in position with the retaining clip.
39  Reconnect the handle wiring connectors (where necessary) and clip the trim cover into position.
40  Fit the handle trim cover and seal to the door and secure it in position by pulling the retaining clip back using a hooked piece of wire. Make sure the trim cover is securely held and install the rubber plug to the door.
41  Install the lock assembly (see steps 30 through 36).

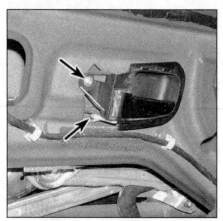

**13.59 To detach the inside door latch from the door, remove the retaining screw (upper arrow) and disengage the latch lever from the operating rod (lower arrow)**

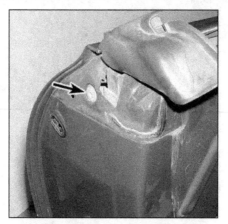

**13.62 To detach the upper end of the window rail guide and the metal guard surrounding the lock mechanism, remove this retaining bolt (arrow)**

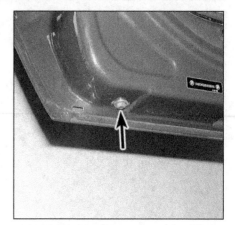

**13.63 To detach the lower end of the window guide rail, remove this bolt (arrow)**

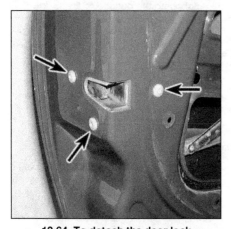

**13.64 To detach the door lock mechanism, remove these three screws (arrows)**

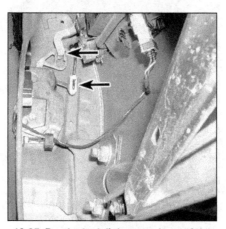

**13.65 Pry the lock linkage rod out of the plastic retainer (lower arrow), and then lift the door lock mechanism slightly to remove the latching pin from the lock gate (upper arrow)**

**13.69 Pry out the oblong access plug from the edge of the door and then, using a screwdriver inserted through the access hole, release the lock plate for the exterior trim**

## Z3 models

57  Remove the door trim panel (see Section 12).
58  Remove the vapor barrier.

### Inside door latch

*Refer to illustration 13.59*

59  Remove the latch retaining screw **(see illustration)**, pull out the latch assembly, disengage the latch lever from the operating rod and remove the latch assembly.
60  Installation is the reverse of removal.

### Door lock mechanism

*Refer to illustrations 13.62, 13.63, 13.64 and 13.65*

61  Remove the door glass and fixed window glass (see Section 16).
62  Remove the bolt **(see illustration)** that retains the upper end of the window rail guide and the metal guard surrounding the lock mechanism.
63  Remove the lower window rail guide bolt **(see illustration)** from the bottom corner of the door. Rotate the window guide rail toward the door hinge and remove it from the door.
64  Peel back the door seal, remove the door lock mechanism screws **(see illustration)** and remove the door lock mechanism.
65  Disconnect the linkage **(see illustration)** from the exterior door handle lock as follows: From inside the door, pry the lock linkage out of the plastic retainer, then lift the door lock mechanism slightly to remove the latching pin from the lock gate. (Rotate the lock mechanism as necessary to disengage the linkage.) Unplug the electrical connector from the handle.
66  Remove the door lock assembly from the door.
67  Installation is the reverse of removal.

### Outside door handle/lock cylinder

*Refer to illustrations 13.69, 13.70, 13.71 and 13.74*

68  Close the window. To prevent accidental operation of the window, unplug the power window harness connector.
69  Pry out the oblong access plug from the edge of the door. Using a screwdriver inserted through the access hole, release the lock plate for the exterior trim **(see illustration)**. Working from the upper end of the trim, carefully pry off the trim.
70  Unplug the electrical connector at the handle **(see illustration)** and remove the

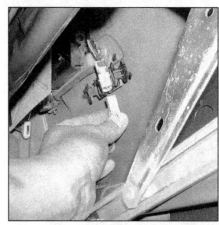

**13.70 Unplug the electrical connector at the handle and remove the plastic connector bracket from the handle**

Chapter 11 Body 11-21

**13.71 Using needle-nose pliers, remove the locking clip (arrow) for the door handle retaining pin and push pin**

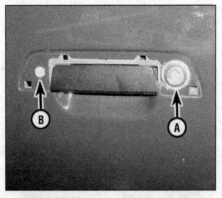

**13.74 While supporting the handle, remove the handle collar nut (A), and then disengage the handle retaining pin (B) from the hole in the door and remove the handle**

**14.4 Unscrew the retaining bolt and remove the window rear guide from the door**

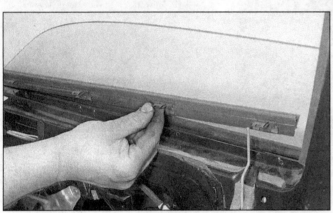

**14.5 Unclip the window inner sealing strip from the top of the door**

**14.6 On models with electric windows disconnect the motor wiring connector to immobilize the motor**

plastic connector bracket from the handle.
71  Using needle-nose pliers, remove the locking clip for the door handle retaining pin and push pin **(see illustration)**.
72  Unplug the door handle linkage from the door lock (see Step 62).
73  Remove the door lock assembly (see Steps 63 and 64).
74  While supporting the handle, remove the handle collar nut **(see illustration)**.

75  Put the door handle at an angle to the handle cavity to disengage the positioning pin from the door is and remove the handle.
76  Installation is the reverse of removal, with the following additions:
a) Make sure the rubber seals are correctly positioned before tightening the outside handle fasteners.
b) Use a hooked tool to pull the locking plate into the locked position.

## 14  Door window glass and regulator (3-Series sedan models) - removal and installation

### Removal

#### Front door window

*Refer to illustrations 14.4, 14.5, 14.6, 14.7a, 14.7b and 14.8*

1  Fully close the window then lower it approximately 1/8-inch.
2  Remove the inner trim panel (see Section 12).
3  Carefully peel the plastic weathershield away from the door panel to gain access to the window components. If the shield is damaged on removal it must be renewed.
4  Loosen and remove the window rear guide rail retaining bolt then unhook the rail and remove it from the door **(see illustration)**.
5  Carefully unclip the window inner sealing strip and remove it from the top of the door **(see illustration)**.
6  On models with electric windows disconnect the wiring connector from the window motor **(see illustration)**.
7  With the aid of an assistant, hold the window in position then slide out the retain-

**14.7a Slide out the retaining clip . . .**

**14.7b . . . and unclip the regulator arms from window glass slides**

# Chapter 11  Body

14.8  Tilt the window glass and remove it from the front door

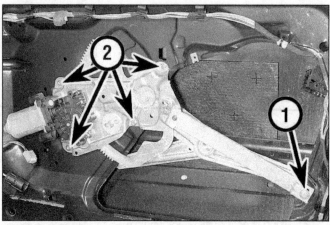

14.11  Front window regulator retaining bolt (1) and rivets (2)

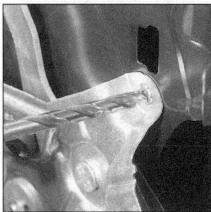

14.12  Drill out the regulator rivets and replace them with 1/4-inch or 6 mm bolts and nuts on installation

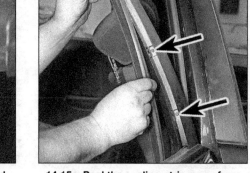

14.15a  Peel the sealing strip away from the front edge of the door then remove the three retaining screws (arrows) . . .

14.15b  . . . and remove the window trim panels from the rear door

ing clips and detach the regulator arms from the glass **(see illustrations)**.

8  Carefully maneuver the window glass upwards and out of the door and recover the slide from the glass guide rails **(see illustration)**.

## Front door window regulator

*Refer to illustrations 14.11 and 14.12*

9  Remove the window glass as described earlier in this Section.

10  Remove the retaining screw securing the interior handle to the regulator then detach the handle from the link rod and remove.

11  Loosen and remove the regulator lower retaining screw **(see illustration)**.

12  Using a drill and 15/64-inch drill bit, carefully drill the heads off the regulator retaining rivets **(see illustration)**. On installa-

tion replace the rivets with 1/4 inch or 6 mm bolts, which are approximately 3/8 inch or 10 mm in length, washers and nuts.

13  Maneuver the regulator assembly out from the door.

## Rear door window glass

*Refer to illustrations 14.15a, 14.15b, 14.17, 14.20 and 14.22*

14  Remove the door inner trim panel (see Section 12).

15  Peel back the window sealing strip from the front edge of the door to gain access to the window trim panel screws. Unscrew the retaining screws and remove the panel from the door **(see illustrations)**.

16  Carefully peel the plastic weathershield away from the door panel to gain access to the window components. If the shield is damaged on removal it must be renewed.

17  Carefully remove the window inner sealing strip from the top of the door **(see illustration)**.

18  Lower the window so that access to the window guide clip can be gained.

19  On models with electric windows, disconnect the wiring connector from the window motor.

20  Carefully ease the window sealing strip out from the door frame **(see illustration)**.

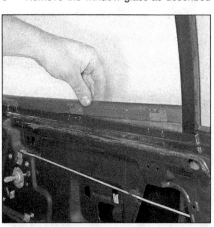

14.17  Unclip the inner sealing strip from the top of the door

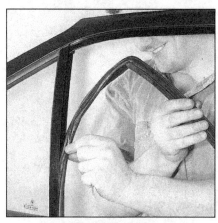

14.20  Ease the window sealing strip out from the door frame

# Chapter 11 Body

14.22 Removing the rear door window glass

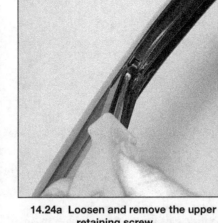

14.24a Loosen and remove the upper retaining screw . . .

14.24b . . . and the lower retaining bolts (arrows) . . .

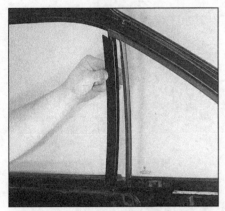

14.24c . . . and position the window guide rail clear of the fixed window

21 Slide out the retaining clip and free the regulator arm from the window guide clip **(see illustrations 14.7a and 14.7b)**.
22 Lift the window out through the top of the door and recover the slide from the window guide **(see illustration)**.

### Rear door fixed window glass

*Refer to illustrations 14.24a, 14.24b and 14.24c*
23 Remove the window glass (see steps 14 through 22).
24 Remove the sealing strip (if not already done so) then loosen and remove the window guide rail retaining screw and bolts and reposition the guide clear of the fixed window **(see illustrations)**.
25 Release the fixed window sealing strip from the door and remove the window.

### Rear door window regulator

*Refer to illustration 14.27*
26 Remove the window (see steps 14 through 22).
27 Remove the regulator (see steps 12 and 13) **(see illustration)**.

## Installation

**Note:** *Upon completion, on models with electric windows, initialize the windows (see Section 19).*

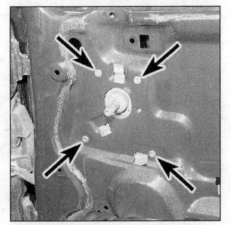

14.27 Rear window regulator retaining rivets (arrows)

### Front door window

28 Installation is the reverse of removal making sure the regulator arms are securely clipped in position. Prior to fitting the weathershield, operate the window and check that it moves easily and squarely in the door frame. If necessary, adjustments can be made by loosening the stop bolt on the regulator mechanism and moving the position of the bolt in its slot. Once the window is correctly adjusted tighten the stop bolt to the specified torque.

### Front door window regulator

29 Installation is the reverse of removal, replacing the regulator rivets with 1/4 inch or 6 mm bolts. Install the window glass and adjust (see step 28).

### Rear door window glass

30 Installation is the reverse of removal. Prior to installing the weathershield, check that the window operates smoothly and easily.

### Rear door fixed window glass

31 Installation is the reverse of removal making sure the window seal is correctly located in the door.

### Rear door window regulator

32 Installation is the reverse of removal. Prior to installing the weathershield, check that the window operates smoothly and easily.

## 15 Door window glass and regulator and rear vent window (3-Series coupe models) - removal and installation

### Front door window glass and regulator

1 Removal of the front door window glass and regulator should be entrusted to a BMW dealer or other qualified repair shop. If the window glass or regulator are disturbed a complex adjustment procedure must be performed on installation. Failure to adjust the window properly will lead to the glass contacting the body when the door is shut, resulting in breakage of the glass.

### Rear vent window

#### Removal

2 Open up the rear vent window and carefully unclip the trim panel from the rear pillar. Disconnect the wiring connector from the interior light and remove the panel.
3 Support the glass then loosen and remove the screw and spacer securing the hinge to the glass and recover the trim cap and seal from outside the glass.
4 Disengage the glass from its front pivot and remove it from the vehicle complete with its sealing strip. Replace the sealing strip if it shows signs of damage or deterioration.
5 If necessary, loosen and remove the retaining bolts and nuts and remove the window hinge from the rear pillar.
6 If the window front locating rail needs replacing, referring to Section 26, remove the rear seat side trim panel, detach the seat belt from its upper mounting and remove the door pillar trim panel. Unscrew the two bolts and remove the seat belt height adjustment

# Chapter 11 Body

**16.3 Remove the bolt (arrow) from the front window channel**

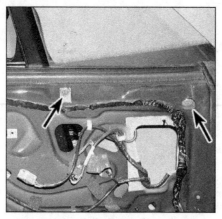

**16.4 To detach the frame that attaches the fixed glass to the door, remove these two retaining screws (arrows)**

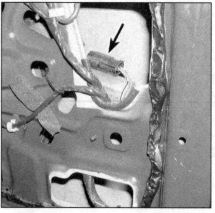

**16.19 Unplug the electrical connector (arrow) for the regulator wiring harness**

mechanism from the door pillar. Loosen and remove the two retaining screws and remove the rail from the pillar.

### Installation

7  If the window locating rail has been removed, ensure that the retaining screw sealing washers are in good condition. Fit the washers to the screws and tighten them lightly. Position the locating rail so that the gap between it and the front door window glass is approximately 3/16-inch along its entire length then tighten the screws to the specified torque. Install the disturbed seat belt and trim components (see Section 26).
8  Ensure the sealing strip is correctly located on the glass and engage the glass with the locating strip.
9  If the hinge was removed, fit the hinge to the pillar and lightly tighten its retaining nuts and bolts.
10  Insert the trim cap and seal from the outside of the glass and fit the spacer and retaining screw to the inside.
11  Close the window and check that it is correctly aligned with the surrounding body panels. Adjust if necessary then tighten the hinge bolts and screw to the specified torque.
12  Check the operation of the window then install the trim panel to the rear pillar making sure it is securely retained by all the relevant clips.

## 16 Door window glass and regulator (Z3 models) - removal, installation and adjustment

1  Remove the door trim panel (see Section 12).
2  Remove the vapor barrier.

## Fixed window glass

*Refer to illustrations 16.3 and 16.4*

3  Remove the bolt from the front window channel, under the front of the door **(see illustration)**.
4  Remove the two retaining screws from the bottom of the frame that attaches the fixed glass to the door **(see illustration)**.
5  To remove the fixed window, lift it straight up.
6  Installation is the reverse of removal. When you're done, check the adjustment of the door window glass (see below).

## Door window
### Removal and installation

7  Remove the fixed door glass (see Steps 3 through 5).
8  With the window closed, raise the widow about 4 to 8 inches.
9  Slide the window forward through the large opening in the door and out of the horizontal window regulator channel at the front.
10  Pivot the front of the window up and out of the door while simultaneously sliding the glass out of the horizontal window regulator channel at the rear.
11  Installation is the reverse of removal. Adjust the window before installing the vapor barrier and the door trim panel.

### Adjustment

12  Verify that the window glass is parallel to its frame. If adjustment is necessary, loosen the regulator rail mounting bolt, reposition the rail and then retighten the regulator rail

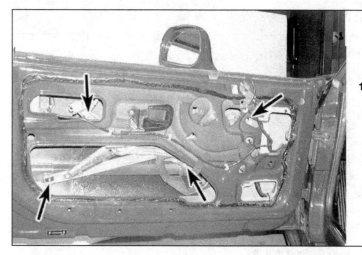

**16.20 To detach the regulator from the door, remove these four mounting bolts (arrows)**

mounting bolt.
13  Verify that the edge of the window glass seats uniformly against the rubber seal/weatherstrip and that the fixed glass is flush to the A-pillar at the windshield frame. If adjustment is necessary, loosen the bolts at the front and rear window channels and reposition the channels as necessary.
14  When you're done, tighten the bolts.
15  Verify that the door window glass, when closed, is parallel to the fixed door glass. If height adjustment is necessary, turn the height adjustment stop screw on the regulator.

### Regulator

*Refer to illustrations 16.19 and 16.20*

16  Remove the door trim panel (see Section 12).
17  Remove the vapor barrier.
18  Remove the fixed door glass and the window glass (see above).
19  Unplug the electrical connector **(see illustration)** for the regulator wiring harness.
20  Remove the regulator-to-door mounting screws **(see illustration)**.
21  Remove the regulator assembly from the door cavity. If necessary, operate the motor to reposition the regulator sector for more clearance.
22  Installation is the reverse of removal.

# Chapter 11 Body

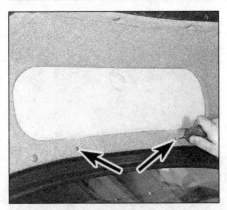

17.2a Unscrew the retaining screws (arrows) and remove the tool kit

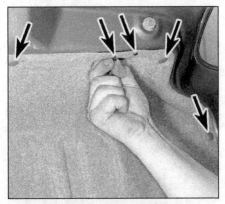

17.2b Remove the retaining clips (arrows) . . .

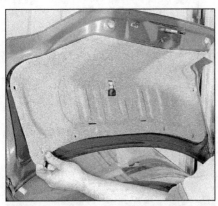

17.2c . . . and remove the trim panel from the trunk lid

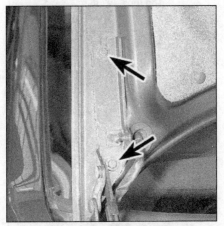

17.5 Trunk lid hinge retaining bolts (arrows)

17.7 Lift the retaining clip and unclip the support strut from the hinge

## 17 Trunk lid and support struts - removal and installation

### 3-Series models

#### Removal

**Trunk lid**

*Refer to illustrations 17.2a, 17.2b, 17.2c and 17.5*

1  Open up the trunk lid and then disconnect the battery negative terminal.
2  Remove the trim caps then remove the retaining screws securing the tool kit to the trunk lid. Remove the tool kit then unclip the trim panel from the trunk lid **(see illustrations)**.
3  Support the trunk lid in the open position and unclip the support struts (see step 7).
4  Disconnect the wiring connectors from the license plate lights, luggage compartment light switch and central locking servo (as applicable) and tie a piece of string to the end of the wiring. Noting the correct routing of the wiring harness, release the harness rubber grommets from the trunk lid and withdraw the wiring. When the end of the wiring appears, untie the string and leave it in position in the trunk lid; it can then be used on installation to draw the wiring into position.
5  Draw around the outline of each hinge with a suitable marker pen then loosen and remove the hinge retaining bolts and remove the trunk lid from the vehicle **(see illustration)**.
6  Inspect the hinges for signs of wear or damage and replace if necessary; the hinges are secured to the vehicle by bolts.

**Support struts**

*Refer to illustration 17.7*

7  Support the trunk lid in the open position. Using a small flat-bladed screwdriver raise the spring clip, and pull the support strut off its upper mounting **(see illustration)**. Repeat the procedure on the lower strut mounting and remove the strut from the vehicle.

#### Installation

**Trunk lid**

8  Installation is the reverse of removal, aligning the hinges with the marks made before removal.
9  On completion, close the trunk lid and check its alignment with the surrounding panels. If necessary slight adjustment can be made by loosening the retaining bolts and repositioning the trunk lid on its hinges. If the paint work around the hinges has been damaged, paint the affected area with a suitable touch-up brush to prevent corrosion.

**Support struts**

10  Installation is a reverse of the removal procedure, ensuring that the strut is securely retained by its retaining clips.

### Z3 models

*Refer to illustrations 17.11, 17.13 and 17.15*

11  Remove the access covers **(see illustration)** from the underside of the trunk lid, unplug all electrical connectors and remove the wiring harness by pulling it out through the opening in the lower left corner of the trunk lid.
12  Remove the retaining clips from the

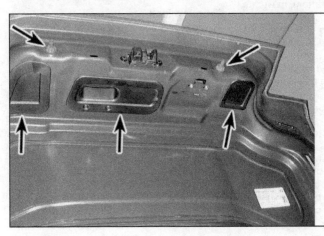

17.11 To unplug the wire harness electrical connectors, remove these three access covers (three lower arrows) from the trunk lid; to adjust the detent buffers (two upper arrows), turn them in or out

# Chapter 11 Body

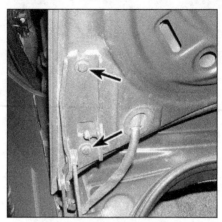

17.13 Be sure to mark the relationship of each hinge to the trunk lid before removing the hinge bolts (arrows)

17.15 To adjust the striker, loosen these bolts (arrows)

18.3 Unclip the lock link rod from the lock cylinder . . .

18.4a . . . then remove the retaining screws (arrows) . . .

18.4b . . . and remove the lock from the boot

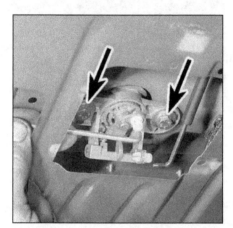

18.6a Unclip the link rods then unscrew the retaining bolts (arrows) . . .

trunk lid support struts **(see illustration 17.7)** and remove the struts.

13   Mark the relationship of the hinges to the trunk lid **(see illustration)** and then, with an assistant supporting the trunk lid, remove the hinge bolts and remove the trunk lid.

14   Installation is the reverse of removal. Make sure that the hinges are aligned with the marks you made during disassembly.

15   If the lock (in the trunk lid) and the striker (on the body) aren't correctly aligned, loosen the striker bolts **(see illustration)**, move the striker slightly and recheck the alignment between the lock and the striker. When the alignment is correct, tighten the striker bolts securely.

16   Close the trunk lid and check the gap between all four sides of the trunk lid and the adjacent bodywork panels. The gap should be even all the way around, and should be about 5 mm wide.

17   Make sure that the trunk lid detent buffers **(see illustration 17.11)** are making contact with the body when the trunk lid is closed. Apply a light coat of grease on the end of each buffer and shut the trunk lid. Open the trunk and note whether there is a grease spot at the point of contact between the buffers and the body. If there isn't, turn out the buffer(s) in 1/4-turn increments until there is.

## 18  Trunk lid lock components - removal and installation

### 3-Series models

#### Removal

**Trunk lid lock**

*Refer to illustrations 18.3, 18.4a and 18.4b*

1   Open up the trunk lid then disconnect the battery negative terminal. **Caution:** *If the radio in your vehicle is equipped with an anti-theft system, make sure you have the correct activation code before disconnecting the battery.*

2   Remove the trim caps then unscrew the retaining screws securing the tool kit to the trunk lid. Remove the tool kit and unclip the trim panel from the trunk lid.

3   Unclip the link rod from the boot lock cylinder **(see illustration)**.

4   Loosen and remove the lock retaining screws and withdraw the lock and link rod from the trunk lid **(see illustrations)**.

**Trunk lid lock cylinder**

*Refer to illustrations 18.6a and 18.6b*

5   Carry out the operations described in paragraphs 1 and 2.

6   Unclip the link rods from the lock cylinder then remove the retaining bolts and remove the lock cylinder from the trunk lid **(see illustrations)**.

18.6b . . . and remove the lock cylinder from the trunk lid

# Chapter 11  Body

## 11-27

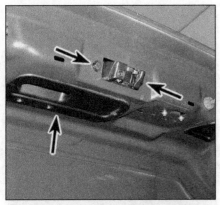

**18.16  To detach the lock from the trunk lid, remove the center access panel (lower arrow), remove the lock retaining bolts (upper arrows), reach through the opening, disengage the link rod from the key lock cylinder and remove the lock assembly**

### Installation

#### Trunk lid lock

7  Install the lock and link rod to the boot and securely tighten its retaining screws.
8  Align the link rod with the lock cylinder and check that it is the correct length so that it clips into the lock without any tension in the rod. If necessary, adjust the rod length prior to installation by screwing it into/out off the threaded adjustment piece.
9  Check the operation of the lock, then clip the trim panel back onto the trunk lid. Install the tool kit and securely tighten its retaining screws.
10  Reconnect the battery then close the trunk lid and check the operation of the lock. If necessary, adjustments can be made by either loosening the retaining bolts and reposition the lock catch or by adjusting the trunk lid rubber buffers.

#### Trunk lid lock cylinder

11  Install the lock cylinder and securely tighten its retaining screws.
12  Clip the central locking solenoid link rod onto the lock cylinder.
13  Carry out the operations described in paragraphs 8 to 10.

### Z3 models

#### Striker

14  To remove the striker, remove the striker retaining bolts **(see illustration 17.15)**.
15  Installation is the reverse of removal. Be sure to adjust the striker (see Step 15 in Section 17).

#### Trunk lid lock

*Refer to illustration 18.16*

16  Remove the center access panel and remove the lock retaining bolts **(see illustration)**.
17  Reach through the access hole and disengage the lock link rod from the lock cylinder **(see illustration 18.3)**.

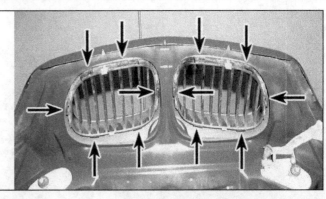

**19.12  To detach either grille from the hood, release these locking tabs and push out the grille from this side**

18  Remove the lock assembly.
19  Installation is the reverse of removal.

#### Trunk lid lock cylinder

20  Refer to Steps 5 and 6 and Steps 11 through 13.

## 19  Grille and front panel - removal and installation

### 3-Series models

#### Grille

**Note:** *This procedure applies to either grille.*
1  Raise the hood and remove the front radiator shroud or air duct (see Chapter 3).
2  Ball up your fist and tap the grille inward with the side of your hand until it pops out of the chrome trim ring.
3  To remove the chrome trim ring, pull it out from the front.
4  To install the grille and chrome trim ring, snap the two parts together, position the assembly in front of its hole in the front panel, then push on the top and bottom of the chrome ring until the assembly snaps into place.

#### Front panel

5  Remove the front bumper (see Section 6).
6  Remove the headlights (see Chapter 12).
7  If you're planning to replace the front panel, remove the two grille assemblies (see Steps 1 through 3). If you're simply removing the front panel assembly to service something, it's not necessary to remove the grilles to remove the front panel.
8  Remove the six upper panel retaining screws (one at each hood lock, on the upper radiator support; two on each end of the radiator).
9  Remove the four lower panel retaining screws from underneath the left and right ends of the front panel (two at each end).
10  Remove the front panel.
11  Installation is the reverse of removal.

### Z3 models (grille only)

*Refer to illustration 19.12*

12  Raise the hood. Working from inside the hood, depress the locking tabs of the grille **(see illustration)** and push out the grille from the back.
13  Installation is the reverse of removal.

## 20  Exterior mirrors and associated components - removal and installation

### 3-Series models

#### Mirror assembly

*Refer to illustrations 20.1a, 20.1b and 20.2*

1  Carefully unclip the mirror interior trim panel from the inside of the door and discon-

**20.1a  Unclip the inner trim panel door . . .**

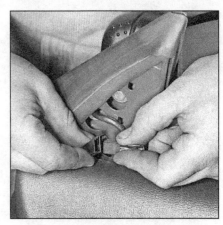

**20.1b  . . . and disconnect the mirror wiring connector**

**20.2 Unscrew the three retaining bolts and remove the mirror assembly from the door**

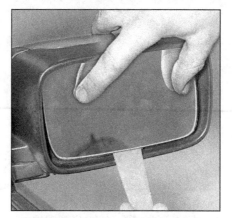

**20.5 Take great care not to break the glass when removing it from the mirror**

**20.10 Exterior mirror motor retaining screws (arrows)**

nect the motor wiring connector **(see illustrations)**.
2  Remove the retaining bolts and remove the mirror from the door. Recover the rubber seal which is fitted between the door and mirror; if the seal is damaged it must be renewed **(see illustration)**.
3  Installation is the reverse of removal, tightening the mirror bolts to the specified torque.

### Mirror glass
*Refer to illustration 20.5*
**Note:** *If the mirror glass is removed when the mirror is cold the glass retaining clips are likely to break.*
4  Tilt the mirror glass fully upwards.
5  Insert a wide plastic or wooden wedge in between the base of the mirror glass and mirror housing and carefully pry the glass from the motor **(see illustration)**. Take great care when removing the glass; do not use excessive force as the glass is easily broken.
6  Remove the glass from the mirror and, where necessary, disconnect the wiring connectors from the mirror heating element.
7  On installation, reconnect the wiring to the glass and clip the glass onto the motor, taking great care not to break it.

### Mirror switch
8  Refer to Chapter 12.

### Mirror motor
*Refer to illustration 20.10*
9  Remove the mirror glass as described above.
10  Unscrew the retaining screws and remove the motor, disconnecting its wiring connector as it becomes accessible **(see illustration)**.
11  On installation reconnect the wiring connector and securely tighten the motor screws. Check the operation of the motor then install the glass as described above.

## Z3 models
### Mirror assembly
*Refer to illustration 20.15*
12  Remove the door trim panel (see Section 12).
13  Detach the electrical harness retaining clips from the mirror.
14  If the vehicle is equipped with heated mirrors, unplug the electrical connector for the heater grid harness.
15  To access the mirror mounting screws, rotate the mirror assembly forward 90 degrees. Remove the two screws **(see illustration)**.
16  Remove the mirror and carefully feed the wiring harness through its hole. If you plan to reuse the same mirror assembly, be careful not to cut or damage the harness.
17  Installation is the reverse of removal.

### Mirror glass
18  Refer to Steps 4 through 7.

### Mirror switch
19  Refer to Chapter 12.

### Mirror motor
20  Refer to Steps 9 through 11.

## 21  Windshield and back glass - general information

These areas of glass are secured by the tight fit of the weatherstrip in the body aperture, and are bonded in position with a special adhesive. Replacement of such fixed glass is a difficult, messy and time-consuming task, which is beyond the scope of the home mechanic. It is difficult, unless one has plenty of practice, to obtain a secure, waterproof fit. Furthermore, the task carries a high risk of breakage; this applies especially to the laminated glass windshield. In view of this, owners are strongly advised to have this sort of work carried out by one of the many specialist windshield fitters.

## 22  Sunroof (3-Series models) - general information

*Refer to illustration 22.2*
Due to the complexity of the sunroof mechanism, considerable expertise is needed to repair, replace or adjust the sunroof components successfully. Removal of the roof first requires the headlining to be removed, which is a complex and tedious operation, and not a task to be undertaken lightly. Therefore, any problems with the sunroof should be referred to a BMW dealer or

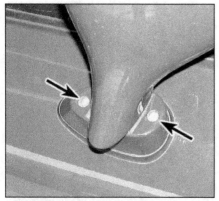

**20.15 To access the mirror mounting screws (arrows), rotate the mirror 90 degrees**

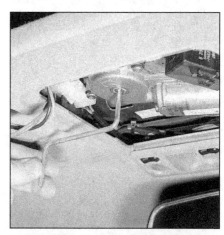

**22.2 If the electric sunroof fails, the sunroof can be moved using the Allen key supplied with the vehicle tool kit**

## Chapter 11  Body

24.2  Front seat rear mounting bolt (arrow)

24.4  Remove the rim cap to gain access to the front seat front mountings

other qualified repair shop.

On models with an electric sunroof, if the sunroof motor fails to operate, first check the relevant fuse. If the fault cannot be traced and rectified, the sunroof can be opened and closed manually using an Allen key to turn the motor spindle (a suitable key is supplied with the vehicle tool kit). To gain access to the motor, unclip the cover from the headlining. Remove the Allen key from the tool kit and insert it into the motor spindle. Disconnect the motor wiring connector and rotate the key to move the sunroof to the required position **(see illustration)**.

## 23  Body exterior fittings - removal and installation

### Wheel arch liners and body under-panels

1  The various plastic covers fitted to the underside of the vehicle are secured in position by a mixture of screws, nuts and retaining clips and removal will be fairly obvious on inspection. Work methodically around the panel, removing its retaining screws and releasing its retaining clips until the panel is free and can be removed from the underside of the vehicle. Most clips used on the vehicle are simply pried out of position. Other clips can be released by unscrewing/prying out the center pins and then removing the clip.

2  On installation, replace any retaining clips that may have been broken on removal, and ensure that the panel is securely retained by all the relevant clips and screws.

### Body trim strips and badges

3  The various body trim strips and badges are held in position with a special adhesive tape. Removal requires the trim/badge to be heated, to soften the adhesive, and then cut away from the surface. Due to the high risk of damage to the vehicle's paintwork during this operation, it is recommended that this task should be entrusted to a BMW dealer or other qualified repair shop.

## 24  Seats - removal and installation

### 3-Series models

#### Removal

**Front seat**

*Refer to illustrations 24.2 and 24.4*

1  Slide the seat fully forwards and raise the seat cushion fully.

2  Loosen and remove the bolts and washers securing the rear of the seat rails to the floor **(see illustration)**.

3  Slide the seat fully backwards and disable the seat belt tensioner (see Section 25). **Warning:** *Failure to disable the tensioner could result in serious injury.*

4  On all models, remove the trim caps from the seat front mounting nuts/bolts then loosen and remove the nuts/bolts and washers **(see illustration)**.

5  On sedan models, unscrew the mounting bolt and free the seat belt lower mounting from the base of the seat.

6  Lift the seat out from the vehicle, where necessary, disconnecting its wiring connectors as they become accessible.

**Rear seat assembly - models with fixed rear seat**

7  Pull down the rear seat armrest (where fitted) and unclip it from the seat back.

8  Pull up on the cushion to release the left- and right-hand retaining clips and remove it from the vehicle.

9  Unclip the top of the seat back then slide it upwards to release its lower retaining pins and remove it from the vehicle.

**Rear seat assembly - models with folding rear seat**

*Refer to illustrations 24.10, 24.12 and 24.13*

10  Pull up on the cushion to release the left and right-hand retaining clips and remove it from the vehicle **(see illustration)**.

11  Loosen and remove the bolts securing the seat belt lower mountings to the body.

12  Fold the seat backs forward and unclip the cover from the seat pivot **(see illustration)**.

13  Release the pivot locking clip by levering it backwards with a screwdriver and remove the seat backs from the vehicle **(see illustration)**.

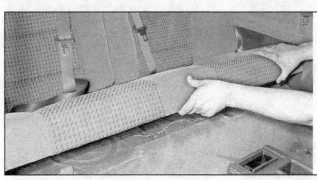

24.10  Unclip and remove the rear seat cushion

24.12  On models with a folding rear seat, unclip the cover from the seat back pivot . . .

24.13  . . . then release the retaining clip and unhook the seat pivot

## Installation

### Front seats

14 Installation is the reverse of removal, noting the following points.
   a) On manually adjusted seats, fit the seat retaining bolts and tighten them by hand only. Slide the seat fully forwards and then slide it back by two stops of the seat locking mechanism. Rock the seat to ensure that the seat locking mechanism is correctly engaged then tighten the mounting bolts to the specified torque.
   b) On electrically adjusted seats, ensure that the wiring is connected and correctly routed then tighten the seat mounting bolts to the specified torque.
   c) On sedan models tighten the seat belt mounting bolt to the specified torque.
   d) Enable the seat belt tensioner mechanism (see Section 25).

### Rear seat assembly - models with fixed rear seat

15 Installation is the reverse of removal making sure the seat back lower locating pegs are correctly engaged with the body.

### Rear seat assembly - models with folding rear seats

16 Installation is the reverse of removal ensuring that the seat pivot is clipped securely in position. Tighten the seat belt lower mounting bolts to the specified torque setting.

## Z3 models

17 Disconnect the cable from the negative battery terminal. **Warning:** *Failure to disconnect the negative battery terminal on 1997 and later models, which use pyrotechnic seat belt tensioners that are powered at all times, is dangerous. When the battery is connected, the tensioners on these models are fully energized and capable of inflicting serious injury if accidentally deployed.*
18 Disconnect the seat belt from the seat guide and unbolt the seat belt from the side of the seat.
19 Move the seat to its rearmost position and remove the two nuts from the front ends of the seat tracks.
20 Move the seat to its full forward position and remove the two nuts from the rear ends of the seat tracks.
21 Lift up the seat slightly and unplug the electrical connectors from the underside of the seat.
22 Remove the seat.
23 On models with a mechanical-spring-type seat belt tensioner (1996 models only), disarm the tensioner (see Section 25).
24 On models with a pyrotechnic-type seat belt tensioner (1997 and later models), unplug the orange connector.
25 Installation is the reverse of removal.

## 25 Front seat belt tensioning mechanism - general information

### 1996 and earlier models

*Refer to illustration 25.5*

1 Most models covered in this manual are fitted with a front seat belt tensioner system. The system is designed to instantaneously take up any slack in the seat belt in the case of a sudden frontal impact, therefore reducing the possibility of injury to the front seat occupants. Each front seat is fitted with its system, the tensioner being situated behind the sill trim panel.
2 The seat belt tensioner is triggered by a frontal impact above a pre-determined force. Lesser impacts, including impacts from behind, will not trigger the system.
3 When the system is triggered, a large spring in the tensioner mechanism retracts and locks the seat belt through a cable which acts on the inertia reel. This prevents the seat belt moving and keeps the occupant firmly in position in the seat. Once the tensioner has been triggered, the seat belt will be permanently locked and the assembly must be renewed.
4 **Warning:** *There is a risk of injury if the system is triggered inadvertently when working on the vehicle. If any work is to be carried out on the seat/seat belt disable the tensioner as follows.*
5 On models with manually adjusted seats, free the seat belt tensioner cable from the front of the seat base by rotating the outer cable retaining clip through 90-degrees and pulling it upwards **(see illustration)**. On completion of work, enable the tensioner by clipping the outer cable back into its bracket, making sure the inner cable is correctly hooked onto the base.
6 On models with electrically adjusted seats, the seat belt tensioner mechanism is disabled by rotating the screw situated in the front outer edge of the seat base. The position of the screw is indicated by the sight glass in the top of the base. When the sight glass is green the tensioner is enabled and when it is red the tensioner is disabled. Rotate the screw to disable the tensioner before carrying out the work and, on completion, enable the tensioner by rotating the screw back again.
7 Also note the following warnings before contemplating any work on the front seat.
**Warning:** *If the tensioner mechanism is dropped, it must be replaced, even it has suffered no apparent damage.*
**Warning:** *Do not allow any solvents to come into contact with the tensioner mechanism.*
**Warning:** *Do not subject the seat to any form of shock as this could accidentally trigger the seat belt tensioner.*

### 1997 and later models

8 The seats on these models are equipped with pyrotechnic (explosive) seat belt tensioners.
9 To disarm a pyrotechnic type tensioner, disconnect the cable from the negative battery terminal. **Warning:** *Failure to disconnect the negative battery terminal on 1997 and later models, which use pyrotechnic seat belt tensioners that are powered at all times, is dangerous. When the battery is connected, the tensioners on these models are fully energized and capable of inflicting serious injury if accidentally deployed.*
10 As insurance, unplug the orange connector under the seat.

## 26 Seat belt components - removal and installation

### 3-Series models

#### Removal

**Front seat belt - sedan models**

*Refer to illustrations 26.3, 26.4, 26.5a, 26.5b, 26.6, 26.7, 26.9 and 26.10*

**Warning:** *On models equipped with seat belt tensioners refer to Section 25 before proceeding.*

1 Remove the front seat (see Section 24) and free the seat belt from the seat.
2 Unclip the rear seat cushion and remove it from the body.
3 Carefully unclip the trim panel from the rear door sill panel **(see illustration)**.
4 Carefully unclip the front sill trim panel

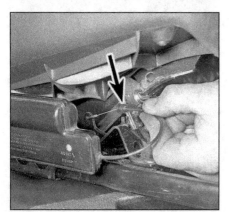

25.5 On models with manually adjusted seats, disable the seat belt tensioner by releasing the outer cable clip from its mounting bracket

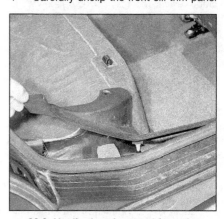

26.3 Unclip the trim panel from the rear door sill . . .

# Chapter 11  Body

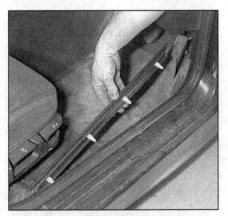

26.4 ... and the panel from the front door sill

26.5a Remove the trim cover ...

26.5b ... then unscrew the mounting nut and free the front seat belt from its upper mount

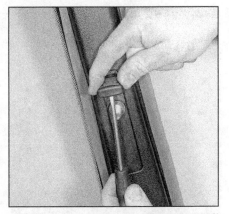

26.6 Release the retaining clip and pull off the height adjustment lever

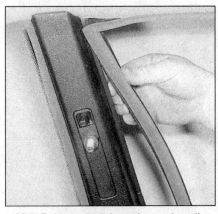

26.7 Release the trim strips and unclip the pillar trim panel

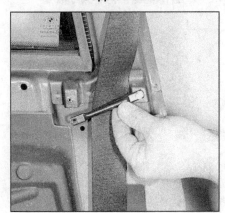

26.9 Unscrew the retaining screws and remove the seat belt guide from the pillar (coupe shown)

and remove it from the vehicle (see illustration).

5  Unclip the trim cover from the seat belt upper mounting. Loosen and remove the seat belt mounting nut and detach the belt from its height adjustment mechanism (see illustrations).

6  Release the retaining lug on the bottom of the height adjustment lever and remove the lever (see illustration).

7  Peel the front and rear door sealing strips away from the door pillar trim panel

(see illustration).

8  Carefully unclip the trim panel from the door pillar and remove it from the vehicle.

9  Unscrew the retaining screw(s) and remove the seat belt guide from the pillar (see illustration).

10  Unscrew the inertia reel retaining bolt and remove the seat belt from the door pillar (see illustration).

11  If necessary, unscrew the retaining bolts and remove the height adjustment mechanism from the door pillar.

### Front seat belt - coupe models

*Refer to illustrations 26.14a, 26.14b, 26.15, 26.16a and 26.16b*

12  Disable the seat belt tensioner (see Section 25).

13  Remove the rear seat (see Section 24).

14  On models with folding rear seats, remove the seat back pivot bushing from the mounting bracket then unclip the top of the seat side section and remove it from the vehicle body (see illustrations).

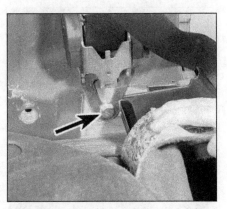

26.10 Unscrew the inertia reel retaining bolt and remove the seat belt (coupe shown)

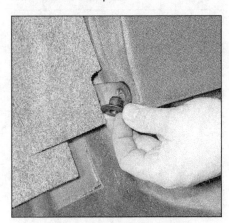

26.14a On coupe models, remove the seat back pivot bushing ...

26.14b ... then unclip the seat side section and remove it from the vehicle

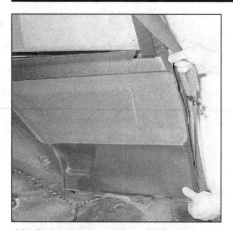

26.15 Unclip the rear seat side trim panel and remove it from the vehicle

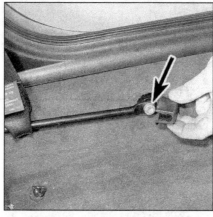

26.16a Remove the trim cap then loosen and remove the seat belt lower mounting rail bolt (arrow) . . .

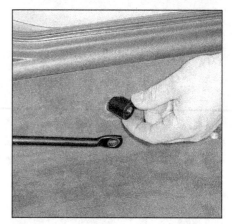

26.16b . . . and recover the spacer fitted between the rail and body

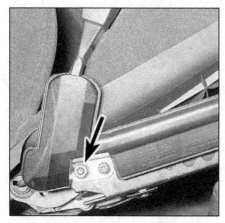

26.19 Front seat belt stalk assembly retaining nut (arrow)

26.22 On models with fixed rear seats, unclip the trim cover from the front of the parcel shelf

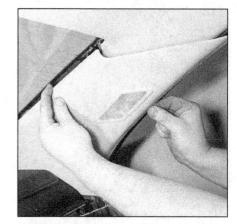

26.23 Unclip the rear pillar trim panels and remove them from the vehicle

15  On all models, open the door and rear window and free the sealing strips from the edge of the rear seat side trim panel. Unclip the trim panel and remove it from the vehicle **(see illustration)**.
16  Remove the trim cap from the front mounting bolt of the seat belt lower mounting rail. Loosen and remove the retaining bolt then recover the spacer which is fitted between the rail and body and slide the belt off from the rail **(see illustrations)**.
17  Remove the seat belt (see steps 5 through 11).

**Front seat belt stalk (incorporating seat belt tensioner mechanism)**
*Refer to illustration 26.19*
18  Remove the seat (see Section 24).
19  Ensuring the tensioner mechanism is disabled, loosen and remove the stalk assembly retaining nut and remove the assembly from the side of the seat **(see illustration)**.

**Rear seat side belts - models with fixed rear seat**
*Refer to illustrations 26.22, 26.23, 26.24 and 26.25*
20  Remove the rear seat (see Section 24).
21  Loosen and remove the bolts and washers securing the rear seat belts to the vehicle body and remove the center belt and buckle.
22  Unclip the trim cover from the front of the parcel shelf and detach it from the seat belts **(see illustration)**.
23  Carefully unclip the left and right-hand trim panels from the rear pillars, disconnecting the wiring from the interior lights as the panels are removed **(see illustration)**.
24  Remove the retaining clips from the front edge of the parcel shelf and slide the shelf forwards and out of position **(see illustration)**. As the shelf is removed, disconnect the wiring connectors from the high-level brake light (where fitted).
25  Unscrew the inertia reel retaining nut and remove the seat belt(s) **(see illustration)**.

26.24 Remove the retaining clips and remove the parcel shelf

26.25 Inertia reel retaining nuts (arrows)

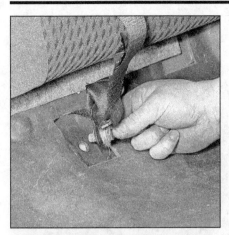

26.27 On models with folding rear seats, loosen and remove the seat belt lower mounting bolt . . .

26.28 . . . then remove the trim cap from seat back and unscrew the inertia reel retaining bolt (arrow)

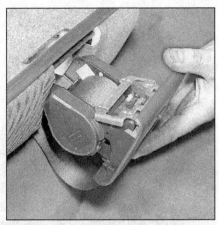

26.29 Unclip the inertia reel from the seat back and remove the seat belt

### Rear seat side belts - models with folding rear seats

*Refer to illustrations 26.27, 26.28 and 26.29*

26  Unclip the rear seat cushion and remove it from the vehicle.
27  Loosen and remove the bolt securing the lower end of the belt to the body **(see illustration)**.
28  Fold the seat forwards and pry out the circular trim cap from the rear of the seat to gain access to the inertia reel bolt **(see illustration)**. Loosen and remove the retaining bolt and washer.
29  Unclip the seat belt trim cover from the top of the seat and withdraw the inertia reel assembly from the seat back **(see illustration)**.

### Rear seat side belt stalk

30  Unclip the rear seat cushion and remove it from the body.
31  On sedan models, carefully unclip the trim panel from the rear door sill.
32  Loosen and remove the bolt and washer and remove the stalk from the vehicle.

### Rear seat center belt and buckle

33  Unclip the rear seat cushion and remove it from the vehicle.
34  Loosen and remove the bolt securing the center belt/buckle to the body and remove it from the vehicle.

### Installation

#### Front seat belt - all models

35  Installation is a reversal of the removal procedure, ensuring that all the seat belt mounting bolts are securely tightened, and all disturbed trim panels are securely retained by all the relevant retaining clips.

#### Front seat belt stalk (incorporating seat belt tensioner mechanism)

36  Ensure the tensioner mechanism is correctly engaged with the seat and tighten its retaining nut to the specified torque. Install the seat (see Section 24) and enable the tensioner mechanism.

### Rear seat side belts - models with fixed rear seat

37  Installation is the reverse of removal ensuring that all seat belt mountings are tighten to the specified torque and all trim panels are clipped securely in position.

### Rear seat side belts - models with folding rear seat

38  Installation is the reverse of removal making sure the inertia reel is clipped securely in position and all seat belt mounting bolts are tightened to the torque.

### Rear seat belt stalk

39  Installation is the reverse of removal, tightening the mounting bolt to the specified torque.

### Rear seat center belt and buckle

40  Installation is the reverse of removal tighten the mounting bolts to the specified torque.

### Z3 models

**Warning:** *Disable the seat belt tensioner(s) before beginning this procedure (see Section 25).*
41  Move the seat to its full forward position and disconnect the seat belt from the seat belt guide (see Section 24).
42  Remove the lower section of the side trim panel behind the seat.
43  Remove the seat belt reel retaining bolts and remove the reel assembly.
44  Remove the seat belt retaining bolt from the side of the seat.
45  Remove the upper belt guide mounting bolt.
46  Remove the seat belt assembly.
47  Installation is the reverse of removal.

## 27 Dash trim panels - removal and installation

**Warning:** *On models equipped with airbags, always disable the airbag system before working in the vicinity of the impact sensors, steering column or instrument panel to avoid the possibility of accidental deployment of the airbag, which could cause personal injury (see Chapter 12).*

### General information

1  The dash trim panels are secured using either screws or various types of trim fasteners, usually studs or clips.
2  Check that there are no other panels overlapping the one to be removed; usually there is a sequence that has to be followed that will become obvious on close inspection.
3  Remove all obvious fasteners, such as screws. If the panel will not come free, it is held by hidden clips or fasteners. These are usually situated around the edge of the panel and can be pried up to release them; note, however that they can break quite easily so replacements should be available. The best way of releasing such clips without the correct type of tool is to use a large flat-bladed screwdriver. Note in many cases that the adjacent sealing strip must be pried back to release a panel.
4  When removing a panel, never use excessive force or the panel may be damaged; always check carefully that all fasteners have been removed or released before attempting to withdraw a panel.

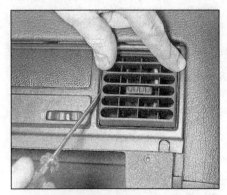

27.6 Where the heater vents are not molded into the glovebox, carefully unclip and remove them

## 11-34    Chapter 11    Body

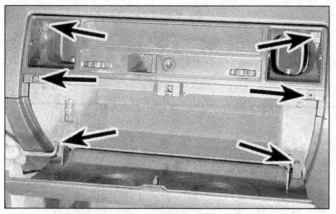

27.7  Loosen and remove the glovebox retaining screws (arrows)

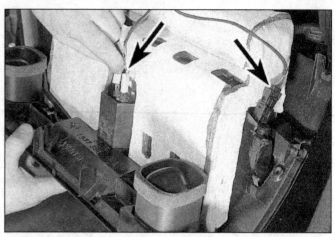

27.9  Slide out the glovebox and disconnect its wiring connectors (arrows)

27.11  To remove the Z3 glove box from the dash, remove these six trim caps (arrows) and the screws behind them (the two lower left arrows indicate the right side center console fasteners)

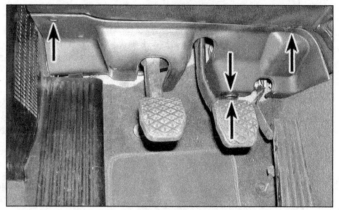

27.16  A typical pedal cover (Z3 shown); to detach it, remove all fasteners (arrows)

5    Installation is the reverse of the removal procedure; secure the fasteners by pressing them firmly into place and ensure that all disturbed components are correctly secured to prevent rattles.

## Glovebox

### 3-Series models

*Refer to illustrations 27.6, 27.7 and 27.9*

6    Where the glovebox heater vents are not molded into the glovebox, unclip and remove them **(see illustration)**.

7    Open up the glovebox lid and pry off the trim caps from the two center retaining screws. Loosen and remove the retaining screws securing the glovebox to the dash. Where the heater vents are part of the glovebox, access to the upper screws is gained through the vents **(see illustration)**.

8    Close the lid and remove the two screws from the base of the glovebox.

9    Slide the glovebox out of position disconnecting the wiring connector(s) as they become accessible **(see illustration)**.

10   Installation is the reverse of removal.

### Z3 models

*Refer to illustration 27.11*

11   Open the glovebox and remove the four trim caps **(see illustration)** located near the glovebox door lock catch, then remove the screws behind the caps.

12   Remove the trim caps and screws located to the left of the glovebox and on the far right end of the dash **(see illustration 27.11)**.

13   Close the glovebox, pull it down and remove it.

14   Installation is the reverse of removal.

## Pedal cover

*Refer to illustration 27.16*

15   Some models are equipped with a pedal cover, which covers the space between the lower instrument panel trim cover and the firewall.

16   To remove the pedal cover, remove the three retaining screws **(see illustration)** and remove the cover.

17   Installation is the reverse of removal.

## Lower dash trim panel

*Refer to illustrations 27.19 and 27.20*

18   The lower dash trim panel covers the under-dash space between the lower edge of the dash and the pedal cover.

19   To detach the left lower dash trim panel, remove the pedal cover (see Steps 15

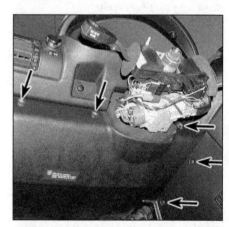

27.19  A typical left lower dash trim panel (Z3 shown); to detach it, remove the two forward center console fasteners (lower right arrows), and then remove all trim panel fasteners (arrows)

and 16), remove the two fasteners from the forward left side of the center console, remove the trim panel fasteners **(see illustration)**, and then pull off the trim panel.

20   To detach the right lower dash trim panel, remove the two fasteners from the forward left side of the center console **(see**

# Chapter 11  Body

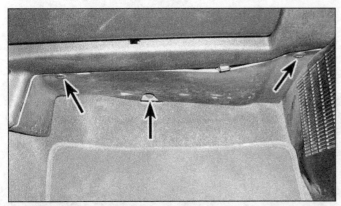

27.20  A typical right lower dash trim panel (Z3 shown); to detach it, remove the two forward center console fasteners (see illustration 27.11), remove the trim panel fasteners (arrows), and then pull down the panel

27.23  A typical knee bolster (Z3 shown); to detach it, remove these fasteners (arrows)

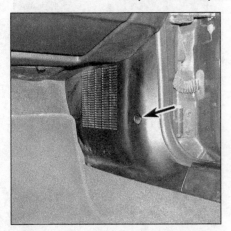

27.26  A typical (right side) kick panel (Z3 shown); to detach it, remove the fastener (arrow), and then pull off the kick panel

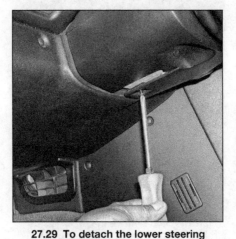

27.29  To detach the lower steering column cover, remove this retaining screw (Z3 model shown, other models similar)

27.30  To detach the upper steering column cover, remove this retaining screw (Z3 model shown, other models similar)

illustration 27.11), remove the trim panel fasteners **(see illustration)** and then pull off the trim panel.
21  Installation is the reverse of removal.

### Knee bolster

*Refer to illustration 27.23*
22  The knee bolster, which is located behind the lower instrument panel trim on the driver's side, affords some extra protection for the driver's legs in the event of a serious head-on crash. You'll need to remove the bolster to get to the upper ends of the pedals, or to troubleshoot or replace any relays, switches or wiring harnesses under the left end of the dash.
23  To detach the knee bolster, remove all fasteners **(see illustration)**.
24  Installation is the reverse of removal.

### Kick panels

*Refer to illustration 27.26*
25  The kick panels cover the space below the ends of the dash, between the front door and the firewall.
26  To detach a kick panel, remove all fasteners **(see illustration)**, and then remove the kick panel.
27  Installation is the reverse of removal.

### Steering column covers

*Refer to illustrations 27.29 and 27.30*
28  The steering column covers protect the ignition switch, key lock cylinder and turn signal/windshield wiper switch assembly on the steering column, between the steering wheel and the instrument cluster.
29  To detach the lower half of the steering column cover, remove the retaining screw **(see illustration)** and pull off the cover.
30  To detach the upper half of the steering column cover, remove the retaining screw **(see illustration)** and lift off the cover.
31  Installation is the reverse of removal.

### Carpets

32  The passenger compartment floor carpet is in one piece and is secured at its edges by screws or clips, usually the same fasteners used to secure the various adjoining trim panels.
33  Carpet removal and installation is reasonably straightforward but very time-consuming because all adjoining trim panels must be removed first, as must components such as the seats, the center console and seat belt lower anchorages.

### Headliner

34  The headliner is clipped to the roof and can only be withdrawn when fittings such as the grab handles, sun visors, sunroof, windshield, rear quarter windows and related trim panels - along with the door, tailgate and sunroof weather-stripping - have been removed.
35  Headliner removal requires considerable skill and experience if it is to be carried out without damage and is therefore best entrusted to an expert.

## 28  Center console - removal and installation

**Warning:** *On models equipped with airbags, always disable the airbag system before working in the vicinity of the impact sensors, steering column or instrument panel to avoid the possibility of accidental deployment of the airbag, which could cause personal injury (see Chapter 12).*

## Chapter 11 Body

28.3 Lift out the rear ashtray then unscrew the two retaining screws (arrows) and remove the ashtray surround

28.6a Loosen and remove the front . . .

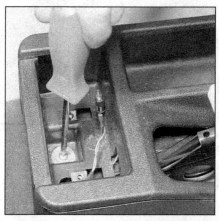

28.6b . . . and rear retaining screws . . .

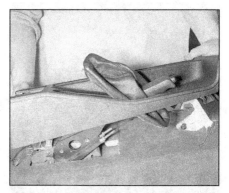

28.6c . . . and remove the rear section of the center console

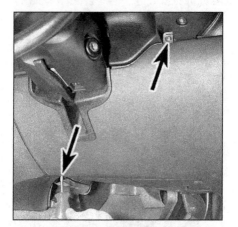

28.9 Unscrew the retaining screws (arrows) and remove the driver's side lower dash panel (coupe shown)

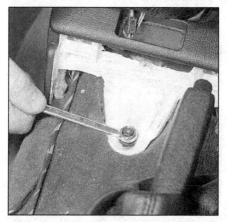

28.11a Unscrew the rear retaining nut . . .

### 3-Series models

#### Removal

*Refer to illustrations 28.3, 28.6a, 28.6b, 28.6c, 28.9, 28.11a and 28.11b*

1  Disconnect the battery negative terminal. **Caution:** *If the radio in your vehicle is equipped with an anti-theft system, make sure you have the correct activation code before disconnecting the battery.*
2  Remove the hazard warning light switch (see Chapter 12).
3  Where ashtrays are fitted, unclip the ashtrays and remove them from the center console. Undo the retaining screws and remove the ashtray surrounds, freeing the illumination bulbholders from them **(see illustration)**.
4  Where no ashtrays are fitted, unclip the storage compartment(s) and remove them from the center console.
5  On all models unclip the parking brake lever boot from the console.
6  Loosen and remove the retaining screws and remove the rear section of the console **(see illustrations)**.
7  If the front section is also to be removed, remove the clock/multi-information display unit (as applicable) (see Chapter 12).
8  Remove the glovebox (see Section 27).
9  Loosen and remove the retaining screws securing the driver's side lower dash panel.

Unclip the panel and remove it from the vehicle **(see illustration)**.
10  Unclip the storage compartment from the front of the center console and disconnect the wiring from the cigarette lighter.
11  Loosen and remove the retaining screws and nut then maneuver the front section of the center console upwards over the gear selector lever and out of the vehicle, disconnecting the various wiring connectors as they become accessible **(see illustrations)**.

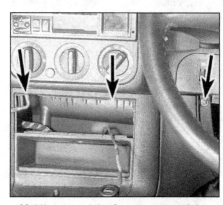

28.11b . . . and the four upper retaining screws (three arrows) and remove the center console front section

#### Installation

12  Installation is the reverse of removal making sure all fasteners are securely tightened.

### Z3 models

*Refer to illustrations 28.15, 28.16, 28.17, 28.19, 28.20, 28.22a, 28.22b, 28.22c, 28.24a and 28.24b*

13  Disconnect the cable from the negative battery terminal. **Warning:** *Do NOT work in the vicinity of the dash with the airbag system*

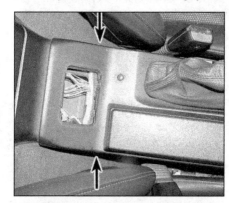

28.15 Remove these two retaining bolts (arrows) from the rear end of the center console

# Chapter 11 Body

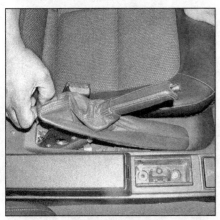

28.16 Pry loose the dust boot for the parking brake lever

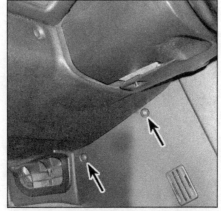

28.17 Remove the retaining screws (arrows) from the left (shown) and right sides of the forward end of the console

28.19 Remove the two mounting screws (arrows) from the radio recess

28.20 Pry back on the shift lever boot until the retaining clips disengage, then remove the shift lever boot

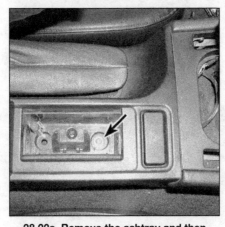

28.22a Remove the ashtray and then remove the ash tray holder retaining screw (arrow)

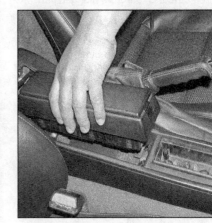

28.22b Remove the cup holder or storage box (shown)

enabled. *The only way to prevent an airbag from accidentally deploying is to disconnect the cable from the negative battery terminal.*

14  Remove the heated seat switch (if applicable) from the console (see Section 4 in Chapter 12).
15  Remove the retaining bolts from the left and right rear corners of the center console **(see illustration)**.
16  Pry loose the rubber boot around the parking brake lever **(see illustration)**.
17  Remove the retaining screws from the left and right lower side panels **(see illustration)**.
18  Remove the radio (see Chapter 12).
19  Remove the two mounting bolts in the recess for the radio, just below the center air vents **(see illustration)**.
20  Remove the shift lever handle (see Chapter 7). Pry back on the shift lever boot (manual transmission) or the shift indicator trim panel (automatic transmission) until the retaining clips disengage, then remove the shift lever boot **(see illustration)** or shift indicator trim panel.
21  Remove the two climate control retaining screws from the radio cavity and detach the climate control assembly from the console (see Chapter 12).
22  Remove the ashtray and then remove the ash tray holder retaining screw **(see illustration)**. Remove the cup holder or storage box **(see illustration)**. Under the ash tray holder remove the console retaining screw **(see illustration)**.
23  Raise the rear end of the console slightly and pull it back far enough to disengage the locking clips.
24  Pull back and lift up the forward end of the console, tilt up the rear end of the console and unplug the cigarette lighter wiring harness **(see illustrations)** and any other electrical connectors plugged into devices attached to the console.

28.22c Under the ash tray holder remove the console retaining screw (arrow)

28.24a Pull back and lift up the forward end of the console . . .

## 11-38　Chapter 11　Body

28.24b ... and then tilt up the rear end of the console and unplug the cigarette lighter wiring harness and any other electrical connectors plugged into devices attached to the console.

25  Slide the center console to the left, rotate it as necessary to clear the parking brake handle and remove the console.
26  Installation is the reverse of removal.

### 29  Dash panel assembly - removal and installation

**Warning:** *On models equipped with airbags, always disable the airbag system before working in the vicinity of the impact sensors, steering column or instrument panel to avoid the possibility of accidental deployment of the airbag, which could cause personal injury (see Chapter 12).*

### 3-Series

#### Removal

*Refer to illustrations 29.6, 29.7, 29.8 and 29.9*

1  Disconnect the battery negative terminal. **Caution:** *If the radio in your vehicle is equipped with an anti-theft system, make sure you have the correct activation code before disconnecting the battery.*
2  Remove the center console (see Section 28).
3  Remove the steering column (see Chapter 10).
4  Remove the instrument panel assembly, cigarette lighter and radio/cassette unit (see Chapter 12).
5  On models equipped with a passenger side airbag, remove the airbag unit (see Chapter 12). Unscrew the airbag mounting frame retaining bolts and remove the frame from the dash.
6  Carefully unclip the left and right-hand windshield pillar trim panels and remove

29.6  Unclip the windshield pillar trim panels and remove them from the vehicle

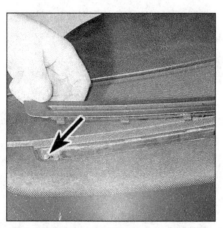

29.8  Unclip the windshield vents from the top of the dash to gain access to the vent screws (arrow)

them **(see illustration)**.
7  Loosen and remove the retaining screws and remove the small section of trim from the passenger end of the dash **(see illustration)**.
8  Unclip the trim covers from around the left and right-hand windshield vents and remove the screws securing the vents to the dash **(see illustration)**.
9  Loosen and remove the retaining bolts from the left and right-hand ends of the dash and recover the spacers **(see illustration)**.
10  Carefully ease the dash assembly away from the firewall. As it is withdrawn, release the wiring harness from its retaining clips on the rear of the dash, while noting its correct routing. **Note:** *Label each wiring connector as it is disconnected from its relevant component. The labels will prove useful on installation, when routing the wiring and feeding the wiring through the dash apertures.* Remove the dash assembly from the vehicle.

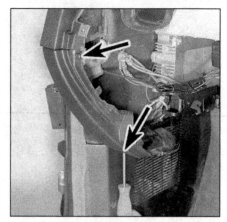

29.7  Unscrew the retaining screws and remove the trim panel from the passenger end of the dash

29.9  Dash retaining bolt (arrow)

#### Installation

11  Installation is a reversal of the removal procedure, noting the following points:
  a) *Maneuver the dash into position and, using the labels stuck on during removal, ensure that the wiring is correctly routed and securely retained by its dash clips.*
  b) *Clip the dash back into position, making sure all the wiring connectors are feed through their respective apertures, then install all the dash fasteners, and tighten them securely.*
  c) *On completion, reconnect the battery and check that all the electrical components and switches function correctly.*

### Z3 models

12  On these models, it's not necessary to remove the dash panel assembly in order to perform any procedure included in this manual.

# Chapter 12
# Chassis electrical system

## Contents

| | Section |
|---|---|
| Airbag system components - removal and installation | 25 |
| Airbag system - general information and precautions | 24 |
| Anti-theft alarm system - general information | 22 |
| Antenna (3-Series models) - general information | 20 |
| Audio unit - removal and installation | 18 |
| Back-up light switch (models equipped with manual transmission) removal and installation | See Chapter 7A |
| Battery check and maintenance | See Chapter 1 |
| Battery - removal and installation | See Chapter 5 |
| Brake light switch - removal and installation | See Chapter 9 |
| Bulbs (exterior lights) - replacement | 5 |
| Bulbs (interior lights) - replacement | 6 |
| Central locking components - removal and installation | 28 |
| Cigarette lighter - removal and installation | 12 |
| Clock/multi-information display/on-board trip computer – removal and installation | 13 |
| Cruise control system components (3-Series models) – removal and installation | 21 |
| Door lock heating system - general information and component removal and installation | 27 |
| Electrical troubleshooting - general information | 2 |
| Electric window components - removal and installation | 29 |

| | Section |
|---|---|
| Exterior light units - removal and installation | 7 |
| Fuses and relays - general information | 3 |
| General information and precautions | 1 |
| Headlight beam alignment - general information | 8 |
| Heated front seat components - removal and installation | 23 |
| Horns - removal and installation | 14 |
| Instrument cluster components - removal and installation | 10 |
| Instrument cluster - removal and installation | 9 |
| Multi-information system - general information and component removal and installation | 26 |
| Speakers - removal and installation | 19 |
| Speedometer drive sender unit - removal and installation | 11 |
| Starter inhibitor/back-up light switch (models with an automatic transmission) - removal and installation | See Chapter 7B |
| Switches - removal and installation | 4 |
| Windshield wiper blade check and replacement | See Chapter 1 |
| Windshield wiper motor and linkage - removal and installation | 16 |
| Windshield/headlight washer system check and adjustment | See Chapter 1 |
| Windshield/headlight washer system components - removal and installation | 17 |
| Wiper arm - removal and installation | 15 |

## Specifications

### Torque specifications

| | |
|---|---|
| Wiper arm to wiper spindle nut | 18 ft-lbs |
| Wiper motor and linkage | |
| Motor-to-linkage bolts | 84 in-lbs |
| Crank-to-motor nut | 19 ft-lbs |
| Wiper spindle nuts | 108 in-lbs |
| Support bracket bolts | 84 in-lbs |
| Airbag system | |
| Driver's side airbag retaining screws | 72 in-lbs |
| Impact sensor mounting bolts | 84 in-lbs |

## 1 General information and precautions

**Warning:** *Before carrying out any work on the electrical system, read through the precautions given in Safety First! at the beginning of this manual and Chapter 5 Part A.*

The electrical system is of the 12 volt negative ground type. Power for the lights and all electrical accessories is supplied by a lead/acid type battery which is charged by the alternator.

This Chapter covers repair and service procedures for the various electrical components not associated with engine. Information on the battery, alternator and starter motor can be found in Part A of Chapter 5.

It should be noted that prior to working on any component in the electrical system, the battery negative terminal should first be disconnected to prevent the possibility of electrical short circuits and/or fires. **Caution:** *If the stereo in your vehicle is equipped with an anti-theft system, make sure you have the correct activation code before disconnecting the battery.*

## 2 Electrical troubleshooting - general information

**Note:** *Refer to the precautions given in Safety first! and in Chapter 5 Part A before starting work. The following tests relate to testing of the main electrical circuits, and must not be used to test delicate electronic circuits (such as anti-lock braking and airbag systems), particularly where an electronic control module (ECM) is used.*

## General

1  A typical electrical circuit consists of an electrical component, any switches, relays, motors, fuses, fusible links or circuit breakers related to that component, and the wiring and connectors which link the component to both the battery and the chassis. To help to pinpoint a problem in an electrical circuit, wiring diagrams are included at the end of this Manual.

2  Before attempting to diagnose an electrical fault, first study the appropriate wiring diagram to obtain a complete understanding of the components included in the particular circuit concerned. The possible sources of a fault can be narrowed down by noting if other components related to the circuit are operating properly. If several components or circuits fail at one time, the problem is likely to be related to a shared fuse or ground connection.

3  Electrical problems usually stem from simple causes, such as loose or corroded connections, a faulty ground connection, a blown fuse, a melted fusible link, or a faulty relay (refer to Section 3 for details of testing relays). Visually inspect the condition of all fuses, wires and connections in a problem circuit before testing the components. Use the wiring diagrams to determine which terminal connections will need to be checked in order to pinpoint the trouble spot.

4  The basic tools required for electrical fault-finding include a circuit tester or voltmeter (a 12-volt bulb with a set of test leads can also be used for certain tests); a self-powered test light (sometimes known as a continuity tester); an ohmmeter (to measure resistance); a battery and set of test leads; and a jumper wire, preferably with a circuit breaker or fuse incorporated, which can be used to bypass suspect wires or electrical components. Before attempting to locate a problem with test instruments, use the wiring diagram to determine where to make the connections.

5  To find the source of an intermittent wiring fault (usually due to a poor or dirty connection, or damaged wiring insulation), a "wiggle" test can be performed on the wiring. This involves wiggling the wiring by hand to see if the fault occurs as the wiring is moved. It should be possible to narrow down the source of the fault to a particular section of wiring. This method of testing can be used in conjunction with any of the tests described in the following sub-Sections.

6  Apart from problems due to poor connections, two basic types of fault can occur in an electrical circuit - open circuit, or short circuit.

7  Open circuit faults are caused by a break somewhere in the circuit, which prevents current from flowing. An open circuit fault will prevent a component from working, but will not cause the relevant circuit fuse to blow.

8  Short circuit faults are caused by a "short" somewhere in the circuit, which allows the current flowing in the circuit to "escape" along an alternative route, usually to ground. Short circuit faults are normally caused by a breakdown in wiring insulation, which allows a feed wire to touch either another wire, or a grounded component such as the bodyshell. A short circuit fault will normally cause the relevant circuit fuse to blow.

## Finding an open circuit

9  To check for an open circuit, connect one lead of a circuit tester or voltmeter to either the negative battery terminal or a known good ground.

10  Connect the other lead to a connector in the circuit being tested, preferably nearest to the battery or fuse.

11  Switch on the circuit, bearing in mind that some circuits are live only when the ignition switch is moved to a particular position.

12  If voltage is present (indicated either by the tester bulb lighting or a voltmeter reading, as applicable), this means that the section of the circuit between the relevant connector and the battery is problem-free.

13  Continue to check the remainder of the circuit in the same fashion.

14  When a point is reached at which no voltage is present, the problem must lie between that point and the previous test point with voltage. Most problems can be traced to a broken, corroded or loose connection.

## Finding a short circuit

15  To check for a short circuit, first disconnect the load(s) from the circuit (loads are the components which draw current from a circuit, such as bulbs, motors, heating elements, etc.).

16  Remove the relevant fuse from the circuit, and connect a circuit tester or voltmeter to the fuse connections.

17  Switch on the circuit, bearing in mind that some circuits are live only when the ignition switch is moved to a particular position.

18  If voltage is present (indicated either by the tester bulb lighting or a voltmeter reading, as applicable), this means that there is a short circuit.

19  If no voltage is present, but the fuse still blows with the load(s) connected, this indicates an internal fault in the load(s).

## Finding a ground fault

20  The battery negative terminal is connected to "ground"- the metal of the engine/transmission and the car body - and most systems are wired so that they only receive a positive feed, the current returning through the metal of the car body. This means that the component mounting and the body form part of that circuit. Loose or corroded mountings can therefore cause a range of electrical faults, ranging from total failure of a circuit, to a puzzling partial fault. In particular, lights may shine dimly (especially when another circuit sharing the same ground point is in operation), motors (such as the wiper motors or the radiator cooling fan motor) may run slowly, and the operation of one circuit may have an apparently unrelated effect on another. Note that on many vehicles, ground straps are used between certain components, such as the engine/transmission and the body, usually where there is no metal-to-metal contact between components due to flexible rubber mountings, etc.

21  To check whether a component is properly grounded, disconnect the battery and connect one lead of an ohmmeter to a known good ground point. Connect the other lead to the wire or ground connection being tested. The resistance reading should be zero; if not, check the connection as follows.

22  If a ground connection is thought to be faulty, disassemble the connection and clean back to bare metal both the bodyshell and the wire terminal or the component ground connection mating surface. Be careful to remove all traces of dirt and corrosion, then use a knife to trim away any paint, so that a clean metal-to-metal joint is made. On reassembly, tighten the joint fasteners securely; if a wire terminal is being reinstalled, use serrated washers between the terminal and the bodyshell to ensure a clean and secure connection. When the connection is remade, prevent the onset of corrosion in the future by applying a coat of petroleum jelly or silicone-based grease or by spraying on (at regular intervals) a proprietary ignition sealer or a water-dispersing lubricant.

## 3  Fuses and relays - general information

### Main fuses

Refer to illustrations 3.1a, 3.1b, 3.1c and 3.3

1  Most of the fuses are located behind the fuse box in the left-hand rear corner of the engine compartment (see illustration). On

3.1a  Most of the fuses and relays are located in the front power distribution box in the left rear corner of the engine compartment (Z3 model shown, other models similar)

# Chapter 12 Chassis electrical system

3.1b Other fuses and relays are located in the E-box, in the right rear corner of the engine compartment (Z3 model shown, other models similar)

3.1c Some fuses and relays are also located above the left kick panel (Z3 model shown, other models similar)

Z3 models, there are also some fuses and relays in the E-box in the right-hand rear corner of the engine compartment **(see illustration)** and above the left kick panel **(see illustration)**.

2   To remove the fuse box cover, release its front retaining clip then unclip the cover.

3   A list of the circuits each fuse protects is given on the label attached to the inside of the fuse box cover. A pair of tweezers for removing the fuses is also clipped to the lid **(see illustration)**.

4   To remove a fuse, first switch off the circuit concerned (or the ignition), then pull the fuse out of its terminals using the tweezers which are clipped to the inside of the fuse box cover. The wire within the fuse should be visible; if the fuse is blown it will be broken or melted.

5   Always replace a fuse with one of an identical rating; never use a fuse with a different rating from the original or substitute anything else. Never replace a fuse more than once without tracing the source of the trouble. The fuse rating is stamped on top of the fuse; note that the fuses are also color-coded for easy recognition.

6   If a new fuse blows immediately, find the cause before replacing it again; a short to ground as a result of faulty insulation is most likely. Where a fuse protects more than one circuit, try to isolate the defect by switching on each circuit in turn (if possible) until the fuse blows again. Always carry a supply of spare fuses of each relevant rating on the vehicle, a spare of each rating should be clipped into the base of the fuse box.

## Relays

*Refer to illustrations 3.7a, 3.7b, 3.7c, 3.7d and 3.9*

7   The majority of relays are located in the fuse box in the left-hand rear corner of the engine compartment. Additional relays can be found in the relay carrier under the left-hand side of the dash or attached to the connector strips under the left and right-hand sides of the dash **(see illustrations)**.

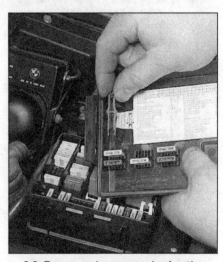

3.3 Fuses can be removed using the tweezers which are clipped to the fuse box lid. Fuse locations are given on the sticker attached to the lid

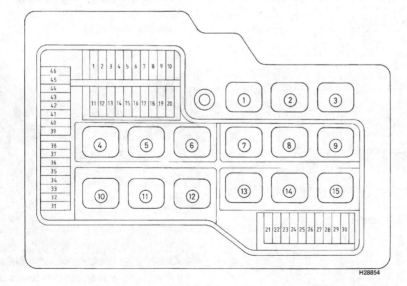

3.7a Typical fuse box relay locations (fuse locations also numbered) (3-Series model shown)

1   Fuel pump relay
2   DME relay
3   Oxygen sensor relay
4   Horn relay
5   Foglight relay
6   Headlight relay
7   Main beam relay
8   Hazard warning light relay
9   Blower motor relay
10  Heated rear window relay
11  ABS overvoltage protection relay
12  ABS pump motor relay
13  Auxiliary cooling fan stage 2 relay
14  Air conditioning compressor relay
15  Auxiliary cooling fan stage 1 relay

*Not all relays present on all models

# Chapter 12 Chassis electrical system

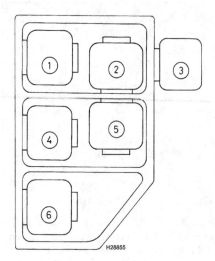

**3.7b Typical relay carrier under behind left-hand side of dash (3-Series model shown)**

1. Independent ventilation relay
2. Double relay module
3. Rear window opening relay - coupe models
4. Crash alarm sensor relay
5. Headlight washer module relay
6. Comfort relay

*Not all relays present on all models

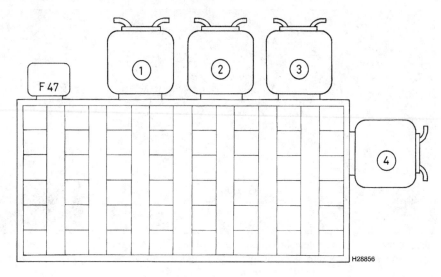

**3.7c Typical connector strip behind the left-hand side of the dash (3-Series model shown)**

1. Unloader relay or starter interlock relay (depending on model)
2. Wiper relay
3. Wiper motor relay
4. Unloader relay or starter interlock relay (depending on model)

*Not all relays present on all models

8  If a circuit or system controlled by a relay develops a fault and the relay is suspect, operate the system; if the relay is functioning it should be possible to hear it click as it is energized. If this is the case the fault lies with the components or wiring of the system. If the relay is not being energized then either the relay is not receiving a main supply or a switching voltage or the relay itself is faulty. Testing is by the substitution of a known good unit but be careful; while some relays are identical in appearance and in operation, others look similar but perform different functions.

9  To replace a relay first ensure that the ignition switch is off. The relay can then simply be pulled out from the socket and the new relay pressed in (see illustration).

## 4 Switches - removal and installation

**Note:** *Disconnect the battery negative cable before removing any switch, and reconnect the lead after installing the switch.* **Caution:** *If the stereo in your vehicle is equipped with an anti-theft system, make sure you have the correct activation code before disconnecting the battery.*

### 3-Series models

#### Ignition switch/steering column lock

1  Refer to Chapter 10.

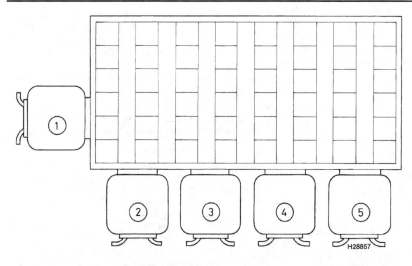

**3.7d Typical connector strip behind the right-hand side of dash (3-Series model shown)**

1. Alarm horn relay
2. Right-hand parking light/license plate light relay
3. Left-hand parking light relay
4. Central locking relay
5. Rear window blower motor relay

*Not all relays present on all models

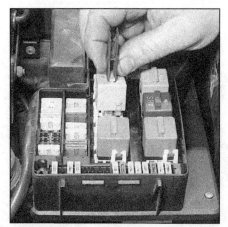

**3.9 Removing a relay from the fuse box**

# Chapter 12 Chassis electrical system

4.3 Removing the driver's side lower dash panel (right-hand drive model shown, left-hand drive similar)

4.4a Remove the screws...

4.4b ... then unclip and remove the steering column shrouds

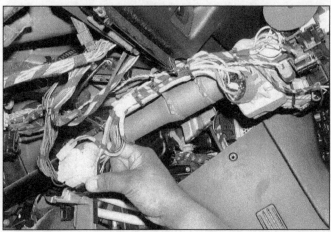

4.5 Disconnect the combination switch wiring connector and free the wiring from the column

4.6 Unscrew and unclip the fastener...

## Steering column combination switches

*Refer to illustrations 4.3, 4.4a, 4.4b, 4.5, 4.6 and 4.7*

2  Remove the steering wheel (see Chapter 10).

3  Remove the driver's side lower dash panel retaining screws then unclip the panel and remove it from the vehicle **(see illustration)**.

4  Unscrew the steering column upper and lower fastener screws and pull out the fasteners. Unclip the two halves of the shroud and remove them from the column **(see illustrations)**.

5  Trace the wiring back from the switch, freeing it from the steering column, and disconnect it at the wiring connector **(see illustration)**.

6  Loosen the switch holder fastener screw and pull out the fastener **(see illustration)**.

7  Release the switch holder from the column then depress the retaining tangs and slide the switch(es) out from the holder **(see illustration)**.

8  Installation is a reversal of the removal procedure ensuring that the wiring is correctly routed.

## Lighting switch

*Refer to illustrations 4.10, 4.11 and 4.12*

9  Remove the driver's side lower dash panel retaining screws then unclip the panel and remove it from the vehicle.

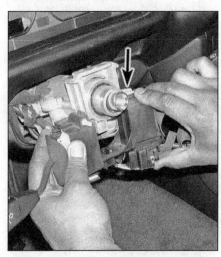

4.7 ... then release the holder from the top of the column and slide out the relevant switch

10  Pull off the light switch control knob to gain access to the switch retaining nut **(see illustration)**.

11  Unscrew the retaining nut then free the switch from the rear of the panel and lower it

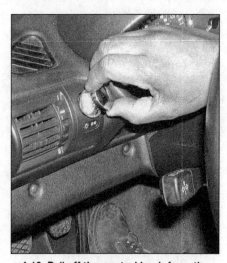

4.10 Pull off the control knob from the lighting switch...

4.11 ... and unscrew the retaining nut

4.12 Lower the lighting switch out of position and disconnect its wiring connector

4.14 On manual transmission models, unclip the boot from the center console and fold it back over the lever

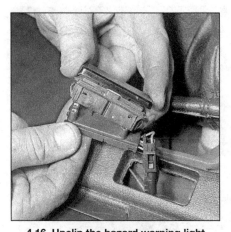

4.16 Unclip the hazard warning light switch and disconnect its wiring connector

4.20 Removing the foglight switch

4.22 Pry the electric mirror switch out from the door trim panel . . .

out of position **(see illustration)**.
12  Disconnect the wiring connector and remove the switch **(see illustration)**.
13  Installation is the reverse of removal making sure the switch groove is correctly engaged with the panel lug.

### Hazard warning light and electric window switches

*Refer to illustrations 4.14 and 4.16*

14  On manual transmission models, carefully unclip the gearchange lever boot from the center console and fold it back over the lever. Where necessary slide the rubber insert up the gearchange lever **(see illustration)**.
15  On models with automatic transmission, remove the retaining screw then depress the detent button and slide the handle off from the top of the selector lever. Unclip the selector lever position display panel from the top of the center console and slide off the selector lever. If the position display panel is a tight fit in the console, remove the storage compartment (see Section 12) and push the panel out from behind.
16  Reach in behind the center console and push the switch out of position and disconnect its wiring connector **(see illustration)**.

17  Installation is the reverse of removal. On automatic transmission models make sure the selector lever handle detent button is correctly engaged with the lever rod before installing the retaining screw.

### Foglight, instrument panel dimmer and headlight leveling switches

*Refer to illustration 4.20*

18  Remove the driver's side lower dash panel retaining screws then unclip the panel and remove it from the vehicle.
19  Reach up behind the dash and disconnect the wiring connector from the switch.
20  Depress the retaining tabs and slide the switch out of position **(see illustration)**.
21  Installation is the reverse of removal.

### Exterior mirror adjustment switch

*Refer to illustrations 4.22 and 4.23*

22  Carefully lever the switch out from the door panel **(see illustration)**.
23  Disconnect the wiring connector and remove the switch **(see illustration)**.
24  Installation is the reverse of removal.

### Cruise control system switch

25  Remove the driver's side lower dash panel retaining screws then unclip the panel and remove it from the vehicle.

26  Unscrew the steering column lower shroud fastener screw and pull out the fastener. Unclip the lower shroud and remove it from the column.
27  Trace the wiring back from the switch, freeing it from the steering column, and disconnect it at the wiring connector.
28  Depress the retaining clips and slide the switch out of position.
29  Installation is the reverse of removal.

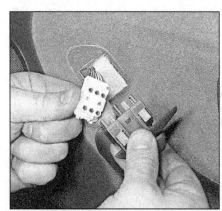

4.23 . . . and disconnect its wiring connector

# Chapter 12  Chassis electrical system

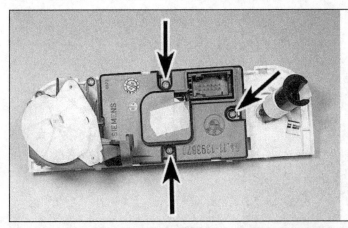

4.33a  Remove the retaining screws (arrows) ...

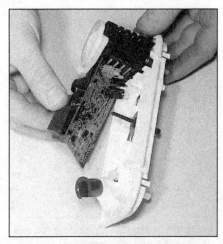

4.33b  ... then unclip the cover and remove the printed circuit board from the heater control panel

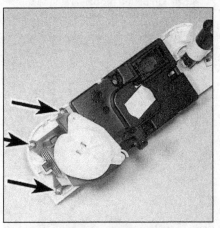

4.37a  Remove the retaining screws (arrows) ...

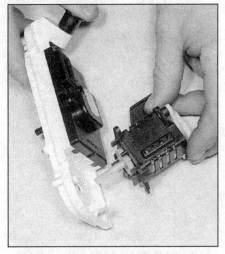

4.37b  ... and remove the blower motor switch from the rear of the heater control panel

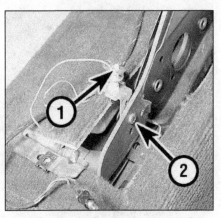

4.44  Parking brake warning light switch wiring connector (1) and retaining screw (2)

## Heated rear window switch

*Refer to illustrations 4.33a and 4.33b*

**Models with an automatic air conditioning system**

30  On these models the switch is an integral part of the control unit and cannot be replaced. If the switch is faulty seek the advice of a BMW dealer.

**Models with manually adjusted air conditioning and models without air conditioning**

31  On these models the switch is an integral part of the heater control panel printed circuit.
32  Remove the climate control or heater control panel (see Chapter 3).
33  Remove the retaining screws then release the retaining clips and detach the printed circuit cover from the rear of the control unit. Lift out the printed circuit board **(see illustrations)**.
34  Installation is the reverse of removal. Check the operation of the switch before installing the control panel to the dash.

## Heater blower motor switch

*Refer to illustrations 4.37a and 4.37b*

**Models with an automatic air conditioning system**

35  On these models the switch is an integral part of the control unit and cannot be replaced. If the switch is faulty seek the advice of a BMW dealer.

**Models with a manually adjusted air conditioning system and models without air conditioning**

36  Remove the heater control panel (see Chapter 3).
37  Remove the retaining screws then unclip the switch from the rear of the control panel and remove it **(see illustrations)**.
38  Installation is the reverse of removal. Check the operation of the switch before installing the control panel to the dash.

## Air conditioning system switches

39  Refer to Steps 35 to 38.

## Heated seat, electrically operated rear sun blind and traction control (ASC+T) switches

40  Remove the storage compartment from the center console (see Section 12).
41  Disconnect the wiring connector then depress the retaining clips and slide the switch out from the panel.
42  Installation is the reverse of removal.

## Parking brake warning light switch

*Refer to illustration 4.44*

43  Remove the rear section of the center console (see Chapter 11) to gain access to the parking brake lever.
44  Disconnect the wiring connector from the warning light switch then remove the retaining screw and remove the switch **(see illustration)**.
45  Installation is the reverse of removal. Check the operation of the switch before installing the center console, the warning light should illuminate between the first and second clicks of the ratchet mechanism.

## Brake light switch

46  Refer to Chapter 9.

## Front door courtesy light switch - sedan models

*Refer to illustrations 4.49a and 4.49b*

47  Open the door. The courtesy light switch is an integral part of the lock catch.
48  Using a suitable marker pen, mark the outline of the catch on the door pillar.
49  Loosen and remove the catch retaining screws then disconnect the wiring connector

# 12-8  Chapter 12 Chassis electrical system

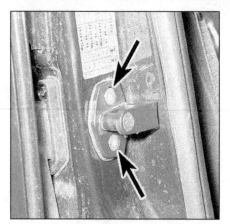

4.49a  On sedan models, remove the two retaining screws (arrows) . . .

4.49b  . . . then remove the catch and disconnect the wiring connector

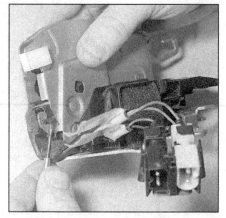

4.52a  On coupe models, release the retaining clip . . .

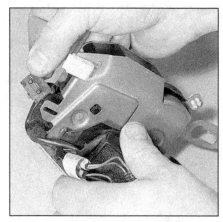

4.52b  . . . and remove the courtesy light switch from the door lock

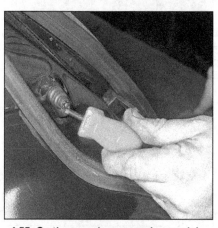

4.55  On the rear door on sedan models, remove the retaining screw . . .

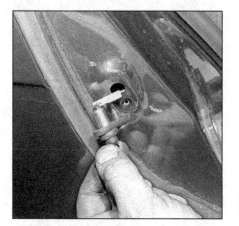

4.56  . . . then remove the switch and disconnect the wiring connector

and remove the catch from the pillar. Take great care not to allow the caged nut plate or wiring connector to fall down inside the door pillar **(see illustrations)**. If either component falls down inside the pillar the trim panel will have to be removed (see Chapter 11).
50  Installation is the reverse of removal, aligning the catch with the marks made prior to removal. Shut the door and check that it locks securely; if necessary adjust by slackening the screws and repositioning the catch.

### Front door courtesy light switch - coupe models

*Refer to illustrations 4.52a and 4.52b*
51  Remove the door lock (see Chapter 11).
52  Release the retaining clip and detach the microswitch from the lock **(see illustrations)**.
53  Install the new switch making sure it is securely retained by the clip.
54  Install the lock (see Chapter 11).

### Rear door courtesy light switch - sedan models

*Refer to illustrations 4.55 and 4.56*
55  Open up the door and loosen the switch retaining screw **(see illustration)**.
56  Withdraw the switch, disconnecting its wiring connector as it becomes accessible.

Tie a piece of string to the wiring to prevent it falling back into the door pillar **(see illustration)**.
57  Installation is a reverse of the removal procedure.

### Luggage compartment light switch

58  Open up the lid then loosen and remove the screw securing the switch to the base of the lid.
59  Disconnect the wiring connector and

4.61  Unclip the switch panel from the headlining . . .

remove the switch.
60  Installation is the reverse of removal.

### Electric sunroof switch

*Refer to illustrations 4.61 and 4.62*
61  Carefully unclip the switch panel from the headlining **(see illustration)**.
62  Disconnect the switch wiring connector then depress the retaining clips and slide the switch out of position **(see illustration)**.
63  Installation is the reverse of removal.

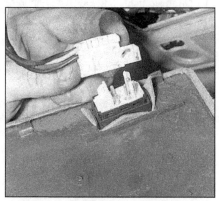

4.62  . . . then disconnect the wiring connector and separate the sunroof switch and panel

# Chapter 12 Chassis electrical system

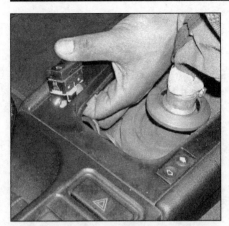

4.72a To remove a power window switch from the center console, remove the shift lever boot, reach under the console, push up the switch and unplug the electrical connector

4.72b To detach the hazard warning switch from the center console, remove the shift lever boot, reach under the console, push up the switch and unplug the electrical connector

4.80 To remove a switch (this one's the air conditioning/recirculation switch) from the upper center console panel, either reach through the radio opening and push out the switch or pry it out and then unplug the electrical connector

## Z3 models

### Ignition switch/steering column lock

64  The ignition switch and key lock cylinder are identical to the switch and lock cylinder used on 3-Series models. Refer to Section 27, Chapter 11, to remove the steering column covers. Refer to Section 19 in Chapter 10 to remove or replace the switch.

### Steering column combination switches

65  The turn signal switch and wiper switch are identical to the switches used on 3-Series models. Refer to Section 27, Chapter 11, to remove the steering column covers. Refer to Steps 2 through 8 to remove or replace the switches.

### Headlight switch

66  Pry off the center trim cap from the headlight switch knob with a small screwdriver and remove the cap **(see illustration 4.10)**.
67  Back off the headlight switch knob locknut **(see illustration 4.11)**, then unscrew and remove the knob.
68  Remove the left lower dash trim panel and the knee bolster (see Section 27 in Chapter 11).
69  Squeeze the two headlight switch locking tabs and push the switch into the dash.
70  Reach up behind the dash, pull down the switch and unplug the electrical connector.
71  Installation is the reverse of removal.

### Hazard warning light and electric window switches

*Refer to illustrations 4.72a and 4.72b*

72  These switches are similar in location and installation to those used on 3-Series models. Refer to Steps 14 through 17. The accompanying illustrations show how to remove two typical switches **(see illustrations)**.

### Automatic transmission program selector switch

73  Remove the gear position indicator trim panel by prying up on the edge. Remove the trim panel. **Note:** *If the gear position indicator trim panel won't fit over the shift lever handle, unscrew the handle retaining bolt with a 3 mm Allen socket driver, then pull the handle straight up.*
74  The remainder of the removal procedure for this switch is similar to the hazard warning light and electric window switches in the console (see previous step). Compress the retaining tabs at both ends of the switch, then push the switch up to remove it.
75  Unplug the electrical connector from the switch.
76  Installation is the reverse of removal.

### Upper center console switches

*Refer to illustration 4.80*

77  Depending on the model, there are up to five switches in the upper center console (models without all five switches have a blank-off plate in place of a switch):

   *Air conditioning compressor control and climate control internal air recirculation switches (dual switches in a single housing)*

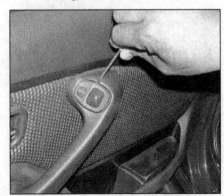

4.91 Pry out the power mirror adjustment button/switch assembly and remove it

   *Driver's side heated seat control switch*
   *Passenger side heated seat control switch*
   *Traction control disable switch*
   *Power convertible top control switch*

These five switch housings are identical, so the removal procedure is the same for each of them:
78  Remove the center console trim panel (see Chapter 11).
79  Remove the radio (see Section 18).
80  Reach through the radio opening and push out the switch or pry it out **(see illustration)** from the center console.
81  Pull the switch out of the console and unplug the electrical connector.
82  Installation is the reverse of removal.

### Interior light switch

83  Remove the trim panels from the A-pillars as follows. Pull each A-pillar trim toward the middle of the windshield, then obliquely upward. Discard any broken trim clips.
84  Remove both sun visor assemblies.
85  Remove the windshield header interior light assembly.
86  Remove the screw under the interior light assembly.
87  Rotate the rear edge of the windshield header trim panel down, pull it back and then remove it.
88  Unplug the electrical connector from the interior light.
89  Remove the retaining clips that attach the light switch assembly to the windshield header trim panel. Remove the interior light switch.
90  Installation is the reverse of removal.

### Power mirror switch

*Refer to illustrations 4.91 and 93*

91  Pry out the power mirror adjustment button and switch assembly **(see illustration)** and remove it.
92  Squeeze the power mirror switch locking tabs and separate the switch from the

# Chapter 12 Chassis electrical system

4.93 Unplug the electrical connector from the switch

4.96 Depress the locking tab (arrow) and remove the cruise control selector lever and switch assembly by pulling it straight out

4.105 Pry the heated seat switch trim panel out of the center console with a small flat-bladed screwdriver

button assembly.
93 Unplug the electrical connector from the switch **(see illustration)**.
94 Installation is the reverse of removal.

### Cruise control system switch

*Refer to illustration 4.96*
95 Remove the steering column covers (see Section 27 in Chapter 11).
96 Depress the locking tab **(see illustration)** and remove the cruise control selector lever and switch assembly by pulling it straight out.
97 Cut the wire harness retaining straps near the column switch assembly and unplug the electrical connectors from the switch assembly.
98 Installation is the reverse of removal.

### Heater blower motor switch

99 Remove the radio (see Section 18).
100 Pull off the blower motor switch knob.
101 Remove the blower switch retaining nut.
102 Push the blower switch into the dash, then reach in through the opening for the radio and pull out the switch.
103 Unplug the switch electrical connector and remove the switch.
104 Installation is the reverse of removal.

### Heated seat switch

*Refer to illustrations 4.105 and 4.107*
105 Pry the edge of the heated seat switch trim panel out of the center console with a small flat-bladed screwdriver **(see illustration)**.
106 Unplug the electrical connector from the back of the heated seat switches and remove the switch/trim panel assembly.
107 Squeeze the locking tangs and push the switch out of the trim panel **(see illustration)**.

## 5  Bulbs (exterior lights) - replacement

### General

1  Whenever a bulb is replaced, note the following points.
 a) Disconnect the battery negative cable before starting work. **Caution:** *If the stereo in your vehicle is equipped with an anti-theft system, make sure you have the correct activation code before disconnecting the battery.*
 b) Remember that if the light has just been in use the bulb may be extremely hot.
 c) Always check the bulb contacts and holder, ensuring that there is clean metal-to-metal contact between the bulb and its live contact(s) and ground. Clean off any corrosion or dirt before fitting a new bulb.
 d) Wherever bayonet-type bulbs are installed, ensure that the live contact(s) bear firmly against the bulb contact.
 e) Always ensure that the new bulb is of the correct rating and that it is completely clean before fitting it; this applies particularly to headlight/foglight bulbs (see below).

### 3-Series models

#### Headlight

*Refer to illustrations 5.2, 5.3, 5.4a and 5.4b*
2  Release the retaining clip and remove the relevant access cover from the rear of the headlight unit **(see illustration)**. To improve access to the left-hand headlight, remove the air cleaner housing (see Chapter 4).
3  Disconnect the wiring connector from the rear of the bulb **(see illustration)**.
4  Unhook and release the ends of the bulb

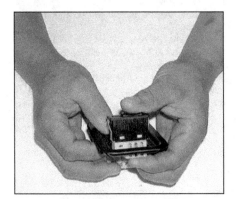

4.107 To separate either heated seat switch from the trim panel, depress the two locking tangs on the side of the switch and push the switch out of the trim panel

5.2 Release the retaining clips and remove the access cover from the rear of the headlight

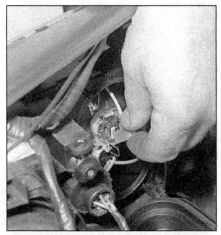

5.3 Disconnect the wiring connector . . .

# Chapter 12 Chassis electrical system

5.4a ... then release the spring clip ...

5.4b ... and withdraw the bulb from the headlight

5.9 Removing the sidelight bulb holder from the headlight

5.11 Release the direction indicator light retaining clip with a screwdriver ...

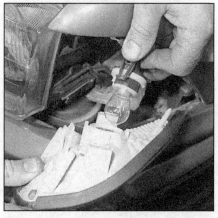

5.12 ... then slide the light out of position and unclip the bulb holder

5.14 Remove the retaining screw ...

retaining clip and release it from the rear of the light unit. Withdraw the bulb **(see illustrations)**.

5  When handling the new bulb, use a tissue or clean cloth to avoid touching the glass with the fingers; moisture and grease from the skin can cause blackening and rapid failure of this type of bulb. If the glass is accidentally touched, wipe it clean using isopropyl alcohol.

6  Install the new bulb, ensuring that its locating tabs are correctly located in the light cutouts, and secure it in position with the retaining clip.

7  Reconnect the wiring connector and install the access cover, making sure it is securely reinstalled.

### Front sidelight

*Refer to illustration 5.9*

8  Depress the retaining clip and remove the access cover from the rear of the headlight unit.

9  Withdraw the sidelight bulb holder from the headlight unit **(see illustration)**. The bulb is of the capless type and is a push fit in the holder.

10  Installation is the reverse of the removal procedure making sure the access cover is securely reinstalled.

### Front turn signal light

*Refer to illustrations 5.11 and 5.12*

11  Using a screwdriver, release the retaining clip and withdraw the direction indicator light from the fender **(see illustration)**.

12  Unclip the bulb holder and remove it from the rear of the light unit **(see illustration)**. The bulb is a bayonet fit in the holder and can be removed by pressing it and twisting in a counterclockwise direction.

13  Installation is a reverse of the removal procedure making sure the light unit is securely retained by its spring.

### Front turn signal side marker light

*Refer to illustrations 5.14, 5.15a and 5.15b*

14  Remove the retaining screw and withdraw the light unit from the fender **(see illustration)**.

15  Twist the bulb holder counterclockwise and remove it from the light. The bulb is of the capless (push-fit) type and can be removed by simply pulling it out of the bulb holder **(see illustrations)**.

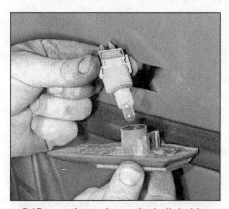

5.15a ... then release the bulb holder from the rear of the lens ...

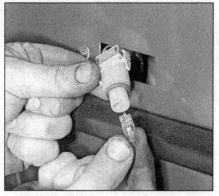

5.15b ... and pull out the bulb

# Chapter 12 Chassis electrical system

5.18 Unclip the foglight and release the bulb cover by twisting it counterclockwise

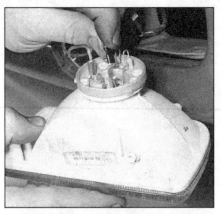

5.19a Disconnect the wiring connector then release the retaining clip . . .

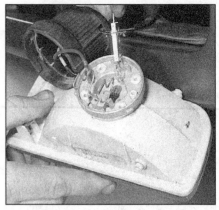

5.19b . . . and withdraw the bulb

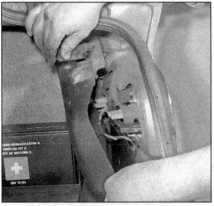

5.24 Release the retaining clip and remove the cover to gain access to the rear light bulbs

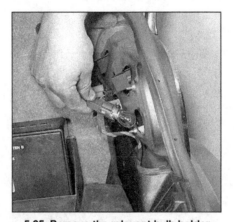

5.25 Remove the relevant bulb holder from the light unit by twisting it counterclockwise

5.29 Removing the trim cover from around the license plate lights

16  Installation is a reverse of the removal procedure. Do not overtighten the retaining screws as the lens is easily cracked.

### Front foglight

*Refer to illustrations 5.18, 5.19a and 5.19b*

17  Insert a flat-bladed screwdriver in through the front bumper grille and release the foglight retaining clip.
18  Withdraw the foglight from the bumper and rotate the foglight cover counterclockwise and release it from the rear of the light unit **(see illustration)**.
19  Disconnect the bulb wiring then release the spring clip and withdraw the foglight bulb **(see illustrations)**.
20  When handling the new bulb, use a tissue or clean cloth to avoid touching the glass with the fingers; moisture and grease from the skin can cause blackening and rapid failure of this type of bulb. If the glass is accidentally touched, wipe it clean using isopropyl alcohol.
21  Insert the new bulb, making sure it is correctly located, and secure it in position with the spring clip.
22  Connect the wiring to the bulb then install the cover to the rear of the unit.
23  Install the foglight to the bumper, making sure it is clipped securely in position.

### Rear taillight cluster

*Refer to illustrations 5.24 and 5.25*

24  From inside the vehicle luggage compartment, rotate the retaining clip 90-degrees and remove the plastic access cover from the rear of the light cluster **(see illustration)**.
25  Rotate the relevant bulb holder counterclockwise and remove it from the light unit **(see illustration)**. The bulbs have bayonet fittings and are removed by pressing in and rotating counterclockwise.
26  Installation is the reverse of the removal sequence.

### High-level brake light

27  From within the luggage compartment, rotate the bulb holder counterclockwise and release it from the light unit. The bulb is a bayonet fit in the holder and can be removed by pressing it and twisting in a counterclockwise direction.
28  Install by reversing the removal procedure.

### License plate light

*Refer to illustrations 5.29 and 5.31*

29  Remove the retaining screws and remove the trim cover from around the lights **(see illustration)**.
30  Press the light unit towards the left and unclip it from the trunk lid.
31  Release the bulb from the contacts and remove it from the light unit **(see illustration)**.
32  Installation is the reverse of removal, making sure the bulb is securely held in position by the contacts.

## Z3 models

### Headlight

*Refer to illustrations 5.33, 5.34 and 5.35*
33  Remove the headlight mounting bolts

5.31 Unclip the license plate light from the trunk lid and remove the bulb from its contacts

# Chapter 12 Chassis electrical system

5.33 Remove the headlight mounting bolts (arrows)

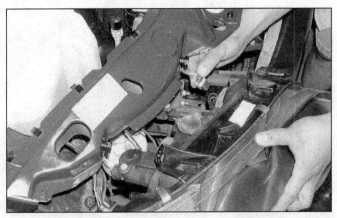

5.34 Pull out the headlight assembly far enough to unplug the electrical connectors for the two headlight bulbs

**(see illustration)**.
34  Pull out the headlight assembly far enough to unplug the electrical connectors for the two headlight bulbs **(see illustration)**.
35  Turn the bulb holder counterclockwise a 1/4-turn and pull it out of the headlight assembly **(see illustration)**. Pull the bulb straight out of the bulb holder.
36  When handling the new bulb, use a tissue or clean cloth to avoid touching the glass with the fingers; moisture and grease from the skin can cause blackening and rapid failure of this type of bulb. If the glass is accidentally touched, wipe it clean using isopropyl alcohol.
37  Install the new bulb in the bulb holder, insert the holder into the headlight assembly and turn it clockwise 1/4-turn to lock it into place. Reconnect the wiring connector and install the headlight assembly.
38  Tighten the headlight mounting bolts securely and then check the headlight adjustment (see Section 8).

### Front turn signal light

*Refer to illustration 5.40*
39  Pull out the headlight assembly (see Steps 33 and 34).
40  Turn the bulb holder counterclockwise 1/4-turn and pull it out **(see illustration)**.
41  To remove the old bulb from the holder,

5.35 Turn the bulb holder counterclockwise a 1/4-turn and pull it out of the headlight assembly

push it in, turn it counterclockwise 1/4-turn and pull it out of the socket.
42  Installation is the reverse of removal.

### Fog light

*Refer to illustrations 5.43, 5.44 and 5.45*
43  Pry out the cover plate next to the fog light lens with a small screwdriver **(see illustration)**.
44  Remove the two fog light retaining screws **(see illustration)**, pull out the fog

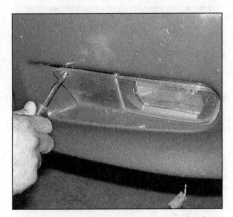

5.43 Pry out the cover plate next to the fog light with a small screwdriver

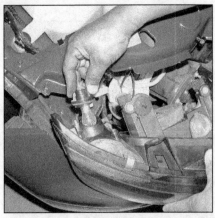

5.40 To detach the front turn signal light from the headlight assembly, turn the bulb holder counterclockwise 1/4-turn and pull it out

light assembly and unplug the electrical connector.
45  Separate the two halves of the fog light assembly, squeeze together the two ends of the wire spring clip that retains the bulb **(see illustration)**, then pull out the bulb and unplug the electrical connector.

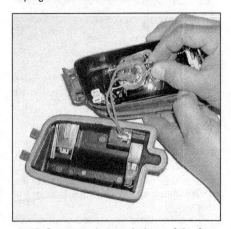

5.45 Separate the two halves of the fog light assembly and then squeeze together the two ends of the wire spring clip that retains the bulb

5.44 To detach the fog light assembly from the body, remove these two retaining screws (arrows)

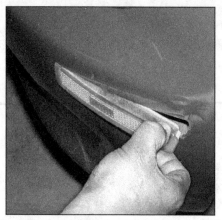

**5.47 Pull the parking light assembly out of the body**

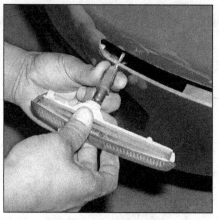

**5.48 Turn the bulb holder counterclockwise and pull it out of the parking light assembly**

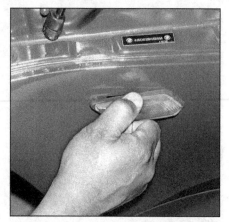

**5.51 To disengage the sidemarker light assembly from the fender, push it to the rear**

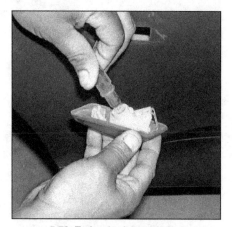

**5.52 Twist the lamp socket counterclockwise and pull it out of the sidemarker light housing**

**5.55a To remove the inside half of the rear taillight housing, unscrew this knurled retaining nut . . .**

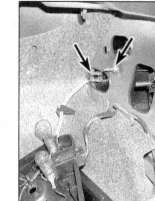

**5.55b . . . the side marker bulb (left arrow), which is located in the exterior part of the housing, can be accessed through the bulb hole; to remove the outer part of the taillight housing, remove the two retaining nuts (upper arrows)**

46  Installation is the reverse of removal. When handling the new bulb, use a tissue or clean cloth to avoid touching the glass with your fingers. Moisture and grease from the skin can cause this type of bulb to fail prematurely. If you accidentally touch the glass part of the bulb, clean it off with isopropyl alcohol.

### Front parking light

*Refer to illustrations 5.47 and 5.48*

47  Pull out the parking light assembly **(see illustration)**,.
48  Turn the bulb holder counterclockwise and pull it out of the parking light assembly **(see illustration)**.
49  Pull the bulb straight out of the holder.
50  Installation is the reverse of removal.

### Front sidemarker light

*Refer to illustrations 5.51 and 5.52*

51  Push the sidemarker light housing to the rear **(see illustration)**, then pull it out of the fender.
52  Twist the lamp socket counterclockwise and pull it out of the sidemarker light housing **(see illustration)**.
53  Pull the old bulb out of the socket.
54  Installation is the reverse of removal.

### Rear back-up, brake, sidemarker, taillight and turn signal bulbs

*Refer to illustrations 5.55a and 5.55b*

**Note:** *The rear back-up, brake, taillight and turn signal bulbs are installed inside a pair of plastic housings that can be easily accessed from the trunk.*

55  Open the trunk, unscrew the knurled retaining nut that secures the tail light cluster assembly and carefully pull the tail light assembly straight out of the body (the tail light bulbs fit through stamped holes in the body) **(see illustrations)**.
56  To remove a bulb from the tail light housing assembly, turn it counterclockwise and pull the bulb out of the housing. (The side marker bulbs are not installed in the inside part of the housing with the other bulbs; they're located in the exterior part of the housing. But they can be accessed through the outer bulb hole in the body.)
57  Installation is the reverse of removal. Before reinstalling the tail light assembly, make sure that the side marker light electrical leads are routed correctly so that they won't be damaged when the knurled nut is tightened.

### Center brake light

*Refer to illustrations 5.58 and 5.59*

58  Pry off the plastic brake light access cover from the inside of the trunk lid **(see illustration)**.

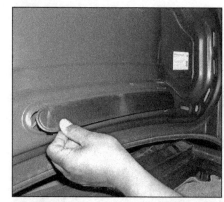

**5.58 To gain access to the center brake light bulbs, pry off this cover from the inside of the trunk lid**

# Chapter 12 Chassis electrical system

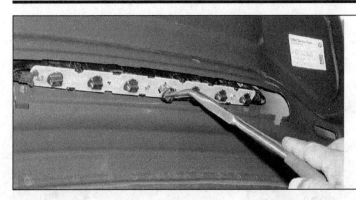

5.59 To remove a center brake light bulb socket, rotate the socket counterclockwise 1/4-turn, then pull the socket straight out

59  Rotate the socket with the bad bulb counterclockwise 1/4-turn, then pull the socket straight out **(see illustration)**.
60  Pull the old bulb straight out of the socket.
61  Installation is the reverse of removal.

## 6  Bulbs (interior lights) - replacement

### General
1  Refer to Section 5.

### Courtesy light
*Refer to illustration 6.2*
2  Using a small, flat-bladed screwdriver, carefully pry light unit out of position and release the bulb from the light unit contacts. On some lights it will be necessary to unclip the reflector to gain access to the bulb **(see illustration)**.
3  Install the new bulb, ensuring it is securely held in position by the contacts, and clip the lens back into position.

### Front seat reading light
*Refer to illustration 6.5*
4  Carefully lever the courtesy light lens out from the headlining.
5  Rotate the reading light bulb counterclockwise and remove it from the light unit

**(see illustration)**.
6  Install the new bulb into position and install the lens to the light unit.

### Luggage compartment light
7  Refer to the information given above in Steps 2 and 3.

### Instrument panel illumination/warning lights
*Refer to illustration 6.9*
8  Remove the instrument panel (see Section 9).
9  Twist the relevant bulb holder counterclockwise and withdraw it from the rear of the panel **(see illustration)**.
10  All bulbs are integral with their holders. Be very careful to ensure that the new bulbs are of the correct rating, the same as those removed; this is especially important in the case of the ignition/battery charging warning light.
11  Install the bulb holder to the rear of the instrument panel then install the instrument panel (see Section 9).

### Glovebox illumination light bulb
12  Open up the glovebox. Using a small flat-bladed screwdriver carefully pry the top of the light assembly and withdraw it. Release the bulb from its contacts.
13  Install the new bulb, ensuring it is

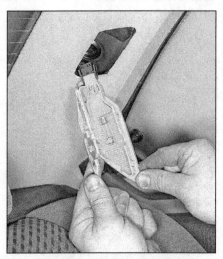

6.2 Replacing the rear pillar courtesy light bulb

securely held in position by the contacts, and clip the light unit back into position.

### Heater control panel illumination bulb
*Refer to illustration 6.17*

#### Models with an automatic air conditioning system
14  Remove the heater control panel from the dash (see Chapter 3) and remove the bulb holder from the rear of the control unit.
15  Installation is the reverse of removal.

#### Models with a manually adjusted air conditioning and models without air conditioning
16  Pull off the heater control panel knobs then remove the retaining screws and unclip the faceplate from the front of the control unit.
17  Using a pair of needle-nose pliers, rotate the bulb holder counterclockwise and remove it from the vehicle **(see illustration)**.
18  Installation is the reverse of removal.

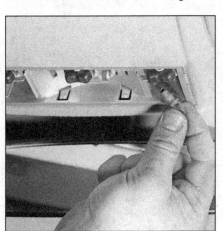

6.5 Replacing the front seat reading light bulb

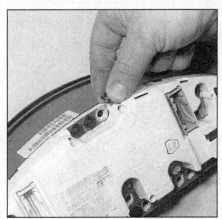

6.9 Instrument panel bulb holders can be removed by twisting them counterclockwise

6.17 Removing the heater control panel illumination bulb

# Chapter 12 Chassis electrical system

6.19 Lighting switch illumination bulb is a push-fit in the switch

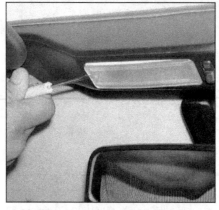

6.20 To access the overhead light in the Z3 windshield frame, pry off the lens

6.21 Remove the overhead light bulb (arrow) by pulling it straight out; make sure the new bulb is fully seated

## Switch illumination bulbs

*Refer to illustration 6.19*

19  All of the switches are equipped with illuminating bulbs; some also have a bulb to show when the circuit concerned is operating. On most switches, these bulbs are an integral part of the switch assembly and cannot be obtained separately. Bulb replacement will therefore require the replacement of the complete switch assembly. The exception to this is the lighting switch illumination bulb which can be removed once the switch has been removed (see Section 4); the bulb is a push-fit in its holder **(see illustration)**.

## Z3 Models

### Overhead light

*Refer to illustrations 6.20 and 6.21*

20  Pry off the overhead light lens **(see illustration)**.
21  Remove the overhead light bulb **(see illustration)**.
22  Installation is the reverse of removal.

### Instrument cluster illumination bulbs

*Refer to illustration 6.24*

23  Remove the instrument cluster (see Section 9).

24  Unscrew and remove the bulb holder **(see illustration)**.
25  Remove the bulb from the holder.
26  Installation is the reverse of removal.

### Climate control assembly illumination bulbs

27  Remove the climate control assembly (see Section 10 in Chapter 3).
28  Pull the illumination bulb out of the backside of the climate control assembly.
29  Installation is the reverse of removal.

### Switch illumination bulbs

30  Switch illumination bulbs on Z3 models are similar to those used on 3-Series models. Refer to Step 19.

## 7  Exterior light units - removal and installation

**Note:** *Disconnect the battery negative cable before removing any light unit, and reconnect the lead after installing the light.* **Caution:** *If the stereo in your vehicle is equipped with an anti-theft system, make sure you have the correct activation code before disconnecting the battery.*

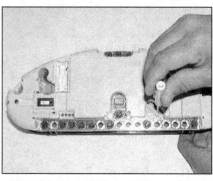

6.24 Using a small screwdriver unscrew the bulb holders on the Z3 instrument cluster

## Headlight

*Refer to illustrations 7.2, 7.4, 7.5, 7.6a, 7.6b and 7.6c*

1  To improve access to the left-hand headlight, remove the air cleaner housing (see Chapter 4).
2  Rotate the retaining ring counterclockwise and detach the main wiring connector from the headlight **(see illustration)**. Where necessary, release the retaining clip and disconnect the other wiring connector.

7.2 Rotate the locking ring and disconnect the main wiring connector from the rear of the headlight

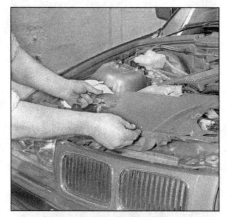

7.4 Remove the retaining screws and remove the plastic cover from above the radiator

7.5 Remove the retaining screws (arrows) . . .

# Chapter 12 Chassis electrical system

7.6a ... and remove the headlight from the vehicle

7.6b Rotate the leveling motor counterclockwise ...

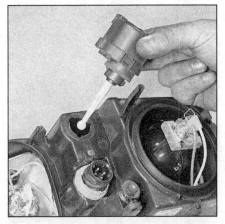

7.6c ... and unclip it from the headlight

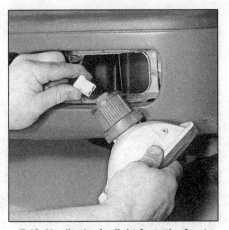

7.13 Unclip the foglight from the front bumper and disconnect its wiring connector

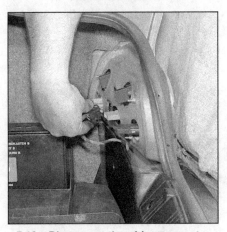

7.16a Disconnect the wiring connector then remove the retaining nuts ...

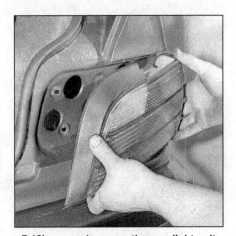

7.16b ... and remove the rear light unit from the vehicle

3  Remove the turn signal light (see step 8).
4  Remove the retaining screws and remove the plastic cover from above the radiator **(see illustration)**.
5  Loosen and remove the headlight retaining screws **(see illustration)**. **Note:** *As the screws are removed, do not allow the plastic retaining clips to rotate. If necessary retain the clips with a suitable open-ended wrench.*
6  Remove the headlight unit from the vehicle. On models equipped with a headlight leveling system, if necessary, remove the motor from the rear of the light unit by rotating it counterclockwise and unclipping its balljoint **(see illustrations)**. **Note:** *The motor can be removed with the headlight unit in position on the vehicle.*
7  Installation is a direct reversal of the removal procedure. Lightly tighten the retaining screws and check the alignment of the headlight with the bumper and hood. If necessary adjust the position of the headlight by screwing the plastic retaining clips in or out of the body (as applicable). Once the light unit is correctly positioned, securely tighten the retaining screws and check the headlight beam alignment using the information given in Section 8.

### Front direction indicator light

8  Using a screwdriver, unhook the retaining spring and withdraw the direction indicator light from the fender. Disconnect the wiring connector from the light unit and remove it from the vehicle.
9  Installation is a reverse of the removal procedure making sure the light unit is securely retained by its spring.

### Front direction indicator side repeater

10  Remove the retaining screw and withdraw the light unit from the fender. Free the bulb holder by rotating it counterclockwise and remove the light unit from the vehicle.
11  Installation is a reverse of the removal procedure. Do not overtighten the retaining screw as the lens is easily cracked.

### Front foglight

*Refer to illustration 7.13*

12  Insert a flat-bladed screwdriver, in through the front bumper grille and release the foglight retaining clip.
13  Withdraw the foglight from the bumper and disconnect its wiring connector **(see illustration)**.
14  Installation is the reverse of removal, making sure the foglight is clipped securely in position. If necessary adjust the foglight aim using the adjusting screw which is accessed through the lower bumper grille.

### Rear light cluster

*Refer to illustrations 7.16a and 7.16b*

15  From inside the vehicle luggage compartment, rotate the retaining clip 90-degrees and remove the plastic access cover from the rear of the light cluster.
16  Disconnect the wiring connector then unscrew the retaining nuts and remove the light unit from the vehicle **(see illustrations)**.
17  Recover the seal which is between the light unit and body and replace it if it shows signs of damage or deterioration.
18  Installation is the reverse of the removal sequence making sure the seal is correctly positioned.

### License plate light

*Refer to illustration 7.20*

19  Remove the retaining screws and remove the trim cover from around the lights.

**7.20 Removing a license plate light**

**8.2a Headlight unit adjustment screws (arrows) (3-Series models)**

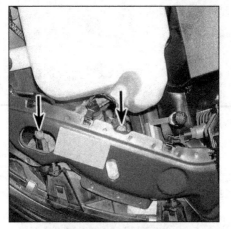

**8.2b Headlight unit adjustment screws (arrows) (Z3 models)**

**9.3a Remove the two retaining screws (arrows) . . .**

**9.3b . . . then remove the instrument panel from the dash**

**9.4a Unplug the electrical connectors from the instrument cluster**

20  Press the light unit towards the left and unclip it from the trunk lid **(see illustration)**.
21  Disconnect the wiring connector and remove the light unit.
22  Installation is the reverse of removal.

### Z3 models

23  Except for the taillight clusters, removal of the exterior light assemblies is part of bulb replacement (see Section 5).

#### Rear taillight cluster

24  Open the trunk, unscrew the knurled retaining nut that secures the tail light cluster assembly and *carefully* pull the tail light assembly straight out of the body (the tail light bulbs fit through stamped holes in the body) **(see illustration 5.55a)**.
25  Unscrew the outer taillight assembly retaining nuts **(see illustration 5.55b)** and remove the taillight assembly.
26  Installation is the reverse of removal.

### 8  Headlight beam alignment - general information

*Refer to illustrations 8.2a and 8.2b*

Accurate adjustment of the headlight beam is only possible using optical beam setting equipment and this work should therefore be carried out by a BMW dealer or qualified repair shop.

For reference the headlights can be adjusted by rotating the adjuster screws on the top of the headlight unit **(see illustrations)**. The outer adjuster alters the horizontal position of the beam while the center adjuster alters the vertical aim of the beam.

Some models are equipped with an electrically operated headlight beam adjustment system which is controlled through the switch in the dash. On these models ensure that the switch is set to the off position before adjusting the headlight aim.

### 9  Instrument cluster - removal and installation

#### Removal

*Refer to illustrations 9.3a, 9.3b, 9.4a and 9.4b*

1  Disconnect the battery negative terminal. **Caution:** *If the stereo in your vehicle is equipped with an anti-theft system, make sure you have the correct activation code before disconnecting the battery.*
2  Remove the steering wheel (see Chapter 10).
3  Loosen and remove the retaining screws from the top of the instrument panel and carefully withdraw the panel from the dash **(see illustrations)**.
4  Lift up the retaining clips then disconnect the wiring connectors **(see illustrations)** and remove the instrument panel from the vehicle.

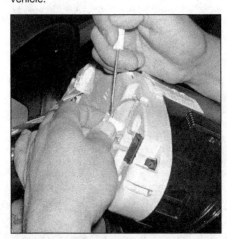

**9.4b Some of the cluster connectors have locking tangs on top that must be pried open with a small screwdriver before they can be unplugged**

# Chapter 12  Chassis electrical system

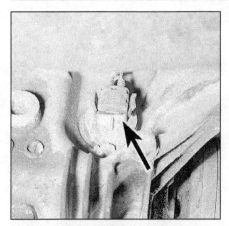

**11.2  The speedometer drive sender unit is mounted in the final drive unit rear cover (arrow)**

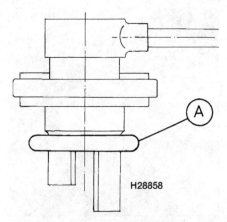

**11.4  Install the new sealing ring (A) to the speedometer drive sender and position it as shown. DO NOT slide it right up to the sender unit face**

**12.2  On models with a clock, unclip and remove the clock mounting frame/storage compartment from the center console**

## Installation

5  Installation is the reverse of removal making sure the instrument panel wiring is correctly reconnected and securely held in position by the retaining clips. On completion reconnect the battery and check the operation of the panel warning lights to ensure that they are functioning correctly.

## 10  Instrument cluster components - removal and installation

No individual components are available for the instrument panel and therefore the panel must be treated as a sealed unit. If there is a fault with one of the instruments, remove the panel (see Section 9) and take it to your BMW dealer for testing. They have access to a special diagnostic tester which will be able to locate the fault and will then be able to advise you on the best course of action.

## 11  Speedometer drive sender unit - removal and installation

### Removal

*Refer to illustration 11.2*

1  Chock the front wheels then jack up the rear of the vehicle and support it on axle stands.
2  Disconnect the wiring connector from the speedometer drive sender unit which is screwed into the rear of the final drive unit **(see illustration)**.
3  Loosen and remove the retaining bolts and withdraw the sender unit. Recover the sender unit sealing ring and discard it; a new should be used on installation.

### Installation

*Refer to illustration 11.4*

4  Apply a smear of oil to the new sealing ring to aid installation and slide the ring up to

**12.4a  Unclip the storage compartment from the console . . .**

the tapered face of the sender unit **(see illustration)**. Do not push the seal right up to the sender unit mating surface.
5  Install the sender to the final drive unit and evenly and progressively tighten its retaining bolts.
6  Reconnect the wiring connector and lower the vehicle to the ground.

## 12  Cigarette lighter - removal and installation

### 3-Series models

#### Removal

*Refer to illustrations 12.2, 12.4a and 12.4b*

1  Disconnect the battery negative terminal. **Caution:** *If the stereo in your vehicle is equipped with an anti-theft system, make sure you have the correct activation code before disconnecting the battery.*
2  Remove the clock/multi-information display unit (see Section 13). On models with a clock, insert a feeler gauge in between the clock mounting frame and storage compartment then release the retaining clip and slide

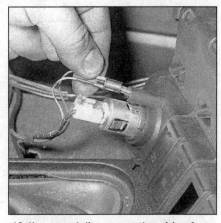

**12.4b  . . . and disconnect the wiring from the cigarette lighter (3-Series models)**

the mounting frame out of position **(see illustration)**.
3  On all models with no clock, carefully pry the upper storage compartment out from the dash center panel.
4  On all models, carefully unclip the storage compartment (which contains the cigarette lighter) and withdraw it from the center dash panel. Disconnect the wiring connectors from cigarette lighter and (if equipped) switches and remove the storage compartment **(see illustrations)**.
5  Unclip the bulb holder from the lighter then depress the retaining tangs and push the lighter out of the panel.

#### Installation

6  Installation is a reversal of the removal procedure, ensuring all the wiring connectors are securely reconnected.

### Z3 models

7  The cigarette lighter in located in the center console. Remove the center console (see Section 28 in Chapter 11) and unplug the wiring harness from the cigarette lighter **(see illustration 28.24b in Chapter 11)**.
8  Installation is the reverse of removal.

# 12-20 Chapter 12 Chassis electrical system

13.3a Unclip and remove the clock as shown . . .

13.3b . . . and disconnect it from the wiring connector (3-Series models)

13.5 To remove the clock/on-board trip computer from the center console on a Z3 model, simply pry it out and unplug it

## 13 Clock/multi-information display/on-board trip computer - removal and installation

### 3-Series models

#### Removal

*Refer to illustrations 13.3a and 13.3b*

1  Disconnect the battery negative terminal. **Caution:** *If the stereo in your vehicle is equipped with an anti-theft system, make sure you have the correct activation code before disconnecting the battery.*
2  Insert a feeler gauge (approximately 0.004-inch thick) in between the base of the clock/multi-information display and mounting.
3  Carefully release the retaining clip then slide the clock/multi-information display unit out from the dash center panel and remove it, disconnecting its wiring connectors as they become accessible **(see illustrations)**.

#### Installation

4  Installation is the reverse of removal.

### Z3 models (clock/on-board trip computer)

*Refer to illustration 13.5*

5  Pry the clock/on-board trip computer from the center console **(see illustration)**, pull out the unit, unplug the electrical connector and remove the unit.
6  Installation is the reverse of removal.

## 14 Horns - removal and installation

### Removal

1  The horns are located behind the left and/or right end(s) of the front bumper.

### 3-Series models

*Refer to illustration 14.3*

2  To gain access to the horn(s) from below, apply the parking brake then jack up the front of the vehicle and support it on axle stands. Remove the retaining screws and remove the access cover from the left-hand underside of the bumper. Unclip and remove the brake disc cooling duct.
3  To gain access to the horns from above, remove the left-hand headlight (see Section 7) **(see illustration)**.
4  Remove the retaining nut/bolt and remove the horn, disconnecting its wiring connectors as they become accessible.

### Z3 models

*Refer to illustration 14.6*

5  Remove the engine splash shield filler panel **(see illustration 6.15a in Chapter 11)**.
6  Unplug the electrical connector and remove the retaining nut **(see illustration)**.

### Installation

7  Installation is the reverse of removal.

## 15 Wiper arm - removal and installation

### Removal

*Refer to illustrations 15.3a and 15.3b*

1  Operate the wiper motor then switch it off so that the wiper arm returns to the park position.
2  Stick a piece of masking tape along the edge of the wiper blade to use as an alignment aid on installation.
3  Pry off the wiper arm spindle nut cover then loosen and remove the spindle nut **(see illustrations)**. Lift the blade off the glass and pull the wiper arm off its spindle. If necessary

14.3 Horns viewed from above with the left-hand headlight unit removed (3-Series models)

14.6 To detach a horn from a Z3 model, unplug the electrical connector and remove this nut (arrow)

15.3a To remove a windshield wiper arm, pry off the wiper arm spindle nut cover . . .

# Chapter 12 Chassis electrical system    12-21

15.3b ... and then loosen and remove the spindle nut (Z3 model shown, 3-Series models similar)

16.9 Unscrew the nut (arrow) from the wiper motor spindle and lift off the washer

16.10 Unscrew the mounting bolts (upper two arrows) and remove the wiper motor support bracket

the arm can be levered off the spindle using a suitable flat-bladed screwdriver. **Note:** *If both windshield wiper arms are to be removed at the same time, mark them for identification; the arms are not interchangeable.*

## Installation

4  Ensure that the wiper arm and spindle splines are clean and dry then install the arm to the spindle, aligning the wiper blade with the tape applied on removal. Install the spindle nut, tightening it to the specified torque setting, and clip the nut cover back in position. **Note:** *BMW recommends that the spindle nut torque should be checked again after 15 minutes.*

## 16  Windshield wiper motor and linkage - removal and installation

### 3-Series models

#### Removal

*Refer to illustrations 16.9, 16.10, 16.12, 16.13a, 16.13b and 16.13c*

1  Disconnect the battery negative terminal. **Caution:** *If the stereo in your vehicle is equipped with an anti-theft system, make sure you have the correct activation code before disconnecting the battery.*

2  Remove the wiper arms (see Section 15).
3  Raise the hood to the vertical position (see Chapter 11).
4  Remove the rubber seal from the top of the heating/ventilation system inlet.
5  Remove the retaining clips securing the wiper motor cover panel in position.
6  Remove the release the grille from the top of the inlet and remove the cover panel(s) (as applicable) from the vehicle.
7  Remove the retaining screws and free the wiring harness duct from the inlet duct.
8  Loosen and remove the retaining screws and retaining plate and remove the inlet duct from the firewall. **Note:** *On 6-cylinder engines it may be necessary to remove the injector and spark plug covers from the engine to enable the inlet to be removed.*
9  Unscrew the large nut(s) from the wiper spindle(s) and remove the washer(s) (as applicable) **(see illustration)**.
10  Loosen and remove the retaining bolts and remove the wiper motor support bracket **(see illustration)**. Recover the spacer and mounting rubbers from the motor.
11  Release the retaining clip and discon-

16.12 Removing the wiper motor assembly from the vehicle

nect the wiring connector from the motor.
12  Maneuver the motor and linkage assembly out of position **(see illustration)**. Recover the rubber grommets from the spindles, inspect them for signs of damage or deterioration, and replace if necessary.
13  If necessary, mark the relative positions of the motor shaft and crank then release the wiper linkage from the motor balljoint **(see illustration)**. Unscrew the retaining nut and free the crank from the motor spindle.

16.13a Unclip the wiper linkage from the balljoint . . .

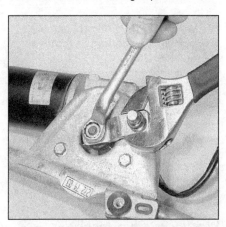

16.13b . . . then remove the retaining nut and remove the crank from the motor spindle

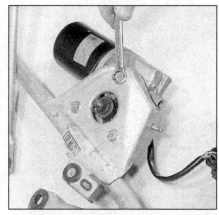

16.13c Remove the three retaining bolts and separate the motor and wiper linkage

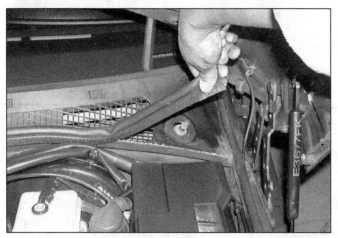

16.15 Peel off the three rubber seals around the protective grille

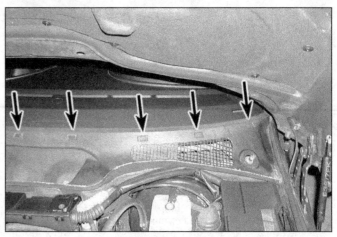

16.16a Pry off the eight retaining clips (arrows) around the edge of the ventilation intake air grille . . .

16.16b . . . with a small screwdriver and remove the grille

16.17 To detach the wiper assembly, remove the left and right mounting bolts (arrows)

bly mounting bolts **(see illustration)**.
18  To disengage the upper locking guide, pull the wiper assembly toward the right side of the vehicle. To remove the wiper assembly, pull it forward.
19  Unplug the electrical connector from the wiper motor. Remove the windshield wiper motor and linkage.
20  To disassemble the windshield wiper motor and linkage, refer to Step 13.
21  Installation is the reverse of removal. Refer to the notes in Step 14.

**17 Windshield/headlight washer system components - removal and installation**

1  The windshield washer reservoir is situated in the engine compartment. On models equipped with headlight washers the reservoir also supplies the headlight washer jets via an additional pump. **Note:** *On some models a second smaller reservoir is also incorporated into the washer system. This reservoir is designed to be filled with concentrated washer fluid for intensive cleaning (refer to your BMW dealer for details).*

### Washer system reservoir

*Refer to illustrations 17.5a and 17.5b*
2  Siphon the windshield washer fluid from

the reservoir or be prepared for fluid spillage.
3  Disconnect the wiring connector from the washer pump. If you're *replacing* the reservoir, remove the pump from the reservoir (see Step 8) and put it aside. (If you're simply removing the reservoir to gain access to or service some other component, it's not necessary to remove the pump.)
4  Disconnect the wiring connector from the reservoir level switch (if equipped).
5  Remove the reservoir fastener and remove the reservoir **(see illustrations)**.

17.5a To detach the windshield washer fluid reservoir, unscrew this fastener (arrow) . . .

Unscrew the motor retaining bolts and separate the motor and linkage **(see illustrations)**.

### Installation

14  Installation is the reverse of removal, noting the following points.
  a) If removed, tighten the motor retaining bolts and crank arm nut to the specified torque.
  b) Ensure that the spindle grommets are correctly installed and maneuver the motor into position. Install the support bracket and tighten both the bracket bolts and spindle nuts to the specified torque.
  c) On completion install the wiper arms (see Section 15).

### Z3 models

*Refer to illustrations 16.15, 16.16a, 16.16b and 16.17*
15  Peel off the left, right and front rubber seals around the protective grille **(see illustration)**.
16  Remove the retaining clips around the edge of the ventilation intake air grille and remove the grille **(see illustrations)**.
17  Remove the left and right wiper assem-

# Chapter 12 Chassis electrical system

17.5b ... and lift the reservoir straight up; if you need to move the reservoir farther than the distance allowed by the pump electrical harness and hose, unplug the connector and detach the hose (arrows) from the pump

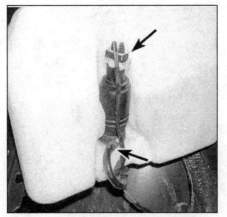

17.7 Before removing the pump, disconnect the wiring connector and washer hose from the pump (arrows)

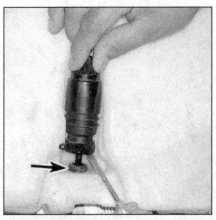

17.8 To remove the pump, carefully pry it out of the reservoir; be sure to inspect the grommet (arrow) for signs of damage or deterioration

Wash off any spilled fluid with cold water.
6 Installation is the reverse of removal. Refill the reservoir and check for leakage.

### Washer pump

*Refer to illustrations 17.7 and 17.8*

7 Disconnect the wiring connector and washer hose from the pump **(see illustration)**.
8 Carefully pry the pump out of the reservoir **(see illustration)** and recover its sealing grommet. Wash off any spilled fluid with cold water. Inspect the pump sealing grommet for signs of damage or deterioration and replace if necessary.
9 Installation is the reverse of removal. Refill the reservoir and check the pump grommet for leaks.

### Washer reservoir level switch

10 Empty the contents of the reservoir or be prepared for fluid spillage as the pump is removed.
11 Disconnect the wiring connector from the level switch and carefully ease the switch out from the reservoir. Recover the sealing grommet and wash off any spilled fluid with cold water.
12 Installation is the reverse of removal, using a new sealing grommet if the original one shows signs of damage or deterioration. Refill the reservoir and check for leaks.

### Windshield washer jets

13 Open up the hood then remove the retaining clips and free the insulation panel from the area around the base of the washer jet.
14 Disconnect the washer hose(s) from the base of the jet. Where necessary, also disconnect the wiring connector from the jet.
15 Release the retaining clip and carefully ease the jet out from the hood, taking great care not to damage the paintwork.
16 On installation, securely connect the jet to the hose and clip it into position in the hood; where necessary also reconnect the wiring connector. Check the operation of the jet. If necessary adjust the nozzles using a pin, aiming one nozzle to a point slightly above the center of the swept area and the other to slightly below the center point to ensure complete coverage.

### Headlight washer jets

17 To improve access, apply the parking brake then jack up the front of the vehicle and support it on axle stands.
18 Remove the retaining screws and remove the small access cover from the bottom of the bumper.
19 Release the retaining clip and remove the brake duct from the bumper.
20 Pull the washer jet assembly off from the supply tube and recover the bushing from the supply tube.
21 Remove the retaining screws and release the washer jet supply tube from the body.
22 Loosen the retaining clip then disconnect the hose and remove the supply tube.
23 On installation, install the bushing and push the jet assembly firmly onto the supply tube. Position the supply tube so the jet is flush with the body panel and securely tighten its retaining screws.

### Wash/wipe system control module

*Refer to illustration 17.26*

24 Remove the two retaining screws then unclip and remove the driver's side lower dash panel.
25 Unclip and remove the undercover to gain access to the module which is clipped in position above the pedals.
26 Release the retaining clips and lower the module out of position **(see illustration)**.
27 Release the retaining clip then disconnect the wiring connector(s) and remove the module from the vehicle.
28 Installation is the reverse of removal.

17.26 Removing the wash/wipe system control module

### Heated washer jet thermostatic switch

29 The heated washer jet thermostatic switch is clipped into the right-hand front brake cooling duct. If necessary, to improve access, apply the parking brake then jack up the front of the vehicle and support it on axle stands.
30 Disconnect the wiring connector then unclip the sensor and remove it from the brake duct.
31 Installation is the reverse of removal ensuring that the switch is clipped securely into position.

---

**18 Audio unit - removal and installation**

*Refer to illustrations 18.2, 18.3 and 18.4*
**Note:** *The following removal and installation procedure is for the range of radio/cassette units installed as standard equipment. Removal and installation procedures of non-standard units may differ slightly.*

1 Disconnect the battery negative cable.
**Caution:** *If the stereo in your vehicle is*

**18.2 Using a small screwdriver, carefully pry open the small access covers from each side of the unit**

**18.3 Using an Allen wrench, unscrew the audio unit mounting screws**

**18.4 Remove the audio unit from the dash and unplug the antenna and electrical connector(s) (arrows)**

equipped with an anti-theft system, make sure you have the correct activation code before disconnecting the battery.

2  Carefully pry open the small access covers from each side of the unit **(see illustration)**.

3  To release the audio unit, use an Allen wrench to loosen (they don't come out) the audio unit mounting screws **(see illustration)**.

4  Remove the audio unit from the dash **(see illustration)** and then unplug the antenna and electrical connector(s).

5  Installation is the reverse of removal.

## 19 Speakers - removal and installation

### Door panel speaker(s)

*Refer to illustration 19.2*

1  Remove the door inner trim panel (see Chapter 11).

2  Unscrew the retaining nut and remove the retaining bracket (where installed) then unclip the speaker from the trim panel **(see illustration)**.

3  Installation is the reverse of removal.

### Drivers side front speaker (situated behind footwell side panel)

*Refer to illustration 19.4*

4  Remove the hood release handle and the left kick panel **(see illustrations 9.9a, 9.9b and 9.9c in Chapter 11)**.

5  Remove the speaker retaining screws **(see illustration)** and pull the speaker off the body.

6  Disconnect the speaker electrical connector.

7  Remove the speaker.

8  Installation is the reverse of removal making sure the speaker is correctly located.

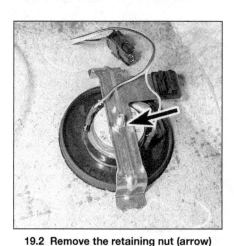

**19.2 Remove the retaining nut (arrow) then remove the mounting bracket and withdraw the speaker from the door trim panel**

### Passenger side front speaker (situated behind footwell side panel)

*Refer to illustrations 19.10a, 19.10b, 19.11a and 19.11b*

9  Remove the two retaining screws and

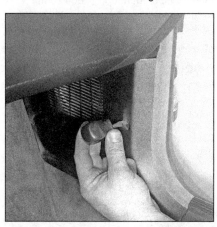

**19.10a Release the fastener by rotating it 90-degrees . . .**

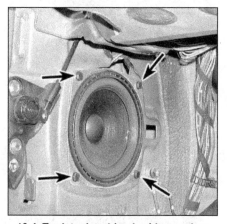

**19.4 To detach a driver's side speaker, remove these four mounting bolts (arrows), pull out the speaker and unplug the electrical connector (Z3 model shown)**

remove the undercover from beneath the dash.

10  Release the trim panel fastener by rotating it 90-degrees then unclip the footwell side trim panel and remove it from the vehicle **(see illustrations)**.

11  Remove the retaining screws and

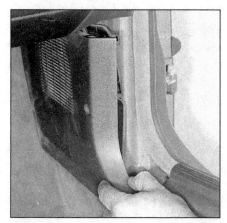

**19.10b . . . then unclip the footwell side trim panel (3-Series model shown)**

# Chapter 12 Chassis electrical system

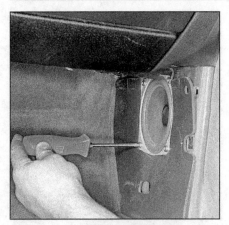

19.11a Remove the four retaining screws...

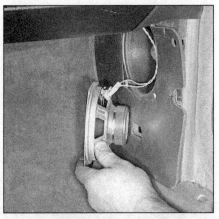

19.11b ...then withdraw the speaker and disconnect its wiring connector (3-Series model shown)

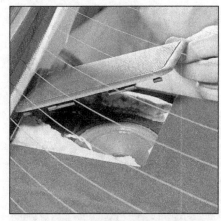

19.13 Unclip the rear speaker grille from the parcel shelf...

19.14a ...and remove the retaining screws (arrows)...

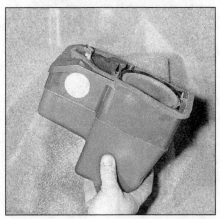

19.14b ...then disconnect the wiring connector and lower the speaker out of position (3-Series model shown)

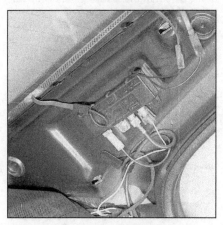

20.2 Radio antenna amplifier is mounted onto the left-hand rear pillar (3-Series models)

remove the speaker, disconnect its wiring connectors as they become accessible **(see illustrations)**.

12 Installation is the reverse of removal making sure the speaker is correctly located.

### Rear speaker

*Refer to illustrations 19.13, 19.14a and 19.14b*

13 Carefully pry the speaker grille out from the rear parcel shelf **(see illustration)**.

14 Disconnect the wiring connector then loosen and remove the speaker retaining screws. Depress the retaining clips and lower the speaker assembly downwards and out through the luggage compartment **(see illustrations)**. If necessary, remove the retaining screws and separate the speaker and mounting box.

15 Installation is the reverse of removal.

## 20 Antenna (3-Series models) - general information

*Refer to illustration 20.2*

The radio antenna is built into the rear window. In order to improve reception an amplifier is installed to boost the signal to the radio/cassette unit. The amplifier unit is located behind the left hand rear pillar trim panel.

To gain access to the antenna amplifier unit, carefully unclip the left hand trim panel from the rear pillar, disconnecting the wiring from the interior light as the panel is removed. Disconnect the antenna lead and wiring then remove the retaining screws and remove the amplifier **(see illustration)**. Installation is the reverse of removal.

## 21 Cruise control system components (3-Series models) - removal and installation

### Electronic control module (ECM)

1 Remove the glovebox (see Chapter 11).
2 Unclip the lower dash panel and remove it from underneath the glovebox aperture.
3 Release the retaining clip, withdraw the ECM then release the retaining clip and disconnect wiring connector.
4 Installation is the reverse of removal.

### Cruise control actuator

5 The cruise control actuator is located in the engine compartment.
6 Disconnect the wiring connector from the actuator.
7 Remove the outer cable retaining clip, then release the cable from the bracket and detach the inner cable from the actuator.
8 Remove the retaining nuts and remove the actuator from the engine compartment.
9 Installation is the reverse of removal.

### System operating switch

10 Refer to Section 4.

### Brake and clutch pedal switches

11 Refer to Chapter 8 (clutch pedal switch) or Chapter 9 (brake switch).

## 22 Anti-theft alarm system - general information

**Note:** *This information is applicable only to the anti-theft alarm system installed by BMW as standard equipment.*

1  Some models in the range are equipped with an anti-theft alarm system as standard equipment. The alarm has switches on all the doors (including the trunk lid), the hood, the glovebox and the ignition switch and also a tilt switch which is sensitive to shocks. If the trunk lid, hood, glovebox or either of the doors are opened or the ignition switch is switched on while the alarm is set, or if the tilt switch senses the vehicle is being tampered with, the alarm horn will sound and the hazard warning lights/headlights will flash. The alarm also has an immobilizer function which makes the ignition system inoperable while the alarm is triggered.

2  The alarm is set using the key in the driver's or passenger front door lock or when the doors are locked using the remote central locking device. The LED in the dash will flash to indicate that the alarm system is operational. If a door or window or the glovebox are not fully closed when the alarm is set, the LED will flash quicker than normal. After a short period the alarm will then operate as normal but with the switch for the open door/window or glovebox switched off.

3  If necessary, the tilt switch sensing function of the alarm can be disabled. To do this turn the alarm on as normal, then turn the alarm on again for a second time (either using the door lock or remote locking device). This will disable the tilt switch but leave all the other switches turned on.

4  Should the alarm system become faulty the vehicle should be taken to a BMW dealer for examination. They will have access to a special diagnostic tester which will quickly trace any fault present in the system.

## 23 Heated front seat components - removal and installation

### Heater mats

On models equipped with heated front seats, a heater pad is fitted to both the seat back and seat cushion. Replacement of either heater mat involves peeling back the upholstery, removing the old mat, sticking the new mat in position and then installing the upholstery. Note that upholstery removal and installation requires considerable skill and experience if it is to be carried out successfully and is therefore best entrusted to your BMW dealer or an automotive upholstery shop. In practice, it will be very difficult for the home mechanic to carry out the job without ruining the upholstery.

### Heated seat switches

Refer to Section 4.

## 24 Airbag system - general information and precautions

1992 and 1993 3-Series models are equipped with a driver's airbag. On all 1994

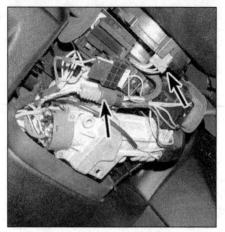

25.4  To disable the airbag system, unplug this orange connector (arrow) under the steering column; if you're removing the steering wheel, unplug this connector (arrow) from the contact ring

and later 3-Series models and on all Z3 models, both a driver's and passenger airbag are standard equipment. Side impact airbags are installed in the front doors of some 1997 4-door 3-Series models and on all 1998 3-Series models. The airbag system comprises the airbag unit(s) (complete with gas generators), two impact sensors (1993 and earlier models) or one impact sensor (integral with the control unit - 1994 and later models), the control unit and a warning light in the instrument panel.

The airbag system is triggered in the event of a heavy frontal impact above a predetermined force, depending on the angle of impact. The airbag is inflated within milliseconds and forms a safety cushion between the driver and steering wheel and (if equipped) the passenger and dash. This prevents contact between the upper body and wheel/dash and therefore greatly reduces the risk of injury. The airbag then deflates almost immediately.

Every time the ignition is switched on, the airbag control unit performs a self-test. The self-test takes approximately 2 to 6 seconds and during this time the airbag warning light on the dash is illuminated. After the self-test has been completed the warning light should go out. If the warning light fails to come on, remains illuminated after the initial period or comes on at any time when the vehicle is being driven, there is a fault in the airbag system. Take the vehicle to a BMW dealer for examination at the earliest possible opportunity.

### Disabling the airbag system

**Warning 1:** *Before working in the vicinity of any airbag system component, be sure to disable the airbag system as follows. Disconnect the negative and positive battery cables, and then wait 10 minutes for the airbag back-up power supply (a capacitor) to discharge before removing the airbag.*

**Warning 2:** *Note that the airbag(s) must not be subjected to temperatures in excess of*

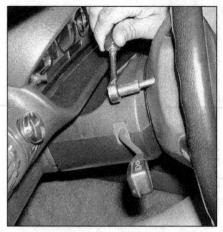

25.5  Loosen and remove the two airbag retaining screws from the rear of the steering wheel, rotating the wheel as necessary to gain access to the screws

90°C (194°F). When the airbag is removed, ensure that it is stored with the trim side facing up.

**Warning 3:** *Do not allow any solvents or cleaning agents to contact the airbag assemblies. They must be cleaned using only a damp cloth.*

**Warning 4:** *The airbags and control unit are both sensitive to impact. If either is dropped or damaged they should be replaced.*

**Warning 5:** *Disconnect the airbag control unit wiring plug prior to using arc-welding equipment on the vehicle.*

**Caution:** *If the stereo in your vehicle is equipped with an anti-theft system, make sure you have the correct activation code before disconnecting the battery.*

## 25 Airbag system components - removal and installation

**Note:** *Refer to the warnings given in Section 24 before carrying out the following operations.*

1  There are two possible types of airbag systems installed on these vehicles: Airbag System I and Airbag System II. The earlier (Airbag I) system uses a control unit and two separate impact sensors, one installed on each front suspension strut tower. On the later (Airbag II) system, a control unit with an integral impact sensor is used, the control unit being mounted underneath the rear seat cushion.

2  Disable the airbag system (see Section 24).

### Driver's side airbag

Refer to illustrations 25.4, 25.5 and 25.6

3  Remove the lower half of the steering column cover (see Section 27 in Chapter 11).

4  Unplug the airbag unit wiring connector **(see illustration)**. If you're planning to remove the steering wheel, also unplug the connector from the contact ring.

# Chapter 12 Chassis electrical system

12-27

25.6 Return the steering wheel to the straight-ahead position then carefully lift the airbag assembly away from the steering wheel and disconnect the wiring connector from the rear of the unit

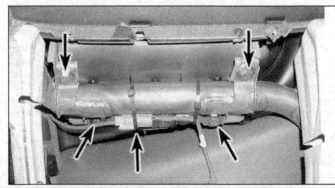

25.9 To detach the airbag from the dash, unplug the electrical connector and remove these four bolts (arrows)

5  Loosen and remove the two airbag retaining screws from the rear of the steering wheel **(see illustration)**, rotating the wheel as necessary to gain access to the screws.
6  Return the steering wheel to the straight-ahead position then carefully lift the airbag assembly away from the steering wheel and disconnect the wiring connector from the rear of the unit **(see illustration)**. **Warning:** *The airbag must not be knocked or dropped and must be stored with the trim side facing up. When carrying the airbag unit, the trim side must face away from your body.*
7  On installation reconnect the wiring connector and seat the airbag unit in the steering wheel, making sure the wire does not become trapped. Install the retaining screws and tighten them to the specified torque setting, noting the right-hand retaining screw should be tightened first. Reconnect the wiring connector and clip it onto the column then install the column shroud and connect the battery.

## Passenger side airbag

*Refer to illustrations 25.9 and 25.10*

8  Remove the glovebox and the lower dash trim panel (see Section 27 in Chapter 11).
9  Unplug the electrical connector and remove the airbag retaining bolts **(see illustration)**. (If you can't unplug the connector while the airbag is still bolted in place, disconnect it after unbolting the airbag.)
10  Carefully pry the airbag access cover (if applicable) from the dash **(see illustration)**. Remove the airbag assembly from the dash.
11  On installation, securely reconnect the wiring connector and seat the airbag in position.
12  Install the airbag retaining screws and tighten them securely. Plug in the electrical connector and install the cover (if applicable).
13  Install the glovebox and the lower dash trim panel (see Chapter 11).

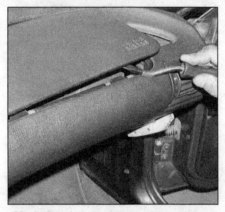

25.10 Carefully pry up the airbag access cover from the dash (Z3 model shown)

## Airbag control unit

### 1993 and earlier models (Airbag system I)

14  Remove the glovebox (see Chapter 11).
15  Release the unit from its mounting bracket and disconnect the wiring connector(s).
16  Installation is the reverse of removal.

### 1994 and later models (Airbag system II)

*Refer to illustrations 25.17 and 25.18*

17  On 3-Series models, unclip the rear seat cushion and remove it from the vehicle. Then disconnect the wiring connector, remove the control unit retaining nuts **(see illustration)** and remove the unit from the vehicle.
18  On Z3 models, remove the center console (see Chapter 11), unplug the electrical connector, remove the control unit retaining nuts **(see illustration)** and remove the unit.
19  When installing the airbag control unit, make SURE that the arrow faces forward. **Warning:** *The control unit MUST be installed with the arrow pointing towards the front of the vehicle. Failure to do so will cause the airbag system to malfunction in the event of an accident.*
20  Installation is otherwise the reverse of removal.

## Airbag wiring contact unit

21  Remove the steering wheel (see Chapter 10).

25.17 Airbag control unit (Airbag system II) - ensure the unit is installed with the arrow pointing towards the front of the vehicle (3-Series models)

### 1993 and earlier models (Airbag system I)

22  Remove the locking tab from the contact unit and lift out the spring.
23  Unscrew the retaining nuts and remove the contact unit from the steering wheel.
24  Install the contact ring to the steering wheel, ensuring that the wiring is correctly routed. Install the retaining nuts, tightening them securely, and lock them in position by

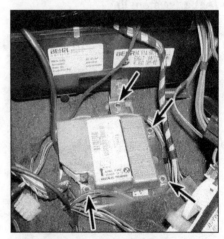

25.18 Airbag control unit retaining nuts (arrows) (Z3 models) - make sure that the arrow points forward when the control unit is installed

applying a dab of varnish/paint to their ends. If a new contact unit is being installed, unscrew the locking screw from the unit.

25  Install the spring to the contact unit and clip the locking tab back into position. Install the steering wheel (see Chapter 10).

26  Reconnect the wiring connector and clip it onto the column then install the column shroud and connect the battery.

### 1994 and later models (Airbag system II)

27  Unclip the trim cover and remove the screw securing the ground lead to the steering wheel.

28  Remove the retaining screws and remove the contact unit from the wheel.

29  Install the new contact unit to the steering wheel, ensuring that the wiring is correctly routed, and securely tighten the retaining screws. If a new contact unit is being installed, depress the retaining tangs and remove the locking peg. Connect the ground lead to the steering wheel, tightening its retaining screw securely, and install the cover.

31  Install the steering wheel (see Chapter 10).

32  Reconnect the wiring connector and clip it onto the column then install the column shroud and connect the battery.

## Impact sensor

### 1993 and earlier models (Airbag system I)

33  There are two impact sensors, one on each side of the engine compartment.

34  Disconnect the wiring connector then remove the retaining bolts and remove the sensor from the side of the suspension strut mounting tower.

35  Installation is the reverse of removal ensuring that the sensor is positioned with its arrow pointing towards the front of the vehicle. Tighten the sensor mounting bolts to the specified torque.

### 1994 and later models (Airbag system II)

36  The impact sensor is an integral part of the control unit.

## 26  Multi-information system - general information and component removal and installation

### General information

1  On some models a multi-information system was standard equipment, and on most other models it is available as an option. There are various types of system all of which are similar in appearance but vary in the amount of functions they carry out. All control units contain a clock along with the following functions:

a) *Outside temperature display - displays the ambient air temperature.*

b) *Check control system - monitors the rear lights, license plate lights and informs the driver of a fault. Also informs the driver if the engine coolant or windshield washer fluid reservoir levels are low. May also display the outside air temperature.*

c) *On-board computer - monitors vehicle speed, distance traveled, fuel consumption etc. as well as air temperature.*

2  Refer to your BMW owner's manual for details on how to use the various functions of each unit. Any problems should be referred to a BMW dealer or other qualified repair shop.

### Component removal and installation

#### Multi-information display unit

3  Refer to Section 13.

#### Check control system electronic control module (ECM)

4  Remove the glovebox (see Chapter 11). Unclip the undercover and remove it from beneath the dash.

5  Release the ECM from its mounting then release the retaining clips and disconnect the wiring connectors and remove the ECM from the vehicle.

6  Installation is the reverse of removal.

#### Ambient air temperature sensor

7  The air temperature sensor is clipped into the left-hand front brake cooling duct. If necessary, to improve access, apply the parking brake then jack up the front of the vehicle and support it on axle stands.

8  Disconnect the wiring connector then unclip the sensor and remove it from the brake duct.

9  Installation is the reverse of removal.

## 27  Door lock heating system - general information and component removal and installation

### General information

1  Some models are equipped with a door lock heating system. The heating element is also built into the handle assembly and is wrapped around the lock housing. The system is switched on by lifting the door handle, a microswitch on the handle switches on the heating element. On later models the system is controlled either by the central locking control module (ECM) or by the central body electronics (ZKE IV) control unit (depending on model and specification).

### Component removal and installation

#### Central locking system control module (ECM)

2  Refer to Chapter 11.

27.7  Door lock heating system microswitch (arrow)

#### Central body electronics (ZKE IV) control module

3  Disconnect the battery negative terminal and remove the glovebox (see Section 27 in Chapter 11). **Caution:** *If the stereo in your vehicle is equipped with an anti-theft system, make sure you have the correct activation code before disconnecting the battery.*

4  Release the retaining clips and slide the control unit out from its retaining bracket. Disconnect the wiring connectors and remove the control unit.

5  Installation is the reverse of removal ensuring that the wiring connectors are securely reconnected.

#### Door lock heating microswitch

*Refer to illustration 27.7*

6  Remove the door exterior handle (see Chapter 11).

7  Release the retaining clip and detach the microswitch from the handle (see illustration).

8  Installation is the reverse of removal ensuring that the switch is clipped securely in position.

#### Door lock heating element

9  At the time of writing it appears the heating element is an integral part of the handle assembly and is not available separately. Consult your BMW dealer for the latest information on parts availability.

## 28  Central locking components - removal and installation

**Warning:** *On models equipped with airbags, always disable the airbag system before working in the vicinity of any airbag system component to avoid the possibility of accidental deployment of the airbag, which could cause personal injury (see Section 24).*

# Chapter 12 Chassis electrical system

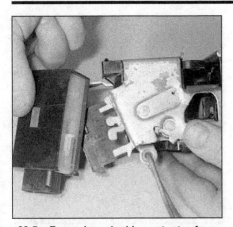

28.5a Removing a locking actuator from the door lock (3-Series models)

28.5b Removing a locking actuator from the door lock (Z3 models)

28.7a On coupe models, remove the retaining screw . . .

## Removal

### Control module

**Note:** *On some models the central locking system is controlled by the central body electronics (ZKE IV) control module (depending on model and specification).*

1  Disconnect the battery negative terminal. **Caution:** *If the radio in your vehicle is equipped with an anti-theft system, make sure you have the correct activation code before disconnecting the battery.*

2  Remove the glovebox (see Section 27 in Chapter 11). Release the retaining clips and remove the dash panel from underneath the glovebox aperture to gain access to the ECM.

3  Disconnect the wiring connector from the control module then unclip it and remove it from the vehicle. On models with anti-lock brakes (ABS) it will be necessary to remove the ABS ECU (see Chapter 9) to gain access to the central locking control module.

### Door lock solenoid

*Refer to illustrations 28.5a and 28.5b*

4  Remove the door lock (see Section 13 in Chapter 11).

5  Release the retaining clip and detach the solenoid from the door lock, noting how it is connected **(see illustrations)**.

### Door lock microswitch - coupe models

*Refer to illustrations 28.7a and 28.7b*

6  Remove the door lock (see Section 13 in Chapter 11).

7  Free the wiring from its retaining clips then loosen and remove the retaining screw and remove the switch from the lock **(see illustrations)**.

### Door handle microswitch - sedan models

*Refer to illustration 28.9*

8  Remove the door exterior handle (see Section 13 in Chapter 11).

9  Remove the retaining clip and detach the switch from the handle **(see illustration)**.

### Trunk lock solenoid

*Refer to illustrations 28.12a, 28.12b and 28.12c*

10  Open up the trunk lid and then, on vehicles with a trunk-mounted battery, disconnect the battery negative terminal. **Caution:** *If the radio in your vehicle is equipped with an anti-theft system, make sure you have the correct activation code before disconnecting the battery.*

11  On 3-Series models, remove the trim

28.7b . . . and withdraw the central locking microswitch from the front door lock

caps then unscrew the retaining screws securing the tool kit to the trunk lid. Remove the tool kit and unclip the trim panel from the trunk lid. On Z3 models, remove the access panels from the trunk **(see illustration 27.11 in Chapter 11)**.

12  Disconnect the wiring connector, unclip the solenoid link rod from the lock cylinder, unscrew the retaining bolts and remove the solenoid and link rod **(see illustrations)**.

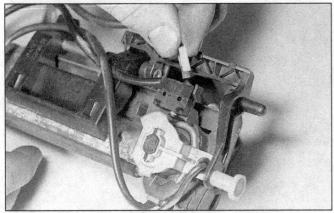

28.9 On sedan models release the retaining clip and remove the central locking microswitch from the handle

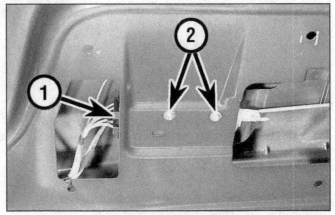

28.12a Trunk locking solenoid wiring connector (1) and retaining screws (2) (3-Series models)

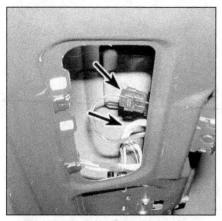
28.12b  To remove the trunk lid locking actuator from a Z3 model, disconnect the wiring connectors (arrows) . . .

28.12c  . . . unclip the solenoid link rod (arrow) from the lock cylinder and remove the retaining bolts (arrows)

28.14  Lift out the first aid box plate and peel back the luggage compartment trim to gain access to the fuel filler flap solenoid

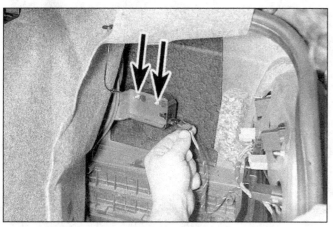
28.15a  Disconnect the wiring connector then loosen the retaining screws (arrows) . . .

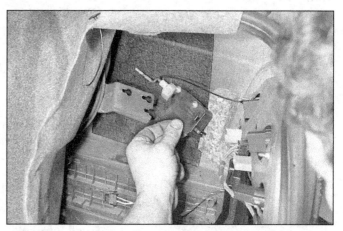
28.15b  . . . and remove the solenoid from its mounting bracket

### Fuel filler flap solenoid

*Refer to illustrations 28.14, 28.15a and 28.15b*

13  Rotate the retaining clip through 90-degrees and remove the access cover from the right-hand rear light unit.
14  Lift out the first aid box plate and peel back the luggage compartment trim to reveal the solenoid **(see illustration)**.
15  Disconnect the wiring connector then loosen the retaining screws and maneuver the solenoid out from the luggage compartment **(see illustrations)**.

### Glovebox lock solenoid

16  Remove the glovebox (see Section 27 in Chapter 11).
17  Unclip the solenoid rod from the lock and guide then unscrew the retaining bolts and remove the solenoid.

### Central locking system impact sensor

*Refer to illustrations 28.19a and 28.19b*

18  Remove the passenger side front speaker (see Section 19). Release the insulation panel from the body and remove it from the vehicle.
19  Loosen the two retaining screws securing the sensor to the body then maneuver it out through the speaker aperture and disconnect it from the wiring connector **(see illustrations)**.

### Installation

20  Installation is the reverse of removal. Prior to installing any trim panels removed for access thoroughly check the operation of the central locking system.

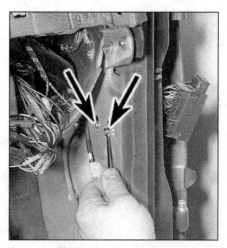
28.19a  Loosen the two screws (arrows) . . .

28.19b  . . . and maneuver the central locking system impact sensor out from behind the pillar

# Chapter 12 Chassis electrical system

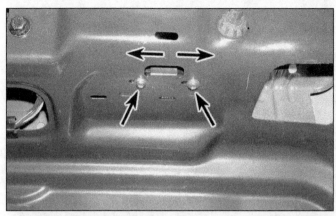

28.21 To adjust the trunk lid locking actuator on a Z3 model, loosen the retaining screws (arrows), slide the actuator to the left or right until the plunger achieves its optimal range of travel, then retighten the retaining screws

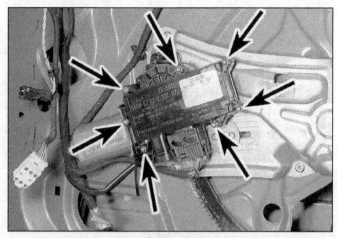

29.13 Electric window ECM retaining screws (arrows) - coupe models

### Adjusting the trunk lid locking actuator (Z3 models)
*Refer to illustration 28.21*

21 The trunk lid locking actuator on Z3 models is adjustable. To adjust it, proceed as follows:

a) Put the central locking system in its unlocked mode.
b) Loosen the actuator retaining screws **(see illustration)**.
c) With the luggage compartment lid open, operate the driver's side door lock with the key and verify that the actuator plunger moves through its full range of motion.
d) Position the actuator to maximize the plunger's range of motion, and then tighten the actuator retaining screws.

## 29 Electric window components - removal and installation

**Warning:** *On models equipped with airbags, always disable the airbag system before working in the vicinity of any airbag system component to avoid the possibility of accidental deployment of the airbag, which could cause personal injury (see Section 24).*
**Note:** *Whenever any component of the electric window electrical system is disconnected, the electronic control unit must be initialized once the battery is reconnected. To do this, close the doors then raise each window to the fully closed position and hold down the switch for approximately 5 seconds. This will enable the one-touch switch facility and the anti-jam window system.*

### Window switches
1 Refer to Section 4.

### Window motors
#### Sedan models
2 Remove the window regulator (see Section 14 in Chapter 11).
3 Loosen and remove the retaining screws and remove the motor from the regulator.
4 On installation, fit the motor to the regulator and securely tighten its retaining screws.
5 Install the regulator assembly (see Section 14, 15 or 16 in Chapter 11) and initialize the windows (see **Note** above).

### Window motors
#### Coupe models
6 Removal and installation of the motors requires the regulator to be removed from the door. This task should be entrusted to a BMW dealer or other qualified repair shop (see Section 15 in Chapter 11).

### Electronic control module
#### Sedan models
**Note:** *On some models the windows are controlled by the central body electronics (ZKE IV) control module (depending on model and specification).*
7 Disconnect the battery and remove the glovebox (see Section 27 in Chapter 11). **Caution:** *If the radio in your vehicle is equipped with an anti-theft system, make sure you have the correct activation code before disconnecting the battery.*
8 Release the retaining clips and remove the lower dash panel from beneath the glovebox aperture.
9 Disconnect the wiring connector then release the retaining clip and withdraw the control module.
10 Installation is a reverse of removal. On completion initialize the windows (see **Note** at the beginning of this Section).

#### Coupe models
*Refer to illustration 29.13*

11 On coupe models there are two control modules, one for each window. Each unit is bolted onto its respective window motor. Prior to removal, lower the window slightly then disconnect the battery negative terminal. **Caution:** *If the radio in your vehicle is equipped with an anti-theft system, make sure you have the correct activation code before disconnecting the battery.*
12 Remove the door inner trim panel (see Section 12 in Chapter 11) and carefully peel off the plastic vapor barrier to reveal the control module.
13 Disconnect the wiring connector(s) then unscrew the retaining screws and remove the ECM squarely from the motor **(see illustration)**. Recover the sealing ring from the motor groove; if the seal shows signs of damage it must be replaced.
14 Prior to installation, ensure that the motor and ECM connecting terminals are free from dirt and corrosion.
15 Fit the sealing ring to the groove in the motor. Align the ECM with the motor terminals and squarely locate it in the motor recess.
16 Install the ECM retaining screws and tighten them securely.
17 Reconnect the battery then initialize the windows (see **Note** at the beginning of this Section) and check the window operation.
18 If all is well, stick the vapor barrier to the door and install the trim panel (see Section 12 in Chapter 11).

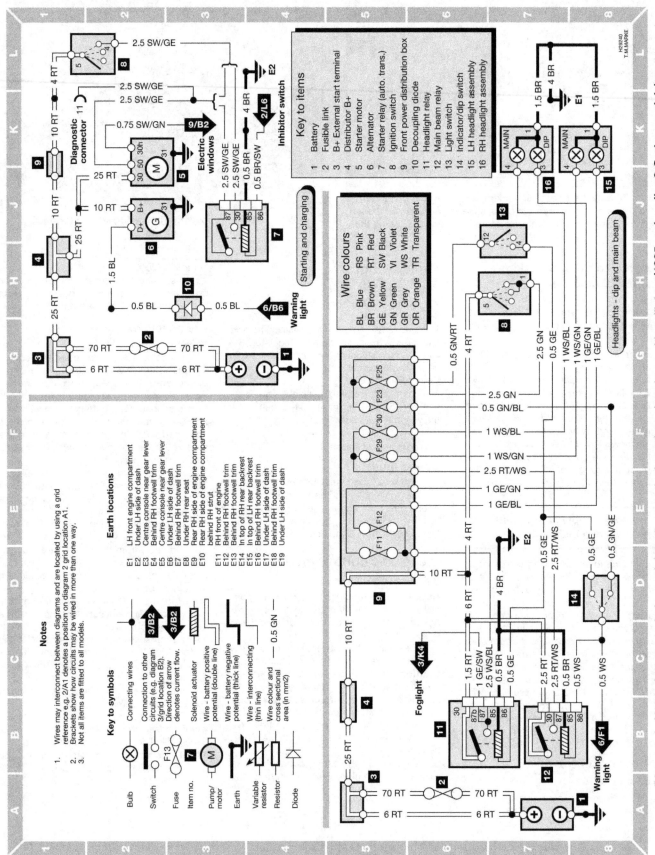

Diagram 1: Notes, key to symbols, earth locations, typical starting, charging and headlights (1995 and earlier 3-Series models)

# Chapter 12 Chassis electrical system

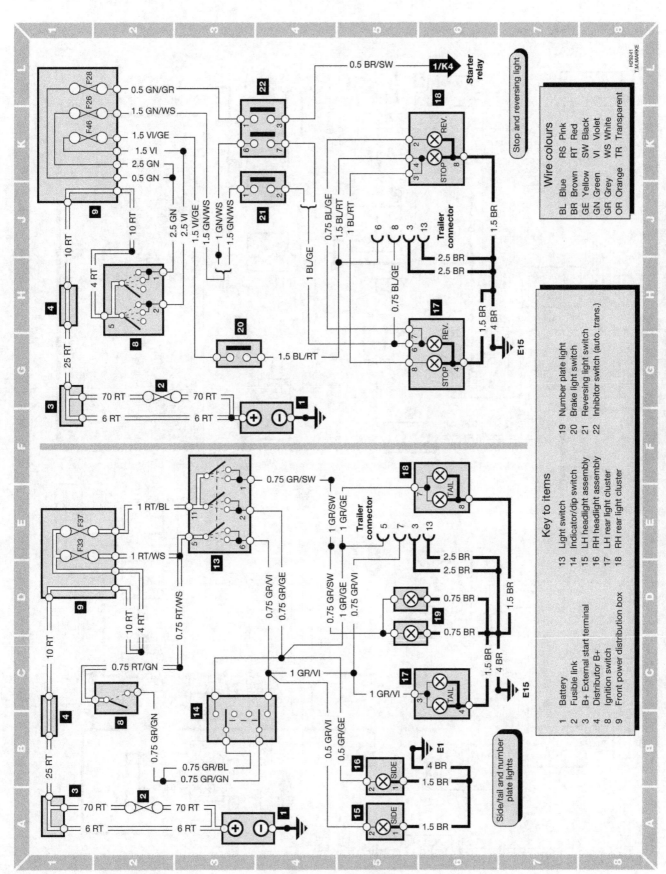

Diagram 2 : Typical side/tail, stop and reversing lights (1995 and earlier 3-Series models)

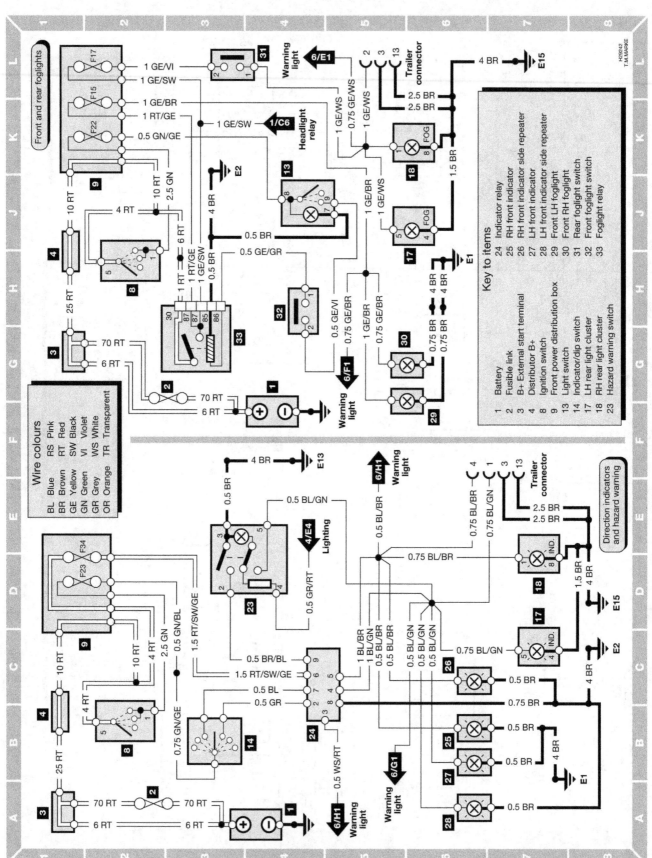

Diagram 3: Typical hazard/direction indicators and foglights (1995 and earlier 3-Series models)

# Chapter 12 Chassis electrical system

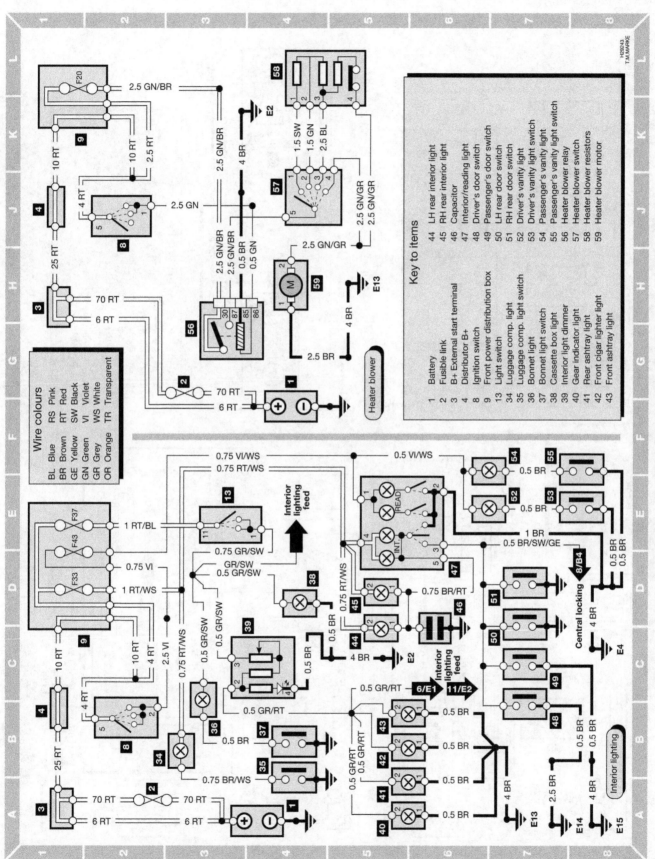

Diagram 4 : Typical interior lighting and heater blower (1995 and earlier 3-Series models)

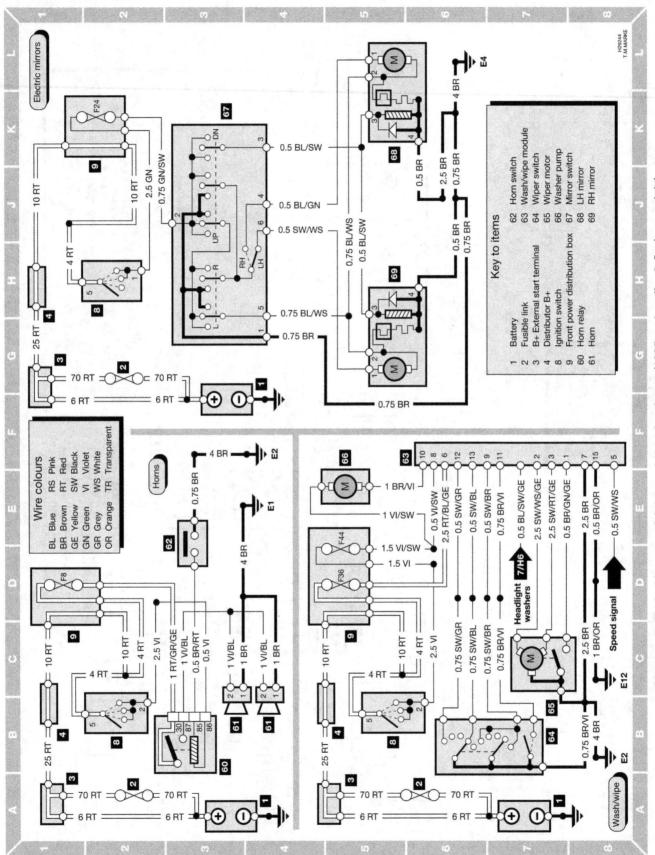

Diagram 5 : Typical horn, wash/wipe and electric mirrors (1995 and earlier 3-Series models)

Chapter 12 Chassis electrical system 12-37

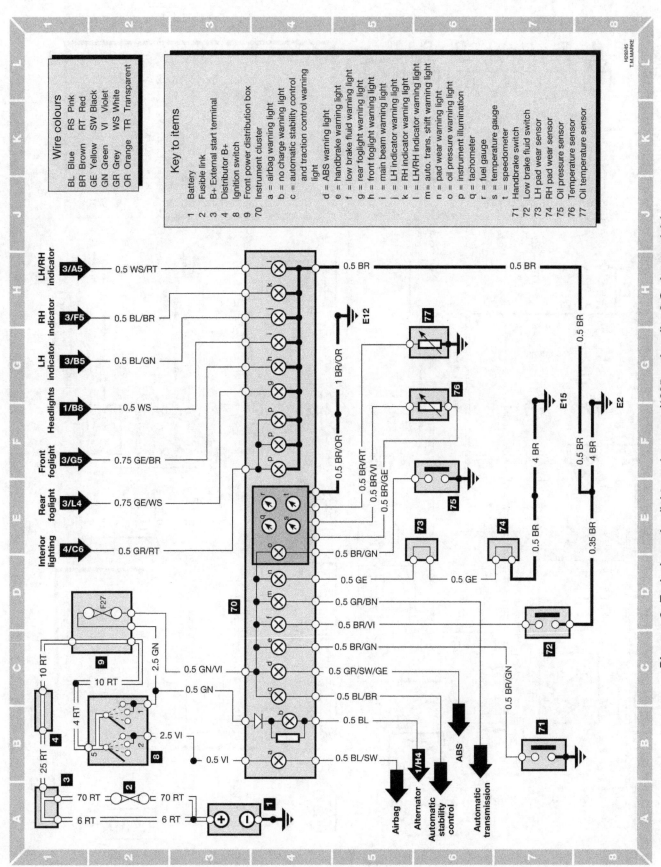

Diagram 6 : Typical warning lights and gauges (1995 and earlier 3-Series models)

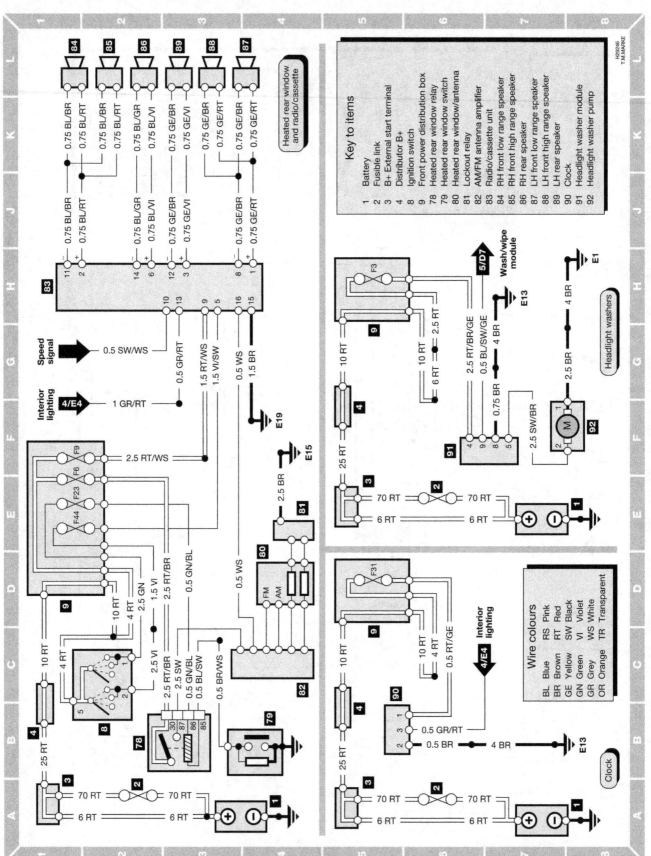

Diagram 7 : Typical heated rear window, clock, radio/cassette and headlight washers (1995 and earlier 3-Series models)

# Chapter 12 Chassis electrical system

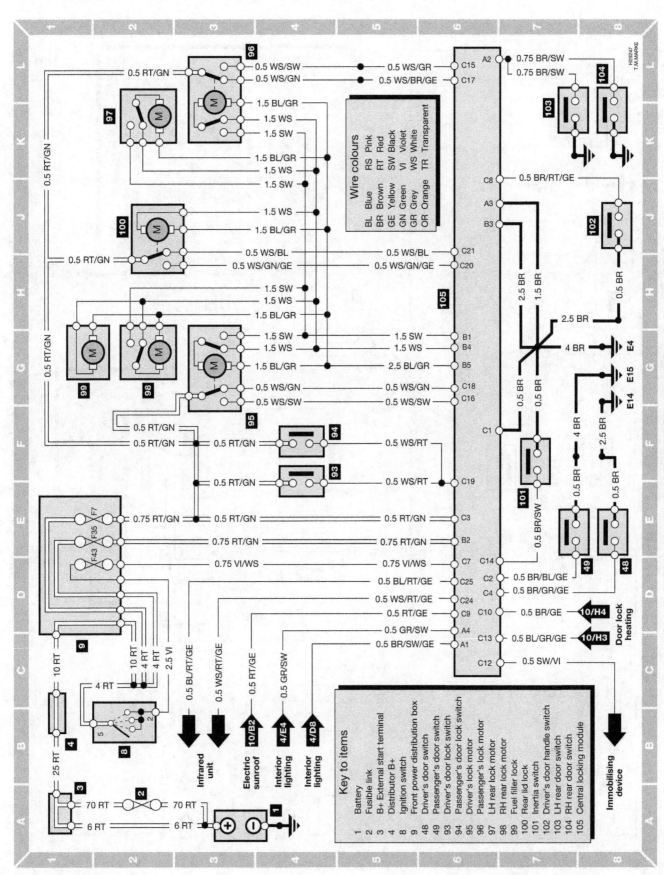

Diagram 8: Typical central locking (1995 and earlier 3-Series models)

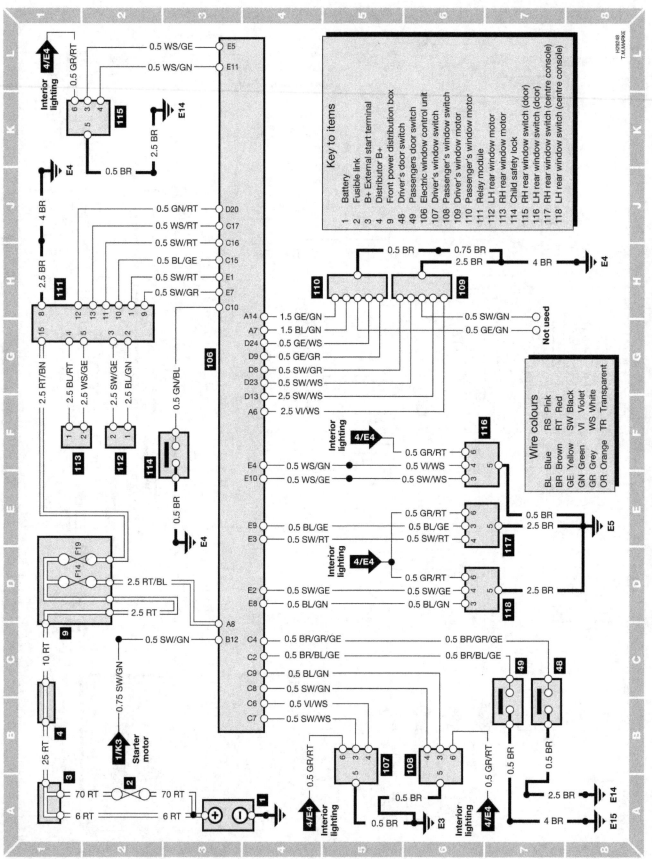

Diagram 9: Typical electric windows (1995 and earlier 3-Series models)

# Chapter 12 Chassis electrical system

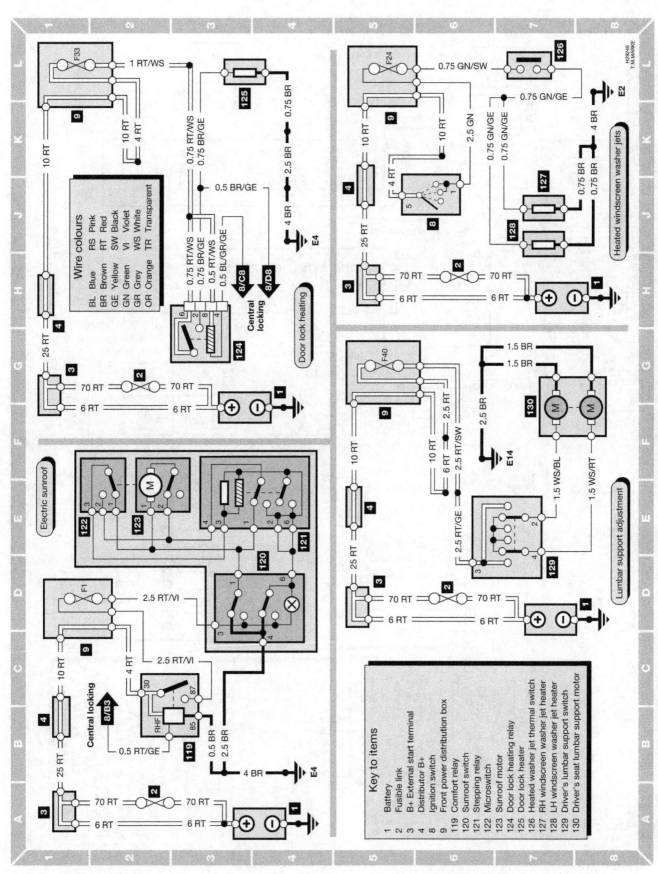

Diagram 10: Typical sunroof, heated door locks, heated washer jets and driver's lumbar support adjustment (1995 and earlier 3-Series models)

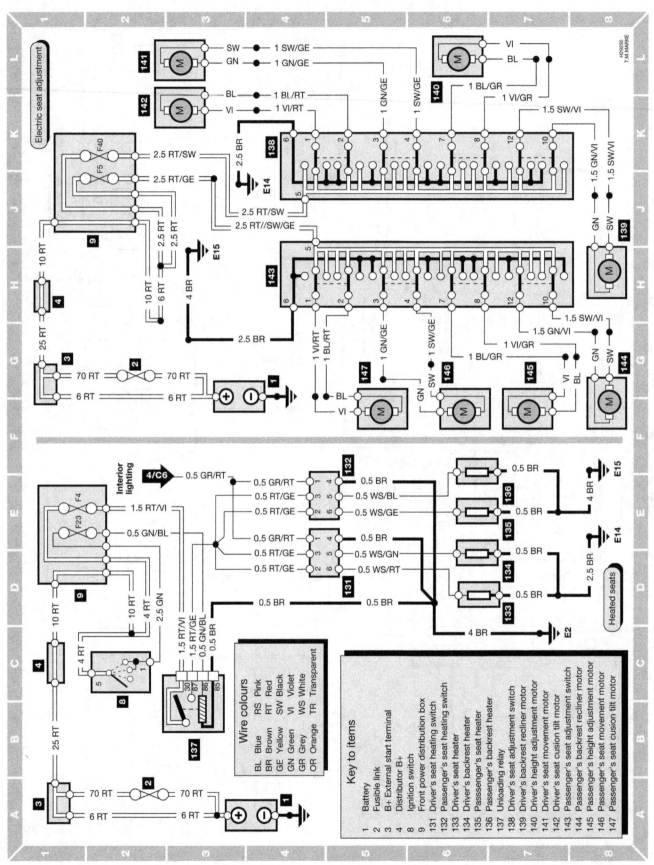

Diagram 11: Typical heated seats and electric seats (1995 and earlier 3-Series models)

## BMW 3 SERIES/Z3 wiring diagrams

**Models from 1996 - Diagram 1**

### Key to symbols

- Bulb
- Switch
- Multiple contact switch (ganged)
- Fuse/fusible link
- Resistor
- Variable resistor
- Item no.
- Pump/motor
- Earth point
- Gauge/meter
- Diode
- Light emitting diode (LED)
- Internal connection (connecting wires)
- Wire splice or soldered joint
- Solenoid actuator
- Plug and socket connector
- Connections to other circuits. Direction of arrow denotes current flow.
- Wire colour (Red wire/white tracer)
- Screened cable
- Denotes alternative wiring variation (brackets)
- Dashed lines denote part of a larger component. Termination identified by connector number and pin number or by pin number in isolation.

### Typical fuse box

| Fuse | Rating | Circuit protected |
|---|---|---|
| F1 | 30A | Power sunroof |
| F2 | 5A | Engine control |
| F3 | 30A | Wash/wipe module |
| F4 | 15A | Heated seats |
| F5 | 30A | Power seats |
| F6 | 20A | Rear window defogger/blower, antenna |
| F7 | 5A | Central locking, convertible roof |
| F8 | 15A | Horn |
| F9 | 20A | Sound system |
| F10 | 30A | ABS/AST |
| F11 | 7.5A | Headlights, foglights, on-board computer |
| F12 | 7.5A | Headlights, foglights, on-board computer |
| F13 | 5A | Not used |
| F14 | 30A | Central body electronics, power windows |
| F15 | 15A | Headlights, foglights |
| F16 | 5A | Engine control, heating and A/C |
| F17 | 10A | Not used |
| F18 | 15A | Engine control |
| F19 | 30A | Power windows |
| F20 | 30A | Blower motor |
| F21 | 5A | ABS/AST |
| F22 | 5A | Instrument illumination |
| F23 | 5A | Antenna, headlights, foglights, heated seats, heating and A/C, instrument cluster, multi-function clock, on-board computer, on-board display, rear window defogger/blower, turn signals/hazard lights |
| F24 | 10A | Convertible roof, power mirrors |
| F25 | 5A | Headlights/foglights, instrument illumination |
| F26 | 10A | Backup lights, electronic transmission control, multi-function clock, on board computer |
| F27 | 5A | Airbag, ABS/AST, convertible roof, instrument cluster, multi-function clock, on-board computer, on-board display |
| F28 | 5A | Central locking, cruise control, electronic transmission control, driveaway protection, starting system |
| F29 | 7.5A | Headlights, foglights |
| F30 | 7.5A | Headlights, foglights |
| F31 | 5A | Central locking, driveaway protection, heating and A/C, instrument cluster, multi-function clock, on-board computer, on-board display, starting system |
| F32 | 30A | Cassette compartment, charging socket, cigar lighter/ashtray lights, glove compartment light |
| F33 | 10A | Ashtray light, cassette compartment, cellular phone, charging socket, cigar lighter, instrument illumination, interior lights glove compartment light, licence plate/luggage light, on-board computer, park/taillights, sound system |
| F34 | 15A | Crash control module, turn signal/hazard warning lights |
| F35 | 25A | Central locking and central body electronics |
| F36 | 30A | Headlight washers, wash/wipe |
| F37 | 10A | Ashtray light, cassette compartment light, charging socket, cigar lighter, glove compartment light, licence plate/luggage compartment lights, instrument illumination, on-board computer, parking/taillights, wash/wipe module |
| F38 | 30A | ABS/AST |
| F39 | 7.5A | Engine control, heating and A/C |
| F40 | 30A | Power seats |
| F41 | 30A | Heating and A/C radiator auxiliary fan |
| F42 | 7.5A | Airbag, convertible roof, roll-over protection |
| F43 | 5A | Anti-theft system, cellular phone, central body electronics, convertible roof, interior lights, rear window defogger/blower, roll over protection, sound system |
| F44 | 15A | Ashtray light, cassette compartment light, charging socket, cigar lighter, glove compartment light, sound system, wash/wipe |
| F45 | 7.5A | Central locking, driveaway protection, instrument cluster, multi-function clock, on-board computer, on-board display, starting system |
| F46 | 15A | ABS/AST, brake lights, cruise control, electronic transmission control, instrument cluster, multi-function clock, on-board computer |
| F47 | 15A | Anti-theft system |
| F48 | 40A | Heating and A/C |
| F49 | 20A | Convertible roof |
| F50 | 5A | Engine control |

Diagram 3, Arrow A — High beam warning light

# Chapter 12 Chassis electrical system

**Models from 1996 - Diagram 2**

### Wire colors
- Bl Blue
- Br Brown
- Ge Yellow
- Gr Gray
- Gn Green
- Or Orange
- Rs Pink
- Rt Red
- Sw Black
- Vi Violet
- Ws White
- Tr Transparent

### Key to items
1. Battery
2. Ignition switch
3. B+ distribution box
4. Front power distribution box
5. Starter motor
6. Alternator
7. Auto. trans. inhibitor switch
8. Immobiliser control unit (EWS II)
9. Data link connector
10. Light switch
11. Combination switch
12. Low beam relay
13. High beam relay
14. Check control module
15. LH high beam
16. RH high beam
17. LH low beam
18. RH low beam
19. Front foglight relay
20. Front foglight switch
21. LH front foglight
22. RH front foglight

H31905

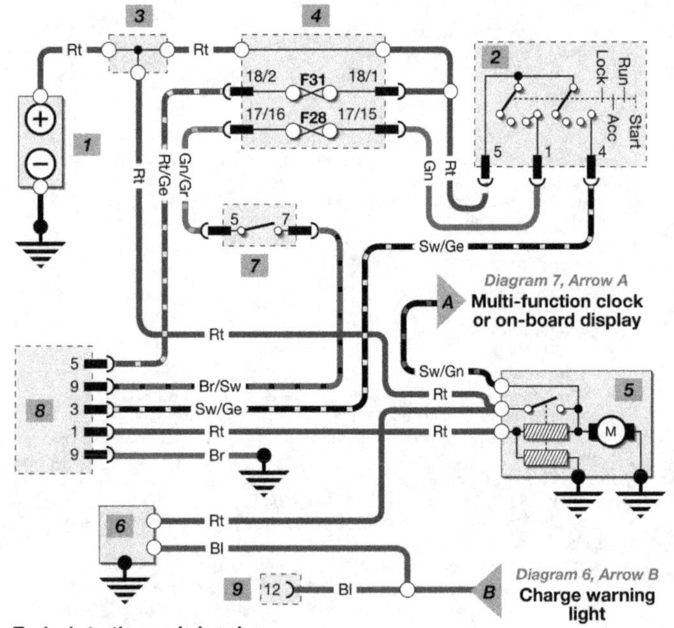

*Typical starting and charging*

*Typical headlights*

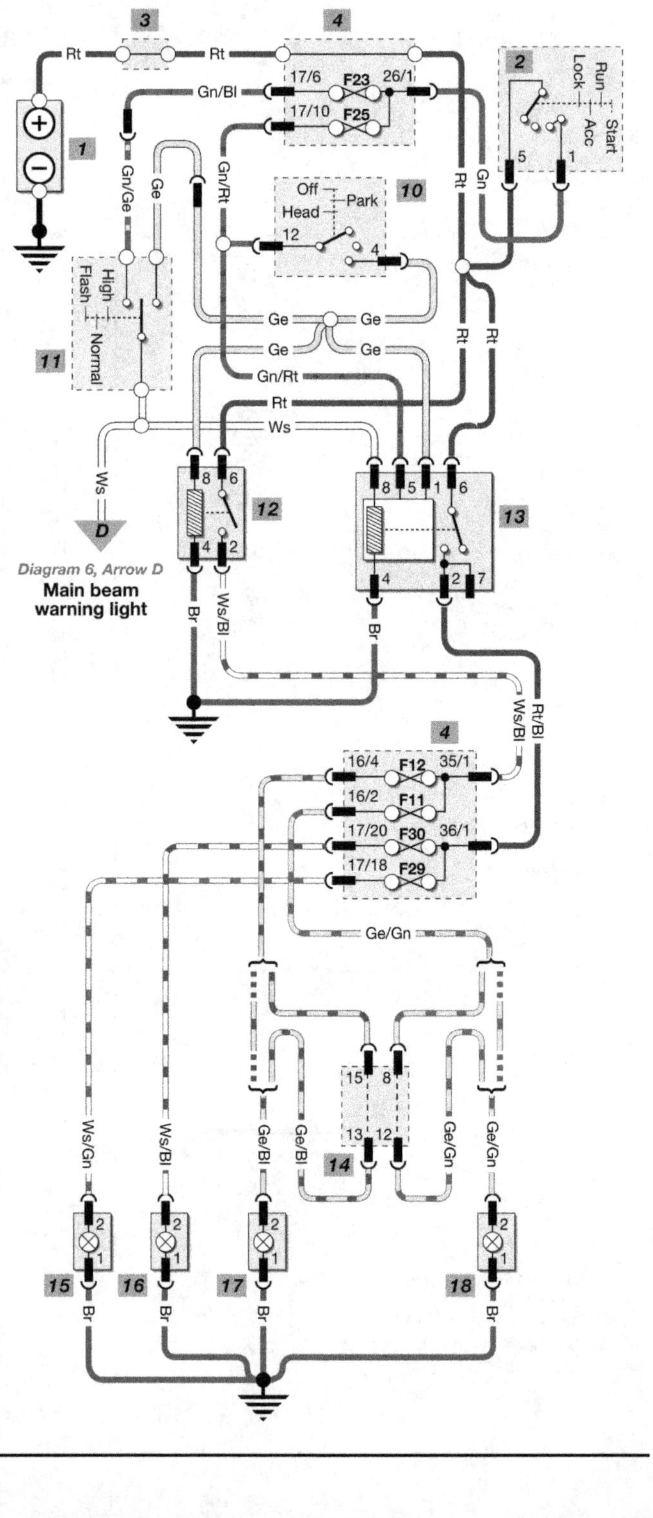

*Typical fog lights*

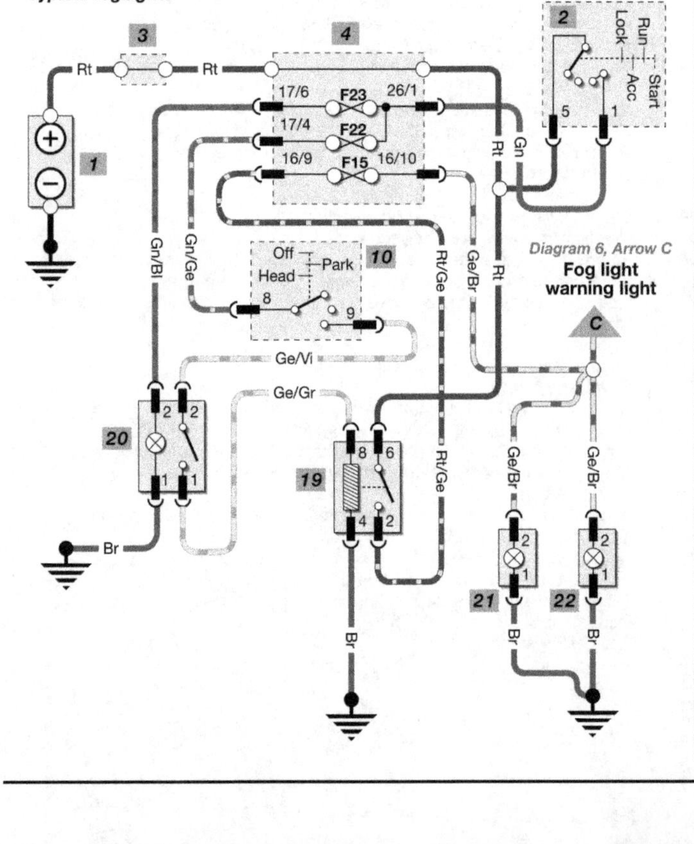

## Chapter 12 Chassis electrical system

**Wire colors**
- Bl  Blue
- Br  Brown
- Ge  Yellow
- Gr  Gray
- Gn  Green
- Or  Orange
- Rs  Pink
- Rt  Red
- Sw  Black
- Vi  Violet
- Ws  White
- Tr  Transparent

**Key to items**
1. Battery
2. Ignition switch
3. B+ distribution box
4. Front power distribution box
10. Light switch
14. Check control module
25. LH front headlight (parking light)
26. RH front headlight (parking light)
27. LH rear light assembly
28. RH rear light assembly
29. LH licence plate light
30. RH licence plate light
31. Brake light switch
32. Backup light switch (manual)
33. Automatic transmission range switch
34. High level brake light

**Models from 1996 - Diagram 3**

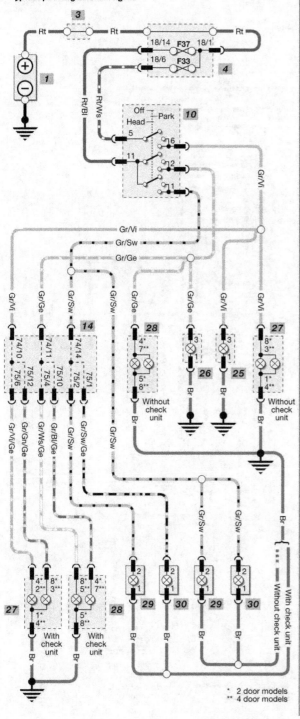

Typical parking and tail lights

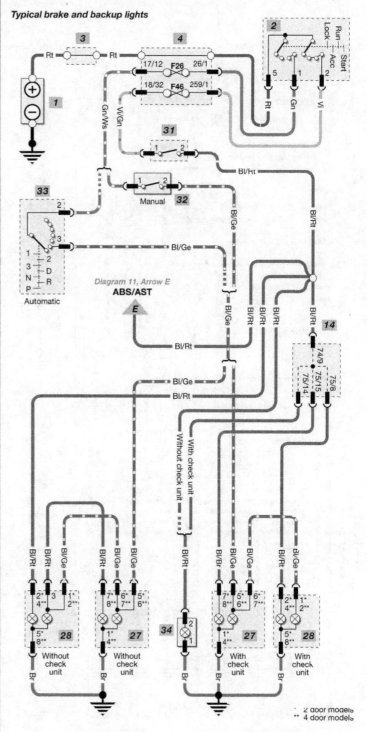

Typical brake and backup lights

* 2 door models
** 4 door models

## 12-46 Chapter 12 Chassis electrical system

**Wire colors**
- Bl Blue
- Br Brown
- Ge Yellow
- Gr Gray
- Gn Green
- Or Orange
- Rs Pink
- Rt Red
- Sw Black
- Vi Violet
- Ws White
- Tr Transparent

**Key to items**
1. Battery
2. Ignition switch
3. B+ distribution box
4. Front power distribution box
11. Combination switch
25. LH front headlight (turn light)
26. RH front headlight (turn light)
27. LH rear light assembly
28. RH rear light assembly
37. Crash control module
38. Hazard flasher relay
39. Hazard switch
40. RH front auxiliary turn light
41. LH front auxiliary turn light
42. Anti-theft control module
43. Body electronics control module
44. Front dome/map reading light
45. LH front footwell light
46. RH front footwell light
47. Glovebox light/switch
48. Trunk compartment light
49. Driver's make up mirror light
50. Driver's make up mirror light switch
51. Passenger's make up mirror light
52. Passenger's make up mirror light switch
53. LH rear interior light
54. RH rear interior light
55. Trunk compartment light switch
56. Passenger door jamb switch
57. Driver's door jamb switch
58. LH rear door switch
59. RH rear door switch
60. RH rear door jamb switch
61. LH rear door jamb switch
62. Trunk lid relay
63. Convertible top storage lid micro switch

**Models from 1996 - Diagram 4**

H31907

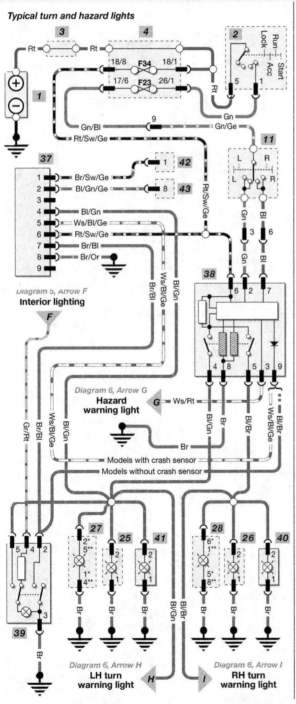

*Typical turn and hazard lights*

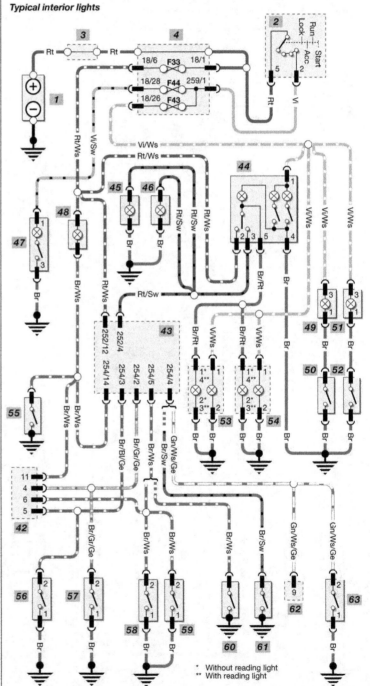

*Typical interior lights*

\* Without reading light
\*\* With reading light

# Chapter 12 Chassis electrical system

12-47

## Wire colors
- Bl Blue
- Br Brown
- Ge Yellow
- Gr Gray
- Gn Green
- Or Orange
- Rs Pink
- Rt Red
- Sw Black
- Vi Violet
- Ws White
- Tr Transparent

## Key to items
1. Battery
2. Ignition switch
3. B+ distribution box
4. Front power distribution box
10. Light switch
43. Body electronics control module
65. Dimmer unit
66. Wash/wipe module
67. Cassette box light
68. Front ashtray light
69. Rear ashtray light
70. Front cigar lighter light
71. Gear shift indicator
72. Transmission position indicator light
73. Horn relay
74. Horn switch
75. LH horn
76. RH horn
77. Rear defogger relay
78. Rear defogger
79. Integrated heating and climate control switch
80. Rear defogger switch
81. Antenna amplifier
82. Mirror control switch
83. Driver's mirror assembly
84. Passenger's mirror assembly

**Models from 1996 - Diagram 5**

H31908

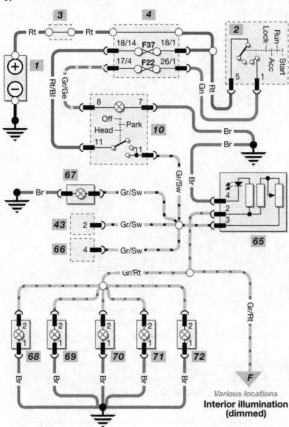

*Typical switch illumination*

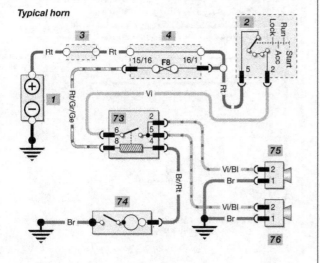

*Typical horn*

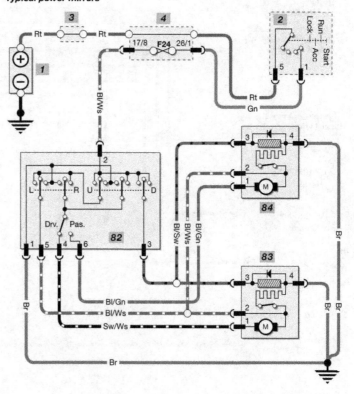

*Typical heated rear window*

*Typical power mirrors*

\* Without electric aerial
\*\* With electric aerial

## Chapter 12 Chassis electrical system

**Models from 1996 - Diagram 6**

### Wire colors
- Bl Blue
- Br Brown
- Ge Yellow
- Gr Gray
- Gn Green
- Or Orange
- Rs Pink
- Rt Red
- Sw Black
- Vi Violet
- Ws White
- Tr Transparent

### Key to items
1. Battery
2. Ignition switch
3. B+ distribution box
4. Front power distribution box
90. Instrument cluster
    - a = high beam ind.
    - b = front fog light ind.
    - c = hazard warning ind.
    - d = LH turn ind.
    - e = RH turn ind.
    - f = ASC + T ind.
    - g = oil ind.
90. Instrument cluster (continued)
    - h = park brake ind.
    - i = check engine ind.
    - j = brake fluid ind.
    - k = charge ind.
    - l = SRS ind.
    - m = instrument illumination
91. Brake fluid level switch
92. Oil pressure switch
93. Park brake switch

H31909

**Typical warning lights and indicators (1 of 2)**

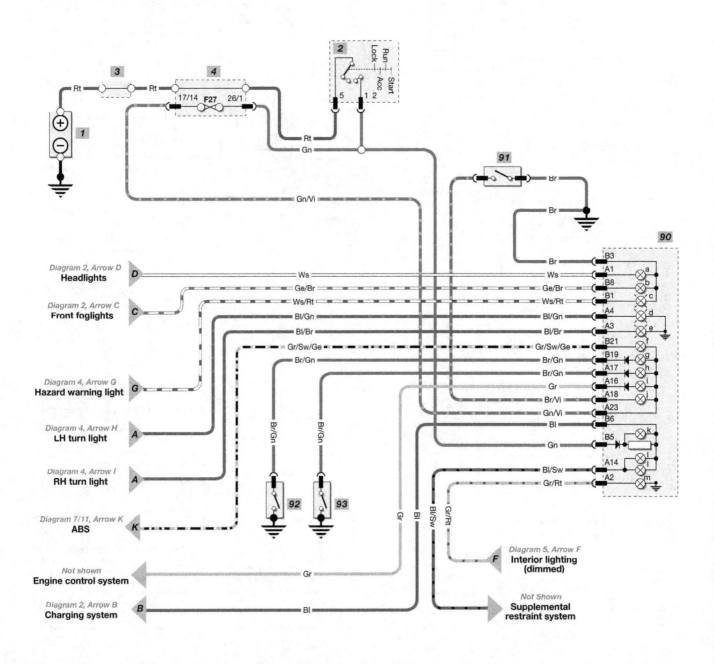

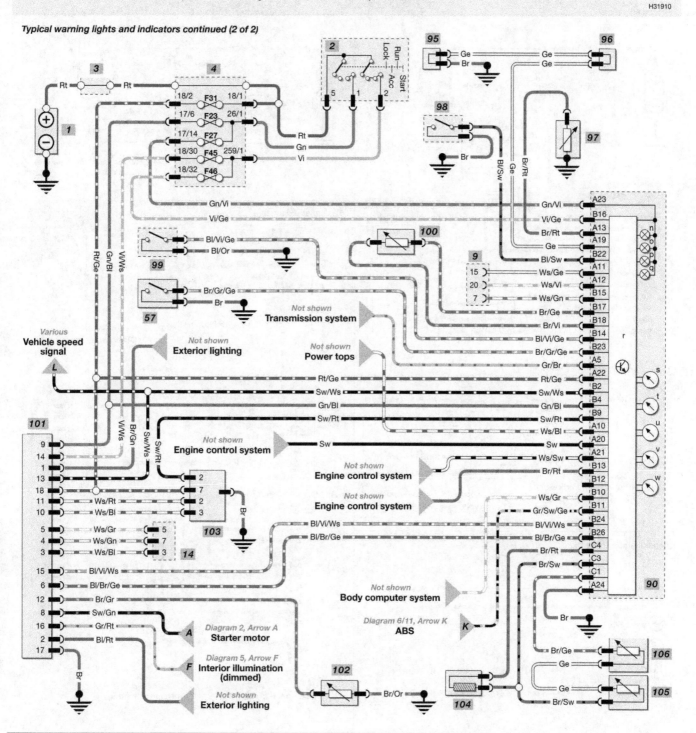

## Chapter 12 Chassis electrical system

**Wire colors**
- Bl Blue
- Br Brown
- Ge Yellow
- Gr Gray
- Gn Green
- Or Orange
- Rs Pink
- Rt Red
- Sw Black
- Vi Violet
- Ws White
- Tr Transparent

**Key to items**
1. Battery
2. Ignition switch
3. B+ distribution box
4. Front power distribution box
43. Body electronics control module
110. Wash/wipe module
111. Wiper motor
112. Wash/wipe switch
113. Washer pump
114. Jet heater thermostat
115. LH washer jet heater
116. RH washer jet heater
117. Radio unit
118. Telephone connector
119. LH front low range speaker
120. LH front high range speaker
121. LH rear speaker
122. RH front low range speaker
123. RH front high range speaker
124. RH rear speaker
125. Sunroof control assembly
126. Sunroof switch
127. Driver's seat lumbar switch
128. Driver's seat lumbar support motor/valve

**Models from 1996 – Diagram 8**

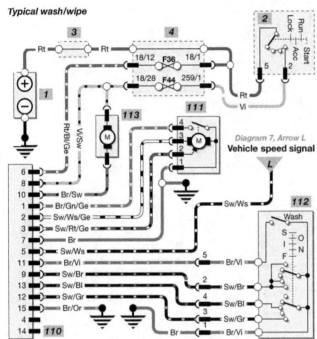
Typical wash/wipe

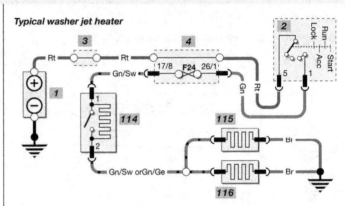
Typical washer jet heater

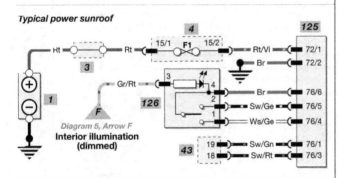
Typical power sunroof

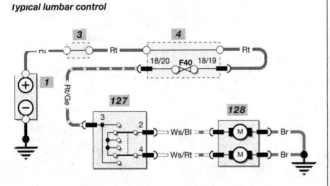
Typical lumbar control

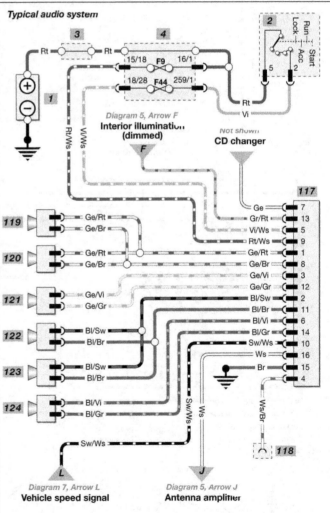
Typical audio system

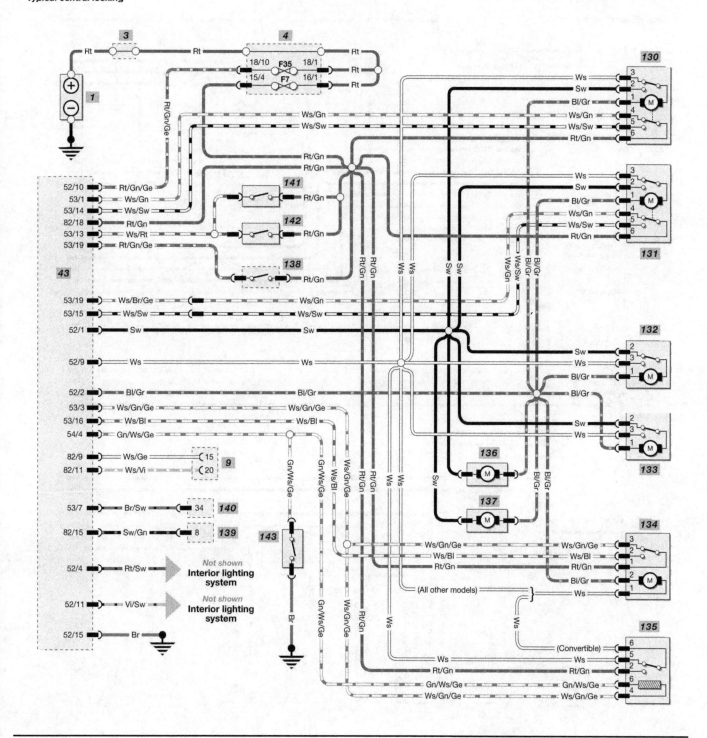

## 12-52  Chapter 12  Chassis electrical system

**Wire colors**
- Bl Blue
- Br Brown
- Ge Yellow
- Gr Gray
- Gn Green
- Or Orange
- Rs Pink
- Rt Red
- Sw Black
- Vi Violet
- Ws White
- Tr Transparent

**Key to items**
1. Battery
3. B+ distribution box
4. Front power distribution box
43. Body electronics control module
56. Passenger's door jamb switch
57. Driver's door jamb switch
145. Relay module
146. Driver's power window motor
147. Passenger's power window motor
148. LH rear window motor
149. RH rear window motor
150. Driver's power window switch
151. Passenger's power window switch
152. LH rear door (console) power window switch
153. RH rear door (console) power window switch
154. LH rear door (door) power window switch
155. RH rear door (door) power window switch
156. Child protect switch

**Models from 1996 - Diagram 10**

H31913

*Typical power windows*

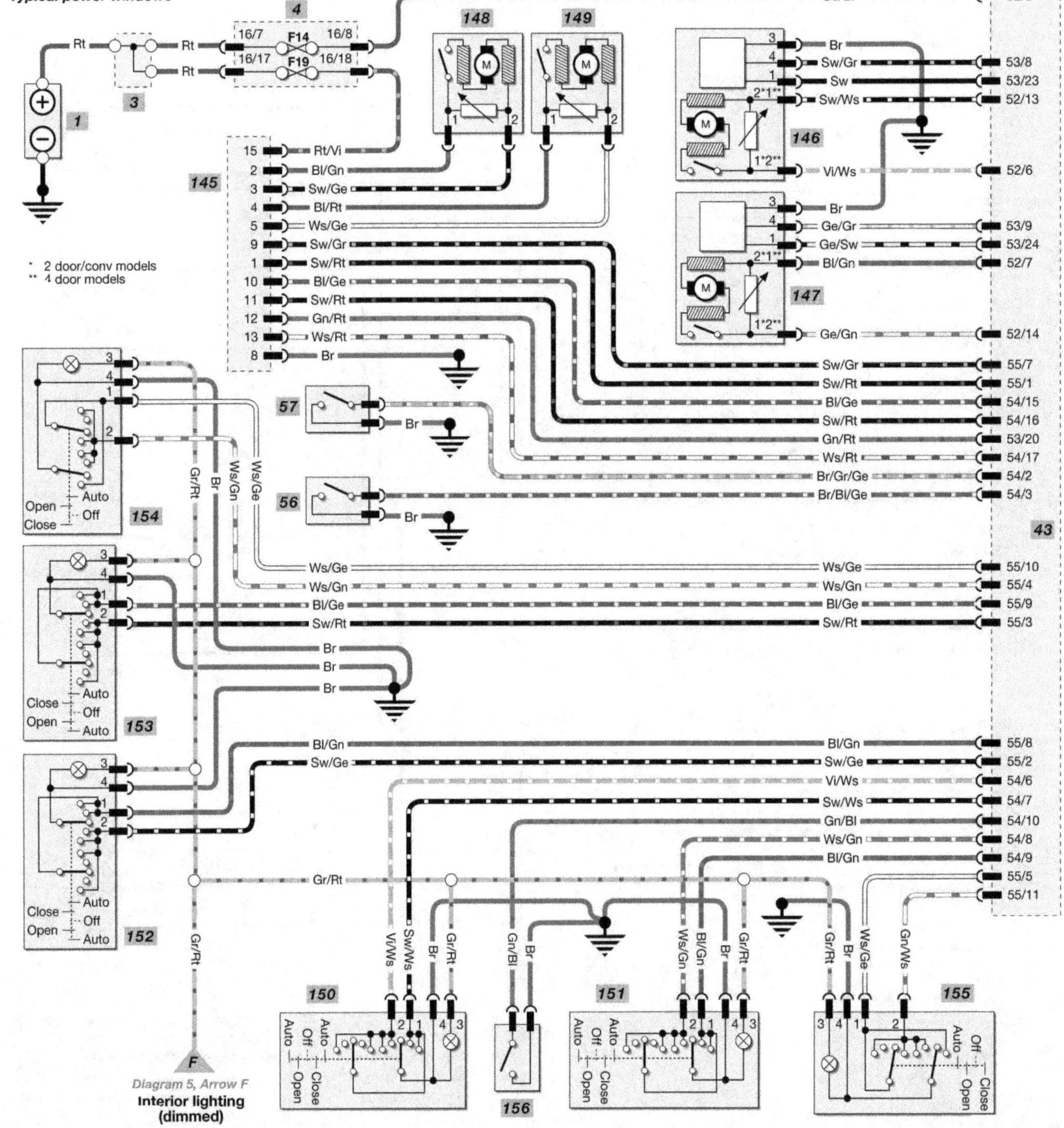

\* 2 door/conv models
\*\* 4 door models

*Diagram 5, Arrow F*
**Interior lighting (dimmed)**

## Chapter 12 Chassis electrical system

**Models from 1996 - Diagram 11**

### Wire colors
- Bl Blue
- Br Brown
- Ge Yellow
- Gr Gray
- Gn Green
- Or Orange
- Rs Pink
- Rt Red
- Sw Black
- Vi Violet
- Ws White
- Tr Transparent

### Key to items
1. Battery
2. Ignition switch
3. B+ distribution box
4. Front power distribution box
9. Data link connector
160. Unloader relay
161. Driver's seat heater switch
162. Passenger's seat heater switch
163. Driver's seat heater
164. Driver's seat back heater
165. Passenger's seat heater
166. Passenger's seat back heater
167. Driver's seat control switch
168. Passenger's seat control switch
169. Driver's seat back recliner motor
170. Driver's seat height motor
171. Driver's seat movement motor
172. Driver's seat cushion tilt motor
173. Passenger's seat back recliner motor
174. Passenger's seat height motor
175. Passenger's seat movement motor
176. Passenger's seat cushion tilt motor
177. ABS control module/hydraulic unit
178. LH front speed sensor
179. LH rear speed sensor
180. RH front speed sensor
181. RH rear speed sensor

*Typical heated seats*

*Typical power seats*

*Typical ABS/AST*

H31914

## Notes

# Index

## A

About this manual, 0-5
Air cleaner assembly, removal and installation, 4A-2, 4B-2
Air conditioning system
    components, removal and installation, 3-12
    general information and precautions, 3-12
Air filter element replacement, 1-20
Air mass meter, 4B-9
Airbag system
    components, removal and installation, 12-26
    inspection, 1-24
Airbag system, general information and precautions, 12-26
Airflow meter, 4A-5
Alternator
    removal and installation, 5A-3
    testing and overhaul, 5A-4
Antenna (3-Series models), general information, 12-25
Anti-lock braking system (ABS)
    components, removal and installation, 9-10
    general information, 9-10
Anti-theft alarm system, general information, 12-25
Anti-theft audio system, 0-13
Audio unit, removal and installation, 12-23
Automatic transmission, 7B-1 through 7B-6
    fluid level check, 1-12
    fluid replacement, 1-21
    fluid seals, replacement, 7B-3
    gear selector cable, removal, installation and adjustment, 7B-3
    gear selector lever, removal and installation, 7B-2
    general information, 7B-1
    overhaul, general information, 7B-5
    removal and installation, 7B-4
Automotive chemicals and lubricants, 0-14

## B

Back-up light switch, testing, removal and installation, 7A-4
Battery
    check, maintenance and charging, 1-17
    jump starting, 0-6
    removal and installation, 5A-2
    testing and charging, 5A-2
Body, 11-1 through 11-38
Body electrical system, 12-1 through 12-31
Body exterior fittings, removal and installation, 11-29
Brakes, 9-1 through 9-12
    Anti-lock braking system (ABS)
        components, removal and installation, 9-10
        general information, 9-10
    booster check valve, removal, testing and installation, 9-7
    booster, general information, 9-7
    caliper, removal, overhaul and installation
        front, 9-4
        rear, 9-6
    disc inspection, removal and installation
        front, 9-5
        rear, 9-6
    fluid level check, 1-8
    fluid replacement, 1-23
    general information, 9-2
    hoses and lines, inspection, 1-15
    hydraulic system, bleeding, 9-2
    light switch, removal and installation, 9-9
    lines and hoses, replacement, 9-3
    master cylinder, general information, 9-6
    pad check, 1-15
    pad replacement
        front, 9-3
        rear, 9-6

# Index

parking
    adjustment, 9-7
    cables, removal and installation, 9-8
    check, 1-15
    lever, removal and installation, 9-8
    shoes, removal and installation, 9-9
pedal, removal and installation, 9-7
shoe check, parking, 1-21
**Bulbs (exterior lights), replacement, 12-10**
**Bulbs (interior lights), replacement, 12-15**
**Bumper, removal and installation, 11-3**
**Buying parts, 0-7**

## C

**Camshaft position sensor, 4A-6**
**Camshafts and followers, removal and installation**
    four-cylinder engines, 2A-10
    six-cylinder engines, 2B-17
**Capacities, 1-2**
**Catalytic converter, general information and precautions, 6-3**
**Center console, removal and installation, 11-35**
**Central locking components, removal and installation, 12-28**
**Charging system, testing, 5A-3**
**Cigarette lighter, removal and installation, 12-19**
**Clock/multi-information display/on-board trip computer, removal and installation, 12-20**
**Clutch and driveline, 8-1 through 8-12**
**Clutch**
    assembly, removal, inspection and installation, 8-2
    check, 1-23
    fluid level check, 1-8
    general information, 8-2
    hydraulic system, bleeding, 8-5
    master cylinder, removal, inspection and installation, 8-5
    pedal, removal and installation, 8-6
    release bearing and lever, removal, inspection and installation, 8-3
    release cylinder, removal, inspection and installation, 8-4
    start switch, check and replacement, 8-6
**Coil spring, rear, removal and installation, 10-9**
**Compression test, description and interpretation, 2A-2, 2B-3**
**Conversion factors, 0-15**
**Coolant pump, removal and installation, 3-6**
**Coolant replacement, 1-23**
**Cooling fan, removal and installation, 3-3**
**Cooling system electrical switches, testing, removal and installation, 3-5**
**Cooling system hoses, disconnection and replacement, 3-2**
**Cooling, heating and ventilation systems, 3-1 through 3-12**
**Crankshaft pilot bearing, replacement, 2B-28**
**Crankshaft position sensor, 4A-5**
**Crankshaft vibration damper/pulley and pulley hub, removal and installation, 2A-5, 2B-7**
**Crankshaft, inspection, 2C-14**

**Crankshaft, installation and main bearing oil clearance check, 2C-16**
**Crankshaft, removal, 2C-10**
**Cruise control system components (3-Series models), removal and installation, 12-25**
**Cylinder block/crankcase, cleaning and inspection, 2C-11**
**Cylinder head**
    cleaning and inspection, 2C-7
    disassembly, 2C-6
    reassembly, 2C-8
    removal and installation
        four-cylinder engines, 2A-13
        six-cylinder engines, 2B-21

## D

**Dash panel assembly, removal and installation, 11-38**
**Dash trim panels, removal and installation, 11-33**
**Diagnosis, 0-18**
**Differential oil replacement, 1-21**
**Differential oil seals, replacement, 8-7**
**Differential, removal and installation, 8-7**
**Door handle and lock components, removal and installation, 11-16**
**Door lock heating system, general information and component removal and installation, 12-28**
**Door, removal, installation and adjustment, 11-13**
**Door trim panel, removal and installation, 11-14**
**Door window glass and regulator and rear vent window (3-Series coupe models), removal and installation, 11-23**
**Door window glass and regulator (3-Series sedan models), removal and installation, 11-21**
**Door window glass and regulator (Z3 models), removal, installation and adjustment, 11-24**
**Driveaxle boots**
    check, 1-21
    replacement, 8-8
**Driveaxle, removal and installation, 8-8**
**Drivebelt(s) check and replacement, 1-13**
**Driveshaft**
    removal and installation, 8-0
    rubber coupling, check and replacement, 8-11
    support bearing, check and replacement, 8-11
    universal joints, check and replacement, 8-12

## E

**Electric window components, removal and installation, 12-31**
**Electrical system, general information and precautions, 12-2**
**Electrical troubleshooting, general information, 12-2**
**Emission control systems, 6-1 through 6-4**
    component replacement, 6-2
    general information, 6-1
**Engine Control Module (ECM), 4A-4, 4B-7**
**Engine coolant level check, 1-8**
**Engine management system check, 1-16**
**Engine oil and filter replacement, 1-11**

# Index

Engine oil level check, 1-7
Engine overhaul
   disassembly sequence, 2C-6
   general information, 2C-3
   reassembly sequence, 2C-15
Engine removal, methods and precautions, 2C-3
Engine, initial start-up after overhaul, 2C-20
Engine, removal and installation, 2C-4
Engine/transmission mounts, inspection and replacement, 2B-28
Exhaust system
   check, 1-15
   general information, removal and installation, 4A-8, 4B-15
Exterior light units, removal and installation, 12-16
Exterior mirrors and associated components, removal and installation, 11-27

## F

Fault finding, 0-18
Fenders, removal and installation, 11-8
Filter replacement
   air, 1-20
   fuel, 1-22
   pollen, 1-16
Fluid leak check, 1-12
Fluid seals, replacement, 7B-3
Flywheel/driveplate, removal and installation, 2B-27
Four-cylinder engines, 2A-1 through 2A-18
Fraction/Decimal/Millimeter Equivalents, 0-16
Front hub assembly, removal and installation, 10-4
Front seat belt tensioning mechanism, general information, 11-30
Front stabilizer bar
   connecting link, removal and installation, 10-7
   removal and installation, 10-7
Front suspension control arm
   balljoint, replacement, 10-7
   removal, overhaul and installation, 10-6
Front suspension strut, removal, overhaul and installation, 10-5
Fuel and exhaust systems, four-cylinder engines, 4A-1 through 4A-10
Fuel and exhaust systems, general information and precautions, 4A-2, 4B-2
Fuel and exhaust systems, six-cylinder engines, 4B-1 through 4B-16
Fuel filter replacement, 1-22
Fuel gauge sender unit, removal and installation, 4B-6
Fuel injection system
   components, removal and installation, 4A-4, 4B-7
   depressurization and priming, 4A-3, 4B-5
   fuel pressure regulator, 4A-4, 4B-9
   fuel pump/fuel gauge sender unit, removal and installation, 4B-5
   fuel rail and injectors, 4A-4, 4B-8
   fuel tank, removal and installation, 4B-3
   general information, 4A-3, 4B-5
   testing and adjustment, 4A-3, 4B-6
Fuses and relays, general information, 12-2

## G

Gear selector cable, removal, installation and adjustment, 7B-3
Gear selector components, removal and installation, 7A-2
Gear selector lever, removal and installation, 7B-2
General engine overhaul procedures, 2C-1 through 2C-20
Grille and front panel, removal and installation, 11-27

## H

Headlight
   beam alignment, check, 1-16
   beam alignment, general information, 12-18
Heated front seat components, removal and installation, 12-26
Heater/ventilation components, removal and installation
   3-Series models, 3-6
   Z3 models, 3-10
Heating and ventilation system, general information, 3-6
Hinge and lock lubrication, 1-16
Hood
   lock(s), removal and installation, 11-13
   release cable, removal and installation, 11-11
   removal, installation and adjustment, 11-10
Horn(s), removal and installation, 12-20
Hose and fluid leak check, 1-12
Hub assembly, removal and installation
   front, 10-4
   rear, 10-8
Hub bearings (rear), replacement, 10-8

## I

Idle speed control valve, 4A-7, 4B-12
Ignition
   coil, removal and installation, 5B-2
   switch, removal and installation, 5A-6
   switch/steering column lock, removal and installation, 10-14
   systems, 5B-1 through 5B-4
   systems
      general information and precautions, 5B-1
      testing, 5B-1
Inlet air temperature sensor, 4B-10
Instrument cluster, removal and installation, 12-18
Instrument cluster components, removal and installation, 12-19
Introduction to BMW 3-series models, 0-5

## J

Jacking and towing, 0-13
Jump starting, battery, 0-6

## K

Knock sensor, removal and installation, 5B-3

## L

Lower control arm, removal, overhaul and installation, rear, 10-12

## M

Main and connecting rod bearings, inspection, 2C-15
Maintenance schedule, 1-6
Maintenance techniques, tools and working facilities, 0-7
Maintenance, bodywork and underframe, 11-1
Maintenance, upholstery and carpets, 11-2
Major body damage, repair, 11-3
Manifolds, removal and installation, 4A-7, 4B-12
Manual transmission, 7A-1 through 7A-6
    back-up light switch, testing, removal and installation, 7A-4
    gear selector components, removal and installation, 7A-2
    general information, 7A-1
    oil level check, 7A-2
    oil replacement, 1-22
    oil seals, replacement, 7A-3
    overhaul, general information, 7A-6
    removal and installation, 7A-4
Master cylinder, general information, 9-6
Minor body damage, repair, 11-2
Mirrors (exterior) and associated components, removal and installation, 11-27
Multi-information system, general information and component removal and installation, 12-28

## O

Oil pan, removal and installation, 2A-14, 2B-23
Oil pump and drive chain, removal, inspection and installation, 2B-26
Oil pump, removal, inspection and installation, 2A-16
Oil seals, replacement, 2B-27, 7A-3
On-board diagnostics, general information, 6-4

## P

Parking brake
    adjustment, 9-7
    cables, removal and installation, 9-8
    check, 1-15
    lever, removal and installation, 9-8
    shoe check, 1-21
    shoes, removal and installation, 9-8
Piston rings, installation, 2C-16
Piston/connecting rod assembly
    inspection, 2C-12
    installation and connecting rod bearing oil clearance check, 2C-18
    removal, 2C-9
Pollen filter replacement, 1-16
Power steering
    pump, removal and installation, 10-16
    system, bleeding, 10-17

## R

Radiator, removal, inspection and installation, 3-4
Rear coil spring, removal and installation, 10-9
Rear hub assembly, removal and installation, 10-8
Rear hub bearings, replacement, 10-8
Rear shock absorber, removal, overhaul and installation, 10-8
Rear stabilizer bar, removal and installation, 10-12
Rear suspension lower control arm, removal, overhaul and installation, 10-12
Rear suspension trailing arm, removal, overhaul and installation, 10-9
Rear suspension upper control arm, removal, overhaul and installation, 10-11
Recommended lubricants and fluids, 1-1
Repair operations possible with the engine in the vehicle, 2A-2, 2B-3
Resetting the service interval display, 1-12
Road test, 1-16

## S

Safety first, 0-17
Seat belt
    check, 1-16
    components, removal and installation, 11-30
    tensioning mechanism (front), general information, 11-30
Seats, removal and installation, 11-29
Service interval display, resetting, 1-12
Shock absorber, removal, overhaul and installation, rear, 10-8
Six-cylinder engines, 2B-1 through 2B-28
Spark plug
    replacement, 1-19
    type and gap, 1-2
Speakers, removal and installation, 12-24
Speedometer drive sender unit, removal and installation, 12-19
Stabilizer bar connecting link (front), removal and installation, front, 10-7
Stabilizer bar, removal and installation
    front, 10-7
    rear, 10-12
Starter motor
    removal and installation, 5A-5
    testing and overhaul, 5A-6
Starting and charging systems, 5A-1 through 5A-6
Starting and charging systems, general information and precautions, 5A-2
Starting system, testing, 5A-4
Steering and suspension check, 1-14
Steering column
    intermediate shaft, removal and installation, 10-15
    removal, inspection and installation, (3-Series models), 10-13
Steering
    gear
        assembly, removal, overhaul and installation, 10-16
        boots, replacement, 10-17
    knuckle, removal and installation, 10-4
    pump, power, removal and installation, 10-16

system, power
  bleeding, 10-17
  steering fluid level check, 1-9
  wheel, removal and installation, 10-13
**Steering knuckle, removal and installation, 10-4**
**Stereo anti-theft system, 0-13**
**Sunroof, (3-Series models), general information, 11-28**
**Suspension and steering, 10-1 through 10-18**
**Suspension and steering, general information, 10-2**
**Suspension control, front, arm balljoint, replacement, front, 10-7**
**Suspension control, front, removal, overhaul and installation, 10-6**
**Suspension strut (front), removal, overhaul and installation, 10-5**
**Suspension trailing arm, rear, removal, overhaul and installation, 10-9**
**Suspension upper control arm, rear, removal, overhaul and installation, 10-11**
**Switches, removal and installation, 12-4**

# T

**Thermostat, removal, testing and installation, 3-2**
**Throttle**
  body, removal and installation, 4A-3, 4B-6
  cable, removal, installation and adjustment, 4A-2, 4B-3
  linkage, lubrication, 1-15
  pedal, removal and installation, 4B-4
  position sensor, 4A-6, 4B-10
**Tie-rod end, removal and installation, 10-17**
**Tie-rod, replacement, 10-17**
**Timing chain**
  covers, removal and installation, 2A-5, 2B-7
  housing, removal and installation, 2A-9
  removal, inspection and installation, 2A-7, 2B-10
  sprockets and tensioner, removal and installation
    four-cylinder engines, 2A-9
    six-cylinder engines, 2B-14
**Tire and tire pressure checks, 1-9**
**Tools, 0-9**
**Top Dead Center (TDC) for No, 1 piston, locating, 2A-3, 2B-3**
**Towing the vehicle, 0-13**
**Transmission, automatic, 7B-1 through 7B-6**
  fluid level check, 1-12
  fluid replacement, 1-21
  fluid seals, replacement, 7B-3
  gear selector
    cable, removal, installation and adjustment, 7B-3
    lever, removal and installation, 7B-2
  general information, 7B-1
  overhaul, general information, 7B-5
  removal and installation, 7B-4
**Transmission, manual, 7A-1 through 7A-6**
  back-up light switch, testing, removal and installation, 7A-4
  gear selector components, removal and installation, 7A-2
  general information, 7A-1
  oil level check, 7A-2
  oil replacement, 1-22
  oil seals, replacement, 7A-3
  overhaul, general information, 7A-6
  removal and installation, 7A-4
**Troubleshooting, 0-18**
**Trunk lid and support struts, removal and installation, 11-25**
**Trunk lid lock components, removal and installation, 11-26**
**Tune-up and routine maintenance, 1-1 through 1-24**

# V

**Valve cover, removal and installation, 2A-4, 2B-5**
**VANOS variable camshaft timing control system**
  general information, 2B-2
  components, removal, inspection and installation, 2B-15
**Vehicle identification numbers, 0-6**

# W

**Weekly checks, 1-7**
**Wheel alignment, general information, 10-17**
**Windshield and back glass, general information, 11-28**
**Windshield washer fluid level check, 1-9**
**Windshield wiper motor and linkage, removal and installation, 12-21**
**Windshield/headlight washer system components, removal and installation, 12-22**
**Windshield/headlight washer system(s) check, 1-16**
**Wiper arm, removal and installation, 12-20**
**Wiper blade check and replacement, 1-18**
**Working facilities, 0-7**

# Notes

# Haynes Automotive Manuals

*NOTE: If you do not see a listing for your vehicle, please visit haynes.com for the latest product information and check out our Online Manuals!*

## ACURA
- **12020** Integra '86 thru '89 & Legend '86 thru '90
- **12021** Integra '90 thru '93 & Legend '91 thru '95
  Integra '94 thru '00 - see HONDA Civic (42025)
  MDX '01 thru '07 - see HONDA Pilot (42037)
- **12050** Acura TL all models '99 thru '08

## AMC
- **14020** Mid-size models '70 thru '83
- **14025** (Renault) Alliance & Encore '83 thru '87

## AUDI
- **15020** 4000 all models '80 thru '87
- **15025** 5000 all models '77 thru '83
- **15026** 5000 all models '84 thru '88
  Audi A4 '96 thru '01 - see VW Passat (96023)
- **15030** Audi A4 '02 thru '08

## AUSTIN-HEALEY
Sprite - see MG Midget (66015)

## BMW
- **18020** 3/5 Series '82 thru '92
- **18021** 3-Series incl. Z3 models '92 thru '98
- **18022** 3-Series incl. Z4 models '99 thru '05
- **18023** 3-Series '06 thru '14
- **18025** 320i all 4-cylinder models '75 thru '83
- **18050** 1500 thru 2002 except Turbo '59 thru '77

## BUICK
- **19010** Buick Century '97 thru '05
  Century (front-wheel drive) - see GM (38005)
- **19020** Buick, Oldsmobile & Pontiac Full-size (Front-wheel drive) '85 thru '05
  Buick Electra, LeSabre and Park Avenue;
  Oldsmobile Delta 88 Royale, Ninety Eight and Regency; Pontiac Bonneville
- **19025** Buick, Oldsmobile & Pontiac Full-size (Rear wheel drive) '70 thru '90
  Buick Estate, Electra, LeSabre, Limited, Oldsmobile Custom Cruiser, Delta 88, Ninety-eight, Pontiac Bonneville, Catalina, Grandville, Parisienne
- **19027** Buick LaCrosse '05 thru '13
  Enclave - see GENERAL MOTORS (38001)
  Rainier - see CHEVROLET (24072)
  Regal - see GENERAL MOTORS (38010)
  Riviera - see GENERAL MOTORS (38030, 38031)
  Roadmaster - see CHEVROLET (24046)
  Skyhawk - see GENERAL MOTORS (38015)
  Skylark - see GENERAL MOTORS (38020, 38025)
  Somerset - see GENERAL MOTORS (38025)

## CADILLAC
- **21015** CTS & CTS-V '03 thru '14
- **21030** Cadillac Rear Wheel Drive '70 thru '93
  Cimarron - see GENERAL MOTORS (38015)
  DeVille - see GENERAL MOTORS (38031 & 38032)
  Eldorado - see GENERAL MOTORS (38030)
  Fleetwood - see GENERAL MOTORS (38031)
  Seville - see GM (38030, 38031 & 38032)

## CHEVROLET
- **10305** Chevrolet Engine Overhaul Manual
- **24010** Astro & GMC Safari Mini-vans '85 thru '05
- **24013** Aveo '04 thru '11
- **24015** Camaro V8 all models '70 thru '81
- **24016** Camaro all models '82 thru '92
- **24017** Camaro & Firebird '93 thru '02
  Cavalier - see GENERAL MOTORS (38016)
  Celebrity - see GENERAL MOTORS (38005)
- **24018** Camaro '10 thru '15
- **24020** Chevelle, Malibu & El Camino '69 thru '87
  Cobalt - see GENERAL MOTORS (38017)
- **24024** Chevette & Pontiac T1000 '76 thru '87
  Citation - see GENERAL MOTORS (38020)
- **24027** Colorado & GMC Canyon '04 thru '12
- **24032** Corsica & Beretta all models '87 thru '96
- **24040** Corvette all V8 models '68 thru '82
- **24041** Corvette all models '84 thru '96
- **24042** Corvette all models '97 thru '13
- **24044** Cruze '11 thru '19
- **24045** Full-size Sedans Caprice, Impala, Biscayne, Bel Air & Wagons '69 thru '90
- **24046** Impala SS & Caprice and Buick Roadmaster '91 thru '96
  Impala '00 thru '05 - see LUMINA (24048)
- **24047** Impala & Monte Carlo all models '06 thru '11
  Lumina '90 thru '94 - see GM (38010)
- **24048** Lumina & Monte Carlo '95 thru '05
  Lumina APV - see GM (38035)
- **24050** Luv Pick-up all 2WD & 4WD '72 thru '82
- **24051** Malibu '13 thru '19
- **24055** Monte Carlo all models '70 thru '88
  Monte Carlo '95 thru '01 - see LUMINA (24048)
- **24059** Nova all V8 models '69 thru '79
- **24060** Nova and Geo Prizm '85 thru '92
- **24064** Pick-ups '67 thru '87 - Chevrolet & GMC
- **24065** Pick-ups '88 thru '98 - Chevrolet & GMC
- **24066** Pick-ups '99 thru '06 - Chevrolet & GMC
- **24067** Chevrolet Silverado & GMC Sierra '07 thru '14
- **24068** Chevrolet Silverado & GMC Sierra '14 thru '19
- **24070** S-10 & S-15 Pick-ups '82 thru '93, Blazer & Jimmy '83 thru '94,
- **24071** S-10 & Sonoma Pick-ups '94 thru '04, including Blazer, Jimmy & Hombre
- **24072** Chevrolet TrailBlazer, GMC Envoy & Oldsmobile Bravada '02 thru '09
- **24075** Sprint '85 thru '88 & Geo Metro '89 thru '01
- **24080** Vans - Chevrolet & GMC '68 thru '96
- **24081** Chevrolet Express & GMC Savana Full-size Vans '96 thru '19

## CHRYSLER
- **10310** Chrysler Engine Overhaul Manual
- **25015** Chrysler Cirrus, Dodge Stratus, Plymouth Breeze '95 thru '00
- **25020** Full-size Front-Wheel Drive '88 thru '93
  K-Cars - see DODGE Aries (30008)
  Laser - see DODGE Daytona (30030)
- **25025** Chrysler LHS, Concorde, New Yorker, Dodge Intrepid, Eagle Vision, '93 thru '97
- **25026** Chrysler LHS, Concorde, 300M, Dodge Intrepid, '98 thru '04
- **25027** Chrysler 300 '05 thru '18, Dodge Charger '06 thru '18, Magnum '05 thru '08 & Challenger '08 thru '18
- **25030** Chrysler & Plymouth Mid-size front wheel drive '82 thru '95
  Rear-wheel Drive - see Dodge (30050)
- **25035** PT Cruiser all models '01 thru '10
- **25040** Chrysler Sebring '95 thru '06, Dodge Stratus '01 thru '06 & Dodge Avenger '95 thru '00
- **25041** Chrysler Sebring '07 thru '10, 200 '11 thru '17 Dodge Avenger '08 thru '14

## DATSUN
- **28005** 200SX all models '80 thru '83
- **28012** 240Z, 260Z & 280Z Coupe '70 thru '78
- **28014** 280ZX Coupe & 2+2 '79 thru '83
  300ZX - see NISSAN (72010)
- **28018** 510 & PL521 Pick-up '68 thru '73
- **28020** 510 all models '78 thru '81
- **28022** 620 Series Pick-up all models '73 thru '79
  720 Series Pick-up - see NISSAN (72030)

## DODGE
- **400 & 600** - see CHRYSLER (25030)
- **30008** Aries & Plymouth Reliant '81 thru '89
- **30010** Caravan & Plymouth Voyager '84 thru '95
- **30011** Caravan & Plymouth Voyager '96 thru '02
- **30012** Challenger & Plymouth Sapporro '78 thru '83
- **30013** Caravan, Chrysler Voyager & Town & Country '03 thru '07
- **30014** Grand Caravan & Chrysler Town & Country '08 thru '18
- **30016** Colt & Plymouth Champ '78 thru '87
- **30020** Dakota Pick-ups all models '87 thru '96
- **30021** Durango '98 & '99 & Dakota '97 thru '99
- **30022** Durango '00 thru '03 & Dakota '00 thru '04
- **30023** Durango '04 thru '09 & Dakota '05 thru '11
- **30025** Dart, Demon, Plymouth Barracuda, Duster & Valiant 6-cylinder models '67 thru '76
- **30030** Daytona & Chrysler Laser '84 thru '89
  Intrepid - see CHRYSLER (25025, 25026)
- **30034** Neon all models '95 thru '99
- **30035** Omni & Plymouth Horizon '78 thru '90
- **30036** Dodge & Plymouth Neon '00 thru '05
- **30040** Pick-ups full-size models '74 thru '93
- **30041** Pick-ups full-size models '94 thru '08
- **30042** Pick-ups full-size models '09 thru '18
- **30045** Ram 50/D50 Pick-ups & Raider and Plymouth Arrow Pick-ups '79 thru '93
- **30050** Dodge/Plymouth/Chrysler RWD '71 thru '89
- **30055** Shadow & Plymouth Sundance '87 thru '94
- **30060** Spirit & Plymouth Acclaim '89 thru '95
- **30065** Vans - Dodge & Plymouth '71 thru '03

## EAGLE
Talon - see MITSUBISHI (68030, 68031)
Vision - see CHRYSLER (25025)

## FIAT
- **34010** 124 Sport Coupe & Spider '68 thru '78
- **34025** X1/9 all models '74 thru '80

## FORD
- **10320** Ford Engine Overhaul Manual
- **10355** Ford Automatic Transmission Overhaul
- **11500** Mustang '64-1/2 thru '70 Restoration Guide
- **36004** Aerostar Mini-vans all models '86 thru '97
- **36006** Contour & Mercury Mystique '95 thru '00
- **36008** Courier Pick-up all models '72 thru '82
- **36012** Crown Victoria & Mercury Grand Marquis '88 thru '11
- **36014** Edge '07 thru '19 & Lincoln MKX '07 thru '18
- **36016** Escort & Mercury Lynx all models '81 thru '90
- **36020** Escort & Mercury Tracer '91 thru '02
- **36022** Escape '01 thru '17, Mazda Tribute '01 thru '11, & Mercury Mariner '05 thru '11
- **36024** Explorer & Mazda Navajo '91 thru '01
- **36025** Explorer & Mercury Mountaineer '02 thru '10
- **36026** Explorer '11 thru '17
- **36028** Fairmont & Mercury Zephyr '78 thru '83
- **36030** Festiva & Aspire '88 thru '97
- **36032** Fiesta all models '77 thru '80
- **36034** Focus all models '00 thru '11
- **36035** Focus '12 thru '14
- **36045** Fusion '06 thru '14 & Mercury Milan '06 thru '11
- **36048** Mustang V8 all models '64-1/2 thru '73
- **36049** Mustang II 4-cylinder, V6 & V8 models '74 thru '78
- **36050** Mustang & Mercury Capri '79 thru '93
- **36051** Mustang all models '94 thru '04
- **36052** Mustang '05 thru '14
- **36054** Pick-ups & Bronco '73 thru '79
- **36058** Pick-ups & Bronco '80 thru '96
- **36059** F-150 '97 thru '03, Expedition '97 thru '17, F-250 '97 thru '99, F-150 Heritage '04 & Lincoln Navigator '98 thru '17
- **36060** Super Duty Pick-ups & Excursion '99 thru '10
- **36061** F-150 full-size '04 thru '14
- **36062** Pinto & Mercury Bobcat '75 thru '80
- **36063** F-150 full-size '15 thru '17
- **36064** Super Duty Pick-ups '11 thru '16
- **36066** Probe all models '89 thru '92
  Probe '93 thru '97 - see MAZDA 626 (61042)
- **36070** Ranger & Bronco II gas models '83 thru '92
- **36071** Ranger '93 thru '11 & Mazda Pick-ups '94 thru '09
- **36074** Taurus & Mercury Sable '86 thru '95
- **36075** Taurus & Mercury Sable '96 thru '07
- **36076** Taurus '08 thru '14, Five Hundred '05 thru '07, Mercury Montego '05 thru '07 & Sable '08 thru '09
- **36078** Tempo & Mercury Topaz '84 thru '94
- **36082** Thunderbird & Mercury Cougar '83 thru '88
- **36086** Thunderbird & Mercury Cougar '89 thru '97
- **36090** Vans all V8 Econoline models '69 thru '91
- **36094** Vans full size '92 thru '14
- **36097** Windstar '95 thru '03, Freestar & Mercury Monterey Mini-van '04 thru '07

## GENERAL MOTORS
- **10360** GM Automatic Transmission Overhaul
- **38001** GMC Acadia '07 thru '16, Buick Enclave '08 thru '17, Saturn Outlook '07 thru '10 & Chevrolet Traverse '09 thru '17
- **38005** Buick Century, Chevrolet Celebrity, Oldsmobile Cutlass Ciera & Pontiac 6000 all models '82 thru '96
- **38010** Buick Regal '88 thru '04, Chevrolet Lumina '88 thru '04, Oldsmobile Cutlass Supreme '88 thru '97 & Pontiac Grand Prix '88 thru '07
- **38015** Buick Skyhawk, Cadillac Cimarron, Chevrolet Cavalier, Oldsmobile Firenza, Pontiac J-2000 & Sunbird '82 thru '94
- **38016** Chevrolet Cavalier & Pontiac Sunfire '95 thru '05
- **38017** Chevrolet Cobalt '05 thru '10, HHR '06 thru '11, Pontiac G5 '07 thru '09, Pursuit '05 thru '06 & Saturn ION '03 thru '07
- **38020** Buick Skylark, Chevrolet Citation, Oldsmobile Omega, Pontiac Phoenix '80 thru '85
- **38025** Buick Skylark '86 thru '98, Somerset '85 thru '87, Oldsmobile Achieva '92 thru '98, Calais '85 thru '91, & Pontiac Grand Am all models '85 thru '98
- **38026** Chevrolet Malibu '97 thru '03, Classic '04 thru '05, Oldsmobile Alero '99 thru '03, Cutlass '97 thru '00, & Pontiac Grand Am '99 thru '03
- **38027** Chevrolet Malibu '04 thru '12, Pontiac G6 '05 thru '10 & Saturn Aura '07 thru '10
- **38030** Cadillac Eldorado, Seville, Oldsmobile Toronado & Buick Riviera '71 thru '85
- **38031** Cadillac Eldorado, Seville, DeVille, Fleetwood, Oldsmobile Toronado & Buick Riviera '86 thru '93
- **38032** Cadillac DeVille '94 thru '05, Seville '92 thru '04 & Cadillac DTS '06 thru '10
- **38035** Chevrolet Lumina APV, Oldsmobile Silhouette & Pontiac Trans Sport all models '90 thru '96
- **38036** Chevrolet Venture '97 thru '05, Oldsmobile Silhouette '97 thru '04, Pontiac Trans Sport '97 thru '98 & Montana '99 thru '05
- **38040** Chevrolet Equinox '05 thru '17, GMC Terrain '10 thru '17 & Pontiac Torrent '06 thru '09

## GEO
Metro - see CHEVROLET Sprint (24075)
Prizm - '85 thru '92 see CHEVY (24060), '93 thru '02 see TOYOTA Corolla (92036)
- **40030** Storm all models '90 thru '93
Tracker - see SUZUKI Samurai (90010)

*(Continued on other side)*

Haynes North America, Inc. • (805) 498-6703 • www.haynes.com

# Haynes Automotive Manuals (continued)

NOTE: If you do not see a listing for your vehicle, please visit haynes.com for the latest product information and check out our **Online Manuals!**

## GMC
- Acadia - see GENERAL MOTORS (38001)
- Pick-ups - see CHEVROLET (24027, 24068)
- Vans - see CHEVROLET (24081)

## HONDA
- 42010 Accord CVCC all models '76 thru '83
- 42011 Accord all models '84 thru '89
- 42012 Accord all models '90 thru '93
- 42013 Accord all models '94 thru '97
- 42014 Accord all models '98 thru '02
- 42015 Accord '03 thru '12 & Crosstour '10 thru '14
- 42016 Accord '13 thru '17
- 42020 Civic 1200 all models '73 thru '79
- 42021 Civic 1300 & 1500 CVCC '80 thru '83
- 42022 Civic 1500 CVCC all models '75 thru '79
- 42023 Civic all models '84 thru '91
- 42024 Civic & del Sol '92 thru '95
- 42025 Civic '96 thru '00, CR-V '97 thru '01 & Acura Integra '94 thru '00
- 42026 Civic '01 thru '11 & CR-V '02 thru '11
- 42027 Civic '12 thru '15 & CR-V '12 thru '16
- 42030 Fit '07 thru '13
- 42035 Odyssey all models '99 thru '10
- Passport - see ISUZU Rodeo (47017)
- 42037 Honda Pilot '03 thru '08, Ridgeline '06 thru '14 & Acura MDX '01 thru '07
- 42040 Prelude CVCC all models '79 thru '89

## HYUNDAI
- 43010 Elantra all models '96 thru '19
- 43015 Excel & Accent all models '86 thru '13
- 43050 Santa Fe all models '01 thru '12
- 43055 Sonata all models '99 thru '14

## INFINITI
- G35 '03 thru '08 - see NISSAN 350Z (72011)

## ISUZU
- Hombre - see CHEVROLET S-10 (24071)
- 47017 Rodeo '91 thru '02, Amigo '89 thru '94 & '98 thru '02 & Honda Passport '95 thru '02
- 47020 Trooper '84 thru '91 & Pick-up '81 thru '93

## JAGUAR
- 49010 XJ6 all 6-cylinder models '68 thru '86
- 49011 XJ6 all models '88 thru '94
- 49015 XJ12 & XJS all 12-cylinder models '72 thru '85

## JEEP
- 50010 Cherokee, Comanche & Wagoneer Limited all models '84 thru '01
- 50011 Cherokee '14 thru '19
- 50020 CJ all models '49 thru '86
- 50025 Grand Cherokee all models '93 thru '04
- 50026 Grand Cherokee '05 thru '19 & Dodge Durango '11 thru '19
- 50029 Grand Wagoneer & Pick-up '72 thru '91 Grand Wagoneer '84 thru '91, Cherokee & Wagoneer '72 thru '83, Pick-up '72 thru '88
- 50030 Wrangler all models '87 thru '17
- 50035 Liberty '02 thru '12 & Dodge Nitro '07 thru '11
- 50050 Patriot & Compass '07 thru '17

## KIA
- 54050 Optima '01 thru '10
- 54060 Sedona '02 thru '14
- 54070 Sephia '94 thru '01, Spectra '00 thru '09, Sportage '05 thru '20
- 54077 Sorento '03 thru '13

## LEXUS
- ES 300/330 - see TOYOTA Camry (92007, 92008)
- ES 350 - see TOYOTA Camry (92009)
- RX 300/330/350 - see TOYOTA Highlander (92095)

## LINCOLN
- MKX - see FORD (36014)
- Navigator - see FORD Pick-up (36059)
- 59010 Rear-Wheel Drive Continental '70 thru '87, Mark Series '70 thru '92 & Town Car '81 thru '10

## MAZDA
- 61010 GLC (rear-wheel drive) '77 thru '83
- 61011 GLC (front-wheel drive) '81 thru '85
- 61012 Mazda3 '04 thru '11
- 61015 323 & Protegé '90 thru '03
- 61016 MX-5 Miata '90 thru '14
- 61020 MPV all models '89 thru '98
- Navajo - see Ford Explorer (36024)
- 61030 Pick-ups '72 thru '93
- Pick-ups '94 thru '09 - see Ford Ranger (36071)
- 61035 RX-7 all models '79 thru '85
- 61036 RX-7 all models '86 thru '91
- 61040 626 (rear-wheel drive) all models '79 thru '82
- 61041 626 & MX-6 (front-wheel drive) '83 thru '92
- 61042 626 '93 thru '01 & MX-6/Ford Probe '93 thru '02
- 61043 Mazda6 '03 thru '13

## MERCEDES-BENZ
- 63012 123 Series Diesel '76 thru '85
- 63015 190 Series 4-cylinder gas models '84 thru '88
- 63020 230/250/280 6-cylinder SOHC models '68 thru '72
- 63025 280 123 Series gas models '77 thru '81
- 63030 350 & 450 all models '71 thru '80
- 63040 C-Class: C230/C240/C280/C320/C350 '01 thru '07

## MERCURY
- 64200 Villager & Nissan Quest '93 thru '01
- All other titles, see FORD Listing.

## MG
- 66010 MGB Roadster & GT Coupe '62 thru '80
- 66015 MG Midget, Austin Healey Sprite '58 thru '80

## MINI
- 67020 Mini '02 thru '13

## MITSUBISHI
- 68020 Cordia, Tredia, Galant, Precis & Mirage '83 thru '93
- 68030 Eclipse, Eagle Talon & Plymouth Laser '90 thru '94
- 68031 Eclipse '95 thru '05 & Eagle Talon '95 thru '98
- 68035 Galant '94 thru '12
- 68040 Pick-up '83 thru '96 & Montero '83 thru '93

## NISSAN
- 72010 300ZX all models including Turbo '84 thru '89
- 72011 350Z & Infiniti G35 all models '03 thru '08
- 72015 Altima all models '93 thru '06
- 72016 Altima '07 thru '12
- 72020 Maxima all models '85 thru '92
- 72021 Maxima all models '93 thru '08
- 72025 Murano '03 thru '14
- 72030 Pick-ups '80 thru '97 & Pathfinder '87 thru '95
- 72031 Frontier '98 thru '04, Xterra '00 thru '04, & Pathfinder '96 thru '04
- 72032 Frontier & Xterra '05 thru '14
- 72037 Pathfinder '05 thru '14
- 72040 Pulsar all models '83 thru '86
- 72042 Roque all models '08 thru '20
- 72050 Sentra all models '82 thru '94
- 72051 Sentra & 200SX all models '95 thru '06
- 72060 Stanza all models '82 thru '90
- 72070 Titan pick-ups '04 thru '10, Armada '05 thru '10 & Pathfinder Armada '04
- 72080 Versa all models '07 thru '19

## OLDSMOBILE
- 73015 Cutlass V6 & V8 gas models '74 thru '88
- For other OLDSMOBILE titles, see BUICK, CHEVROLET or GENERAL MOTORS listings.

## PLYMOUTH
- For PLYMOUTH titles, see DODGE listing.

## PONTIAC
- 79008 Fiero all models '84 thru '88
- 79018 Firebird V8 models except Turbo '70 thru '81
- 79019 Firebird all models '82 thru '92
- 79025 G6 all models '05 thru '09
- 79040 Mid-size Rear-wheel Drive '70 thru '87
- Vibe '03 thru '10 - see TOYOTA Corolla (92037)
- For other PONTIAC titles, see BUICK, CHEVROLET or GENERAL MOTORS listings.

## PORSCHE
- 80020 911 Coupe & Targa models '65 thru '89
- 80025 914 all 4-cylinder models '69 thru '76
- 80030 924 all models including Turbo '76 thru '82
- 80035 944 all models including Turbo '83 thru '89

## RENAULT
- Alliance & Encore - see AMC (14025)

## SAAB
- 84010 900 all models including Turbo '79 thru '88

## SATURN
- 87010 Saturn all S-series models '91 thru '02
- Saturn Ion '03 thru '07- see GM (38017)
- Saturn Outlook - see GM (38001)
- 87020 Saturn L-series all models '00 thru '04
- 87040 Saturn VUE '02 thru '09

## SUBARU
- 89002 1100, 1300, 1400 & 1600 '71 thru '79
- 89003 1600 & 1800 2WD & 4WD '80 thru '94
- 89080 Impreza '02 thru '11, WRX '02 thru '14, & WRX STI '04 thru '14
- 89100 Legacy all models '90 thru '99
- 89101 Legacy & Forester '00 thru '09
- 89102 Legacy '10 thru '16 & Forester '12 thru '16

## SUZUKI
- 90010 Samurai/Sidekick & Geo Tracker '86 thru '01

## TOYOTA
- 92005 Camry all models '83 thru '91
- 92006 Camry '92 thru '96 & Avalon '95 thru '96
- 92007 Camry, Avalon, Solara, Lexus ES 300 '97 thru '01
- 92008 Camry, Avalon, Lexus ES 300/330 '02 thru '06 & Solara '02 thru '08
- 92009 Camry, Avalon & Lexus ES 350 '07 thru '17
- 92015 Celica Rear-wheel Drive '71 thru '85
- 92020 Celica Front-wheel Drive '86 thru '99
- 92025 Celica Supra all models '79 thru '92
- 92030 Corolla all models '75 thru '79
- 92032 Corolla all rear-wheel drive models '80 thru '87
- 92035 Corolla all front-wheel drive models '84 thru '92
- 92036 Corolla & Geo/Chevrolet Prizm '93 thru '02
- 92037 Corolla '03 thru '19, Matrix '03 thru '14, & Pontiac Vibe '03 thru '10
- 92040 Corolla Tercel all models '80 thru '82
- 92045 Corona all models '74 thru '82
- 92050 Cressida all models '78 thru '82
- 92055 Land Cruiser FJ40, 43, 45, 55 '68 thru '82
- 92056 Land Cruiser FJ60, 62, 80, FZJ80 '80 thru '96
- 92060 Matrix '03 thru '11 & Pontiac Vibe '03 thru '10
- 92065 MR2 all models '85 thru '87
- 92070 Pick-up all models '69 thru '78
- 92075 Pick-up all models '79 thru '95
- 92076 Tacoma '95 thru '04, 4Runner '96 thru '02 & T100 '93 thru '08
- 92077 Tacoma all models '05 thru '18
- 92078 Tundra '00 thru '06 & Sequoia '01 thru '07
- 92079 4Runner all models '03 thru '09
- 92080 Previa all models '91 thru '95
- 92081 Prius all models '01 thru '12
- 92082 RAV4 all models '96 thru '12
- 92085 Tercel all models '87 thru '94
- 92090 Sienna all models '98 thru '10
- 92095 Highlander '01 thru '19 & Lexus RX330/330/350 '99 thru '19
- 92179 Tundra '07 thru '19 & Sequoia '08 thru '19

## TRIUMPH
- 94007 Spitfire all models '62 thru '81
- 94010 TR7 all models '75 thru '81

## VW
- 96008 Beetle & Karmann Ghia '54 thru '79
- 96009 New Beetle '98 thru '10
- 96016 Rabbit, Jetta, Scirocco & Pick-up gas models '75 thru '92 & Convertible '80 thru '92
- 96017 Golf, GTI & Jetta '93 thru '98, Cabrio '95 thru '02
- 96018 Golf, GTI, Jetta '99 thru '05
- 96019 Jetta, Rabbit, GLI, GTI & Golf '05 thru '11
- 96020 Rabbit, Jetta & Pick-up diesel '77 thru '84
- 96021 Jetta '11 thru '18 & Golf '15 thru '19
- 96023 Passat '98 thru '05 & Audi A4 '96 thru '01
- 96030 Transporter 1600 all models '68 thru '79
- 96035 Transporter 1700, 1800 & 2000 '72 thru '79
- 96040 Type 3 1500 & 1600 all models '63 thru '73
- 96045 Vanagon Air-Cooled all models '80 thru '83

## VOLVO
- 97010 120, 130 Series & 1800 Sports '61 thru '73
- 97015 140 Series all models '66 thru '74
- 97020 240 Series all models '76 thru '93
- 97040 740 & 760 Series all models '82 thru '88
- 97050 850 Series all models '93 thru '97

## TECHBOOK MANUALS
- 10205 Automotive Computer Codes
- 10206 OBD-II & Electronic Engine Management
- 10210 Automotive Emissions Control Manual
- 10215 Fuel Injection Manual '78 thru '85
- 10225 Holley Carburetor Manual
- 10230 Rochester Carburetor Manual
- 10305 Chevrolet Engine Overhaul Manual
- 10320 Ford Engine Overhaul Manual
- 10330 GM and Ford Diesel Engine Repair Manual
- 10331 Duramax Diesel Engines '01 thru '19
- 10332 Cummins Diesel Engine Performance Manual
- 10333 GM, Ford & Chrysler Engine Performance Manual
- 10334 GM Engine Performance Manual
- 10340 Small Engine Repair Manual, 5 HP & Less
- 10341 Small Engine Repair Manual, 5.5 HP to 20 HP
- 10345 Suspension, Steering & Driveline Manual
- 10355 Ford Automatic Transmission Overhaul
- 10360 GM Automatic Transmission Overhaul
- 10405 Automotive Body Repair & Painting
- 10410 Automotive Brake Manual
- 10411 Automotive Anti-lock Brake (ABS) Systems
- 10420 Automotive Electrical Manual
- 10425 Automotive Heating & Air Conditioning
- 10435 Automotive Tools Manual
- 10445 Welding Manual
- 10450 ATV Basics

Over a 100 Haynes motorcycle manuals also available

Haynes North America, Inc. • (805) 498-6703 • www.haynes.com